Engineering Materials and Processes

Springer
London
Berlin
Heidelberg
New York
Hong Kong
Milan
Paris
Tokyo

Series Editor

Professor Brian Derby, Professor of Materials Science
Manchester Science Centre, Grosvenor Street, Manchester, M1 7HS, UK

Other titles published in this series:

Fusion Bonding of Polymer Composites
C. Ageorges and L. Ye

Composite Materials
D.D.L. Chung

Titanium
G. Lütjering and J.C. Williams

Corrosion of Metals
H. Kaesche

Phase Diagrams and Heterogeneous Equilibria
B. Predal, M. Hoch and M. Pool

Orientations and Rotations
A. Morawiec
Publication due October 2003

Corrosion and Protection
E. Bardal
Publication due November 2003

Computational Mechanics of Composite Materials
M.M. Kaminski
Publication due March 2004

Liming Dai

Intelligent Macromolecules for Smart Devices

From Materials Synthesis to Device Applications

Liming Dai, PhD
Department of Polymer Engineering, The University of Akron, Akron, OH
44235-0301, USA

British Library Cataloguing in Publication Data
Dai, Liming
  Intelligent macromolecules for smart devices. -
  (Engineering materials and processes)
  1.Smart materials 2.Macromolecules
  I.Title
  620.1'1
  ISBN 1852335106

Library of Congress Cataloging-in-Publication Data
A catalog record for this book is available from the Library of Congress

Apart from any fair dealing for the purposes of research or private study, or criticism or review, as permitted under the Copyright, Designs and Patents Act 1988, this publication may only be reproduced, stored or transmitted, in any form or by any means, with the prior permission in writing of the publishers, or in the case of reprographic reproduction in accordance with the terms of licences issued by the Copyright Licensing Agency. Enquiries concerning reproduction outside those terms should be sent to the publishers.

Engineering Materials and Processes ISSN 1619-0181
ISBN 1-85233-510-6 Springer-Verlag London Berlin Heidelberg
Springer-Verlag is a part of Springer Science+Business Media
springeronline.com

© Springer-Verlag London Limited 2004

Whilst we have made considerable efforts to contact all holders of copyright material contained in this book, we have failed to locate some of these. Should holders wish to contact the Publisher, we will be happy to come to some arrangement with them.

The use of registered names, trademarks etc. in this publication does not imply, even in the absence of a specific statement, that such names are exempt from the relevant laws and regulations and therefore free for general use.

The publisher makes no representation, express or implied, with regard to the accuracy of the information contained in this book and cannot accept any legal responsibility or liability for any errors or omissions that may be made.

Typesetting: Electronic text files prepared by authors
Printed and bound in the United States of America
69/3830-543210 Printed on acid-free paper SPIN 10833984

Preface and Acknowledgements

All matter is made up of *atoms*. The term *molecule* is used to describe groups of atoms that tend to exist together in a stable form. As the term implies, *macromolecule* refers to a molecule of large size. *Intelligent macromolecules* are those large molecules, including certain polymers, dendrimers, fullerenes and carbon nanotubes, with the capability to change their conformational structures and/or properties in response to an external stimulus.

Polymers have long been used as electrically insulating materials: after all, metal cables are coated in plastic to insulate them. The visit of MacDiarmid to Shirakawa at Tokyo Institute of Technology in 1974 and, later, Shirakawa to MacDiarmid and Heeger at University of Pennsylvania, however, led to the discovery of conducting polyacetylene in 1977 - a prototype conducting polymers with the so-called *conjugated* structure of alternating single and double bonds for delocalization of π-electrons. This finding opened up the important new field of polymers for electronic applications and was recognized by the 2000 Nobel Prize in Chemistry. The subsequent discovery of the electroluminescent light emission from conjugated poly(p-phenylene vinylene), by Friend's group at Cavendish Laboratory in 1990, revealed the significance of the use of conjugated polymers in photonic devices. Various conjugated polymers can now be synthesized to show the processing advantages of plastics and the optoelectronic properties of inorganic semiconductors or metals.

With DNA, proteins, and polysaccharides comprising the most important components of a living cell, life is polymeric in its essence. The self-assembly of lipids contributes to even larger *supramolecular* structures. Nature handles these biomacromolecular and supramolecular in a mysterious way to create various nanostructures/nanomachines that enable creatures to function. The salient feature of biomacromolecules is that they can adopt unique, well-defined conformations according to specific environmental conditions. Human beings have done remarkable work and learned a lot from nature. Recent decades have witnessed the appearance of synthetic polymers, including dendrimers, which can respond reversibly to a change in solvent, pH, temperature, electric or optical fields, or other stimuli. These *stimuli-responsive* biomacromolecules and synthetic polymers are of significance to various potential applications, ranging from sensors/actuators to controlled release of drugs.

Sometimes history repeats itself. The visit made by Kroto in 1985 to Smalley and Curl at Rice University led also to the Noble-Prize-winning discovery of *buckminsterfullerene* C_{60} - a conjugated molecule with a soccerball-like structure consisting of 12 pentagons and 20 hexagons facing symmetrically. Just as the discovery of C_{60} has created an entirely new branch of carbon chemistry, the subsequent discovery of *carbon nanotubes* by Iijima in 1991 opened up a new era in material science and nanotechnology. These elongated nanotubes consist of carbon hexagons arranged in a concentric manner with both ends normally capped by fullerene-like structures containing pentagons. Having a conjugated all-carbon structure with unusual molecular symmetries, fullerenes and carbon nanotubes also show interesting electronic, photonic, magnetic, and mechanical properties, which make them attractive for various applications, including optical limiters, photovoltaic cells and field emitting displays.

This book deals with the synthesis, fundamentals, and device applications of a wide range of intelligent macromolecules. The diversity both in the synthesis/fabrication of intelligent macromolecules and in the development/application of various smart devices appeals to a broad group of students, teachers, scientists and engineers with backgrounds in physical, chemical, material, biological, and medical sciences and engineering. The main objective of this book is to bridge the gap between the synthesis of intelligent macromolecules and the development of smart devices through a broad treatment of the field by covering not only the materials' synthesis and structure study, but also the device construction and application. In order to cover the multidisciplinary field of such diversity, the text has been divided into three major parts after the introductory chapter. The first part from Chapter 2 to Chapter 5 deals with the basic science of various intelligent macromolecular systems. Chapters 6 and 7 follow the first part of the text to present an overview of various methodologies developed for assembling these intelligent macromolecules into multi-dimensional structures, which have been so crucial to the development of smart devices. A large variety of smart devices based on these intelligent macromolecules and functional structures are then described in the final part of the book. The above approach, I hope, will allow the reader to first review the scientific basis for intelligent macromolecules. The basic knowledge is then extended to functional devices; many of them are of practical importance. While many self-explanatory illustrations could provide an overview understanding even to those who are new to the field, the large number of updated references cited in each of the chapters should enable advanced readers to quickly review the multidisciplinary and challenging field with information on the latest developments.

At this place I wish to express my sincere thanks to Mr Oliver Jackson for his suggesting and encouraging me to write this book and also his very kind and patient cooperation during its completion, without which this book would never have been appeared. I would also like to thank my colleagues who contributed in one way or the other to the text and authors whose work was cited, in particular: Vardhan Bajpai, Yoseph Bar-Cohen, Ray Baughman, Bruce Crissinger, Mei Gao, Zhixin, Guo, Hans Griesser, Sinan Li, Albert Mau, Darrell Reneker, Chris Strauss, Chris Toprakcioglu, Alfred Uhlherr, Gordon Wallace, Zhonglin Wang, John White, Paul White, Berthold Winkler, Junbing Yang and Zhi Yu. Last, but not the least, I thank my parents, wife Lin Zhu and two sons, Quanbin Kevin Dai, and Quanzhe Alvin Dai, for their love, unceasing patience, and continuous support. This book is dedicated to my father, who passed away from a heart attack in July, 2000.

Liming Dai
Akron, Ohio

February 2003

Table of Contents

1 The Concepts of Intelligent Macromolecules and Smart Devices 1
   1.1   Introduction 1
   1.2   The Concept of Intelligent Macromolecules 4
          1.2.1   Synthetic Macromolecules 4
                 1.2.1.1   Chain Structure and Classification 4
                 1.2.1.2   Synthesis 6
                 1.2.1.3   Chain Conformation 9
                 1.2.1.4   Macromolecular Structure in Solution 12
                 1.2.1.5   Primary, Secondary, Tertiary and Quaternary
                                 Structure 18
          1.2.2   Biological Macromolecules 19
                 1.2.2.1   Structure of DNA 19
                 1.2.2.2   Structure of Proteins 21
                 1.2.2.3   Structure of Polysaccharides 25
          1.2.3   Carbon Nanomaterials 27
          1.2.4   Intelligent Macromolecules 28
   1.3   The Concept of Smart Devices 29
          1.3.1   Self-assembling and Micro-/Nano-fabrication 29
          1.3.2   Functional Structures and Smart Devices 31
   1.4   References 35

Part I.   Intelligent Macromolecules

2 Conducting Polymers 41
   2.1   Introduction 41
   2.2   Conjugated Conducting Polymers 42
          2.2.1   Structure and Properties 42

		2.2.1.1 π-π* Conjugation 42
		2.2.1.2 Doping 43
	2.2.2	Synthesis 50
		2.2.2.1 Syntheses of Soluble Conjugated Polymers 51
		2.2.2.2 Syntheses of Conjugated Polymer Films 60
2.3	Charge Transfer Polymers 63
	2.3.1	Organic Charge Transfer Complexes 63
	2.3.2	Polymeric Charge Transfer Complexes 64
	2.3.3	Charge Transfer Between Fullerene C_{60} and Polymers 67
2.4	Ionically Conducting Polymers 68
	2.4.1	Structural Features of Polymer Electrolytes 68
	2.4.2	Transport Properties and Chain Dynamics 69
2.5	Conductively Filled Polymers 72
	2.5.1	Polymers Filled with Conductive Solids 72
	2.5.2	Polymers Filled with Conjugated Conducting Polymers 76
2.6	References 76

3 Stimuli-responsive Polymers 81
3.1 Introduction 81
3.2 Solvent-responsive Polymers 82
3.3 Temperature-responsive Polymers 86
    3.3.1 Temperature-responsive Polymers in Solution 86
    3.3.2 Temperature-responsive Polymers on Surface 91
3.4 pH-responsive Polymers 95
3.5 Ionically Responsive Polymers 97
3.6 Electrically Responsive Polymers 98
3.7 Photoelectrochromism 103
3.8 Photoresponsive Polymers 104
3.9 Biochromism 111
3.10 Photomodulation of Enzyme Activity 113
3.11 References 113

4 Dendrimers and Fullerenes 117
4.1 Introduction 117
4.2 Dendrimers 119
    4.2.1 Synthesis 119
        4.2.1.1 Divergent Approach 119
        4.2.1.2 Convergent-growth Approach 122
        4.2.1.3 Other Miscellaneous Approaches 123
    4.2.2 Structure 127
        4.2.2.1 Dendrimers with a Metal Core 127
        4.2.2.2 Dendrimers with a Hollow Core 127
        4.2.2.3 Dendrimers with a Hydrophobic Interior and
                  Hydrophilic Exterior Layer 129
        4.2.2.4 Dendrimers with Guest Molecules Trapped in
                  their Cavities 131

		4.2.2.5	Dendrimers with Different Terminal Groups – Dendritic Block Copolymers 132
4.3	Fullerene C_{60} 133
	4.3.1	Chemistry of C_{60} 134
		4.3.1.1	Addition Reactions 134
		4.3.1.2	Dimerization and Polymerization 139
	4.3.2	Polymeric Derivatives of C_{60} 140
		4.3.2.1	Fullerene Charm Bracelets 141
		4.3.2.2	Fullerene Pearl Necklaces 144
		4.3.2.3	Flagellenes 146
4.4	References 150

5 Carbon Nanotubes 157
- 5.1 Introduction 157
- 5.2 Structure 159
- 5.3 Property 160
- 5.4 Synthesis 165
  - 5.4.1 Multi-wall Carbon Nanotubes (MWNTs) 165
  - 5.4.2 Single-wall Carbon Nanotubes (SWNTs) 166
- 5.5 Purification 167
- 5.6 Microfabrication 168
  - 5.6.1 Opening, Filling and Closing 168
  - 5.6.2 Filling 170
  - 5.6.3 Tip-closing 172
- 5.7 Chemical Modification 172
  - 5.7.1 End-functionalization 173
    - 5.7.1.1 Oxidation of Carbon Nanotubes 173
    - 5.7.1.2 Covalent-Coupling via the Oxidized Nanotube Ends 174
  - 5.7.2 Modification of Nanotube Outerwall 179
    - 5.7.2.1 Sidewall Fluorination of Carbon Nanotubes ... 179
    - 5.7.2.2 The Attachment of Dichlorocarbene to the Sidewall 181
    - 5.7.2.3 Modification via 1,3-Dipolar Cycloaddition of Azomethine Ylides 181
    - 5.7.2.4 The Reaction Between Aniline and Carbon Nanotubes 182
  - 5.7.3 Functionalization of Carbon Nanotube Innerwall 184
  - 5.7.4 Other Physical Chemistries of Carbon Nanotubes 185
    - 5.7.4.1 Modification of Carbon Nanotubes via Mechanochemical Reactions 185
    - 5.7.4.2 Modification of Carbon Nanotubes via Electrochemical Reactions 186
    - 5.7.4.3 Modification of Carbon Nanotubes via Photochemical Reactions 187
- 5.8 Non-covalent Chemistry of Carbon Nanotubes 188

	5.8.1	Non-covalent Attachment of Small Molecules onto the Nanotube Sidewall 188
	5.8.2	Non-covalent Wrapping of Polymer Chains onto the Nanotube Sidewall 190
5.9	Modification of Aligned Carbon Nanotubes 191	
	5.9.1	Plasma Activation of Aligned Carbon Nanotubes 192
	5.9.2	Acid Oxidation with Structural Protection 194
	5.9.3	Electrochemical Modification of Aligned Carbon Nanotubes 195
5.10	References 196	

Part II. From Intelligent Macromolecules to Smart Devices

6 Ordered and Patterned Macromolecules 203
   6.1   Introduction 203
   6.2   Oriented and Patterned Conjugated Polymers 204
           6.2.1   The Necessity 204
                  6.2.1.1   For Electronic Applications 204
                  6.2.1.2   For Non-linear Optical Applications 205
           6.2.2   Oriented Conjugated Polymers 206
                  6.2.2.1   Synthesis-induced Orientation 206
                  6.2.2.2   Liquid Crystalline Conjugated Polymers 208
                  6.2.2.3   Post-synthesis Orientation 213
           6.2.3   Patterned Conjugated Polymers 215
                  6.2.3.1   Photolithographic Patterning 217
                  6.2.3.2   Pattern Formation by Self-assembling 220
                  6.2.3.3   Pattern Formation by Polymer Phase Separation 223
                  6.2.3.4   Plasma Patterning of Conjugated Polymers 225
   6.3   Aligned and Patterned Carbon Nanotubes 228
           6.3.1   The Necessity 228
                  6.3.1.1   Molecular Computing 229
                  6.3.1.2   Electron Emitters 229
                  6.3.1.3   For Membrane Applications 229
           6.3.2   Horizontally Aligned and Micropatterned Carbon Nanotubes 230
                  6.3.2.1   Horizontally Aligned Carbon Nanotubes 230
                  6.3.2.2   Micropatterns of Horizontally Aligned Carbon Nanotubes 232
           6.3.3   Perpendicularly Aligned and Micropatterned Carbon Nanotubes 236
                  6.3.3.1   Perpendicularly Aligned Carbon Nanotubes 236
                  6.3.3.2   Micropatterns of Perpendicularly Aligned Carbon Nanotubes 239
                  6.3.3.3   Perpendicularly Aligned and Micropatterned Carbon Nanotubes by Self-assembly 249
   6.4   Aligned Non-carbon Nanotubes 252

		6.4.1 Aligned B:C:N Nanotubes 252
		6.4.2 Aligned Inorganic Nanotubes 252
		6.4.3 Aligned Polymer Nanotubes 253
		6.4.4 Aligned Peptide Nanotubes 253
	6.5	References 254

7 Macromolecular Nanostructures 265

- 7.1 Introduction 265
- 7.2 Polymer Nanoparticles 266
  - 7.2.1 Polymer Nanospheres by Polymerization.. 266
  - 7.2.2 Dispersion of Pre-formed Polymers... 268
    - 7.2.2.1 Polymer Nanosphere by Emulsifying Dispersion 269
    - 7.2.2.2 Polymer Nanospheres by Supercritical Fluid Method 269
- 7.3 Self-assembling of Pre-formed Polymers... 269
  - 7.3.1 Shell-core Polymer Nanoparticles 270
- 7.4 Polymer Nanowires, Nanotubes and Nanofibers.................................... 275
  - 7.4.1 Tip-assisted Syntheses of Polymer Nanowires 275
  - 7.4.2 Template Syntheses of Polymer Nanowires, Nanotubes and Nanofibers 279
  - 7.4.3 Electrospinning of Polymer Nanofibers.................................... 282
- 7.5 Polymer Nanofilms... 286
  - 7.5.1 Polymer Nanofilms by Solution Casting 286
  - 7.5.2 Polymer Nanofilms by Plasma Polymerization 287
  - 7.5.3 Polymer Nanofilms by Langmuir-Blodgett Deposition............ 288
  - 7.5.4 Polymer Brushes by End-adsorption 289
    - 7.5.4.1 Polymer Mushrooms 296
    - 7.5.4.2 Polymer Brushes 297
- 7.6 Nanostructured Polymers with Special Architectures 300
  - 7.6.1 Self-assembly of Ordered Nanoporous Polymers 300
  - 7.6.2 Coaxial Polymer Nanowires and Nanofibers........................... 302
  - 7.6.3 Multilayered Polymer Nanofilms.. 305
  - 7.6.4 Nanostructured Polymers by Phase Separation 309
- 7.7 References 311

Part III. Smart Devices

8 Electronic Devices 321

- 8.1 Introduction 321
- 8.2 Conjugated Polymer Devices 321
  - 8.2.1 Electromagnetic Shielding 322
  - 8.2.2 Schottky Barrier Diodes and Field-effect Transistors............... 326
    - 8.2.2.1 Schottky Barrier Diodes... 326
    - 8.2.2.2 Field-effect Transistors 329
- 8.3 C_{60} Superconductivity 330
- 8.4 Polymer Batteries and Carbon Nanotube Supercapacitors..................... 333

	8.4.1	Conducting Polymer Batteries 333
	8.4.2	Biofuel Cells 336
	8.4.3	Carbon Nanotube Supercapacitors... 337
8.5	Carbon Nanotube Nanoelectronics.. 338
	8.5.1	Carbon Nanotube Nanowires.. 338
	8.5.2	Carbon Nanotube Superconductors 340
	8.5.3	Carbon Nanotube Rings.. 340
	8.5.4	Carbon Nanotube Nanocircuits... 342
	8.5.5	Carbon Nanotube-based Random Access Memory (RAM) for Molecular Computing.. 345
8.6	DNA Molecular Wires and DNA Computing 346
	8.6.1	DNA Molecular Wires... 346
	8.6.2	DNA Computing on Chips 351
8.7	References 353

9 Photonic Devices 357

9.1	Introduction 357
9.2	Light-emitting Polymer Displays 359
	9.2.1 Device Construction 359
	9.2.2 Quantum Efficiency 361
	9.2.3 Interface Engineering... 364
	9.2.3.1 Chemical Derivatization of the Metal Electrodes......... 364
	9.2.3.2 Polymer-polymer Interface 366
	9.2.4 Modification of the Charge Injection Characteristics 368
	9.2.5 Light-emitting Electrochemical Cells (LECs)........................... 368
	9.2.6 Color Tuning... 372
	9.2.7 Patterned Emission 375
9.3	Laser Action of Conjugated Polymers 376
9.4	Carbon Nanotube Displays... 378
9.5	Bucky Light Bulbs and Optical Limiters.. 381
	9.5.1 C_{60} Light Bulbs... 381
	9.5.2 C_{60} Optical Limiters 381
9.6	Photovoltaic Cells 383
	9.6.1 Polymer Photovoltaic Cells Containing Fullerenes 383
	9.6.2 Polymer Photovoltaic Cells Containing Carbon Nanotubes 385
9.7	Light-harvesting Dendrimers 387
9.8	Electronic Windows, Electrochromic Displays and Electronic Papers 392
	9.8.1 Electrochromic Windows 392
	9.8.2 Electrochromic Displays.. 393
	9.8.3 Electronic Papers 395
9.9	References 397

10 Sensors and Sensor Arrays 405

10.1	Introduction 405
10.2	Conjugated Polymers Sensors 406
	10.2.1 Conjugated Polymer Sensors with Electrical Transducers 406

		10.2.1.1 Conjugated Polymer Conductometric Sensors............407
		10.2.1.2 Conjugated Polymer Potentiometric Sensors...............412
		10.2.1.3 Conjugated Polymer Amperometric Sensors 412
		10.2.1.4 Conjugated Polymer Voltammetric Sensors.................413
	10.2.2	Conjugated Polymer Sensors with Optical Transducers...........413
		10.2.2.1 Conjugated Polymer Fluorescent Ion Chemosensors..413
		10.2.2.2 Conjugated Polymer Fluorescent TNT Sensors...........418
		10.2.2.3 Conjugated Polymer Light-harvesting "Turn-on" Sensors 420

10.3 Charge Transfer Polymer Sensors..422
10.4 Ionically Conducting Polymer Sensors 422
10.5 Conductively Filled Polymers Sensors..423
    10.5.1   Conductively Filled Polymer Humidity Sensors........................424
    10.5.2   Conductively Filled Polymer Gas Sensors..................................425
    10.5.3   Conducting Polymer-coated Fabric Sensors: Smart Textiles...425
10.6 Dendrimer Sensors 426
    10.6.1   Dendrimer Gas Sensors 426
              10.6.1.1 Dendrimer Iodine (Vapor) Sensor................................426
              10.6.1.2 Dendrimer SO_2 Gas Sensors.......................................426
              10.6.1.3 Dendrimer CO Gas Sensors 427
    10.6.2   Dendrimer Sensors for Carbonyl Compounds 428
10.7 Fullerene C_{60} Sensors 428
    10.7.1   Fullerene Humidity Sensors..428
    10.7.2   Fullerene Gas Sensors 430
10.8 Carbon Nanotube Sensors 431
    10.8.1   Carbon Nanotube Gas Sensors 431
              10.8.1.1 Carbon Nanotube Ammonia and Nitrogen Dioxide Sensors 431
              10.8.1.2 Carbon Nanotube Hydrogen Sensors 433
              10.8.1.3 Carbon Nanotube Oxygen Sensors 434
              10.8.1.4 Carbon Nanotube Thermoelectric Nanonose 435
              10.8.1.5 Carbon Nanotube Carbon Dioxide Sensors..................439
    10.8.2   Carbon Nanotube Pressure and Temperature Sensors 440
    10.8.3   Carbon Nanotube Chemical Force Sensors 442
    10.8.4   Carbon Nanotube Resonator Mass Sensors 443
    10.8.5   Carbon Nanotube Glucose Sensors...444
10.9 DNA Sensors...446
    10.9.1   DNA Sensors Based on Oligonucleotide-functionalized Polypyrroles..447
    10.9.2   DNA Diagnostic Biosensors..447
    10.9.3   DNA Sensor for Detection of Hepatitis B Virus 447
    10.9.4   DNA Fluorescent Sensor for Lead Ions.......................................448
    10.9.5   DNA Molecular Break Lights..449
    10.9.6   DNA Quartz Oscillators and Cantilevers....................................450
10.10 Sensors Arrays..451

       10.10.1 Conducting Polymer "Electronic Noses" 452
       10.10.2 DNA Arrays 454
       10.10.3 Protein Arrays 455
  10.11 References 456

11 Actuators and Nanomechanical Devices 461
  11.1 Introduction 461
  11.2 Conducting Polymer Actuators 462
       11.2.1 Self-powered Actuators 465
       11.2.2 Conducting Polymer Microtweezers 466
  11.3 Actuators Based on Composites of Ion-exchange Polymers and Metals 468
  11.4 Responsive Polymer Actuators 471
  11.5 Carbon Nanotube Actuators 475
  11.6 Smart Electromechanical Devices Based on Carbon Nanotubes 478
       11.6.1 Carbon Nanotube Quantum Resistors and Nanoresonators 478
       11.6.2 Carbon Nanotube Nanoprobes 481
       11.6.3 Carbon Nanotube Nanotweezers 483
       11.6.4 Carbon Nanotube Bearings, Switches and Gears 484
  11.7 C_{60} Abacus and Fullerene Vehicles 487
  11.8 Smart Devices Based on Biomolecules 488
       11.8.1 Flagellar Motors 488
       11.8.2 DNA Switches 488
  11.9 References 489

Index 491

Chapter 1

The Concepts of Intelligent Macromolecules and Smart Devices

1.1 Introduction

All matter is made up of *atoms*. The term *molecule* is used to describe groups of atoms that tend to exist together in a stable form. As the term implies, *macromolecule* refers to a molecule of extraordinarily large size (Flory, 1953). The macromolecules discussed in this book include both conventional large molecules (*e.g.* various synthetic/natural polymers, fullerene C_{60} and carbon nanotubes), in which the atoms are held together by covalent bonds, and supramolecules formed by the association of macromolecules or low molecular weight compounds by physical forces (*e.g.* H-bonding, electrostatic forces, donor-acceptor interactions, *etc.*).

The properties of macromolecules depend strongly not only on the chemical nature of the constituent atoms but also on their arrangement within a single macromolecule and in space. Macromolecules offer a unique class of materials with the natural length of polymer chains and their morphologies in the bulk, lying precisely at the nanometer-length scale, indicating considerable room for creating new properties and functions even without any change in their chemical composition. Indeed, the systematic organization of matter at the nanometer-length scale is a key feature of biological systems. Every living cell is filled with natural nanomachines of DNA, RNA, proteins, *etc.*, which interact to produce tissues and organs (Clegg and Clegg, 1987). All creatures great and small are simply made up of some nitrogen, oxygen, hydrogen, carbon, calcium, a little sulfur, iron, phosphorus, and some other elements. In a mysterious way, nature handles these inexpensive building blocks, which man has known for centuries, to create various nanostructures that enable creatures to function.

At the nanometer level, the wave-like properties of electrons inside matter and atomic interactions are influenced by the size of the material (Goldstein, 1997). As a consequence, changes in melting points, as well as magnetic, optic and electronic properties can be observed as the material takes on nanodimensions (Goldstein,

1997). Due to the high surface-to-volume ratio associated with nanometer-sized materials, a tremendous improvement in chemical properties is also achievable through a reduction in size (Goldstein, 1997). Besides, new phenomena, such as the confinement-induced quantization effect, also occur when the size of materials becomes comparable to the deBroglie wavelength of charge carriers inside (Goldstein, 1997). By creating nanostructures, therefore, it is possible to control the fundamental properties of materials (including macromolecular materials, of course) even without changing the materials' chemical composition. This should, in principle, allow us to develop new intelligent materials and smart devices of desirable properties and functions for numerous applications.

The recent discoveries of buckminsterfullerene C_{60} (Kroto *et al.*, 1985) and the carbon nanotube (Iijima, 1991) have opened up a new era in material science and nanotechnology. Nanotechnology is the creation of useful materials, devices and systems through the control of matter at the nanometer-length scale and the exploitation of novel properties and phenomena developed at that scale (Timp, 1998). Humans had had experience in nanotechnology centuries ago, though we were not aware of it due to our limited ability to see and control matter at a billionth of a meter scale for a long time. For instance, catalysts allow thousands of chemical reactions to go on producing the rearrangement of atoms in ways imitating the elegance and economy of nature. Photography allows us to produce pictures with unlimited resolution that even the most advanced digital camera cannot match. Particles on the photographic film have sizes well below the wavelength of the light, allowing every detail of an object to be recorded on the film. However, the manufacturing methods we have been using for a long time, including casting, grinding and milling, would be considered very crude from the perspective of the molecular level. Recent developments in nanoscience and nanotechnology have resulted in a large variety of tools and methods, including the scanning tunneling microscope (STM) (Magonov and Whangbo, 1996), the atomic force microscope (AFM) (Wickramasinghe, 1992), laser tweezers (Sheetz, 1998), soft-lithographic (Jackman and Whitesides, 1999) and self-assembling (Varner, 1988) techniques, which have been devised for building materials and devices atom-by-atom or molecule-by-molecule through the so-called "bottom-up" and/or "top-down" approaches (*vide infra* Krummenacker and Lewis, 1995). With so many intelligent macromolecules and nanotools having already been reported and with yet more to be developed, there will be vast opportunities for developing numerous new devices and systems of potentially useful characteristics. These concepts can be better illustrated by, but not limited to, the following possibilities:

1) Our bodies naturally have intricate molecular machines functioning within them at all times. If any of these molecular machines is damaged, major health problems can arise. Currently we are unable to treat medical disorders on a molecular level, therefore leaving us unable to fully combat many of the illnesses and disorders which currently plague our population. The field of nanomedicine has the ability to breach such boundaries by developing many ways for doctors to treat patients on a molecular level. For instance, in the future, we will have smart machines that mimic those naturally existing in the body to serve in place of those damaged or defective parts in the body.

2) Solar energy has been considered to be the ultimate source of electricity for this world. At present the conversion of solar energy to electricity is feasible with

the help of solar cells (also termed photovoltaic cells). The intensive ongoing research into polymeric photovoltaic cells could provide the world with very cost-efficient and clean energy from sunlight, which will change our daily lives forever. Imagine if you had a car that ran on a photovoltaic cell during the day and stored energy for night. Imagine also if you had a house equipped with such energy systems. You would never have any kind of bills any more for electricity and gas.

3) In everyday life, we could experience the innovations too. For instance, intelligent macromolecules would help us to solve problems, such as bulky clothes for the winter. Imagine if nice, good-looking jackets which would be very light and slim but more protective and tough could be manufactured. This could be done if some thermo-responsive macromolecules could be attached to nanotube fibers that run along the fabric in the outer part of the jacket. The polymer would shrink when the fibers detect cold conditions from the ambient, keeping the inside layers from moisture and cold air.

These are just some ideas which could become reality. The day is not far away when our imagination will become reality, as the technology in intelligent macromolecules and smart devices is developing so rapidly that it will affect every aspect of our lives in the long term. However, most of these developments will come from an interdisplinary effort combining knowledge from the fields of physics, chemistry, biochemistry, materials science, and electronic engineering. Due to its advanced nature, the subject is not considered in conventional textbooks, and, as a consequence of the interdisciplinary approach, it has not been treated in any monograph. The purpose of this book is to 'fill the gap' by providing an up-to-date summary of the field to a broad group of scientists and engineers with interests in intelligent materials and associated smart devices. In order to cover the multidisciplinary field of such diversity, I have divided the text into three major parts. The first part, under the heading of *Intelligent Macromolecules,* covers the basic science of conducting polymers (Chapter 2), stimuli-responsive polymers (Chapter 3), dendrimers and fullerene C_{60} (Chapter 4) and carbon nanotubes (Chapter 5). This is followed by Chapters 6 and 7 in the second part of the book, *From Intelligent Macromolecules to Smart Devices*, dealing with the construction of oriented polymers/carbon nanotubes and polymer nanostructures, which fulfill a particularly useful bridge between the development of intelligent macromolecules and the construction of smart devices. Various devices based on intelligent macromolecules, including sensors, actuators, polymer photovoltaic cells, polymer displays, and DNA computers, are then described in the final part of the book concerning *Smart Devices* (Chapters 8–11). The above approach allows the reader to first review the scientific basis for intelligent macromolecules. The basic knowledge is then extended to functional structures and actual devices, many of which are of practical importance.

In this chapter, the concepts of intelligent macromolecules and smart devices are demonstrated by outlining the distinct structural features of synthetic macromolecules, biomacromolecules, and carbon nanomaterials, followed by describing the self-assembling and scanning probing approaches toward functional structures and smart devices.

1.2 The Concept of Intelligent Macromolecules

1.2.1 Synthetic Macromolecules

1.2.1.1 Chain Structure and Classification

Polymer is a combination of the two words *poly* and *monomer*, and literally means many monomers or *macromolecule* (for convenience, polymer and macromolecule are used interchangeably in this book). The molecular weights of most natural and synthetic polymers range from 10^4 to 10^7 daltons. Although these polymers are so large, they can be defined by the basic repeat unit often termed as the monomer unit. Many of the overall properties of a particular polymer depend on the constituent monomer units and the way they are arranged in the macromolecule. The simplest form of polymers is the unbranched chain, the so-called chain structure, because they consist of many links of the monomer units. If only one kind of monomer unit is involved in a polymer chain, it is called a *homopolymer*. However, there are a number of polymers consisting of two or more different monomer units with many different ways in which they can be arranged in the macromolecule to form the so-called *copolymer*. To describe a linear homopolymer, it is sufficient to specify the number of monomeric units forming the chain. For a copolymer consisting of two monomeric units A and B, however, it is necessary to specify both the ratio of the numbers of the two monomeric units (*i.e.* composition of the copolymer) and the rule of sequencing A and B along the polymer chain. Table 1.1 shows a number of copolymers consisting of two monomer units, but with different sequences. Only one of them shows that the A and B units alternate in a completely regular fashion, and so this block copolymer is said to be an alternating block copolymer.

Table 1.1. Types of block copolymers (After Vollmert, 1973)

∿∿A—B—B—A—B—A—A—A—B—A—A—B—A—B—B—B—A∿∿
1. Statistical or random sequence

∿∿A—B—A—B—A—B—A—B—A—B—A—B—A—B—A—B—A—B∿∿
2. Alternating sequence

∿∿A—A—A—A ··· A—A—A—A—B—B—B—B ··· B—B—B—A—A—A∿∿
3. Segment polymers = block copolymers

Table 1.2 lists some of the most commonly known polymers, along with the monomer structure.

Table 1.2. Repeat units of some common polymers (After Hamley, 2000)

Structure	Name	Acronyms
$-[-\underset{\underset{CH_3}{|}}{C}=CHCH_2CH_2-]_n-$	Poly(isoprene)	PI
$-[-CH=CHCH_2CH_2-]_n-$	Poly(butadiene)	PB
$-[-CH_2CH_2-]_n-$	Poly(ethylene)	PE
$-[-CF_2CF_2-]_n-$	Poly(tetraflouroethylene)	PTFE
$-[-\underset{\underset{CH_3}{|}}{C}HCH_2-]_n-$	Poly(propylene)	PP
$-[-\underset{\underset{Cl}{|}}{C}HCH_2-]_n-$	Poly(vinyl chloride)	PVC
$-[-\underset{\underset{C_6H_5}{|}}{C}HCH_2-]_n-$	Poly(styrene)	PS
$-[-\underset{\underset{COOCH_3}{|}}{\overset{\overset{CH_3}{|}}{C}}CH_2-]_n-$	Poly(methyl methacrylate)	PMMA
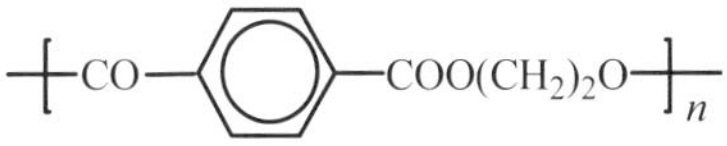 $-[-CO-C_6H_4-COO(CH_2)_2O-]_n-$	Poly(ethylene teraphthalate)	PET
$-[-CH_2CHCN-]_n-$	Poly(acrylonitrile)	PAN
$-[-OCH_2CH_2-]_n-$	Poly(ethylene oxide)	PEO
$-[-CO(CH_2)_5NH-]_n-$	Poly(α-caprolactam)	Nylon-6

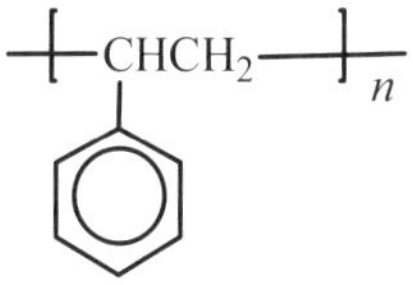

Apart from the block copolymer, various other copolymers, including graft (branched), star, and cross-linked copolymers, can be synthesized (Vollmert, 1973). Figure 1.1 shows the major classes of macromolecules (Dvornic and Tomalia, 1996). Also included in Figure 1.1 is a relatively new class of macromolecules called dendrimers. The dendritic polymers are molecular species with a large number of hyperbranched chains composed of two or more monomer units surrounding a central core (Chapter 4). The more recently discovered buckminsterfullerene C_{60}, a carbon macromolecule with a soccerball-like structure consisting of 12 pentagons and 20 hexagons facing symmetrically, and carbon nanotubes, consisting of carbon hexagons arranged in a concentric manner with both ends normally capped by fullerene-like structures containing pentagons, can be regarded as the newest members of the polymer family.

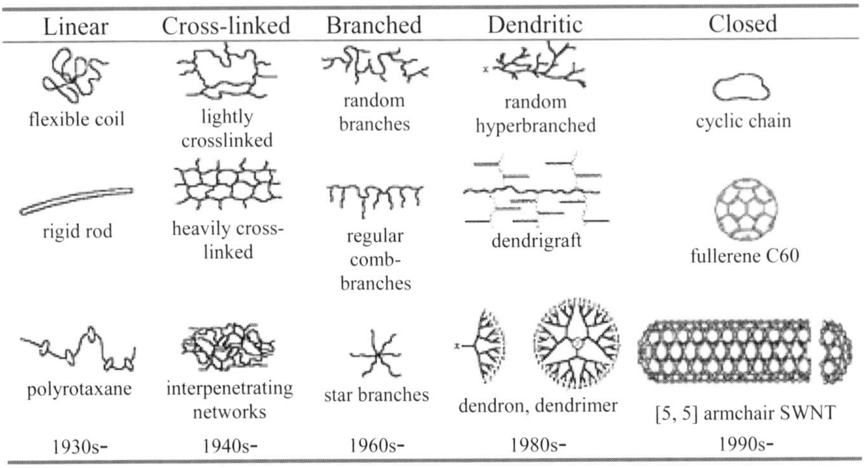

Linear	Cross-linked	Branched	Dendritic	Closed
flexible coil	lightly crosslinked	random branches	random hyperbranched	cyclic chain
rigid rod	heavily cross-linked	regular comb-branches	dendrigraft	fullerene C60
polyrotaxane	interpenetrating networks	star branches	dendron, dendrimer	[5, 5] armchair SWNT
1930s-	1940s-	1960s-	1980s-	1990s-

Figure 1.1. Major classes of macromolecular architecture arranged by increasing order of complexity from left to right (classes) and from top to bottom (subclasses) (After Dvornic and Tomalia, 1996, copyright 1996 Elsevier)

1.2.1.2 Synthesis

To form a polymer, a monomer must be capable of being linked to two or more other monomers. Although many methods can be used to polymerize a given monomer species into macromolecules, polymerization methods may be divided into two broad categories: step-growth polymerization and chain-growth polymerization.

Step-growth polymerization, as the name suggests, refers to stepwise reactions between any monomer species of two or more functionalities. A typical example is the condensation polymerization of diol and diacid monomers to form polyesters with the release of water molecules [Equation (1.1)].

$$n\text{HO-R-OH} + n\text{HOOC-R'-COOH} \rightarrow \text{HO-[R-OCO-R']}_n\text{-COOH} + n\text{H}_2\text{O} \qquad (1.1)$$

The formation of highly branched or network-like polymer chains often occurs when the step-growth polymerization involves a monomer species with more than two functionalities.

Chain-growth polymerization, also called addition polymerization, can be further divided into free radical addition polymerization and ionic polymerization. Free radical polymerizations normally involve three steps known as initiation, chain propagation, and chain termination [Equations (1.2–1.7)].

Initiation:

$$I \rightarrow 2R^{\bullet} \qquad (1.2)$$
$$R^{\bullet} + M \rightarrow RM^{\bullet} \qquad (1.3)$$

Chain propagation:

$$RM^{\bullet} + M \rightarrow RMM^{\bullet} \qquad (1.4)$$
$$RM_{n-1}^{\bullet} + M \rightarrow RM_n^{\bullet} \qquad (1.5)$$

Chain termination:

(a) Termination by combination

$$RM_n^{\bullet} + RM_m^{\bullet} \rightarrow RM_{n+m}R \qquad (1.6)$$

(b) Termination by disproportionation

$$RM_n^{\bullet} + RM_m^{\bullet} \rightarrow RM_n + RM_m \qquad (1.7)$$

At the initiation stage, the polymerization is initiated by the creation of free radicals ($R^{\bullet}$) from initiators (*e.g.* organic peroxides or azo compounds) to produce monomeric free radicals ($RM^{\bullet}$) [Equations (1.2)–(1.3)]. Chain growth or propagation proceeds by the addition of monomer (M) molecules to monomeric free radicals to form the so-called polymeric free radicals [Equation (1.5)]. Chain termination of the polymerization generally results from the combination of two polymeric free radical active centers [Equation (1.6)]. Alternatively, chain termination can occur through the abstraction of a hydrogen atom from one growing chain by another [Equation (1.7)]. The overall rate of polymerization and the length of the polymer chains formed in the free radical polymerization are determined by the rates of the above individual processes (Flory, 1953; Vollmert, 1973).

Ionic polymerizations can be divided into two types: anionic and cationic. While anionic polymerizations proceed by the addition of a monomer to a reactive anionic chain end, cationic polymerizations are polymerizations that proceed by addition of monomer to a cationic chain end. Therefore, catalysts for cationic polymerizations are usually electron-deficient [*e.g.* protonic acids (H_2SO_4 and H_3PO_4), lewis acids ($AlCl_3$, BF_3, ZnO, $FeCl_3$ and PF_5), and halogens (I_2)], and molecules of electron-rich double bonds are also good monomers. In contrast, anionic polymerizations are often initiated by basic species, such as organoalkali compounds (*e.g.* BuLi) and alkali metals (*e.g.* Na, Li). Good monomers for anionic polymerizations are vinyl molecules ($CH_2=CR_1R_2$), with R_1 and R_2 being electron-withdrawing groups or unsaturated functions with the capability to stabilize a negative charge through resonance or inductive effects. The propagation steps can be summarized as:

Anionic $M + I^+ \rightarrow M^-I^+$ (1.8)

Cationic $M + I^- \rightarrow M^+I^-$ (1.9)

As can be seen in Equations (1.8)–(1.9), the reactive chain end is always associated with a counter ion of the opposite charge in ionic polymerizations of either type. Since the propagating species have the same charge, the combination termination by coupling two active centers cannot occur. Particularly, there is no formal termination step in anionic polymerization with the carbanion end groups remaining continuously reactive, hence the name of *living polymerization*. Polymer chains with the living carbanion end groups can be used for the synthesis of block copolymers by growing new block sequence(s) from monomer(s) sequentially added (Odian, 1970). Alternatively, the living carbanion can be used for end-functionalization of the polymer chain by deactivating the living chain end with specific molecular species (*e.g.* proton donors) and subsequent derivatization (Morton, 1983).

Table 1.3 listed some common monomers for free radical, cationic, and/or anionic polymerizations.

Table 1.3. Polymerization of vinyl monomers (After Kaufman and Falcetta, 1977)

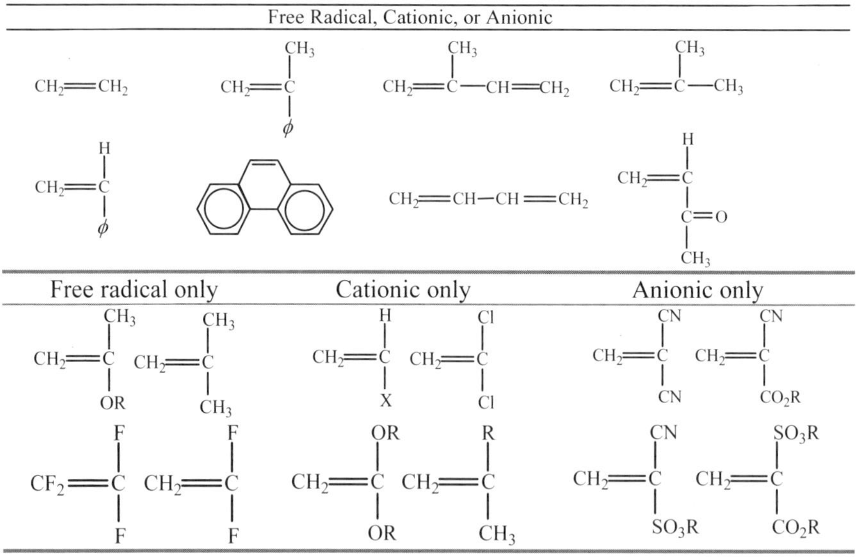

Free Radical, Cationic, or Anionic			
$CH_2\!=\!CH_2$	$CH_2\!=\!C(CH_3)\phi$	$CH_2\!=\!C(CH_3)\!-\!CH\!=\!CH_2$	$CH_2\!=\!C(CH_3)\!-\!CH_3$
$CH_2\!=\!CH\phi$	(phenanthrene)	$CH_2\!=\!CH\!-\!CH\!=\!CH_2$	$CH_2\!=\!CH\!-\!C(=\!O)\!-\!CH_3$

Free radical only	Cationic only	Anionic only
$CH_2\!=\!C(CH_3)OR$ $CH_2\!=\!C(CH_3)CH_3$	$CH_2\!=\!CH(H)X$ $CH_2\!=\!C(Cl)Cl$	$CH_2\!=\!C(CN)CN$ $CH_2\!=\!C(CN)CO_2R$
$CF_2\!=\!CF_2$ $CH_2\!=\!CF_2$	$CH_2\!=\!C(OR)OR$ $CH_2\!=\!C(R)CH_3$	$CH_2\!=\!C(CN)SO_3R$ $CH_2\!=\!C(SO_3R)CO_2R$

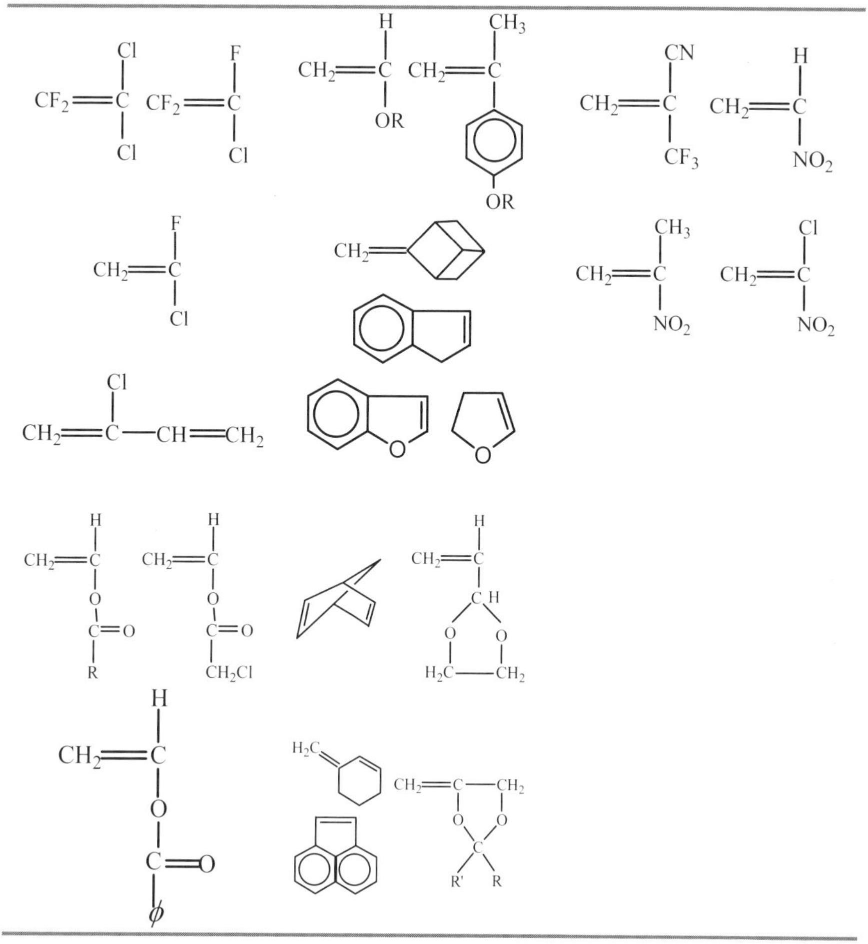

1.2.1.3 Chain Conformation

In the preceding sections we have seen how macromolecules are made by various polymerization methods. Here, we start to look at the conformational structures (*i.e.* the arrangements in three dimensions) and properties for individual macromolecular chains. The unique feature for all polymer chains is their usually long length compared to the dimensions of their cross-section. As a consequence, most polymer chains are highly flexible. At a local scale, the conformation of an individual polymeric macromolecule depends on rotations of the C-C bonds that make up the backbone of most synthetic macromolecules. To study the rotation of C-C bonds in a polymer chain, let us first consider a short segment of polyethylene, $-(CH_2)_n-$, whose chain contains only four methylene groups (Figure 1.2).

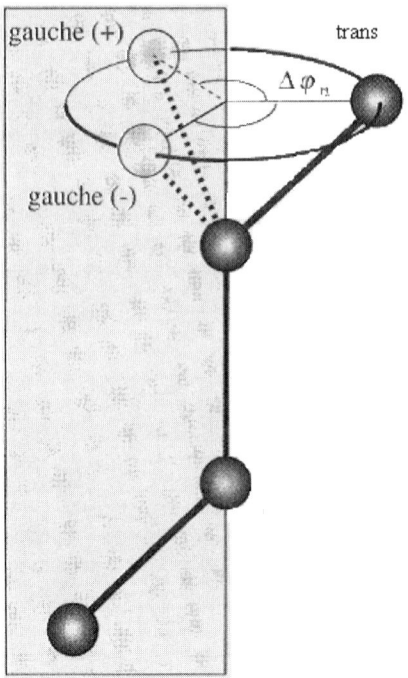

Figure 1.2. Configuration parameters along a chain of ethylene groups (Courtesy of Kleman and Lavrentovich, 2003, copyright 2003 Springer-Verlag. Reproduced with permission)

As can be seen in Figure 1.2, a plane (the shadow area) is defined by three of the carbon atoms in this segment. If we allow free rotation about the carbon-carbon bond, the fourth atom can then be anywhere on the circle indicated in the figure, though some positions (*e.g.* trans, gauche$^+$, gauch$^-$) are more likely than others since completely free rotation is impossible due to steric hindrances. Each successive carbon atom along the chain can randomly take any of several positions on a similar circle based on the position of the preceding atom, the total number of possible conformations in a polymer chain thus being formidably large. If we consider a chain with 1000 links (which corresponds to a polyethylene chain with a modest molecular weight of 14000) and consider the three most favorable conformations (t, g$^+$, g$^-$) only for each link, for example, the number of possible conformations is about 3^{1000}!

Among all of the possible conformations for a given chain, one of them is of particular interest. In this case, each successive carbon atom lies in the same plane in the trans location with respect to adjacent atoms in the chain (Figure 1.3). This fully-extended all-trans conformation has the *fully extended length* of $nl\sin(\theta/2)$, whereas the *counter length* of the chain along all the zigzag path from atom to atom is nl. These do not, however, give a realistic measure of the size of the polymer chain that often coils up in the molten or in a dilute solution. This, together with the conformational fluctuation due to thermal motion, prompts us to take a statistical average when considering the size of polymer coils.

Figure 1.3. The fully extended all-trans configuration of a carbon-carbon chain (After Kaufman and Falcetta, 1977, copyright 1997 John Wiley & Sons, Inc. Reproduced with permission)

There are two average parameters that can be used to characterize the size of polymer coils, namely the root-mean-square end-to-end distance, $<r^2>_0^{1/2}$, and the radius of gyration, $<R_g^2>_0^{1/2}$, where $<>$ indicates a thermal average. The root-mean-square end-to-end distance is the average separation between chain ends while the radius of gyration serves as a measure of the average distance of a chain segment from the center of mass of the coil. To simplify the calculation of polymer chain dimensions, the chain is supposed to consist of n volumeless links of length l that can rotate freely in space with no restrictions at all on the angles between successive links. The chain coils back, and even crosses itself, many times, leading to a dense 'clumped up' structure. This model is then called the freely jointed chain model, in which the polymer coil effectively executes a random walk, as schematically shown in Figure 1.4.

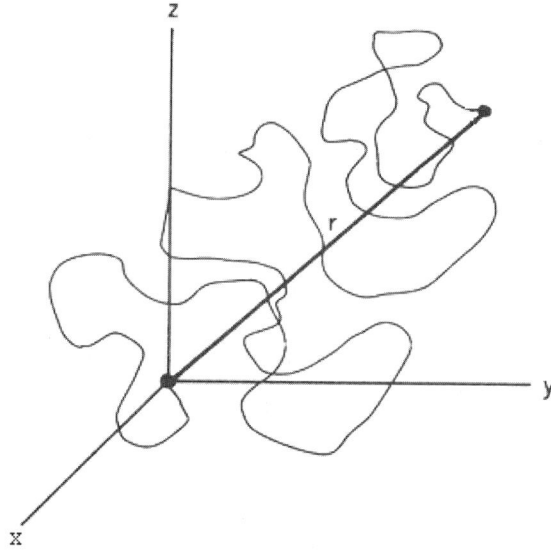

Figure 1.4. A schematic representation of a random-coil polymer chain, with one end at the original of a coordinate system and the other at a distance r from the origin (After Kaufman and Falcetta, 1977, copyright 1997 John Wiley & Sons, Inc. Reproduced with permission)

The calculation of the end-to-end distance of a freely jointed chain is then the same as the *random-flight* problem associated with Brownian motion. The statistics of random walks had been worked out for Brownian motion by Einstein, and it turns out that the mean-square end-to-end distance is:

$$<r^2>_0 = nl^2 \qquad (1.10)$$

Random polymer coils are often also called Gaussian chains. This is because the probability of finding one end of the chain in a given volume element follows a Gaussian function. For linear Gaussian chains, the root-mean-square radius of gyration is:

$$<R_g^2>_0^{1/2} = <r^2>_0^{1/2} / 6^{1/2} \qquad (1.11)$$

Thus, simple calculation shows that is $<r^2>_0^{1/2}$, which is quite small compared to the contour length of the chain. For a polyethylene chain of 1000 carbon atoms with C-C bonds 1.54 Å long, the fully extended length is calculated to be 1270 Å, whereas $<r^2>_0^{1/2} = 49$ Å and 20 Å.

1.2.1.4 Macromolecular Structure in Solution

Coil-globule Transition
In order to understand the structure of single polymer chains, one needs to separate chain-chain interactions from the intra-chain properties. In a good solvent, the repulsive interaction between polymer segments forces polymer chains to adopt a swollen coil-like conformation at a diluted concentration. When the solvent quality becomes worse, however, the unfavorable interactions between solvent and polymer segments lead to a contraction of the coil. By investigating polymers in dilute solution and as a function of their concentration, the intermolecular interference can be eliminated. Also the interaction between solvent and polymer becomes effectively zero when the measurements are carried out at the θ-point (where the second virial coefficient, $A_2 = 0$). In a solvent worse than the θ-solvent, the attractions between monomers are so strong that the polymer coil collapses into a compact globule conformation. The coil-globule transition has generated a great deal of interest. For detailed discussion of the current state of research on coil-globule transition, interested readers are referred to specialized reviews and monographs (Cantor and Schimmel, 1980; Shmakov, 2002).

From a thermodynamic point of view, in order for solution to take place, it is essential that the free energy $\Delta G = \Delta H - T\Delta S$, which is the driving force in the solution process, should be less than 0, where ΔH and ΔS represent the change in enthalpy and the change in entropy in the solution process, respectively.

The ideal solution law, as embodied by Raoult's law, provides the basis for the treatment of a simple molecule solution. Ideal solution behavior requires that the following conditions be fulfilled: a) the entropy of mixing must be given by

(Moelwyn-Hughes, 1965)

$$\Delta S = -k(n_1 \ln N_1 + n_2 \ln N_2) \tag{1.12}$$

where n_1 and n_2 are the numbers of molecules of solvent and solute, respectively, and N_1 and N_2 their mole fractions; b) the heat of mixing ΔH must equal 0. Deviations from ideality may arise from the failure of either of these conditions. Numerous investigations (Meyer and Luhdemann, 1935; Morawetz, 1975; Yamakawa, 1971) on polymer solutions have revealed that polymer solutions do deviate greatly from the ideal solution behavior due to failures in both condition a) and b). It, therefore, is necessary to derive expressions of the thermodynamic functions (entropy, heat of mixing, *etc.*) for mixed polymer and solvent.

Flory-Huggins Liquid Lattice Theory
Flory and Huggins derived expressions of entropy and enthalpy for polymer solutions on the basis of a liquid lattice model (Flory, 1953), in which the molecules in the pure liquids and in their solution are considered to be arranged with enough regularity to be represented approximately by a lattice, as schematically illustrated in Figure 1.5.

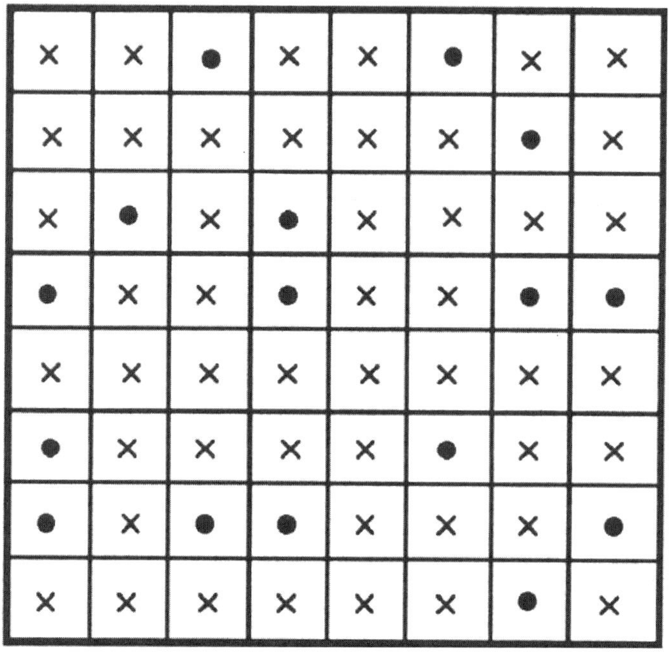

Figure 1.5. Molecules of low molecular weight solute distributed over the lattice used to describe a binary solution: • solute molecule; × solvent molecule

Although the Flory-Huggins liquid lattice theory has its limitations (Flory, 1953; Napper, 1983), it has served as a fundamental framework in treating the thermodynamics of polymer solutions over several decades and will be briefly stretched here.

The Entropy of Mixing

In a simple liquid consisting of nearly spherical molecules, the pure solvent may be arranged in the lattice in only one way. For a solution, the total number of ways of arranging the n_1 identical molecules of the solvent and n_2 identical molecules of solute on the lattice comprising $n_0 = n_1 + n_2$ cells gives the total number of arrangements, W, as

$$W = n_0! / n_1! n_2! \tag{1.13}$$

By following the Boltzmann relation, therefore, the entropy of mixing should be given by

$$\Delta S = k \ln W \tag{1.14}$$

where k is the Boltzmann constant.

In view of Stirling's approximation, *i.e.*

$$\ln n! = n \ln n - n \tag{1.15}$$

by suitable rearrangement, Equation (1.15) could be reduced to Equation (1.16).

$$\Delta S = k[(n_1 + n_2)\ln(n_1 + n_2) - n_1 \ln n_1 - n_2 \ln n_2] \tag{1.16}$$

The above liquid lattice model treatment cannot possibly hold for polymer solutions, where the solute molecule may be a thousand or more times the size of the solvent. The assumed approximate interchangeability between solvent and solute molecules in the liquid lattice is thus unrealistic. However, the long chain polymer may be considered to consist of r chain segments, each of which is equal in size to a solvent molecule. Each of the segments occupies one lattice cell, and a segment and a solvent molecule may replace one another in the liquid lattice, as schematically shown in Figure 1.6. According to the situation illustrated in Figure 1.6, Flory and Huggins (Flory, 1942; Flory and Krigbaum, 1951; Huggins, 1942a, b) calculated the total configurational entropy, ΔS, of a polymer solution with N_1 solvent molecules and N_2 macromolecules as to be

$$\Delta S = -k(N_1 \ln \phi_1 + N_2 \ln \phi_2) \tag{1.17}$$

where ϕ_1 and ϕ_2 are the volume fractions of solvent and solute, *i.e.*

$$\phi_1 = N_1 / N_1 + rN_2; \qquad \phi_2 = rN_2 / N_1 + rN_2 \tag{1.18}$$

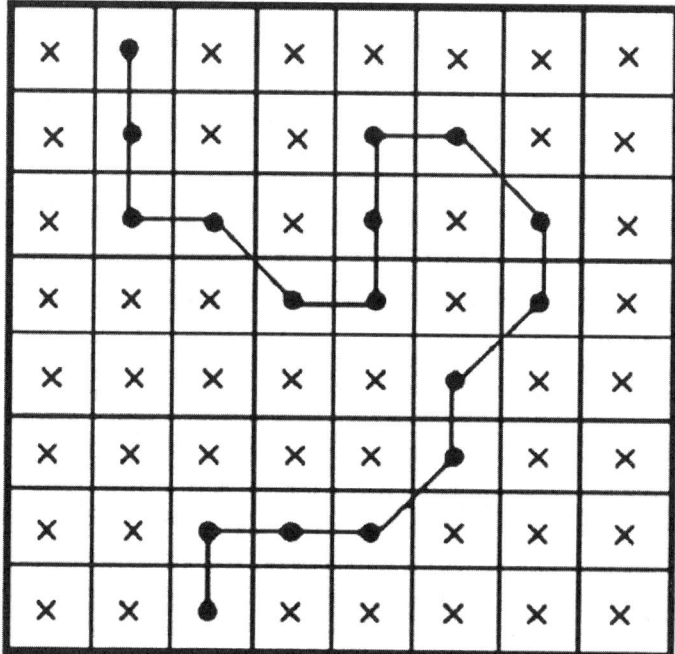

Figure 1.6. Segments of a polymer chain located in the liquid lattice

The Heat and Free Energy of Mixing
If $\Delta\varepsilon$ is the energy change associated with the pair contact between a solvent molecule and a chain segment in a polymer solution, the heat change of mixing N_1 molecules of solvent with N_2 polymer molecules of r_1 chain segments is given:

$$\Delta H = Z\Delta\varepsilon N_1 \phi_1 \tag{1.19}$$

where Z is the total number of contacts between a chain segment and its neighbors; ϕ_2 is the volume fraction of the solute [see Equation (1.18)].

We shall find it advantageous to recast Equation (1.19) in the following form:

$$\Delta H = kT\chi N_1 \phi_2 \tag{1.20}$$

where $X = Z\Delta\varepsilon / kT$ is the so-called Huggins parameter, which is a dimensionless quantity and characterizes the interaction energy per solvent molecule divided by kT. The quantity $kT\chi$ represents merely the difference in energy of a solvent molecule immersed in the pure polymer ($\phi_2 = 1$) compared with one surrounded by molecules of its own kind, *i.e.* the absolute strength of the interaction between polymer segments and solvent molecules. The solvent power of a given solvent for a polymer is also measured by the second virial coefficient, which is originally defined through a power series expression of the osmotic pressure, π, as a function of the polymer concentration, C, as follows (Brandrup and Immergut, 1966):

$$\pi/C = RT(1/M + A_2C + A_3C^2 + \cdots) \tag{1.21}$$

where R is the gas constant, T the absolute temperature, M the number average molecular weight of the polymer and A_2, A_3, ... are the second, third, ... virial coefficients.

The χ is related to the second virial coefficient, A_2, by (Napper, 1983)

$$A_2 = (1/2 - \chi)\, v_2^2/V_1 \tag{1.22}$$

where v_2 is the specific volume of the polymer and V_1 is the molar volume of the solvent (Cohen, 2000).

The free energy of mixing can be easily obtained by combining Equation (1.17) with Equation (1.20). That is

$$\Delta G = \Delta H - T\Delta S = kT[N_1\ln\phi_1 + N_2\ln\phi_1 + \chi N_1\phi_2] \tag{1.23}$$

Critical Solution Temperatures

To consider phase separation in polymer-solvent or polymer-polymer mixtures, it is often more convenient to write the χ parameter as follows:

$$\chi = A + B/T \tag{1.24}$$

where A and B are constants with the sign of B determining whether mixing is favored enthalpically at a high temperature or low temperature.

Knowing the temperature dependence of χ allows the equilibrium behavior of a polymer-solvent or polymer-polymer mixture to be predicted by the Flory-Huggins theory, as illustrated in Figure 1.7.

As can be seen from Figure 1.7, the limit of thermodynamic stability of the homogenous phase is given by the condition:

$$\partial^2 \Delta G_m/\partial \phi^2 = 0 \tag{1.25}$$

The second derivative of the Gibbs energy with respect to composition is equal to zero [*i.e.* Equation (1.25)] at points of inflection on the Gibbs energy curves for $T < T_c$ (Figure 1.7). The locus of such points defines the spinodal curve. The two sides of the spinodal curve meet a bimodal point at the critical solution temperature, which is then defined by the conditions:

$$\partial \Delta G_m/\partial \phi = \partial^2 \Delta G_m/\partial \phi^2 = \partial^3 \Delta G_m/\partial \phi^3 = 0 \tag{1.26}$$

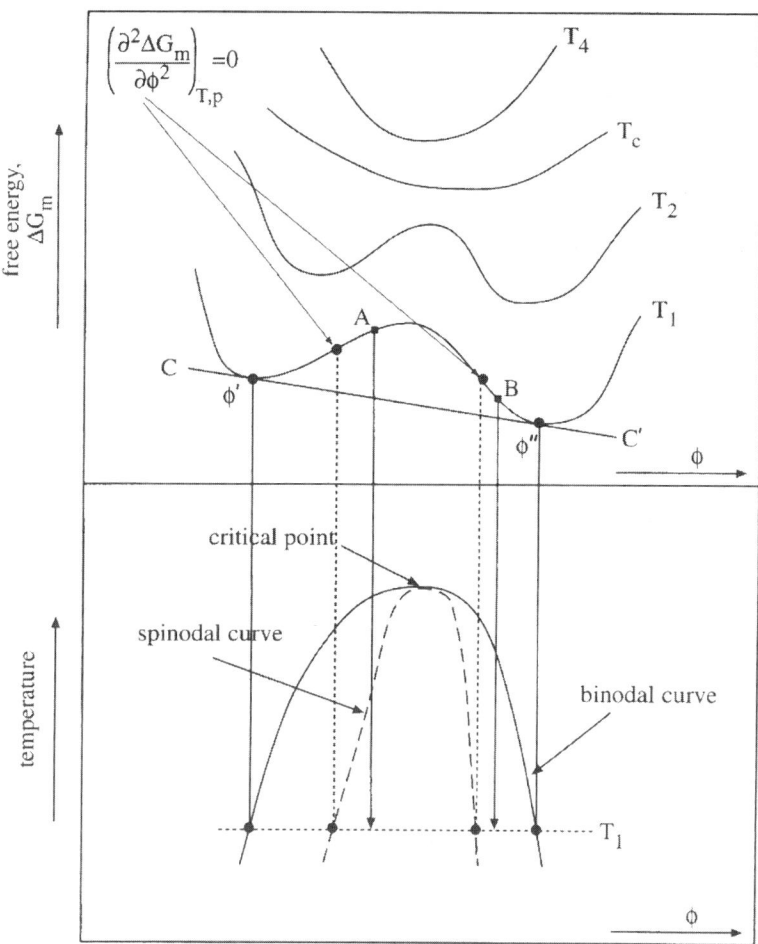

Figure 1.7. Analysis of phase behavior of a binary blend of polymer and solvent or two polymers exhibiting an upper critical solution temperature, T_c. Top: variation of Gibbs free energy with composition, ϕ ($\phi = \phi_1$ or ϕ_2) at four temperatures. The tie line CC' defines the compositions on the bimodal curve. The locus of points defined by the points of inflection $(\partial^2 \Delta G / \partial \phi^2)_{T,P} = 0$ define the spinodal curve. At point A (inside the spinodal curve), the mixture will spontaneously phase separate (into domains with compositions ϕ' and ϕ'') via spinodal decomposition. However, at point B (outside the spinodal curve) there is an energy barrier to phase separation, which then occurs by nucleation and growth (After Hamley, 2000, copyright 2000 John Wiley & Sons, Inc. Reproduced with permission)

Figure 1.7 shows that the spinodal and bimodal curves meet at a common maximum, which is called an *upper critical solution temperature* (UCST). In contrast to the UCST (Figure 1.8(a)), there are also systems to exhibit a *lower critical solution temperature* (LCST, Figure 1.8(b)).

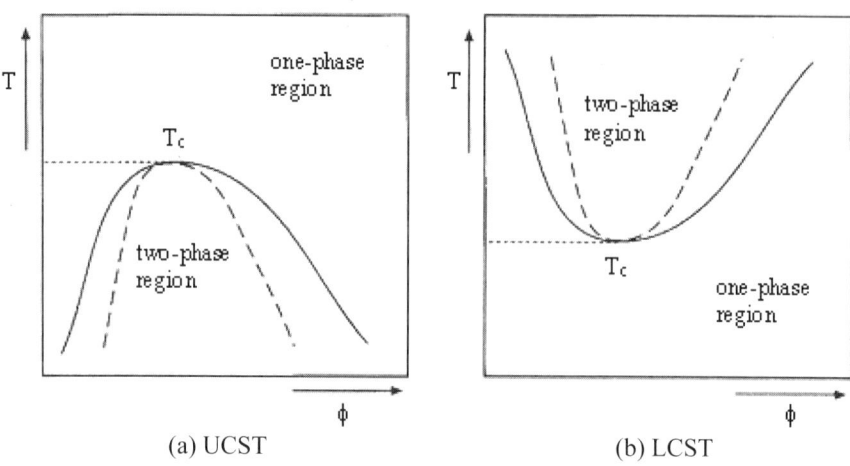

Figure 1.8. Phase diagrams for polymer blends showing (a) upper critical solution temperature (LCST); (b) lower critical solution temperature (LCST) behavior. The solid lines are binodals and the dashed lines are spinodals (After Hamley, 2000, copyright 2000 John Wiley & Sons, Inc. Reproduced with permission)

In the case of UCST, solubility is enhanced by an increase in temperature, which is often observed with non-polar macromolecules. This is because in these systems thermal motion increases contacts between unlike segments or solvent molecules and reduces that between similar species. In systems with specific interactions between molecules, such as hydrogen bonding, the less common LCST may occur, as the solvent quality decreases with increasing temperature. As we shall see in Chapter 3, poly(N-isopropylacrylamide) in water is an example of a system showing an LCST.

1.2.1.5 Primary, Secondary, Tertiary and Quaternary Structure

As has been seen from the above discussion, a polymer chain usually consists of a large number of mobile structural units, which can adopt various conformational structures. The random coil discussed above is the predominating type of *secondary structure*. While the *primary* structure refers to the kind of structural units a chain consists of and to the steric arrangement of neighboring structural units (*i.e.* chemical constitution and configuration), the secondary structure relates to the arrangement of the polymer chain within the range of a single macromolecule (*i.e.* conformation). The tertiary structure is concerned with the arrangement of the macromolecules to form more complex aggregates. The association of two or more tertiary aggregates can lead to quaternary structures. Figure 1.9 shows that a polymer chain can exist in a large number of different structures, including the extended chain, random coil, folded chains.

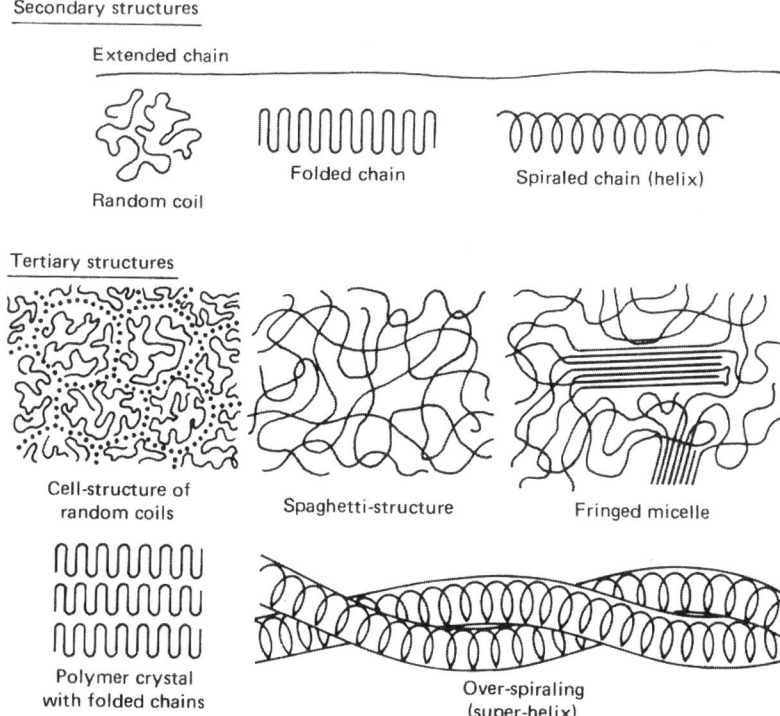

Figure 1.9. Schematic representations of secondary and tertiary structures of macromolecules (After Vollmert, 1973, copyright 1973 Springer-Verlag. Reproduced with permission)

As we shall see below, the helix plays a most important role with proteins and nucleic acids (*e.g.* DNA), in which the helical structures are stabilized by hydrogen bonds.

1.2.2 Biological Macromolecules

1.2.2.1 Structure of DNA

Deoxyribonucleic acid (DNA), the carrier of genetic information in all living species, serves as a typical example of a biomacromolecule with an extremely high molecular weight (typically, in the order of 10^6 to 10^9). DNA is a polydeoxyribonucleotide that contains many monodeoxyribonucleotides covalently linked by 3′,5′-phosphodiester bonds. Figure 1.10 shows the structure of a linear DNA strand.

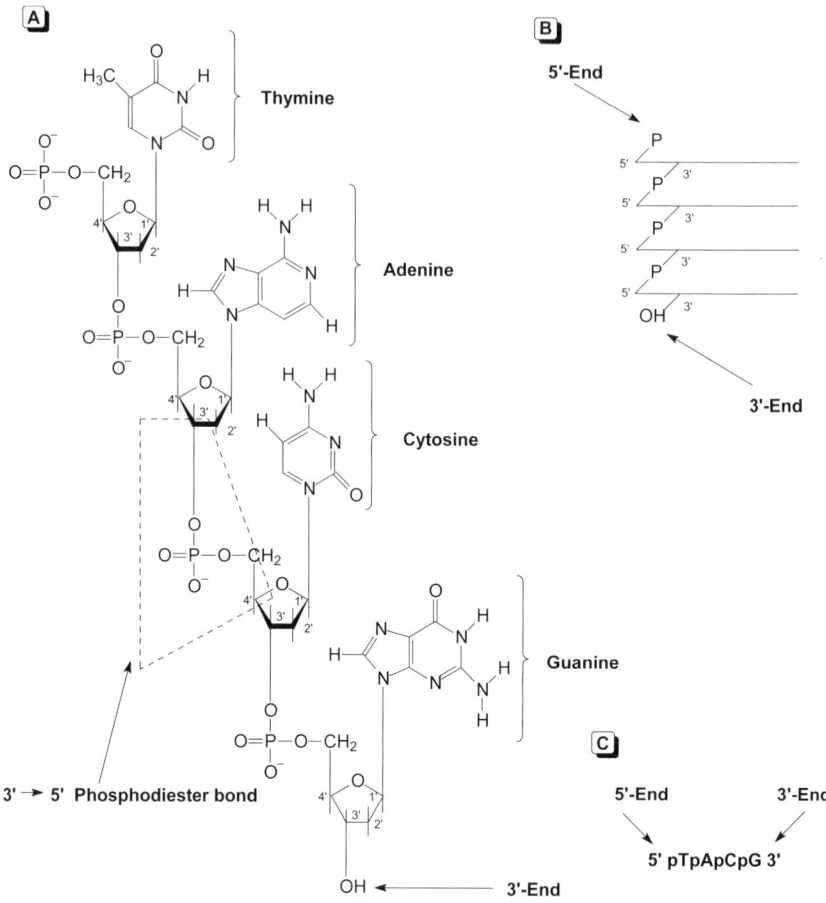

Figure 1.10. (A) Chemical structure of DNA with the nucleotide sequence shown written in the 5′→3′ direction; (B) the DNA structure written in a more stylized form, emphasizing the ribose-phosphate backbone; (C) a simplified representation of the nucleotide sequence (After Champe and Harvey, 1994, copyright 1994 J.B. Lippincott Co.)

As can be seen in Figure 1.10, phosphodiester bonds join the 5′-hydroxyl group of the deoxypentose of one nucleotide to the 3′-hydroxyl group of the deoxypentose of another nucleotide through a phosphate group. The resulting macromolecular chain with unbound 5′- and 3′-ends tends to coil with another chain through hydrogen bonding (Figure 1.11(a)) to form the double-helix structure (Figure 1.11(b)). In most DNA helices, the 5′-end of one strand is paired with the 3′-end of the other strand (Calladine and Drew, 1997). While the hydrophilic deoxyribose-phosphate backbone of each chain is on the outside of the double helical molecule, the hydrophobic bases are stacked inside normal to the helix (Figure 1.11(b)).

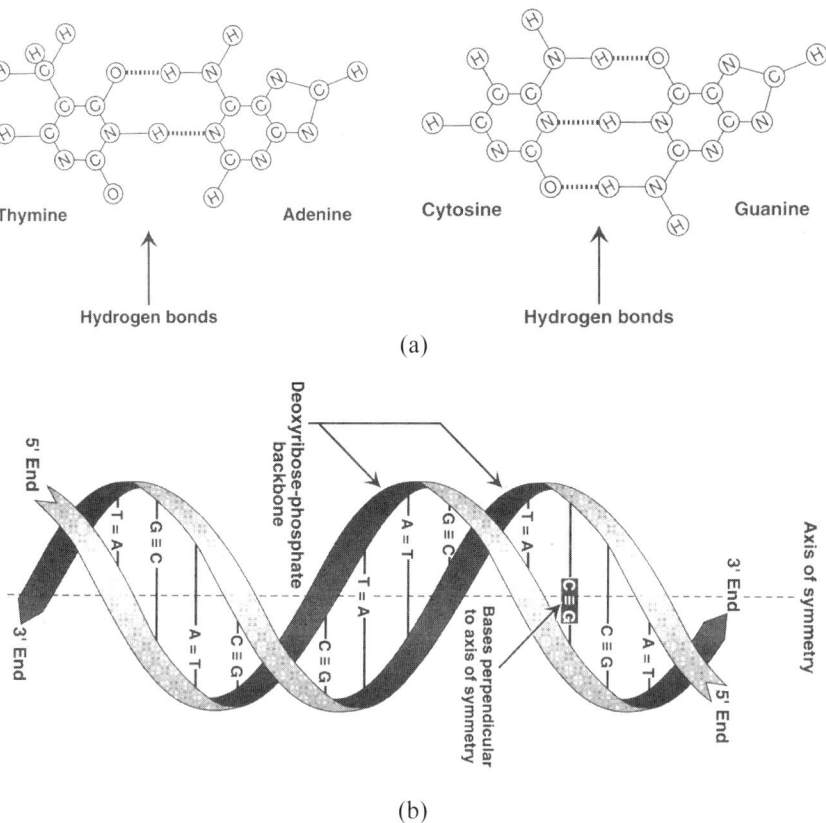

Figure 1.11. (a) Hydrogen bonds between complementary bases; (b) the Watson-Crick model of the DNA double helix illustrating some of its major structural features (After Champe and Harvey, 1994, copyright 1994 J.B. Lippincott Co.)

As illustrated in Figure 1.11, the base of one strand is paired with the bases of the second strand in such a delicate way that adenine (A) is always paired with thymine (T), whereas cytosine (C) always pairs with guanine (G). For each single strand of the DNA molecule, therefore, there is always a complementary strand. The specific pairing between complementary DNA strands has played an important role in regulating self-assembled supramolecular structures and DNA sensors. More importantly, it also provides a basis for replication of DNA chains, as each of the single DNA strands separated out from the double helix (Figure 1.11(b)) can serve as a template for the growth of a new complementary strand (Guschlbauer, 1976).

1.2.2.2 Structure of Proteins

Proteins represent another major class of biomacromolecules with amino acids covalently joined together by peptide bonds through the amide linkages between the α-carboxyl group of one amino acid and the α-amino group of another. The 20 amino acids commonly found in proteins, listed in Table 1.4, all contain a central carbon atom C (often denoted Cα), a carboxyl group (COOH), and an amino group (NH_2). They differ from each other in that the side group R can be hydrophilic (*e.g.* R = CH_2-OH, serine; R = CH_2-SH, cysteine), hydrophobic [R = CH_3, alanine; R = CH_2-$(CH_3)_2$, valine], negatively charged (R = CH_2-COO^-, aspartic acid), or positively charged [R = $(CH_2)_4$-NH_2^+, lysine].

Table 1.4. 20 Amino acids commonly found in proteins

Aliphatic	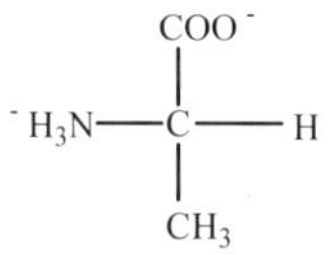 Alanine (Ala)	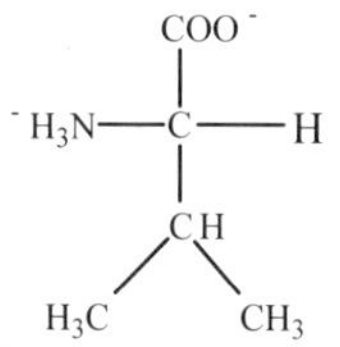 Valine (Val)	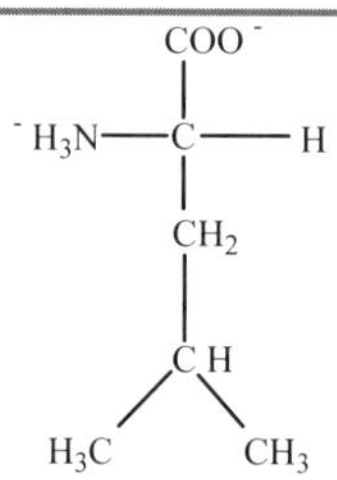 Leucine (Leu)	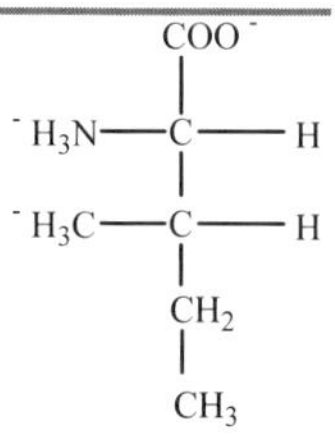 Isoleucine (Ile)
Non-polar	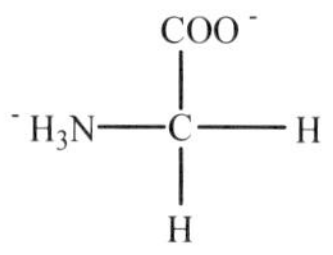 Glycine (Gly)	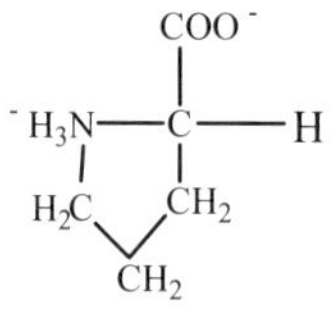 Proline (Pro)	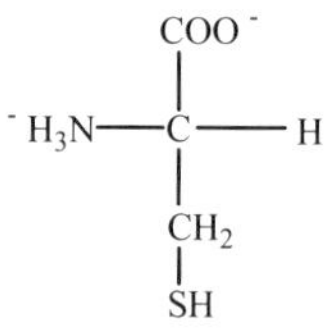 Cysteine (Cys)	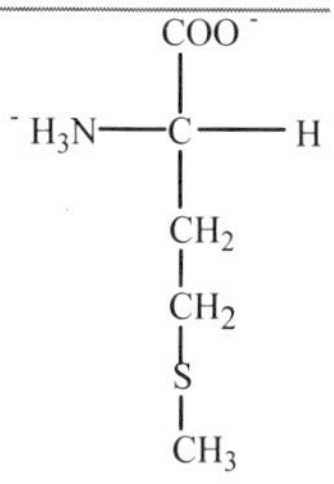 Methionine (Met)
Aromatic	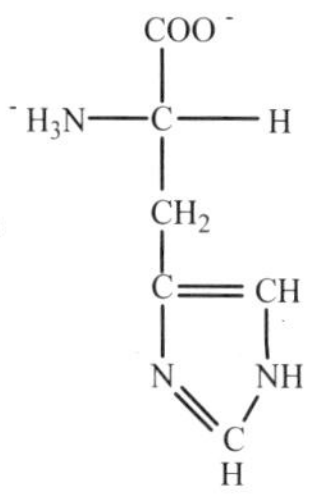 Histidine (His)	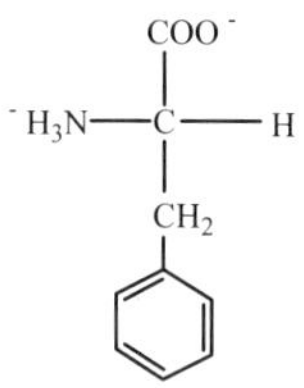 Phenylalanine (Phe)	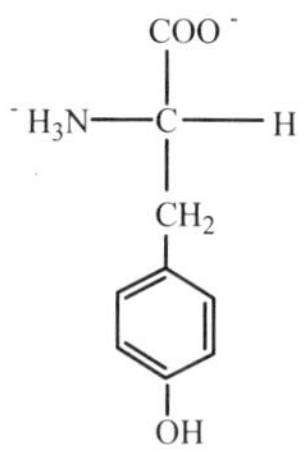 Tyrosine (Tyr)	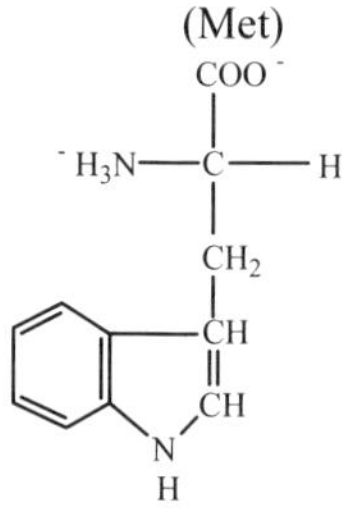 Tryptophan (Trp)

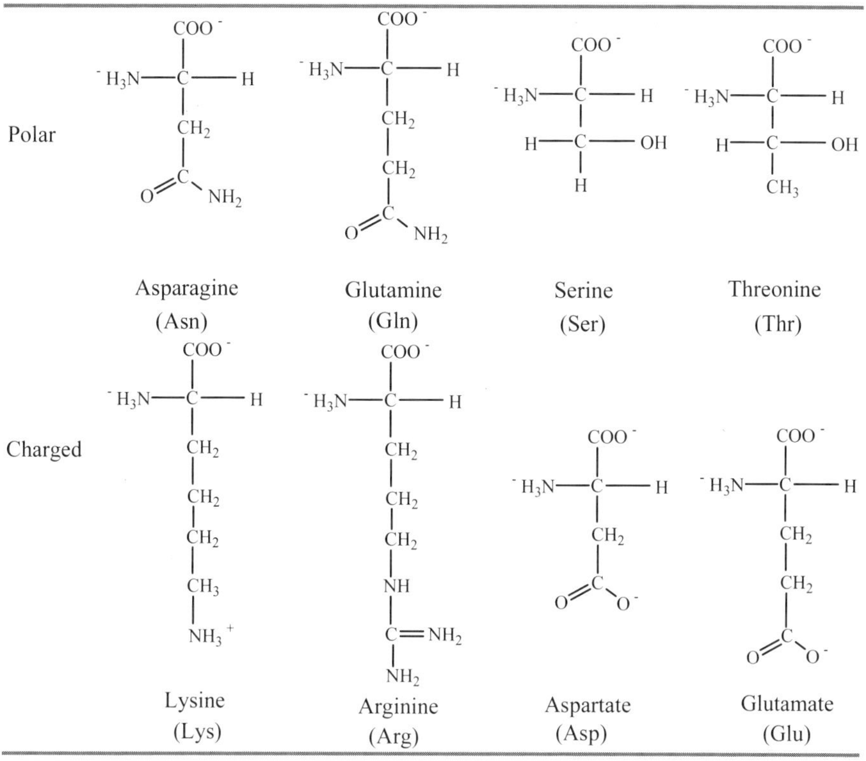

Polar	Asparagine (Asn)	Glutamine (Gln)	Serine (Ser)	Threonine (Thr)
Charged	Lysine (Lys)	Arginine (Arg)	Aspartate (Asp)	Glutamate (Glu)

The formation of peptides (biomacromolecules containing no more than 100 amino acid subunits) and proteins (biomacromolecules with a large number of amino acid subunits) involves the release of one molecule of water per amino acid to form the so-called peptide bond [Equation (1.27)].

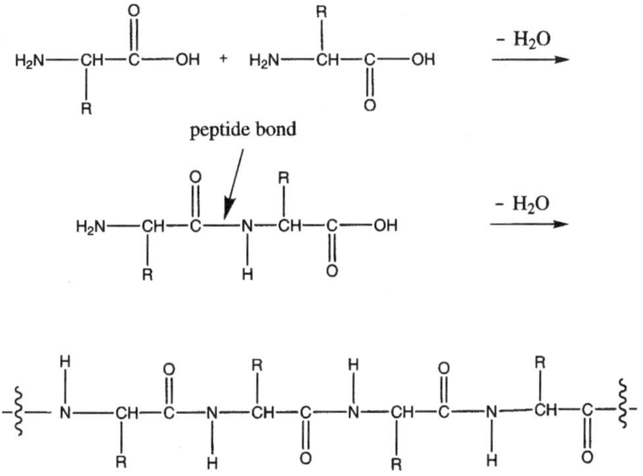

(1.27)

To break down the peptide bonds non-enzymically requires prolonged exposure to a strong acid or base at elevated temperatures.

Just as in the case of most synthetic macromolecules, the conformational structure of a protein is characterized by several organizational levels (*e.g.* primary, secondary, tertiary, and quaternary) (Figure 1.12).

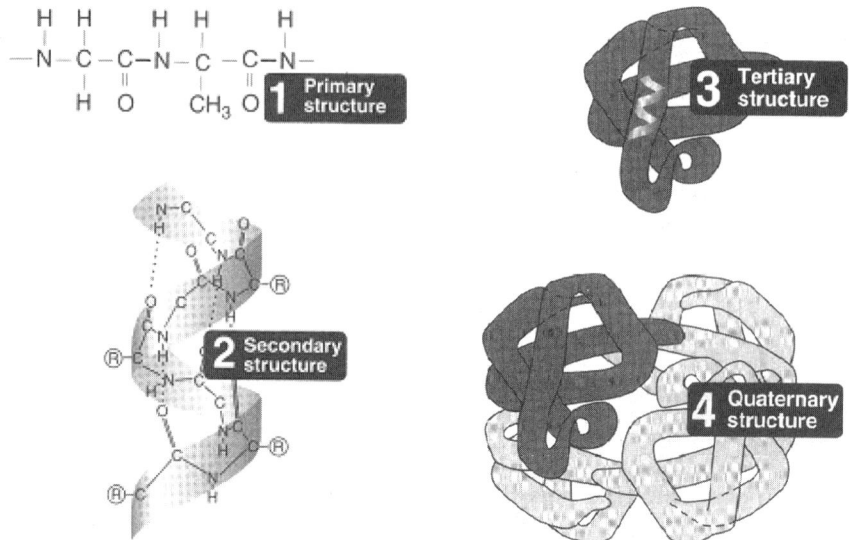

Figure 1.12. Four levels of protein structure (After Champe and Harvey, 1994, copyright 1994 J.B. Lippincott Co.)

The rigid, planar peptide unit (CO-NH) with the partial double bond character between C and N hinders free rotation of atoms around the main chain, enabling proteins to maintain a well-defined three-dimensional structure. On the other hand, the main-chain C-Cα and N-Cα single bonds with free rotations allow the protein chain to fold in many different ways. The linear sequence of the linked amino acids (*primary structure*) contains the information necessary to generate a protein macromolecule with a unique three-dimensional structure. Although the structural complexity of proteins increases with their organizational levels, certain structural elements (*secondary structure*) are repeated in a wide variety of proteins. The secondary structure forms through hydrogen bonding between the NH and C=O groups from different peptide units, leading to two building elements, α-helices and β-sheets. As shown in the secondary structure of Figure 1.12, the C=O group of the n-th amino acid residue ("residue" or "moiety" refers to each of the monomer units) is bound to the NH group of the (n + 4)-th residue in the α-helix. The main chain is coiled in a helicoidal fashion. In contrast to this, the main chain remains almost fully stretched out in the β-sheets, forming hydrogen bonds between NH and C=O groups of different polypeptide strands from the same or different protein chains (Figure 1.13). Both α-helix and β-sheets can usually be found in a single protein macromolecule.

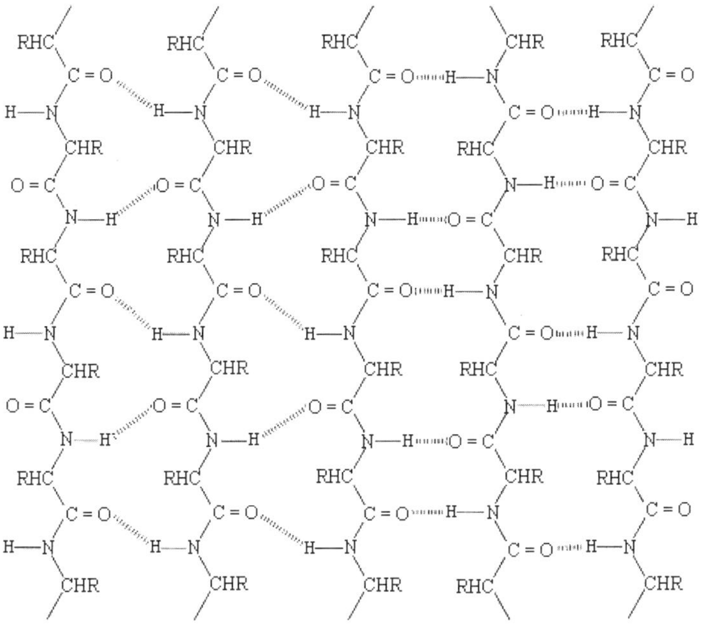

Figure 1.13. Schematic structure of a β-sheet formed by hydrogen bonds between protein strands. The strands can be either parallel (three strands on the left side) or antiparallel (three strands on the right side) (After Champe and Harvey, 1994, copyright 1994 J.B. Lippincott Co.)

As mentioned above, the side groups (R) in all of the amino acids listed in Table 1.4 are different, polar or non-polar. The secondary structure, in conjugation with a complex interplay of hydrophobic, hydrophilic, hydrogen bond, and van der Waals interactions result in the formation of the so-called *tertiary* and *quaternary* higher levels of conformational structures (Neurath and Hill, 1975). It is these complexities of the protein structure that makes possible their specific responses to the surrounding molecules and external stimuli, and hence the biological functions (Dickerson and Geis, 1969).

1.2.2.3 Structure of Polysaccharides

Polysaccharides are polymers of monosaccharides, representing yet another class of biomacromolecules composed of monomer units other than amino acids or nucleotides. As shown in Figure 1.14, monosaccharides can be linked by glycosidic bonds to form a larger structure. Dissaccharides, including lactose, maltose and sucrose, contain two monosaccharides, while oligosaccharides contain from 3 to about 12 monosaccharide units. Polysaccharides contain more than 12 of the same or different monosaccharides.

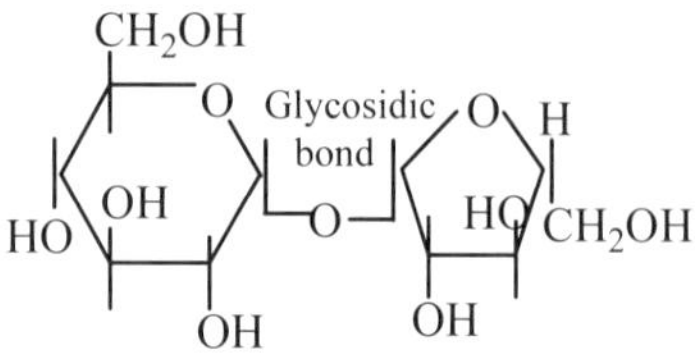

Sucrose: α-D-glucopyranosyl-(1,2)-
β-D-fructofuranoside

Figure 1.14. Two hexoses are linked into a disaccharide through a glycosidic bond

Examples of the linear polysaccharide homopolymers include amylose (a component of starch that serves as a storage reservoir of nutrients in plants), cellulose (a major component of wood or other plant fibers), and chitin (a major component of the exoskeletons of insects and other arthropods) (Figure 1.15). Some other linear polysaccharides may have a disaccharide as the repeating unit (*e.g.* hyaluronic acid, Figure 1.15). Some more complex polysaccharides contain less regular sequence of residues, as exemplified by heparin in Figure 1.15.

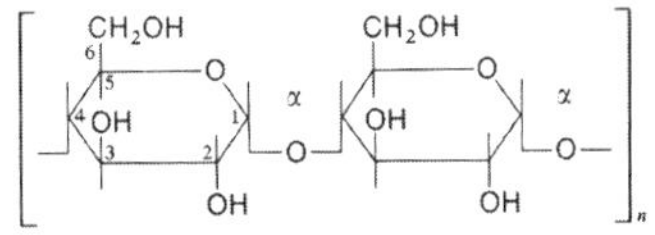

Amylose

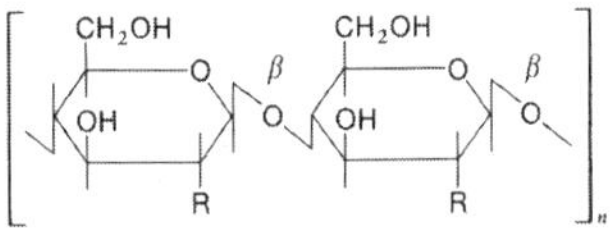

Cellulose, R = OH

Chitin, R = NH—C(=O)—CH_3

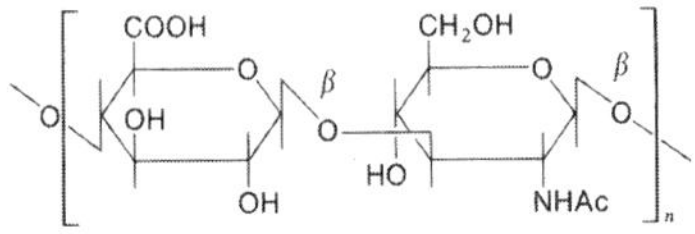

Hyaluronic acid

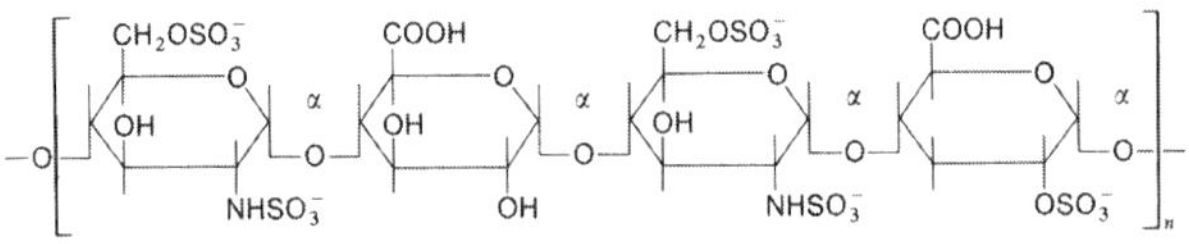

Heparin

Figure 1.15. Chemical structure of some common polysaccharide chains. The numbering scheme for hexose sugars is shown in the amylose structure. Note that linkages start from the anomeric configuration of a sugar in which the C-1 bond to oxygen is on the opposite side of the sugar plane from the C-6 bond to oxygen. In *β*-linkages, the C-1 and C-6 bonds to oxygen are on the same side of the sugar plane (After Cantor and Schimmel, 1980, copyright 1980 W.H. Greeman and Company)

Unlike proteins and nucleic acids that always have linear or circular unbranched backbones, some polysaccharides even have branched side chains (Kirkwood, 1974). Although polysaccharides possess a large variety of conformational structures, the various levels of structure described earlier for proteins and nucleic acids can also be applied to polysaccharides. The primary structures of polysaccharides seem to be very regular with a single residue or a single sequence repeating over the chain length. Very rigid secondary structures may be found in some fibrous polysaccharides, such as chitin and cellulose. While some of the polysaccharide fibers contain ribbons made from two chains stabilized by inter-residue hydrogen bonding, other polysaccharides have helical secondary structures. The folding of the helical secondary structures leads to the formation of tertiary structures in polysaccharides. A quaternary structure is then formed through the association of individual helical or folded polysaccharides. Apart from their chemical nature, the structures at various organizational levels also play important roles in regulating the macroscopic shapes, mechanical properties and bio-functions of polysaccharide materials.

1.2.3 Carbon Nanomaterials

Carbon has long been known to exist in three forms, namely amorphous carbon, graphite and diamond (Marsh, 1989). Certain polymers (*e.g.* polyacetylene) may be regarded as modified amorphous carbon with hydrogen atom(s) bound to each of the carbon atoms that are chemically linked in a specific manner (Figure 1.16(a)). Depending on how the carbon atoms are arranged, their properties vary. For example, graphite is soft and black and the stable, common form of carbon (Figure 1.16(b)). Diamond is hard and transparent due to the unusual form of carbon (Figure 1.16(c)). In diamond each carbon atom is bound to four other carbon atoms in a regular repetitive pattern. The density is 3.51 g/cm^3. In graphite the carbon atoms are located at the corners of regular and fused hexagons arranged in parallel layers. Its density is considerably lower, 2.26 g/cm^3.

However, the Noble-Prize-winning discovery of buckminsterfullerene C_{60} (Kroto *et al.*, 1985) – a *conjugated* molecule with a soccerball-like structure consisting of 12 pentagons and 20 hexagons facing symmetrically (Figure 1.16(d)) – has created an entirely new branch of carbon chemistry (Hirsch, 1994; Taylor, 1995). The subsequent discovery of carbon nanotubes by Iijima (1991) opened up a new era in materials science and nanotechnology (Dresselhaus *et al.*, 1996; Harris, 2001). These elongated nanotubes consist of carbon hexagons arranged in a concentric manner with both ends normally capped by fullerene-like structures containing pentagons (Figure 1.16(e)). Having a *conjugated* all-carbon structure with unusual molecular symmetries, fullerenes and carbon nanotubes show interesting electronic, photonic, magnetic and mechanical properties attractive for various applications, including optical limiters, photovoltaic cells, and field emitting displays, much like the conjugated polymers to be discussed below.

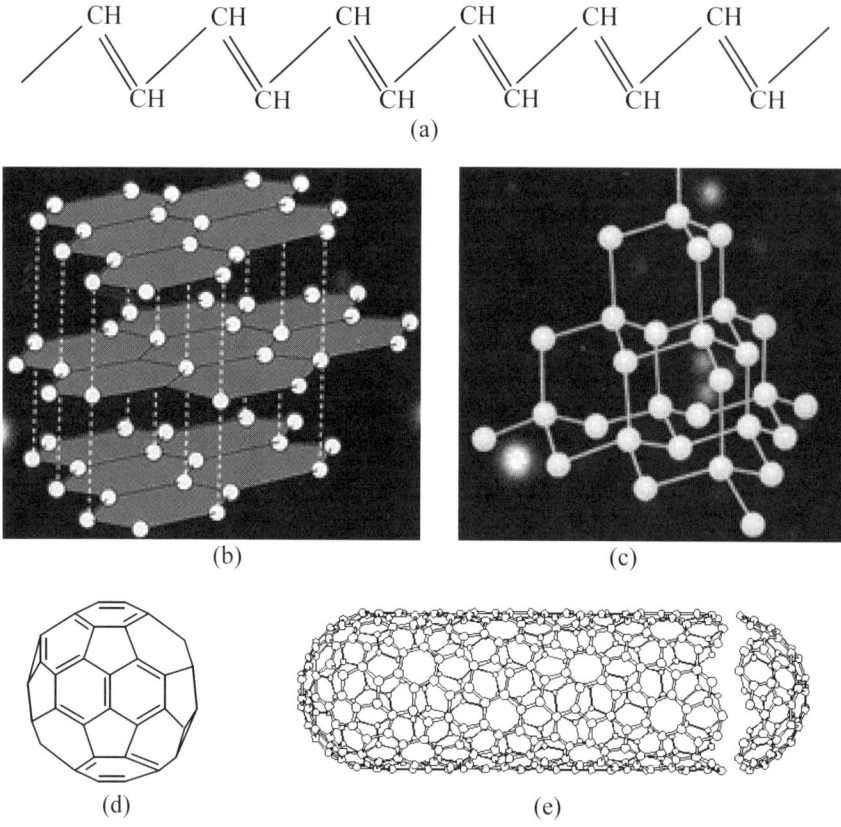

Figure 1.16. One-, two- and three-dimensional carbon materials: (a) polyacetylene; (b) graphite; (c) diamond; (d) buckminsterfullerene C_{60}; (e) [5,5] armchair single wall carbon nanotube

1.2.4 Intelligent Macromolecules

Polymers have been traditionally used as electrically insulating materials. Over the last three decades or so, however, various polymers have been synthesized with unusual electronic, photonic and magnetic properties. For example, the visit of MacDiarmid to Shirakawa at Tokyo Institute of Technology in 1974 and, later, Shirakawa to MacDiarmid and Heeger at The University of Pennsylvania, led to the discovery of conducting polyacetylene (Shirakawa *et al.*, 1977; Dai, 2001). This finding opened up the important new field of polymers for electronic applications and was recognized by the 2000 Nobel Prize in Chemistry (Jones, 2000). The subsequent discovery of the electroluminescent light emission from conjugated poly(*p*-phenylene vinylene), by Friend's group at Cavendish Laboratory in 1990 (Burroughes *et al.*, 1990), revealed the significance of the use of conjugated polymers in photonic devices.

Apart from the above-mentioned electrically active conjugated polymers, buckminsterfullerene C_{60} and carbon nanotubes (Kwon *et al.*, 1991; Osada *et al.*, 1992; Skotheim *et al.*, 1998; Sutani *et al.*, 2001), various other stimuli-responsive macromolecules have been demonstrated to undergo fast and reversible changes in conformation and/or properties in response to small changes in their environment, for example, a change in temperature (Bae *et al.*, 1987; Chen and Hoffman, 1995; Hoffman, 1987; Tanaka, 1979), pH (Gan *et al.*, 2000; Siegel and Firestone, 1988), optical (Feringa, 2001; Liu, 2001) and magnetic field (Miller and Epstein, 1994, 1995). Due to the similarity between the stimuli-responsive macromolecules and biopolymers, these stimuli-responsive macromolecules are also termed *intelligent macromolecules* (or *intelligent polymers*, *smart polymers*). Recent research advances have led to the development of a large variety of smart devices (*e.g.* polymer/carbon nanotube flat panel displays and chemical/biological nanosensors) from the broadly defined intelligent macromolecules.

1.3 The Concept of Smart Devices

1.3.1 Self-assembling and Micro-/Nano-fabrication

By definition, a machine is an assembly of parts that transmits and modifies forces, motion, and energy from one to another in a predetermined manner. Similarly, an assembly of molecules or macromolecules in a predetermined manner may act as a molecular or supramolecular machine. Through non-covalent self-assembling, nature provides supramolecular architectures and/or molecular machines with a complexity and efficiency that none of today's synthetic systems can attain (Mann *et al.*, 2000). For instance, proteins can self-assemble into working molecular machines, objects that *do* something, such as cutting and splicing other molecules or making muscles to contract. They also join with other molecules to form huge assemblies like the ribosome.

Supramolecules are aggregates of macromolecular or low-molecular weight compounds. They are characterized both by the spatial arrangement of their components, their architecture, and by the nature of the intermolecular bonds that hold these components together. Supramolecules possess well-defined structural and other physicochemical properties. Just like intelligent macromolecules, some of the supramolecular assemblies can perform machine-like movement and/or other smart performances in response to an appropriate external stimulus.

As schematically shown in Figure 1.17, Lehn and co-workers (Breuning *et al.*, 2001) have recently reported a new class of polypyridine-derived, tetranuclear metal complexes of the $[2 \times 2]M_4^{II}$ grid-type (M represents transition metal) supramolecular structures formed by the self-assembling of ligands containing two tridentate, terpyridine-like, binding subunits with metal ions of octahedral coordination geometry. Because their constituents are linked by reversible connections, metal coordination and hydrogen bonds, for example, certain molecules undergo a spontaneous and continuous assembly and de-assembly processes in a given set of conditions. The spontaneous but controlled generation of well-defined, functional supramolecular architectures of nanometric size through self-organization represents a means of performing programmed engineering and the processing of intelligent supramolecular structures and functional smart devices.

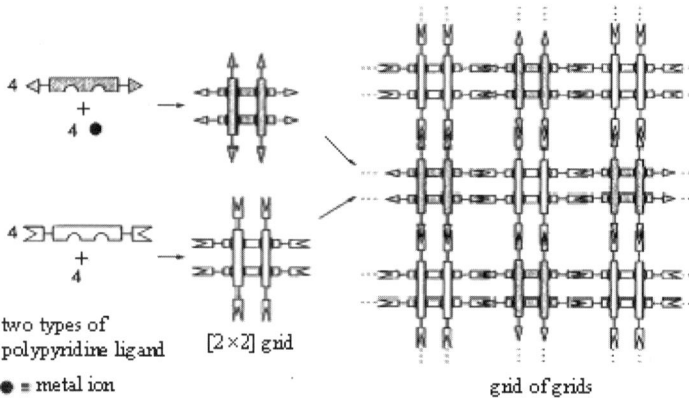

Figure 1.17. A grid of grids formed from [2 × 2] grids (After Breuning *et al.*, 2001, copyright 2001 Wiley-VCH Verlag)

Just as the self-assembling approach was developed largely by chemists, physicists have contributed new tools of great promise for molecular engineering. These are the proximal probes, including the scanning tunneling microscope, STM, and the atomic force microscope, AFM (Wickramasinghe, 1992), and lithographic techniques such as photolithography and soft-lithography (Dai and Mau, 2001). The lithography can be regarded as a "top-down approach", with which large structures (machines) could be used to make smaller structures (machines) that can then build even smaller structures (machines), and so on. Similar to today's semiconductor technology in that standard materials would be sculpted into a desired shape, the "top-down approach" would not possess control on a molecular scale. A proximal-probe device places a sharp tip in proximity to a surface and uses it to probe (and sometimes modify) the surface and any molecules that may be stuck to it. Therefore, the proximal-probed approach would extend the control to the molecular level by building things atom-by-atom, the so-called "bottom-up approach". For example, an IBM group (Eigler and Schweizer, 1990) has spelled out the name of IBM with 35 precisely placed xenon atoms by STM (Figure 1.18). The precision here is complete, like the precision of molecular assembly: each atom sits in a dimple on the surface of a nickel crystal. The atoms can rest either in one dimple or in another, but never somewhere in between.

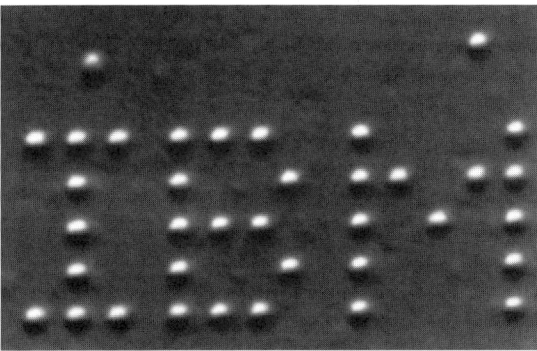

Figure 1.18. STM image of a patterned array of xenon atoms on a nickel surface (After Eigler and Schweizer, 1990, copyright 1990 Macmillan Magazines Limited)

1.3.2 Functional Structures and Smart Devices

Recently, the self-assembling of macromolecules, especially intelligent macromolecules, to create functional materials and/or smart devices has attracted a great deal of interest (Brunsveld *et al.,* 2001; Caulder and Raymond, 1999; Conn and Rebek, 1997; Klok and Lecommandoux, 2001; Tsukruk, 1997). Tools originally developed by biochemists and biotechnologists to deal with biomolecular machines found in nature can be redirected to make new functional structures and smart devices. Based on biochemical molecular recognition principles, for example, Mirkin and co-workers (Mirkin, 2000; Taton *et al.,* 2000) have used DNA chains as linkages for the assembly of nanoscale inorganic building blocks into multi-dimensional functional architectures. To demonstrate this concept, these authors used two sets of gold nanoparticles. While one set of the nanoparticles carries one kind of DNA single-strand chains, the other set of the nanoparticles is bound with complementary single-strand DNA chains. The pairing action between the two complementary DNA chains could link the nanoparticles into both two- and three-dimensional architectures with a predetermined supramolecular structure by controlling the sequence-specific DNA hybridization events and surface-grafting chemistry.

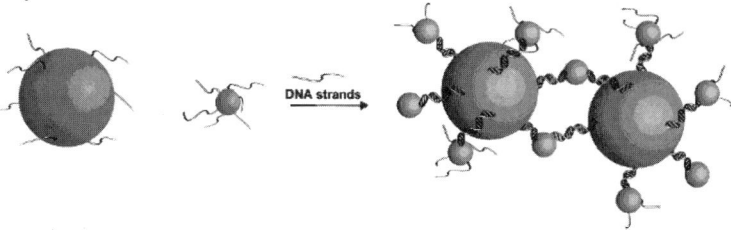

Figure 1.19. Multi-dimensional functional architectures formed by self-assembling through the hybridization events between the complementary DNA single-strand chains bound onto nanoparticles (After Mirkin, 2000, copyright 2000 The American Chemical Society. Reproduced with permission)

Figure 1.19 schematically shows two sets of nanoparticles of different sizes via the DNA hybridization (Mirkin, 2000). Depending on the inter-particle distances, which can be regulated by changing the DNA linker length, these agglomerations could cause the solution to change color, suggesting considerable room for the use of the supramolecular assembly based on the DNA-modified nanoparticles in optical sensors (Storhoff *et al.*, 2000), biodiagnostics (Nam *et al.*, 2002), and molecular electronics (Park *et al.*, 2000).

The potential of the self-assembling of synthetic macromolecules, especially block copolymers, into functional architectures and/or devices has also been demonstrated recently (Klok and Lecommandoux, 2001). By the self-assembling of tailor-made triblock copolymer chains (Figure 1.20 *top*), Stupp and co-workers (Zubarev *et al.*, 1999) have demonstrated a beautiful mushroom-shaped supramolecular structure (Figure 1.20, *bottom left*), which resembles the toxin α-hemolysin (a protein assembled from seven polypeptides, Figure 1.20 *bottom right*).

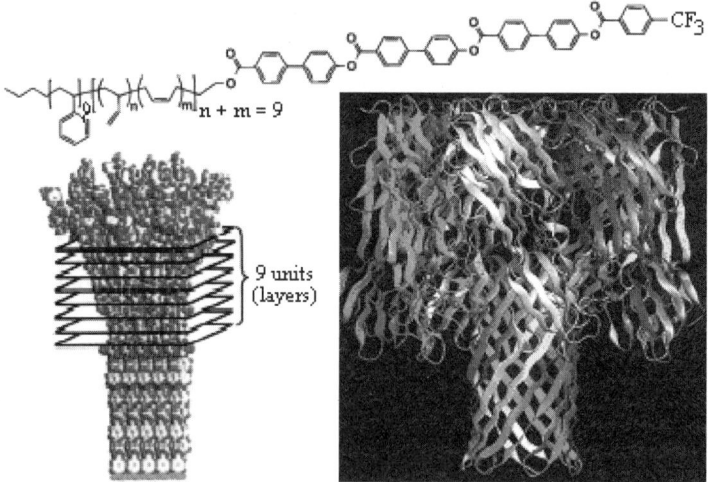

Figure 1.20. A self-assembly of block copolymers mimicking the size and shape of α-hemolysin (After Zubarev *et al.*, 1999; Roubi, 1999)

The distinct characteristics of the mushroom-shaped macromolecular assembly could allow it to break through cells with its hydrophobic stem and to move molecules and ions across cell membranes through its hydrophilic hollow core, in much the same way as α-hemolysin.

On the other hand, self-assembling at the liquid/solid interface has also been exploited for the development of various smart surfaces. In this context, Israelachvili and co-workers (Jeppesen *et al.*, 2001) have demonstrated that poly(ethylene glycol) tethered chains (*see*: polymer brushes in Chapter 7) bearing biotin at their free ends can latch onto the biotin receptor streptavidin and pull it back, like a harpoon (Figure 1.21), though polymer chains normally don't like to adopt a highly stretched, high-energy conformation. Given that it is ubiquitous for ligands to be tethered to the cell surface through proteins and carbohydrates, the above finding should have an important implication for biological systems (*e.g.* designing smart drug carriers).

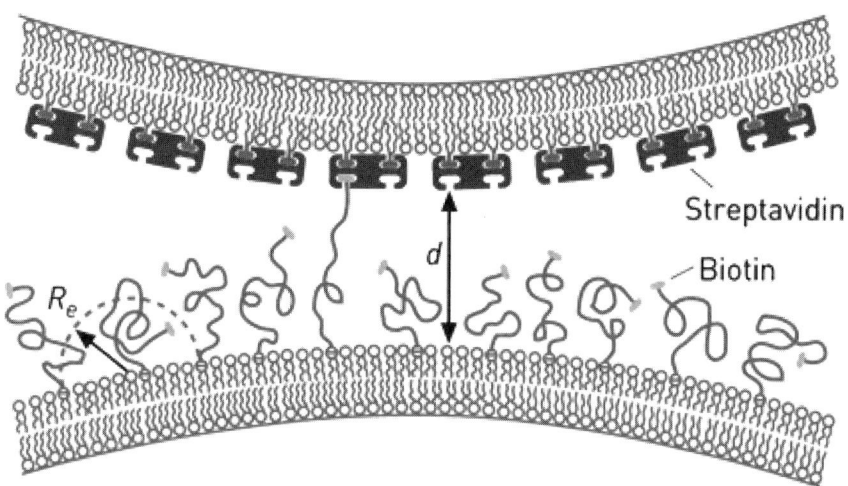

Figure 1.21. Surface force measurements (Chapter 7) show that poly(ethylene glycol) tethers bearing biotin at their tips can extend beyond their average reach (R_e) to allow biotin to bind with streptavidin even when $d > R_e$ (After Jeppesen *et al.*, 2001, copyright 2001 American Association for the Advancement of Science)

In a somewhat related study, Klibanov and co-workers (Tiller *et al.*, 2001; Borman, 2001) have recently reported that assemblies of *N*-alkylated poly(4-vinylpridine) or alkylated polyethyleneimines on a wide range of substrates, including glass, polyethylene, polypropylene, nylon, and poly(ethylene terephthalate) plastics can kill wild-type and antibiotic-resistant bacteria (Lin *et al.*, 2002; Tew *et al.*, 2002; Tiller *et al.*, 2002), leading to new materials and/or devices (*e.g.* smart clothing and drugs) with strong antimicrobial properties.

In conjunction with lithographic patterning methods, the self-assembling approaches discussed above could provide very effective means to produce a large variety of smart devices. In this context, Whitesides and co-workers (Jacobs *et al.*, 2002) have successfully fabricated flexible, cylindrical light-emitting diode (LED) displays through patterned assembly (Figure 1.22(a)).

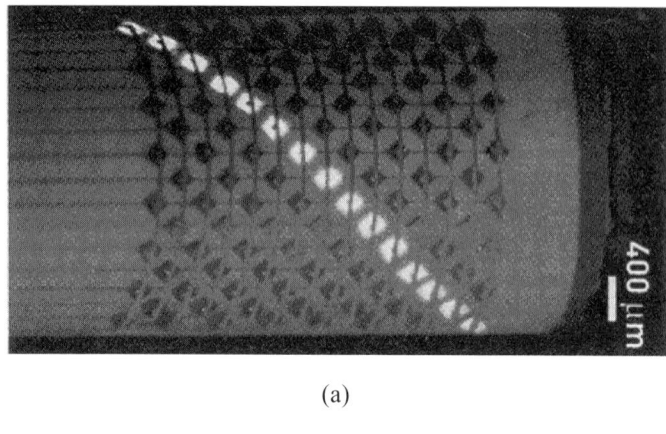

(a)

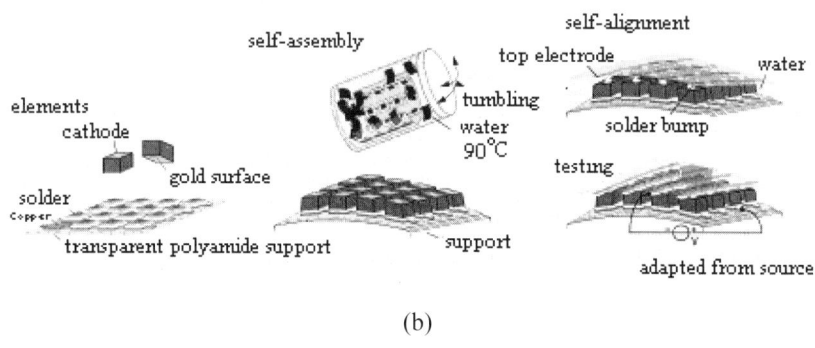

(b)

Figure 1.22. (a) A diagonal line of LEDs lights up on the non-planar, fully addressable array of 113 LEDs; (b) Schematic representation of the self-assembling process for the production of the array of LEDs shown in (a) (After Jacobs *et al.*, 2002; Dagani, 2002, copyright 2002 American Association for the Advancement of Science)

In particular, these authors first photolithographically produced an array of 113 280-μm-wide solder-coated copper squares connected by parallel copper wires on a transparent polyimide film (Figure 1.22(b)). They then added LEDs with a layer of gold (anode) on the bottom and a small circular cathode on the top, which have the same size as the copper squares. The subsequent addition of hot water melts the solder and a gentle tumbling of the vial causes the gold bottoms of the LEDs to grab onto the liquid solder due to an effect associated with the minimization of the solder's free surface area. Finally, they deposited solder to the top cathode of each LED through dip-coating, followed by placing a polyimide-supported parallel copper wire of a top electrode onto the LED array and fused it to the individual diodes by melting the solder on the LEDs. Applying a potential between the top and substrate electrodes in a proper manner can light up any number of the 113 LEDs thus produced individually or collectively. The whole process can be completed within a short period, giving an indication of the potential for mass-fabrication of flexible optoelectronic devices.

1.4 References

Bae, Y.H., Okano, T., Hsu, R., Kim, S.W. (1987) *Makromol. Chem., Rapid Commun.* **8**, 481.
Borman, S. (2001) *C&E News* **May 28**, 13.
Brandrup, J., Immergut, E.H. (eds) (1966) *Polymer Handbook*, John Wiley & Sons, Inc., New York.
Breuning, E., Ziener, U., Lehn, J.-M., Wegelius, E., Rissanen, K. (2001) *Eur. J. Inorg. Chem.* 1515.
Brunsveld, L., Folmer, B.J.B., Meijer, E.W., Sijbesma, R.P. (2001) *Chem. Rev.* **101**, 4071.
Burroughes, J.H., Bradley, D.C.C., Brown, A.R., Mackay, M.K., Friend, R.H., Burn, P.L. (1990) *Nature* **347**, 539.
Calladine, C.R., Drew, H.R. (1997) *Understanding DNA: The Molecule & How It Works*, Academic Press, New York.
Cantor, C.R., Schimmel, P.R. (1980) *Biophysical Chemistry*, W.H. Freeman and Company, New York.
Caulder, D.L., Raymond, K.N. (1999) *Acc. Chem. Res.* **32**, 975.
Champe, P.C., Harvey, R.A. (1994) *Lippincott's Illistrated Reviews: Biochemistry*, 2nd edn, J.B. Lippincott Co., Philadelphia.
Chen, G.H., Hoffman, A.S. (1995) *Macromol. Rapid Commun.* **16**, 175.
Cohen, C. (2000) in *Scattering in Polymeric and Colloidal Systems*, Brown, W., Mortensen, K. (eds), Gordon and Breach Science Publishers, Amsterdam.
Clegg, P.C., Clegg, A.G. (1987) *Biology of the Mammal*, Heinemann, London.
Conn, M.M., Rebek, J. (1997) *Chem. Rev.* **97**, 1647.
Dai, L. (2001) *Aust. J. Chem.* **54**, 11.
Dai, L., Mau, A.W.H. (2001) *Adv. Mater.* **13**, 899, and references cited therein.
Dickerson, R., Geis, I. (1969) *The Structure and Action of Proteins*, Benjamin/Cummings, Menlo Park.
Dresselhaus, M.S., Dresselhaus, G., Eklund, P. (1996) *Science of Fullerenes and Carbon Nanotubes*, Academic, San Diego.
Dvornic, P.R., Tomalia, D.A. (1996) *Curr. Opin. Colloid Interf. Sci.* **1**, 221.
Eigler, D.M., Schweizer, E.K. (1990) *Nature* **344**, 524.
Feringa, B.L. (2001) *Acc. Chem. Res.* **34**, 504.
Flory, P.J. (1942) *J. Chem. Phys.* **10**, 51.
Flory, P.J. (1953) *Principle of Polymer Chemistry*, Cornell University Press, New York.
Flory, P.J., Krigbaum, W.R. (1951) *Ann. Rev. Phys. Chem.* **2**, 383.
Gan, L.H., Gan, Y.Y., Deen, G.R. (2000) *Macromolecules* **33**, 7893.
Goldstein, A.N. (1997) *Handbook of Nanophase Materials*, Marcel Dekker, Inc., New York.
Guschlbauer, W. (1976) *Nucleic Acid Structure*, Springer-Verlag, New York.
Hamley, I.W. (2000) *Intoduction to Soft Matter: Polymers, Colloids, Amphiphiles and Liquid Crystals*, John Wiley & Sons, Inc., New York.
Harris, P.J.F. (2001) *Carbon Nanotubes and Related Structures - New Materials for the Twenty-First Century*, Cambridge University Press, Cambridge.
Hirsch, A. (1994) *The Chemistry of the Fullerenes*, Thieme, Stuttgart.
Hoffman, A.S. (1987) *J. Controlled Release* **6**, 297.
Huggins, M.L. (1942a) *J. Phys. Chem.* **46**, 151.
Huggins, M.L. (1942b) *J. Am. Chem. Soc.* **64**, 1712.
Iijima, S. (1991) *Nature* **354**, 56.
Jackman, R.J., Whitesides, G.M. (1999) *CHEMTECH*, **May**, 18.
Jacobs, H.O., Tao, A.R., Schwartz, A., Gracias, D.H., Whitesides, G.M. (2002) *Science* **296**, 323.
Jeppesen, C., Wong, J.Y., Kuhl, T.L., Israelachvili, J.N., Mullah, N., Zalipsky, S., Marques, C.M. (2001) *Science* **293**, 465.

Jones, N. (2000) *New Sci.* **Oct. 21**, 14.
Kaufman, H.S., Falcetta, J.J. (1977) *Introduction to Polymer Science and Technology: An SPE Textbook*, John Wiley & Sons, Inc., New York.
Kirkwood, S. (1974) *Ann. Rev. Biochem.* **43**, 401.
Kleman, M., Lavrentovich, O.D. (2003) *Soft Matter Physics: An Introduction*, Springer-Verlag, New York.
Klok, H.-A., Lecommandoux, S. (2001) *Adv. Mater.* **13**, 1217.
Kroto, H.W., Heath, J.R., O'Brien, S.C., Curl, R.F., Smalley, R.E. (1985) *Nature* **318**, 162.
Krummenacker, M., Lewis, J. (1995) *Prospects in Nanotechnology: Toward Molecular Manufacturing*, John Wiley & Sons, Inc., New York.
Kwon, I.C., Bae, Y.H., Kim, S.W. (1991) *Nature* **354**, 291.
Lin, J., Tiller, J.C., Lee, S.B., Lewis, K., Klibanov, A.M. (2002) *Biotechnol. Lett.* **24**, 801.
Liu, R.S.H. (2001) *Acc. Chem. Res.* **34**, 555.
Magonov, S.N., Whangbo, M.-H. (1996) *Surface analysis with STM and AFM: experimental and theoretical aspects of image analysis*, Weinheim, New York.
Mann, S., Shenton, W., Li, M., Connolly, S., Fitzmaurice, D. (2000) *Adv. Mater.* **12**, 147.
Marsh, H. (1989) *Introduction to Carbon Science*, Butterworths, London.
Meyer, K.H., Luhdemann, R. (1935) *Heelv. Chim. Acta* **18**, 307.
Miller, J.S., Epstein, A.J. (1994) *Angew. Chem. Int. Ed.* **33**, 385.
Miller, J.S., Epstein, A.J. (1995) *C&E News* **Oct. 2**, 30.
Mirkin, C.A. (2000) *Inorg. Chem.* **39**, 2258.
Moelwyn-Hughes, E.A. (1965) *Physcical Chemistry*, 2nd edn., Pergamon Press, Oxford.
Morawetz, H. (1975) *Macromolecules in Solution*, 2nd edn., John Wiley & Sons, Inc., New York.
Morton, M. (1983) *Anionic Polymerization: Principles and Practice*, Academic Press, New York.
Nam, J.M., Park, S.J., Mirkin, C.A. (2002) *J. Am. Chem. Soc.* **124**, 3820.
Napper, D.H. (1983) *Polymeric Stabilization of Colloidal Dispersions*, Academic Press, London.
Neurath, H., Hill, R.C. (eds), (1975) *The Protein*, 3rd edn., Academic Press, New York.
Odian, G. (1970) *Principles of Polymerization*, McGraw-Hill, New York.
Osada, Y., Okuzaki, H., Hori, H. (1992) *Nature* **355**, 242.
Park, S.L., Lazarides, A.A., Mirkin, C.A., Brazis, P.W., Kannewurf, C.R., Letsinger, R.L. (2000) *Angew. Chem. Int. Ed.* **39**, 3845.
Roubi, M. (1999) *C&E News* **January 25**, 9.
Sheetz, M.P. (ed.) (1998) *Laser Tweezers in Cell Biology*, Academic Press, San Diego.
Shirakawa, H., Louis, E.J., MacDiarmid, A.G., Chiang, C.K., Heeger, A.J. (1977) *J. Chem. Soc., Chem. Commun.* 578.
Shmakov, S.L. (2002) *Polym. Sci. Ser. B* **44**, 310.
Siegel, R.A., Firestone, B.A. (1988) *Macromolecules* **21**, 3254.
Skotheim, T.A., Elsenbaumer, R.L., Reynolds, J.R. (ed.) (1998) *Handbook of Conducting Polymers*, 2nd edn., Marcel Dekker, Inc., New York.
Storhoff, J.J., Lazarides, A.A., Mucic, R.C., Mirkin, C.A., Letsinger, R.L., Schatz, G.C. (2000) *J. Am. Chem. Soc.* **122**, 4640.
Sutani, K., Kaetsu, I., Uchida, K. (2001) *Rad. Phys. Chem.* **61**, 49.
Tanaka, T. (1979) *Polymer* **20**, 1404.
Taton, T.A., Mucic, R.C., Mirkin, C.A., Letsinger, R.L. (2000) *J. Am. Chem. Soc.* **122**, 6305.
Taylor, R. (ed.) (1995) *The Chemistry of Fullerenes*, World Scientific, Singapore.
Tew. G.N., Liu, D.H., Chen, B., Doerksen, R.J., Kaplan, J., Carroll, P.J., Klein, M.L., DeGrado, W.F. (2002) *Proc. Natl. Acad. Sci. USA* **99**, 5110.
Tiller, J.C., Lee, S.B., Lewis, K., Klibanov, A.M. (2002) *Biotechnol. Bioeng.* **79**, 466.
Tiller, J.C., Liao, C., Lewis, K., Klibanov, A.M. (2001) *Proc. Natl. Acad. Sci. USA* **98**, 5981.
Timp, G. (ed.) (1998) *Nanotechnology*, Springer-Verlag, New York.

Tsukruk, V.V. (1997) *Prog. Polym. Sci.* **22**, 247.
Varner, J.E. (ed.) (1988) *Self-assembling Architecture*, Alan R. Liss, New York.
Vollmert, B. (1973) *Polymer Chemistry*, Springer-Verlag, Berlin.
Wickramasinghe, H.K. (1992) *Scanned Probe Microscopy*, American Institute of Physics, New York.
Yamakawa, H. (1971) *Modern Theory of Polymer Solutions*, Harper & Row Publishers, New York.
Zubarev, E.R., Pralle, M.U., Li, L.M., Stupp, S.I. (1999) *Science* **283**, 523.

Part I
Intelligent Macromolecules

Chapter 2

Conducting Polymers

2.1 Introduction

Polymers have long been used as insulating materials. For example, metal cables are coated in plastic to insulate them. However, there are at least four major classes of semiconducting polymers that have been developed so far. They include conjugated conducting polymers, charge transfer polymers, ionically conducting polymers and conductively filled polymers.

The conductively filled conducting polymers were first made in 1930 for the prevention of corona discharge. The potential uses for conductively filled polymers have since been multiplied due to their ease of processing, good environmental stability and wide range of electrical properties. Being a multi-phase system in nature, however, their lack of homogeneity and reproducibility has been an inherent weakness for conductively filled polymers. Therefore, controlling the quality of dispersion to obtain homogeneous conducting polymer composites is critically important.

The report of electrical conductivity in ionic polymers in 1975 (Wright, 1975) attracted considerable interest. Since then, various ionically conducting polymers or polymer electrolytes have been prepared for a wide range of applications ranging from rechargeable batteries to smart windows. Polymer electrolytes are also highly processable. The ionic conduction mechanism requires the dissociation of opposite ionic charges and the subsequent ion migration between coordination sites, which are generated by the slow motion of polymer chain segments. Consequently, polymer electrolytes normally show a low conductivity and high sensitivity to humidity. They often become electrically non-conducting upon drying.

The discovery of electrical conductivity in molecular charge transfer (CT) complexes in the 1950s (Akamatu et al., 1954) promoted the development of conducting CT polymers, and led to subsequent findings of superconductivity with molecular CT complexes in 1980 (Jerome et al., 1980) and with fullerene in 1986 (Iqbal et al., 1986). The conductivity in CT complexes arises from the formation of appropriate segregated stacks of electron donor and acceptor molecules and a certain degree of charge transfer between the stacks. A desired crystal structure is, therefore, essential for good conductivity in the molecular CT complexes. However, the resultant materials are often brittle and unprocessable. To overcome this problem, attempts have been made to attach electron donor and/or acceptor moieties onto polymer backbones to produce charge transfer polymers with good processability and stacking properties.

Along with all of the activities described above, various conjugated polymers have been synthesized during the past 25 years or so which show excellent electrical properties (Skotheim *et al.*, 1986). Owing to the delocalization of electrons in a continuously overlapped π-orbital along the polymer backbone, certain conjugated polymers also possess interesting optical and magnetic properties. These unusual optoelectronic properties allow conjugated polymers to be used for a large number of applications, including protecting metals from corrosion, sensing devices, artificial actuators, all-plastic transistors, non-linear optical devices and light-emitting displays. Due to the backbone rigidity intrinsically associated with the delocalized conjugated structure, however, most *unfunctionalized* conjugated polymers are intractable *(i.e.* insoluble, infusible and brittle). Some of them are even unstable in air.

The present chapter gives an overview of various methods developed for the preparation of advanced conjugated conducting polymers, charge transfer polymers, ionically conducting polymers and conductively filled polymers. The structure, conduction mechanism and electrical properties for each of these are also discussed, which underpin their applications in a wide range of smart devices as presented in later chapters.

2.2 Conjugated Conducting Polymers

2.2.1 Structure and Properties

2.2.1.1 π-π* Conjugation

Table 2.1 lists the repeat units and conductivities for some common conjugated polymers (Dai, 1999). As can be seen in Table 2.1, the conjugated structure with alternating single and double bonds or conjugated segments coupled with atoms providing *p*-orbitals for a continuous orbital overlap (*e.g.* N, S) seems to be necessary for polymers to become intrinsically conducting. This is because just as metals have high conductivity due to the free movement of electrons through their structure, in order for polymers to be electronically conductive they must possess not only charge carriers but also an orbital system that allows the charge carriers to move. The conjugated structure can meet the second requirement through a continuous overlapping of π-orbitals along the polymer backbone. Due to its simple conjugated molecular structure and fascinating electronic properties, polyacetylene has been widely studied as a prototype for other electronically conducting polymers (Chien, 1984).

Table 2.1. Some conjugated conducting polymers

Polymer (date conductivity discovered)	Structure	π-π* gap (eV)	Conductivity[#] (S/cm)
I. Polyacetylene and analogues			
Polyacetylene (1977)	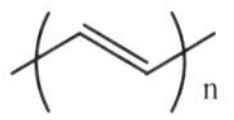	1.5	$10^3 - 1.7\times10^5$
Polypyrrole (1979)	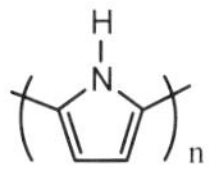	3.1	$10^2 - 7.5\times10^3$
Polythiophene (1981)	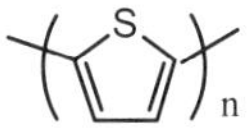	2.0	$10 - 10^3$
II. Polyphenylene and analogues			
Poly(paraphenylene) (1979)	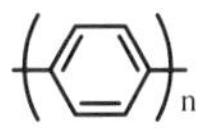	3.0	$10^2 - 10^3$
Poly(p-phenylene vinylene) (1979)	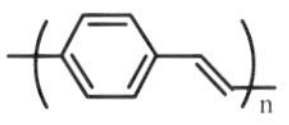	2.5	$3 - 5\times10^3$
Polyaniline (1980)	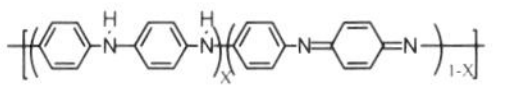	3.2	$30 - 200$

[#] The range of conductivities listed is from that originally found to the highest values obtained to date (after Dai, 1999, copyright 1999 Marcel Dekker, Inc.)

2.2.1.2 Doping

Since most organic polymers do not have intrinsic charge carriers, the required charge carriers may be provided by partial oxidation (*p*-doping) of the polymer chain with electron acceptors (*e.g.* I_2, AsF_5) or by partial reduction (*n*-doping) with electron donors (*e.g.* Na, K). Through such a doping process, charged defects (*e.g.* polaron, bipolaron and soliton) are introduced, which could then be available as the charge carriers. In the case of I_2-doped *trans*-polyacetylene, it was estimated that nearly 85% of the positive charge is delocalized over 15 CH units to give a positive soliton. In fact, the insulator-to-metal transition in conjugated polymers is not so simple, and the way in which charges can be stabilized on the polymer chains and the nature of the charge transport process are still a matter of debate. Nevertheless, the simple band theory can provide some useful information about the doping-induced changes in electronic structure.

According to band theory (Harrison, 1979), the electrical properties of direct gap inorganic semiconductors are determined by their electronic structures, and the electrons move within discrete energy states called *bands*. By analogy, the bonding and antibonding π-orbitals of the sp^2 hybridized π-electron materials (*e.g.* polyenes) generate energy bands, which are fully occupied (π-band) and empty (π*-band). The highest occupied band is called the *valence band*, and the lowest unoccupied band is the *conduction band*. The energy difference between them is called the *band gap*. Electrons must have certain energy to occupy a given band and need extra energy to move from the valence band to the conduction band. Moreover, the bands should be partially filled in order to be electrically conducting, as neither empty nor full bands can carry electricity. Owing to the presence of partially filled energy bands, metals have high conductivities (Figure 2.1(a)). The energy bands of insulators and semiconductors, however, are either completely full or completely empty. For instance, most conventional polymers have full valence bands and empty conduction bands, which are separated from each other by a wide energy gap (Figure 2.1(b)). In contrast, conjugated polymers have narrower band gaps (Figure 2.1(c)) and doping can change their band structures by either taking electrons from the valence band (*p*-doping) or adding electrons to the conduction band (*n*-doping).

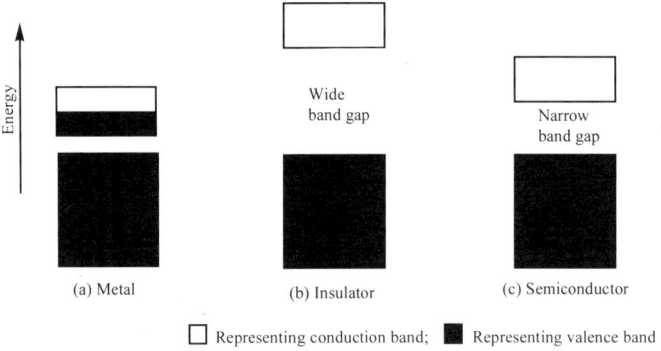

Figure 2.1. A schematic representation of energy gaps in (a) metal; (b) insulator; (c) semiconductor (After Dai, 1999, copyright 1999 Marcel Dekker, Inc.)

When an electron is added (removed) to the bottom of the conduction band (from the top of the valence band) of a conjugated polymer (Figure 2.2(a)), the conduction (valence) band ends up being partially filled and a radical anion (cation), commonly termed as a polaron (Brédas and Street, 1985), is created (Figure 2.2(b)). The formation of polarons causes the injection of states from the bottom of the conduction band and top of the valence band into the band gap. A polaron carries both spin (1/2) and charge (±1e). Addition (removal) of a second electron on a chain already having a negative (positive) polaron results in the formation of a bipolaron (spinless) through dimerization of two polarons, which can lower the total energy (Figure 2.2(c)). In conjugated polymers with a degenerate ground state *(i.e.* two equivalent resonance forms), like *trans*-polyacetylene, the bipolarons can further lower their energy by dissociating into two spinless solitons at one-half of the gap energy (Figure 2.2(d)). Solitons do not form in conjugated polymers with non-degenerate ground states, such as in polypyrrole, polythiophene and polyaniline (Brédas and Street, 1985). The population of polarons, bipolarons, and/or solitons increases with the doping level. At high doping levels, the localized polarons, bipolarons or solitons near to individual dopant ions could overlap, leading to new energy bands between and even overlapping the valence and conduction bands, through which electrons can flow.

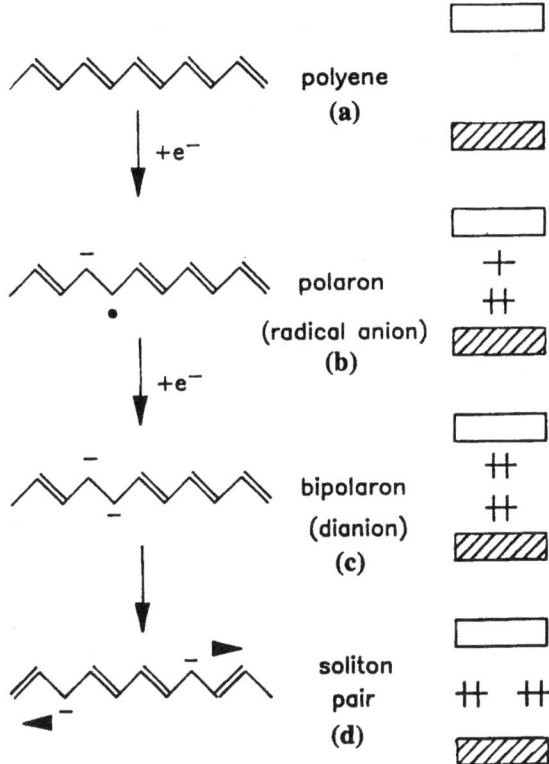

Figure 2.2. A schematic description of the formation of polaron, bipolaron, and soliton pair on a *trans*-polyacetylene chain by doping (After Dai, 1999, copyright 1999 Marcel Dekker, Inc.)

The bulk conductivity of conducting polymers should, in principle, consist of contributions from intra-chain, inter-chain and inter-domain electron transportations (Dai *et al.*, 1999). While the details for each of the transporting processes and their relative importance are still not fully understood, some of the factors that influence conductivity have been recognized. As can be seen from the foregoing discussion, the doping process is the most obvious factor which influences conductivity of conjugated polymers. Other factors include the orientation, crystallinity and purity of the conjugated polymers. Some of the most commonly used doping methods are described below.

Chemical Doping

Almost all conjugated polymers, including those listed in Table 2.1, can be either partially oxidized (*p*-type redox doping) or partially reduced (*n*-type redox doping) by electron acceptors or electron donors. For example, the treatment of *trans*-polyacetylene with an oxidizing agent such as iodine leads to the doping reaction [Equation (2.1)] and a concomitant increase in conductivity of about 10^{-5} to 10^2 S/cm (Figure 2.3).

$$\textit{trans-}(CH)_x + 3/2xyI_2 \rightarrow [CH^{+y}(I_3)_y^-]_x \qquad (2.1)$$

Similarly, most conjugated polymers like *trans*-polyacetylene can be doped with electron donors (*n*-doping) to gain high conductivities.

$$\textit{trans-}(CH)_x + [Na^+(C_{10}H_8)^-] \rightarrow [(Na^+)_y(CH)^{-y}]_x + C_{10}H_8 \qquad (2.2)$$

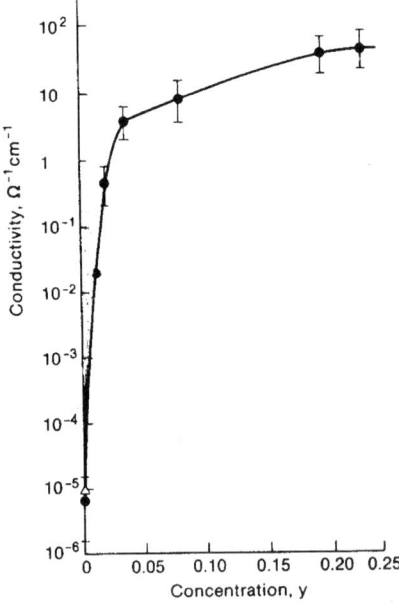

Figure 2.3. Electrical conductivity of I_2-doped *trans*-polyacetylene as a function of iodine concentration (After Chiang *et al.*, 1978, copyright 1978 American Institute of Physics)

More interestingly, it was found that the *n*- and *p*-type dopants could compensate each other. This compensation process is illustrated by the undoping of an Na-doped polyacetylene film with a composition of $(CHNa_{0.27})_x$ by I_2. As can be seen in Figure 2.4, the electrical conductivity of the Na-doped sample gradually decreases upon I_2-doping and increases, after reaching a minimum, with further *p*-doping (Chiang *et al.*, 1978).

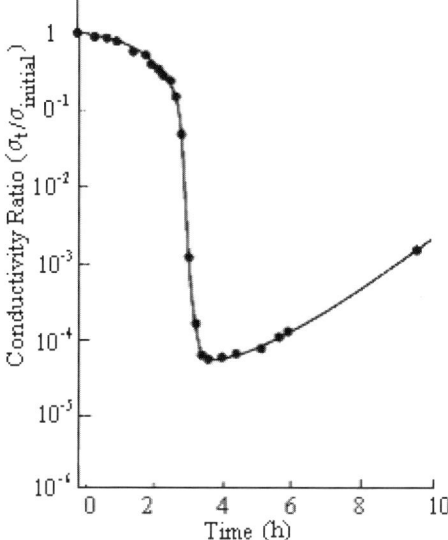

Figure 2.4. Electrical conductivity ratio of an Na-doped polyacetylene as a function of time of exposure to iodine (After Chiang *et al.*, 1978, copyright 1978 American Institute of Physics)

The relatively low conductivities attainable by the I_2-doping can be attributed to Na^+ ions remaining in the initially *n*-doped polyacetylene film in the form of NaI (0.5NaI/CH unit). Although the minimum conductivity for the fully compensated sample is still higher than that of the pristine polyacetylene film prior to Na-doping, the observed compensation process clearly demonstrates the feasibility to reversibly change the electrical properties of conjugated polymers in a controllable fashion.

Electrochemical Doping
Owing to the extensive conjugation of π-electrons, conjugated polymers can also be easily oxidized (*p*-doping) or reduced (*n*-doping) electrochemically with the conjugated polymer acting as either an electron source or an electron sink. In particular, the doping reaction [Equation (2.3)] can be accomplished by applying a DC power source between a *trans*-polyacetylene-coated positive electrode and a

negative electrode. Both of them are immersed in a solution of LiClO$_4$ in propylene carbonate (Chien, 1984).

$$\textit{trans-}(CH)_x + (xy)(ClO_4)^- \rightarrow [CH^{+y}(ClO_4)_y^-]_x + (xy)e^- \quad (y \leq 0.1) \quad (2.3)$$

Compared with chemical doping, electrochemical doping has several distinct advantages. Firstly, a precise control of the doping level can be achieved simply by monitoring the amount of current passed. Secondly, doping-undoping is highly reversible with no chemical products requiring removal. Finally, both *p*- or *n*-type doping can be achieved even with dopant species that cannot be introduced by conventional chemical means. In both cases, however, counter "dopant" ions are introduced for stabilizing the charge along the polymer backbone. The incorporation of counter ions can be both a hindrance and an advantage. While the counter ions may cause an undesirable structural distortion and a deteriorated effect on conductivity, they facilitate conjugated conducting polymers for actuation applications (Chapter 11).

In order to eliminate the incorporation of counter ions, "photo-doping" and "charge-injection doping" methods have been exploited to achieve the redox doping effects. Besides, some conjugated polymers, such as polyaniline and poly(heteroaromatic vinylenes), can also acquire high conductivities through the protonation of imine nitrogen atoms without any electron transfer between the polymer and "dopants" occurring - *i.e.* "non-redox doping" (Epstein and MacDiarmid, 1995).

Photo-doping
The irradiation of a conjugated polymer (*e.g. trans*-polyacetylene) macromolecule with a light beam of energy greater than its band gap could promote electrons from the valence band into the conduction band, as schematically shown in Equation (2.4) (Heeger *et al.*, 1988). Although the photogenerated charge carriers may disappear once the irradiation ceases, the application of an appropriate potential during irradiation could separate electrons from holes, leading to photoconductivity.

$$\textit{trans-}(CH)_x \xrightarrow{h\upsilon} \diagup\diagdown\diagup\diagdown + \diagup\diagdown\diagup\diagdown - \diagup\diagdown\diagup\diagdown \quad (2.4)$$

Charge-injection Doping
Using a field-effect transistor (FET) geometry (Chapter 8), charge carriers can be injected into the band gap of conjugated polymers [*e.g.* polyacetylene, poly(3-hexylthiophene), P3HT] by applying an appropriate potential on the metal/insulator/polymer multilayer structure (Garnier *et al.*, 1994). Just like photo-doping, the charge-injection doping does not generate counter ions, allowing a systematic study of the electrical properties as a function of the charge carrier density with a minimized distortion of the material structure. Using the charge-injection doping method, Schön *et al.* (2001) have recently demonstrated that a thin P3HT self-assembled film exhibits a metal-insulator transition with a metallic-like

temperature dependence. At temperatures below 2.35 K, these authors observed superconductivity when the charge density exceeds 2.5×10^{-14} cm^{-2}. This observation of superconductivity appears to be closely related to two-dimensional charge transport in the self-assembled polymer film. Although much of Schön's work is currently in question (Jacoby, 2002) and the detailed superconducting mechanism remains to be found (Tinkham, 1975), the above observation may suggest that the conductivity of conjugated polymers could be tuned over the largest possible range from insulating to superconducting.

Non-redox Doping
Unlike redox doping, the non-redox doping does not cause any change in the number of electrons associated with the polymer backbone, but merely a rearrangement of the energy levels. The most studied doping process of this type is the protonic doping of polyaniline emeraldine base (PANI-EB) with aqueous protonic acids, such as HCl, *d,l*-camphorsulfonic acid (HCSA), *p*-CH$_3$-(C$_6$H$_4$)SO$_3$H and (C$_6$H$_5$)SO$_3$H, to produce conducting polysemiquinone radical cations via the reaction shown in Equation (2.5) (Dai *et al.*, 1998a).

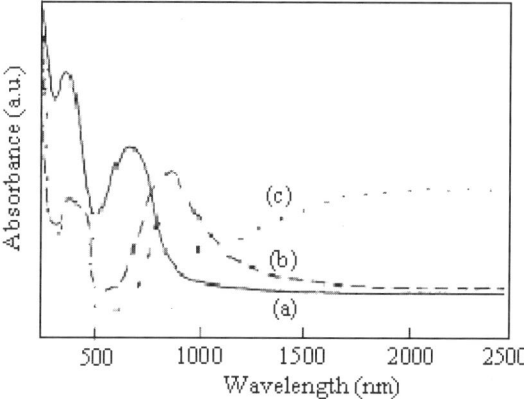

(2.5)

Figure 2.5 shows the evolution in time of the ultraviolet/visible/near-infrared (UV/vis/NIR) spectrum for the as-prepared PANI-EB after doping by HCSA (Dai *et al.*, 1998a).

Figure 2.5. UV/vis/NIR spectra of PANI-EB: (a) pristine PANI-EB; (b) after having been doped with HCSA; (c) the HCSA-doped PANI-EB after treatment with *m*-cresol (*vide infra*) (After Dai *et al.*, 1998a, copyright 1998 The American Chemical Society. Reproduced with permission)

The absorption peaks at 322 and 625 nm for the pristine PANI-EB (curve (a)) are attributable to the π-π* transition of the benzenoid rings and exciton absorption of the quinoid rings, respectively (Xia *et al.*, 1995). The absorption peaks at about 420 and 800 nm seen in curve (b) appeared at the expense of the interband transitions and can be taken as evidence for the presence of localized polarons produced by the HCSA doping (Cheng and Lin, 1995).

Secondary Doping
The interaction of an HCSA-doped PANI-EB with *m*-cresol was found to cause the absorption band characteristic of the localized polarons to largely disappear at 800 nm (curve (c) in Figure 2.5), while a very intense free carrier tail commencing at *ca.* 1000 nm developed (Dai *et al.*, 1998a). These spectroscopic changes have been attributed to the so-called "secondary doping" process, which causes a conformational transition of the polymer chain from a "compact coil" to an "expanded coil" due to molecular interactions between the HCSA-doped polyaniline and *m*-cresol (MacDiarmid and Epstein, 1994, 1995). The observed free carrier tail in the near-infrared region was considered to arise from delocalization of electrons in the polaron band, leading to a concomitant increase in conductivity by up to several orders of magnitude. Indeed, an increase in conductivity for a HCSA-doped polyaniline film from *ca.* 0.1 S/cm when cast from chloroform up to 400 S/cm when cast from *m*-cresol has been recorded (Cao *et al.*, 1992a, b).

The "compact coil" to "expanded coil" transition has long been ascribed to the "secondary doping" process. Some evidence from spectroscopic and viscosity measurements has been reported (MacDiarmid and Epstein, 1994, 1995). More recently, Dai *et al.* (2000, 2001) investigated the secondary doping of HCSA-doped PANI with *m*-cresol within the gallery of clay particles. They observed a significant increase in the inter-sheet distance of the polymer-intercalated clay particles (Figure 2.6), arising from the "compact coil" to "expanded coil" transition for polyaniline chains associated with the secondary doping.

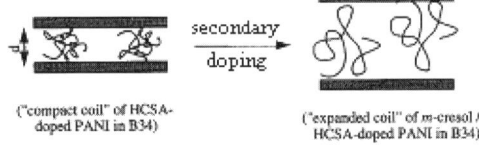

("compact coil" of HCSA-doped PANI in B34)   ("expanded coil" of m-cresol / HCSA-doped PANI in B34)

Figure 2.6. Schematic representation of the secondary doping of HCSA-doped PANI-EB with *m*-cresol within the clay gallery of Bentone 34 (B34). Note, d represents the height of the clay gallery and the diagrams are not drawn to scale (After Dai *et al.*, 2000, copyright 2000 Kluwer Academic Publishers. Reproduced with permission)

2.2.2 Syntheses

Although various catalyst systems (radiation, cationic, transition/rare-earth metal, *etc.*) can be used to initiate acetylene polymerization, the Ziegler-Natta catalyst remains the major one for chemical polymerization of acetylene (Shirakawa, 2001). Particularly, the Ti(OBu)$_4$/AlEt$_3$ is a preferred catalyst system for producing linear polymers of high molecular weight, high crystallinity and high conductivity upon doping (Chien, 1984). In this context, Ito *et al.* (1974) have succeeded in polymerizing acetylene directly in the form of a thin film, which is often termed as *Shirakawa's polyacetylene*, on the wall of a reaction vessel by adding about 1000 times more Ti(OBu)$_4$/AlEt$_3$ catalyst than the reaction stoichiometrically required. Meanwhile, the Luttinger catalyst (*i.e.* NaBH$_4$/Co(NO$_3$)$_2$·6H$_2$O) has also been used for acetylene polymerization (Chien, 1984; Luttinger, 1962). In both cases, acetylene is usually bubbled into a stirred catalyst solution for a homogeneous polymerization. However, the final product is often an intractable (*i.e.* insoluble and infusible) black powder (Chien, 1984). In fact, before the early 1980s no conjugated conducting polymer had been shown to be soluble in any solvent without decomposition (Dai, 1999), due to backbone rigidity associated with the delocalized conjugated structure. Furthermore, most unfunctionalized conjugated polymers were found to be unstable in air (Baker, 1988). Since then, a number of techniques have been developed to overcome these problems. Relatively stable conducting polymeric materials have been made, for example, by physically blending conjugated polymers with certain non-conjugated macromolecules (Baker, 1988) and chemically synthesizing conducting polymer colloids (Armes, 1996) or conjugated oligomers (de Meijere, 1999). On the other hand, many clever synthetic routes have now been devised for preparing processable conjugated polymers, as described below.

2.2.2.1 Syntheses of Soluble Conjugated Polymers

By Substitution

As is well known, both chemical and physical properties of a polymer material may change with substitution. Consequently, soluble forms of various conjugated polymers have been prepared by grafting suitable side groups and/or side chains along their conjugated backbones. Examples include polyacetylene, poly(*p*-phenylene vinylene) (PPV), polythiophene, polypyrrole and polyaniline grafted derivatives (Ferraro, 1987).

Polyacetylene derivatives: soluble poly(methylacetylene) and poly(phenylacetylene) have been synthesized (Yoshino, 1997). These polymers can be regarded as polyacetylene grafted with methyl or phenyl side groups. More recently, various soluble di-substituted polyacetylene derivatives, including poly(diphenylacetylene), poly(1-alkyl-2-phenylacetylene), poly(1-chloro-2-phenylacetylene) and poly[1-phenyl-2-*p*-(triphenylsilyl)phenylacetylene] have been synthesized to show electroluminescent properties (Chapter 9).

Poly(p-phenylene vinylene) derivatives: PPV is one of the most studied polymers for electroluminescent applications. The most common method to prepare PPV is the so-called Wessling route (Wessling and Zimmerman, 1968, 1972) from the sulfonium precursor polymer [Equation (2.6)].

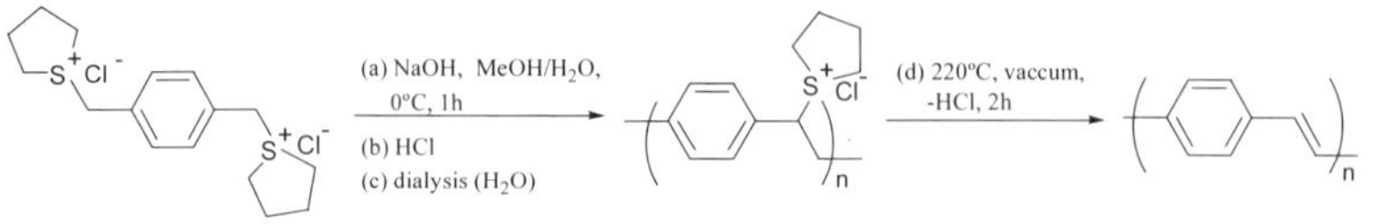

[Scheme 2.6 shown]

(a) NaOH, MeOH/H$_2$O, 0°C, 1h
(b) HCl
(c) dialysis (H$_2$O)
(d) 220°C, vaccum, -HCl, 2h

(2.6)

Although the final product from the Wessling route is insoluble and infusible with a rigid-rod microcrystalline structure, various soluble forms of PPV have been reported. For example, the Wessling method has been extended to synthesize soluble derivatives of PPV, such as poly[2-methoxy-5-(2'-ethyl-hexyloxy)-*p*-phenylene vinylene] (*i.e.* MEH-PPV, Burn *et al.* 1993), poly(2-butyl-5(2'-ethylhexyl)-1,4-phenylene vinylene) (*i.e.* BuEH-PPV, Gettinger *et al.*, 1994), and poly(2,5-dimethoxy-*p*-phenylene vinylene) (*i.e.* PDMeOPV, Bradley, 1992; Brédas *et al.*, 1996), by introducing various side groups/chains onto the phenylene units. Without involving the water-soluble sulfonium salts, soluble PPVs have also been obtained from the substituted dichloro-*p*-xylene in organic solvents via the so-called Gilch route (Gilch and Wheelwright, 1966), as shown in Equation (2.7).

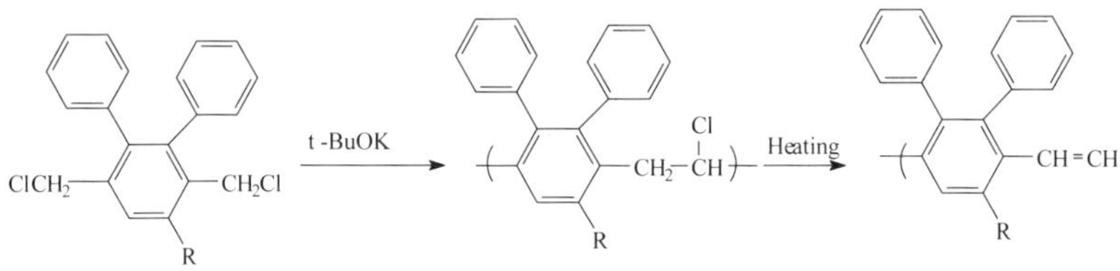

[Scheme 2.7 shown]

(2.7)

soluble chlorine   R = H   (DP-PPV)
precursor polymers R = 4-biphenyl (DPbp-PPV)

Compared with the Wessling method, the Gilch route allows easier access to a large range of substituted PPV derivatives soluble in organic solvents (Hörhold *et al.*, 1997; Hsieh *et al.*, 1998). Using either the Wessling or Gilch route, Winkler *et al.* (1999) have recently synthesised PPV derivatives grafted with oligo(ethylene oxide) side chains of different chain lengths [Equation (2.8)].

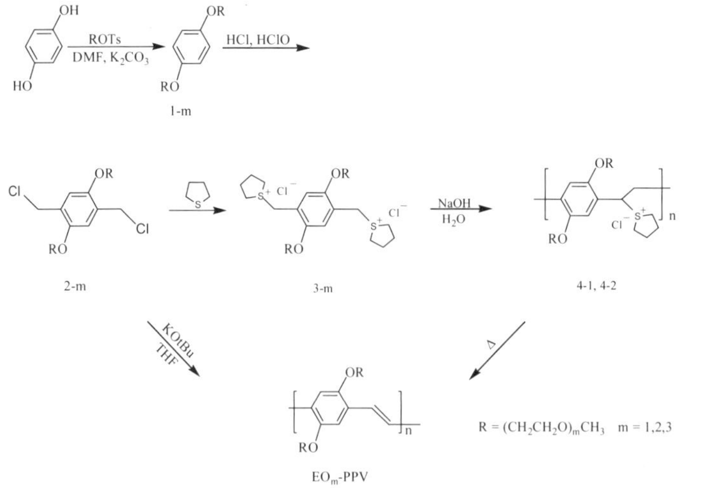

[Scheme 2.8 shown]

(2.8)

R = (CH$_2$CH$_2$O)$_m$CH$_3$ m = 1,2,3

EO$_m$-PPV

The same authors have also used the Wittig or Horner-Wadsworth-Emmons reaction to prepare phenylenevinylene oligomers/polymers substituted with crown ether rings via Equation (2.9) (Winkler et al., 2000).

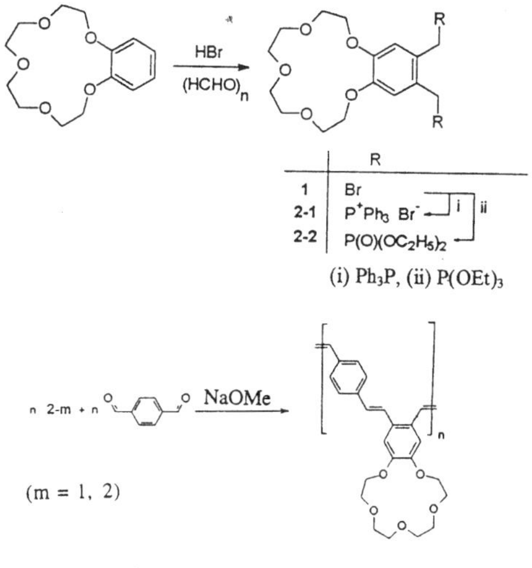

$$\tag{2.9}$$

CE-OPV

Other interesting PPV derivatives, including cyano-substituted PPVs, have been produced by a Knoevenagel condensation reaction between a diacetonitrile and a dialdehyde (Feast et al., 1996), which can also be used for synthesis of hetero-aromatic cyanopolymers such as poly(thienylenephenylene vinylene) and poly(thienylene vinylene).

Polythiophenes: Equation (2.10) represents the most common route to poly(3-alkylthiophenes), P(3AT)s, in which 3-bromothiophene is used as the starting material. P(3AT)s thus prepared are meltable and soluble in most common organic solvents (Tamao et al., 1982). Color tuning of photo- and electro-luminescence has been demonstrated for P(3AT)s with well-defined regioregularities (Gill et al., 1994). The regioregularity also has a profound influence on crystallinity and electrical conductivity (Sentein et al., 1996).

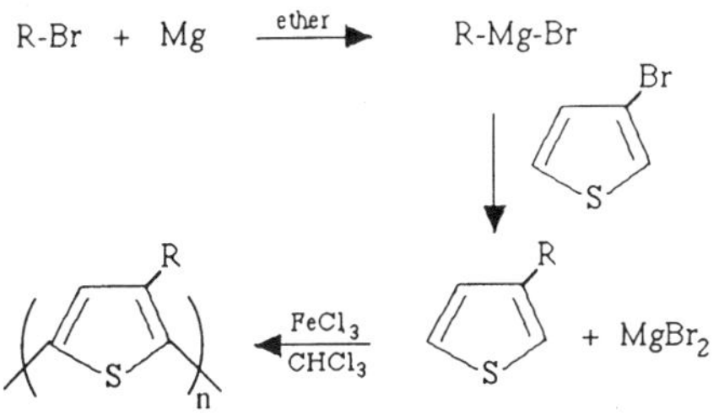

$$\tag{2.10}$$

Soluble polypyrrole and polyaniline have also been synthesised by introducing alkyl side groups (Patil *et al.*, 1987; Tour, 1994). Grafting with hydrophilic side groups/chains (-SO_3H, -COOH, -OH, *etc.*) even allows the formation of water-soluble conducting polymers (Brédas and Silbey, 1991). Generally speaking, the high solubility of the substituted conjugated polymers is often gained at the expense of significantly reduced conductivity. For instance, the conductivities of substituted polyacetylenes, after doping, were shown to be lower than that of doped polyacetylene by several orders of magnitude (Chien, 1984). Polyaniline coatings with a relatively high electrical conductivity, however, have been produced from soluble polyaniline emeraldine salt generated by the protonic doping of polyaniline emeraldine-base with organic sulfonic acids containing large organic groups *(e.g.* HCSA) (Cao *et al.*, 1992a, b).

By Copolymerization

Copolymerization of conjugated polymers with various soluble segments provides an alternative way to circumvent the intractability of conjugated polymers. The combination of optoelectronic properties characteristic of conjugated structures and the solubility of the soluble polymeric segments into a single copolymer chain should, in principle, lead to a material with properties characteristic of both the constituent components.

Random and alternating copolymerization: in 1981, Chien and co-workers synthesised random copolymers of acetylene and methylacetylene (Chien *et al.,* 1981). They found that the resulting random copolymers are more tractable than polyacetylene homopolymer, being swellable in toluene and pentane when the acetylene content is less than 55%(wt). Both the intractability and conductivity, in doped form, decrease with increasing percentage content of methylacetylene, and the highest obtainable conductivity is about 45 S/cm for an AsF_5-doped sample. Random copolymers of acetylene and phenylacetylene have also been made (Deits *et al.*, 1982), which are soluble in methylene chloride when the acetylene content is 26%(wt). These random copolymers, after doping, also show much lower conductivities than that of the polyacetylene homopolymer. Recently, Yang *et al.* (1998) have synthesised soluble copolymers with alternating conjugated and non-conjugated segments via a Wittig-type coupling reaction between the appropriate dialdehyde and 1,4-xylylenebis(triphenylphosphonium chloride) [Eq. (2.11)].

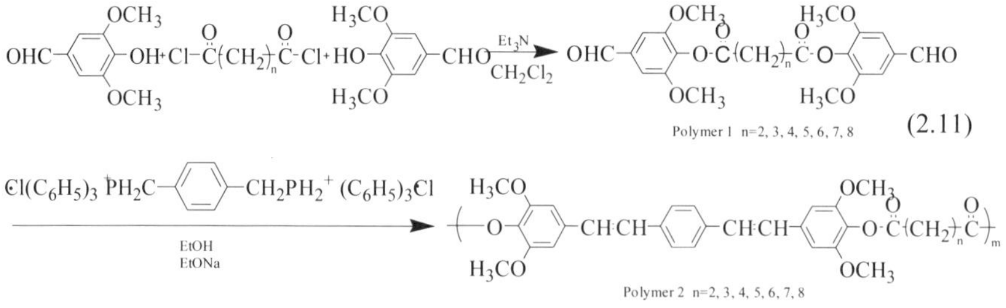

$$\text{(2.11)}$$

Polymer 1 n=2, 3, 4, 5, 6, 7, 8

Polymer 2 n=2, 3, 4, 5, 6, 7, 8

The copolymers prepared according to Equation (2.11) were found to have conductivities in the range of 10^{-3}–10^{-2} S/cm after having been cast into films from chloroform and doped with iodine (Yang et al., 1994).

In 1988, Thakur (1988) claimed that a conjugated structure is not always necessary for a polymer to be electronically conducting by demonstrating high conductivities up to 10^{-1} S/cm for "I_2-doped" cis-1,4-polyisoprene. In view of the ease with which conducting rubber can be made and the potential challenge of Thakur's claim to modern theory of conducting polymers (Section 2.2.1.1), Thakur's work received complimentary comments from several scientific journals (Borman, 1990; Calvert, 1988; Rothman, 1988). Meanwhile, various other polymers with isolated (non-conjugated) double bonds, such as trans-1,4-polyisoprene, 1,4-poly-(2,3-dimethylbutadiene) and trans-1,4-polybutadiene, have also been found to become dark in color and conductive when "doped" with iodine in the solid state. Dai and co-workers (Dai, 1992; Dai et al., 1994; Dai and White, 1991) have demonstrated that the treatment of certain polydiene chains of isolated double bonds with iodine, both in solution and the solid state, produces conjugated sequences in the polydiene backbone through polar addition of I_2 into the isolated double bonds in the polymer chain, followed by HI elimination. The final products resemble random copolymers consisting of polyacetylene and polydiene segments. The reaction mechanism for the I_2-induced conjugation is shown in Equation (2.12), and the conjugated sequences thus produced have been shown to be responsible for the high conductivity of the "I_2-doped" polyisoprene.

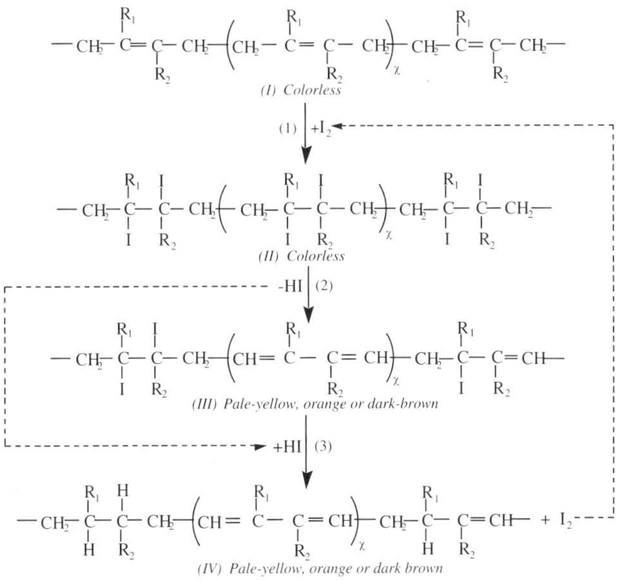

$$\text{(2.12)}$$

R_1, R_2 refer to either H or CH_3 group.

Block and graft copolymerization: just as in the case of graft copolymers *vs.* substituted polymers, block copolymers differ from random and alternating copolymers in that relatively long conjugated polymer block(s) are attached onto a soluble polymer chain (or *vice versa*). Therefore, the integrity of conjugated structures can largely be preserved in block or graft copolymers, suggesting they possess better electrical properties than their substituted or random/alternating counterparts. In this context, Dandreaux *et al.* (1983) and Kminek and Trekoval (1984) demonstrated that ethylene oxide or methylmethacrylate (MMA) could be grafted onto a sodium-doped polyacetylene homopolymer [Equation (2.13)].

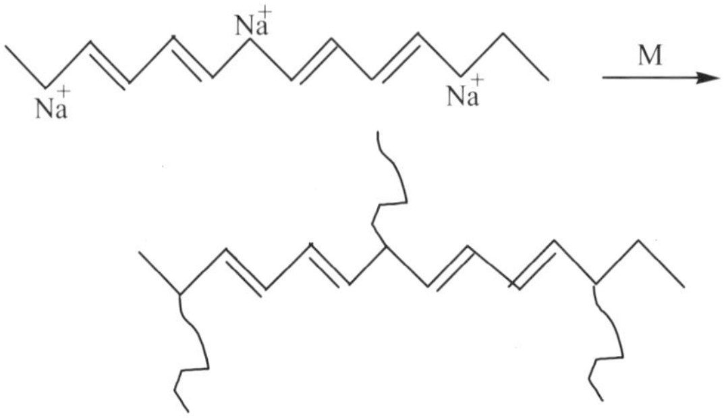

(2.13)

M = methyl methacrylate or ethylene oxide

The soluble graft copolymers of acetylene and ethylene oxide thus prepared were reported to have conductivities of up to 1 S/cm after I_2-doping in the solid state. This value of conductivity is still relatively low with respect to that of a polyacetylene homopolymer (Table 2.1). Possible explanations include: a) polyethylene oxide grafted chains are non-conducting; b) the grafting reaction itself introduces defects along the polyacetylene conjugated backbone.

It seems therefore more desirable to graft conjugated polymer chains onto a soluble polymer backbone. Using $Ti(OBu)_4/AlEt_3$ as a catalyst, Bates and Baker (1983), Aldissi (1986), and Galvin and Wnek (1985) have successfully copolymerized acetylene onto polystyrene, 1,4-polybutadiene, and 1,4-polyisoprene chains [Equations (2.14) & (2.15)], yielding either a block copolymer through the so-called anionic to Ziegler-Natta route [Equation (2.14)] or a graft copolymer by polymerizing acetylene in the presence of a second functionalized polymer [Equation (2.15)].

$$nBu\{CH_2\underset{Ph}{CH}\}_n CH_2\underset{Ph}{C^-}HLi^+ \xrightarrow[-\,LiOBu]{Ti(OBu)_4} nBu\{CH_2\underset{Ph}{CH}\}_n CH_2\underset{Ph}{CH}Ti(OBu)_3 \xrightarrow{Ti^{IV} \rightarrow Ti^{III}}$$

$$nBu\{CH_2\underset{Ph}{CH}\}_n CH_2\underset{Ph}{CH}Ti(OBu)_2 \xrightarrow{C_2H_2} nBu\{CH_2\underset{Ph}{CH}\}_{n+1}\{CH{=}CH\}_m CH{=}CH_2 \qquad (2.14)$$

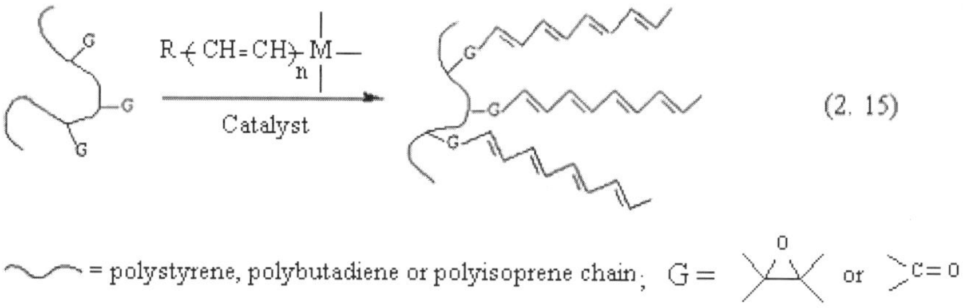

$$R\text{-}(CH=CH)_n\text{-}M\text{-} \xrightarrow{\text{Catalyst}} \quad (2.15)$$

∿ = polystyrene, polybutadiene or polyisoprene chain; G = (epoxide) or >C=O

Bolognesi *et al.* (1986) have used the same catalyst system for grafting copolymerization of acetylene onto 1,2-polybutadiene. Moreover, these authors have also grafted polyacetylene chains onto the 1,4-polydiene backbone by first lithiating polybutadiene chains with *sec*-butyllithium in the presence of tetramethylethylenediamine (TMEDA), followed by complexation with Ti(OBu)$_4$ to form a new polymeric catalyst for graft copolymerization of acetylene (Bolognesi *et al.*, 1985). It has been noted that polyacetylene chains in all of the block and graft copolymers synthesized by the Shirakawa-type titanium catalyst exist in *trans*- form. Using a novel Luttinger-type cobalt catalyst system, Armes *et al.* (1986), and Dai and White (Dai, 1997; Dai and White, 1997) have successfully synthesised both block [Equation (2.16)] and graft [Equation (2.17)] copolymers of 1,4-polydiene and polyacetylene, in which polyacetylene chains are mostly in the *cis*- form due to the low-temperature nature of the copolymerization reaction.

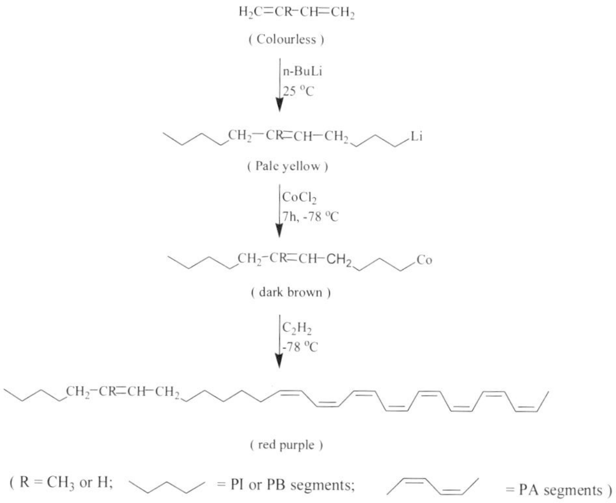

$$H_2C=CR\text{-}CH=CH_2$$
(Colourless)
$$\downarrow \text{n-BuLi, 25 °C}$$
∿CH$_2$–CR=CH–CH$_2$∿Li
(Pale yellow)
$$\downarrow \text{CoCl}_2, \text{7h, -78 °C}$$
∿CH$_2$–CR=CH–CH$_2$∿Co
(dark brown)
$$\downarrow \text{C}_2\text{H}_2, \text{-78 °C}$$
∿CH$_2$–CR=CH–CH$_2$∿══════
(red purple)

(2.16)

(R = CH$_3$ or H; ∿ = PI or PB segments; ══ = PA segments)

Furthermore, it was demonstrated that isomerization of the *cis*-polyacetylene blocks in a solution of the copolymers of polyisoprene and polyacetylene (PI-PA) prepared by the cobalt catalyst was retarded by some intramolecular barriers *(e.g.*

the chemically bonded polyisoprene chains), making the *cis-* isomer very stable at room temperature (Dai, 1997).

$$\begin{array}{c}
\sim\!\!\sim\!\!CH_2\text{-}CR\!=\!CH\text{-}CH_2\!\sim\!\!\sim\!\!CH_2\text{-}CR\!=\!CH\text{-}CH_2\!\sim\!\!\sim \\
\Big\downarrow \text{\textit{Sec}-BuLi, TMEDA} \\
\sim\!\!\sim\!\!\underset{\text{Li}}{CH}\text{-}CR\!=\!CH\text{-}CH_2\!\sim\!\!\sim\!\!CH_2\text{-}CR\!=\!CH\text{-}\underset{\text{Li}}{CH}\!\sim\!\!\sim \\
\Big\downarrow \text{CoCl}_2 \; (7h, -78^\circ C) \\
\sim\!\!\sim\!\!\underset{\text{Co}}{CH}\text{-}CR\!=\!CH\text{-}CH_2\!\sim\!\!\sim\!\!CH_2\text{-}CR\!=\!CH\text{-}\underset{\text{Co}}{CH}\!\sim\!\!\sim \\
\Big\downarrow C_2H_2 \; (-78^\circ C) \\
\sim\!\!\sim\!\!\underset{\text{\textasciitilde}}{CH}\text{-}CR\!=\!CH\text{-}CH_2\!\sim\!\!\sim\!\!CH_2\text{-}CR\!=\!CH\text{-}\underset{\text{\textasciitilde}}{CH}\!\sim\!\!\sim
\end{array} \qquad (2.17)$$

(R = CH_3 or H; $\sim\!\!\sim$ = PI or PB segments; $\sim\!\!\sim\!\!\sim$ = PA segments)

The syntheses of soluble copolymers of polystyrene and polythiophene or polyparaphenylene have also been reported (Francois and Olinga, 1993; Zhong and Francois, 1991). Polystyrene-polythiophene block copolymer films were found to convert into pure polythiophene films on heating at 380°C under vacuum or argon through selective depolymerization of polystyrene sequences (Francois and Olinga, 1993). The polythiophene films thus prepared can be controlled to have a spheric or fibrillar morphology, and show conductivity of up to 60 S/cm after doping with $FeCl_3$.

Another interesting area closely related to the block and/or graft conjugated copolymers is the synthesis of hyperbranched/dendritic (starburst) supramolecules with conjugated moieties (Müller, 1997). Dendritic macromolecules are molecular species with a large number of hyperbranched chains of precise length and constitution surrounding a central core (Tomalia and Durst, 1993). Dendritic polymers differ from conventional linear polymers in that the former possess much more free volume and terminal end groups, and hence become more soluble than the latter (see Chapter 4 for details). Although synthesis of fully conjugated dendritic systems is difficult, Duan *et al.* (1995) have prepared electrically conducting dendrimers by modifying the periphery of a poly(amidoamine) dendrimer with naphthalene diimide anion radicals, which can aggregate into a π-stacked network to provide the pathway for conductivity. Dendritic oligo(phenylene ethynylene)s up to 125 Å in diameter (Figure 2.7(a)) (Xu and Moore, 1993), along with dendritic oligophenylenes (Morgenroth *et al.*, 1997) and conjugated dendrimers of the type shown in Figure 2.7(b & c), respectively, have also been synthesized (Kuwabara *et al.*, 1994). A multilayer electroluminescence device incorporating the conjugated dendrimer shown in Figure 2.7(c) has been demonstrated to have reasonably good long-term stability.

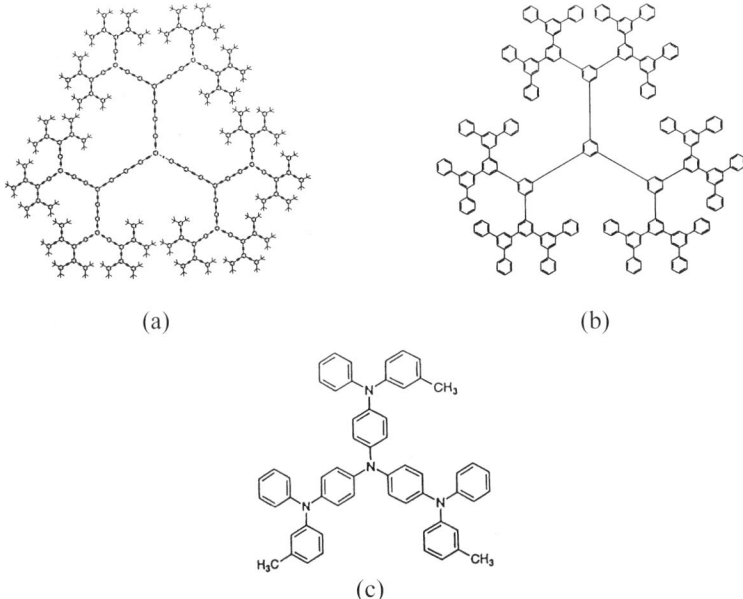

Figure 2.7. Conjugated dendritic polymers: (a) dendritic oligo(phenylacetylene); (After Xu and Moore, 1993); (b) dendritic oligophenylene; (After Morgenroth *et al.*, 1997) (c) 4,4′,4″-tris(3-methylphenyl-phenylamino)-triphenylamide (After Kuwabara *et al.*, 1994) (Copyrights Wiley-VCH Verlag)

Besides, Meier *et al.* (2000) have recently synthesized some highly soluble stilbenoid dendrimers with alkoxy chains substituted at the peripheral benzene rings [Equation (2.18)].

(2.18)

Wang et al. (1997) have also reported a route to conjugated star polymers with a hyperbranched 1,3,5-polyphenylene core surrounded by regioregular poly(3-hexylthiophene) arms [Equation (2.19)].

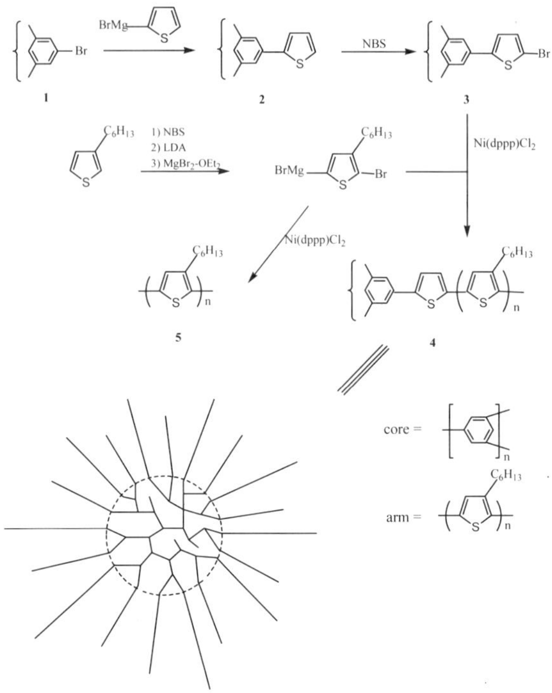

(2.19)

Other related examples include poly(2,5-bisphenyl-1,3,4-oxadiazole) (Herrema et al., 1995), polyquinoxaline (Jandke et al., 1998), hyperbranched poly-Schiff bases (Dai et al., 1998b) and hyperbranched PPV (Lin et al., 2000).

2.2.2.2 Syntheses of Conjugated Polymer Films

The syntheses of soluble conjugated polymers discussed above offer excellent possibilities for advanced device fabrication using various solution-processing methods (e.g. spin coating). However, solution-processing techniques may cause some problems related to the quality of polymer films thus formed, including possible deterioration of the mechanical properties of conjugated polymer films by chemical modification of their structures, the reduction of device stability by trapped impurities (e.g. O_2, residual solvent) and difficulty in choosing suitable solvent(s) for preparing pinhole-free multilayer polymer thin films. On the other hand, some *in situ* film-forming polymerization methods, including the precursor route and electropolymerization, can be used to circumvent the above-mentioned problems associated with post-synthesis solution processing.

Precursor Route
The formation of conjugated polymers by, for example, thermal conversion of soluble non-conjugated precursor polymers has also been tried in order to circumvent the intractability of conjugated polymers. To mention but a few examples:

Durham polyacetylene: the Durham route to polyacetylene developed by Edwards and Feast has attracted a considerable amount of attention (Edwards and Feast, 1980). As seen in Equation (2.20), it involves the ring-opening metathesis polymerization (ROMP) of cyclobutene derivatives, followed by a thermal conversion into polyacetylene with the evolution of 1,2-bis(trifluoromethyl)benzene. Thermal conversion at a relatively low temperature (< 50 °C) results in the formation of a high quality polyacetylene with a low crystallinity.

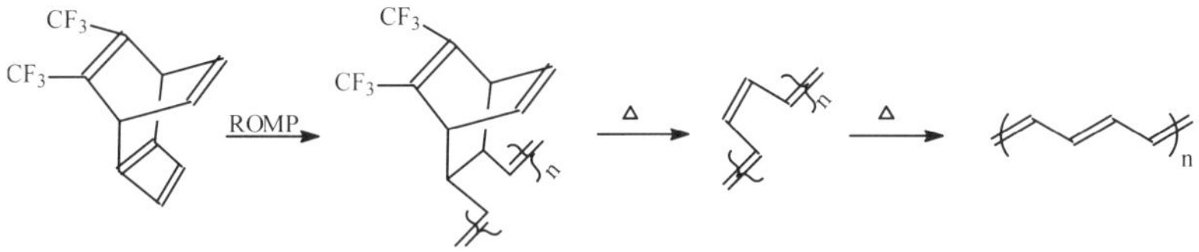

$$(2.20)$$

The material thus prepared could be useful for non-linear optical (NLO) applications, where the scattering of light by crystallites needs to be minimized. Changing the reaction conditions can vary the morphology of the final product.

With a refined Durham route, Knoll and Schrock (1989) were able to make polyacetylene prepolymers with a well-controlled molecular weight and its distribution close to unity. Furthermore, it is now possible to synthesize end-functionalized polyacetylene chains and to produce block copolymers with well-defined polyacetylene phase structure through the Durham route (Schrock, 1990). Based on the ROMP technique, Swager *et al.* (1988) reported an alternative precursor route to polyacetylene of low crystallinity [Equation (2.21)]. This method differs from the Durham route in that there is no mass loss upon thermal conversion. However, the resulting polyacetylene was shown to have a low conductivity after I_2-doping, and to possess a significant amount of sp^3 defects.

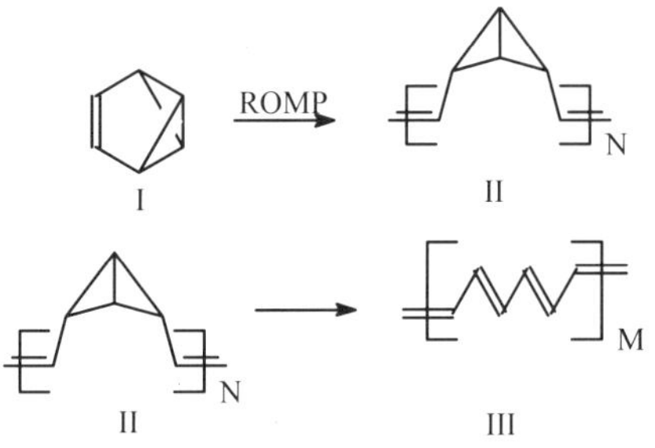

$$(2.21)$$

From poly(phenylvinylsulfoxide): polyacetylene-like polymers including polyacetylene-polystyrene diblock copolymers have also been made from poly(phenylvinylsulfoxide), PPVS (Kanga, 1990) and its polystyrene copolymers (Leung and Tan, 1994) by thermal elimination of phenylsulfenic acid (Reibel, 1992), as shown in Equation (2.22).

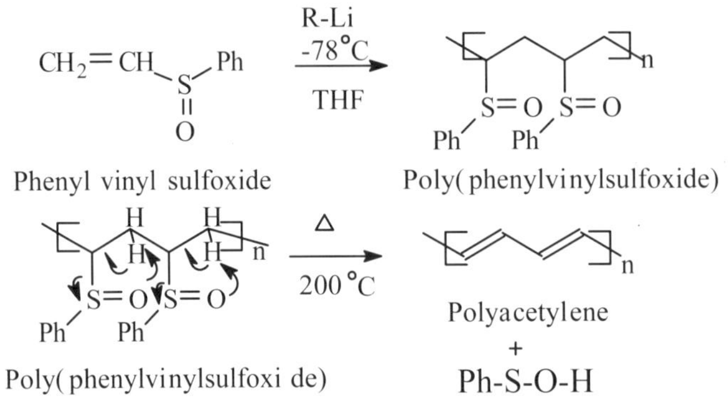

$$\text{Phenyl vinyl sulfoxide} \xrightarrow[\text{THF}]{\text{R-Li}, -78^\circ\text{C}} \text{Poly(phenylvinylsulfoxide)}$$

$$\text{Poly(phenylvinylsulfoxide)} \xrightarrow{\Delta, 200\,^\circ\text{C}} \text{Polyacetylene} + \text{Ph-S-O-H} \qquad (2.22)$$

Electropolymerization

Apart from the electrochemical doping, both the potentiostatic method with a constant potential and potentiodynamic technique by scanning the potential within a certain range of voltages have been used for electrochemical polymerization of electronically conducting polymers (Chien, 1984). For instance, pyrrole can be oxidatively polymerized on a suitable anode using a simple two-electrode electrochemical cell. Equation (2.23) shows the mechanism for electropolymerization of pyrrole, which is also valid for many other conjugated conducting polymers (Skotheim et al., 1986).

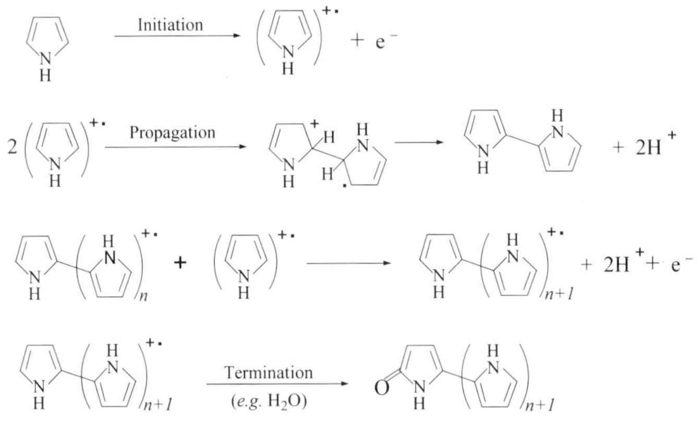

(2.23)

As can be seen from Equation (2.23), the electrochemical oxidation initially produces radicals. This is followed by the formation of dimers via the radical-radical recombination. Subsequent electrochemical oxidation of the dimer intermediates results in the formation of "oligomeric" radicals with a higher molecular weight, which could also combine with monomer radicals. Repeating the above steps yields a polypyrrole film on the anode electrode. The growth of polypyrrole macromolecules is believed to be governed by the radical-radical coupling and can be terminated via the exhaustion of reactive radical species in the vicinity of the electrode or by other chain termination processes. To balance the charge on the polymer backbone, counter ions are normally incorporated into the polymer film during the chain growth process.

2.3 Charge Transfer Polymers

2.3.1 Organic Charge Transfer Complexes

In 1973, Heeger and co-workers (Coleman *et al.*, 1973; Garito and Heeger, 1974) discovered that an organic ionic salt consisting of tetrathiofulvalene (TTF – electron donor, Figure 2.8(a)) and 7,7,8,8-tetracyano-*p*-quinodimethane (TCNQ – electron acceptor, Figure 2.8(b)) has a low-temperature conductivity (*ca.* –220 °C) as good as that of copper at room temperature. The observed high conductivity was attributed to a "herring bone"-type crystal structure formed by the flat TTF and TCNQ (Figure 2.8(c)), in which π-orbitals on adjacent molecules are overlapped to form continuous one-dimensional bands. The charge carriers in this system are provided by the electron transfer between the electron donor (D) and acceptor (A) molecules.

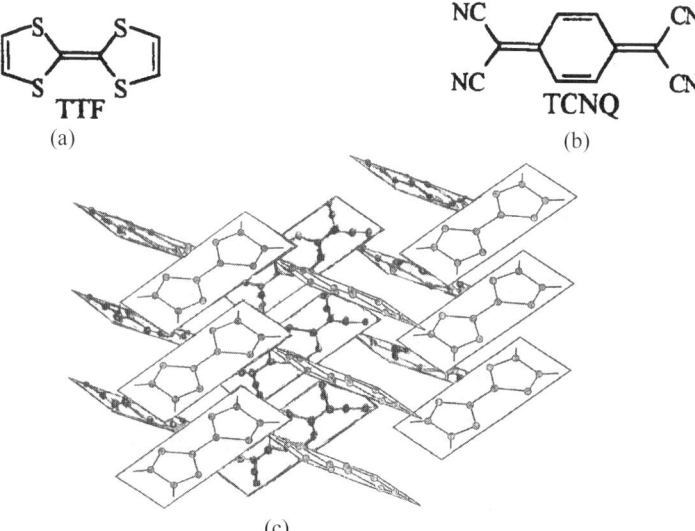

Figure 2.8. The molecular structure of (a) the electron donor TTF; (b) the electron acceptor TCNQ; (c) the crystal structure of TTF-TCNQ charge-transfer complex, in which π-orbitals on adjacent molecules can overlap to form continuous one-dimensional bands. Transfer of electrons from TTF to TCNQ partially empties the valence band of the former and partially fills the conduction band of the latter (After Ball 1994, copyright 1994 Princeton University Press).

If there were one electron transfer from each of the TTF molecules to each TCNQ molecule, the valence band (conduction band) associated with the TTF (TCNQ) molecules would then be completely emptied (filled), forming an insulator. However, it was found that only three fifths of an electron on average is transferred per molecule in TTF-TCNQ, leading to partially filled energy bands for charge carriers to move around (Section 2.2.1.2). The TTF-TCNQ salt is a typical example of a general class of organic conductors, normally called charge transfer (CT) complexes. As these systems involve two or more different molecules, there is plenty of room for regulating the electronic properties by interchanging different donors or acceptors, or by modifying their chemical structures. Indeed, a large number of CT complexes with various TTF (TCNQ) derivatives as the electron donors (acceptors) have been reported since 1973, as have many other systems with different CT moieties. Several CT salts have been found to even have superconducting properties, as exemplified by (BEDT-TTF)$_2$·X with BEDT being [bis(ethylenedithio)tetrathiafulvalene] and X = ClO_4^-, PF_6^-, AsF_6^-, I_3^-, IBr_2^-, BrI_2^-, *etc.* The major drawback for most CT complexes is that they are brittle and unprocessable.

2.3.2 Polymeric Charge Transfer Complexes

In the CT complexes discussed above, the conductivity can be attributed to two factors: the formation of appropriate segregated stacks of donors and acceptors and a certain degree of charge transfer between the stacks. Consequently, the formation of a desired crystal structure, composed of individual donor and acceptor molecules, is essential, but it is not readily susceptible to chemical control. In this regard, organic molecules or macromolecules with D and/or A moieties separated by chemically bonded spacers have been studied (Jiang *et al.*, 1998). These types of molecules possess several important advantages. Firstly, the ionization potential of D, together with the electron affinity of A, can be tuned by adjusting the nature and the mode of the chemical substitution. Hence, it is possible to prepare molecules with D and/or A in either a neutral or charged state by controlling the electron transfer from D to A. Molecules with the CT moieties in the neutral ground state should, in principle, behave like photoconductors or semiconductors, whereas their counterparts with the CT units in the charged state could be metallic conductors or superconductors. Secondly, the barrier to electron hopping along the molecules and their stacking columns may be reduced by introducing heavy atoms (*e.g.* transition metals) into the molecular spacer as certain metal atoms can overlap with the molecules through their large atomic orbitals. Finally, the incorporation of A and/or D moieties into polymer chains could increase the chances for packing segregated stacks with order (Becker *et al.*, 1989) and gain good processability and film-forming properties.

So far, various D-and/or A-containing polymers, with the CT moieties either as pendant groups or as constituent components of the polymer backbones, have been prepared. For example, Litt and Summers (1973) reported the synthesis of poly-*N*-acylethylenimines grafted with D molecules, such as 10-methylphenothiazine (MP) and 4-(methylthio)anisole (MTA) (Figure 2.9).

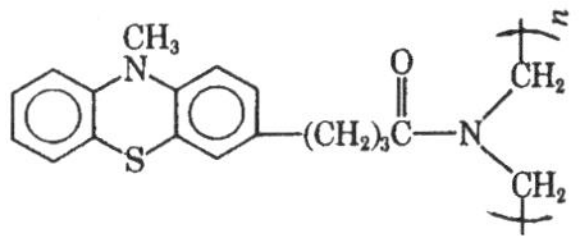

PMP

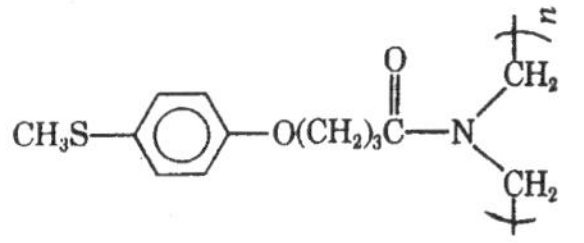

PMTA

Figure 2.9. The structure of PMP and PMTA (After Litt and Summers, 1973, copyright 1973 John Wiley & Sons, Inc.)

In these polymeric electron donors, the distance between two adjacent donors along the polymer chain allows the insertion of just one electron acceptor molecule (Figure 2.10), leading to good complexing with a stack of alternating D and A.

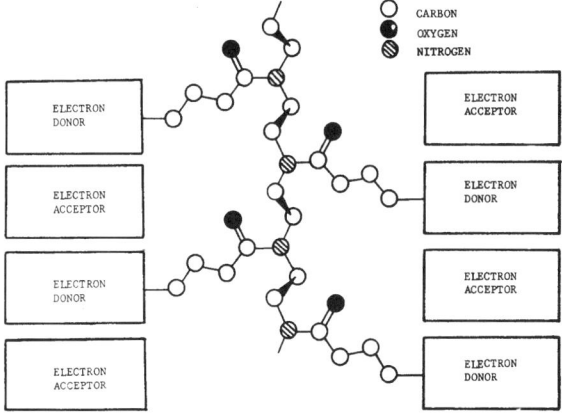

Figure 2.10. Schematic representation of the regular CT complex of polymeric electron donor with electron acceptors (After Litt and Summers, 1973, copyright 1973 John Wiley & Sons, Inc.)

Upon complexation with electron acceptors, such as dichlorodicyanoquinone (DDQ), TCNQ, tetracyanoethylene (TCNE) and tetranitrofluorenone (TNF), the aforementioned polymeric electron donors can become two hundred times as conducting as the corresponding model organic molecular complexes. The polymeric donor-TNF complex even showed photoconductivity (Summers and Litt, 1973).

More recently, Tamura et al. (1999) and Frenzel et al. (2001) have synthesized conjugated polymers containing TTF in the polymer backbone (Figure 2.11). The resulting TTF-containing polymers are soluble and have good film-forming properties. Conductivities up to about 5.5×10^{-1} S/cm were obtained through the formation of CT complexes with iodine.

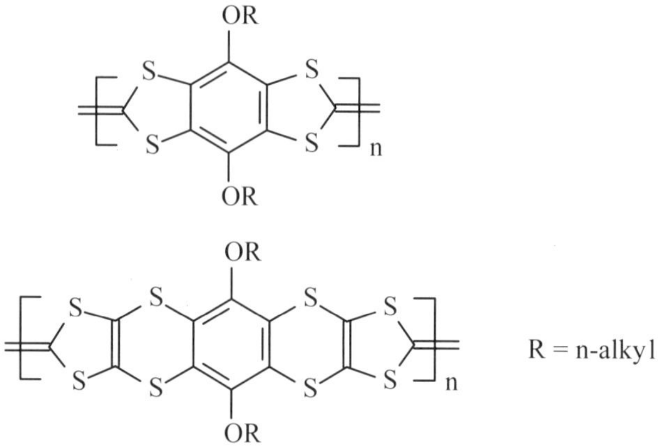

Figure 2.11. Some TTF-containing polymers with extended π-conjugation (After Frenzel *et al.*, 2001, copyright 2001 Elsevier. Reproduced with permission)

The use of metal-containing linkage(s) to facilitate the electron transfer from a D to an A was demonstrated by the synthesis of highly conductive metal-containing polymers based on the TTF-metal bis-dithiolene type of compounds (Figure 2.12) (Rivera and Engler, 1979).

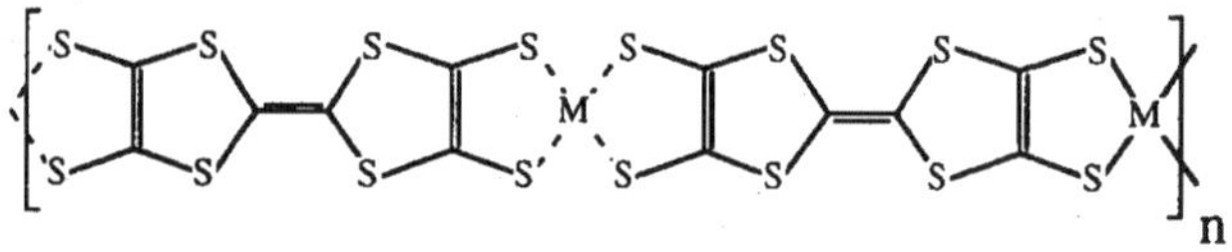

Figure 2.12. TTF-metal bis-dithiolene organometallic polymers (After Rivera and Engler, 1979, copyright 1979 The Royal Society of Chemistry. Reproduced with permission)

These compounds can show conductivities as high as 3×10^4 S/cm when the metal (M) atom is nickel. The conductivity in these metal-containing polymers arises from the charge transfer and the ability of the transition metal atom to provide the orbital overlap, implying that the conjugated structure and subsequent doping may not always be necessary for the metal-containing conducting polymers. A series of conducting polymers with porphyrin or metal-to-ligand CT porphyrin chromophores/electrophores incorporated into their backbones have been studied for potential applications ranging from artificial photosynthesis to light-emitting diodes (Jiang *et al.*, 1998).

2.3.3 Charge Transfer Between Fullerene C_{60} and Polymers

Buckminsterfullerene C_{60} has a soccerball-like structure with a diameter of 7.1 Å, consisting of 12 pentagons and 20 hexagons facing symmetrically (see Chapter 4 for details). Owing to its unusual molecular structure, fullerene C_{60} has a unique electronic structure. In contrast to benzene, double-bond resonant structures in five-membered rings of buckyballs are not favored, and the curvature of the C_{60} molecule leads to a greater intermolecular orbital overlap than graphite (Dai, 1999). As such, fullerenes have chemical characteristics different from the classical aromatic compounds and, unlike insulating diamond, they are semiconductors. The distribution in energy of the 60 π-electrons in C_{60} has been determined, for example, by Hückel calculations (Haddon, 1992). The theoretically determined low-lying LUMO near to the zero of the Hückel energy axis suggests a high electron affinity for the C_{60} molecule. Therefore, C_{60} can be used as an electron acceptor in the preparation of CT compounds. Complexation of C_{60} with strong electron donors (*e.g.* cobaltocene or tetrakis(dimethylamino)ethylene, TDAE) resulted in the formation of electrically insulating CT compounds (Allemand *et al.*, 1991a), which show *ferromagnetism* below the critical temperature T_c = 17 - 24 K (Allemand *et al.*, 1991b). On the other hand, doping fullerene C_{60} with alkali metals (Hammond and Kuck, 1992) has led to a dramatic increase in conductivity. Surprisingly, the K_xC_{60} (typically, x = 3) combination even showed *superconductivity* (SC) below 18 K (Hebard *et al.*, 1991).

While the ability for C_{60} to provide both of the cooperative phenomena (namely, superconductivity and ferromagnetism) in its solid state CT complexes is fascinating, C_{60} has also been used as a *p*-type dopant for conjugated conducting polymers. When doping conducting polymers with C_{60}, however, the LUMO of C_{60} is located between the top of the valence band and the bottom of the conduction band (*i.e.* in the forbidden gap) of most conducting polymers (Schlebusch *et al.*, 1996), indicating that fullerenes are weak dopants (Zakhidov, 1991). Consequently, the dark conductivities of the conducting polymers could only be marginally improved by doping with fullerene C_{60} (Dai *et al.*, 1998a). Due to photo-induced CT between the fullerene molecules and conjugated conducting polymers (Figure 2.13), however, various interesting optoelectronic phenomena, including quenching of photoluminescence, photo-induced adsorption, photo-induced dichroism and enhancement of photoconductivity, have been observed (Sariciftci, 1995). For instance, C_{60}-doped photoconducting films with performances comparable to some of the best commercial photoconductors *(e.g.* thiapyryliumdye aggregates) (Yu *et al.*, 1994) have been prepared by liquid-phase mechanical mixing of PVK with fullerenes (a mixture of C_{60} and C_{70}) (Wang, 1992). The improved photoconductivity of C_{60}-doped PVK was attributed to a photo-induced charge separation of the electron transfer complex between C_{60} and PVK. Furthermore, doping a poly(3-alkylthiophene)-C_{60}, P(3AT)-C_{60}, composite with an alkali metal, A (*e.g.* K vapor, *n*- doping), was suggested to cause extrinsic CT from A to both C_{60} and conducting polymer, leading to superconductivity for the CP-$(C_{60})_yA_x$ ternary composite (T_c = 17 K for 5 mol% C_{60}) (Sariciftci, 1995).

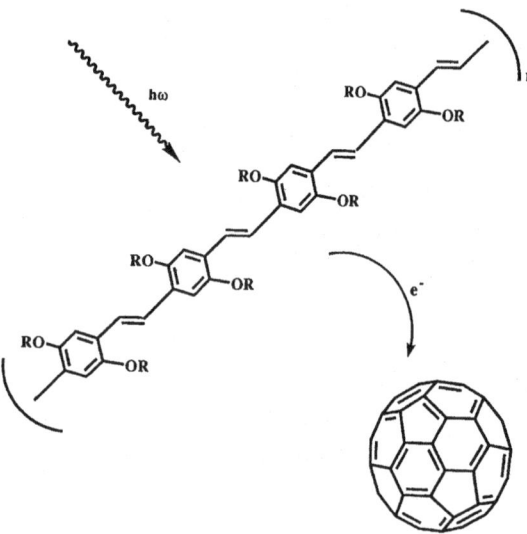

Figure 2.13. Schematic representation of the photo-induced electron transfer from conjugated polymers onto fullerene C_{60} (After Sariciftci, 1995, copyright 1995 Elsevier. Reproduced with permission)

2.4 Ionically Conducting Polymers

Just like the electronically conducting polymers (*i.e.* conjugated polymers and CT polymers) discussed above, the development of solid-state ionic conductors has been considered important because of the many uses to which they have been applied, for example, in electrochromic devices, supercapacitors, thermoelectric generators and secondary lithium batteries. The development of solid-state ionic materials based on ceramics, glasses or inorganic crystals has for decades been an active research area, while the ionically conducting polymers or polymer electrolytes represent a relatively new class of solid ionics. Since the inorganic solid-state ionics have received excellent coverage elsewhere (Vashishta *et al.*, 1979), we focus our attention here on polymer electrolytes.

2.4.1 Structural Features of Polymer Electrolytes

Polymer electrolytes are normally prepared by dissolving a salt species in a solid polymer host. Conductivity in polymer electrolytes is believed to arise from the ion migration between coordination sites repeatedly generated by the local motion of polymer chain segments. Therefore, a desirable polymer host must possess: a) electron-donating atoms or groups for the coordinate bond formation with cations, b) low bond rotation barriers for an easy segmental motion of the polymer chain, and c) an appropriate distance between coordinating centers for multiple intra-polymer bonding with cations. Although a large number of macromolecules can meet the above requirements, poly(ethylene oxide), PEO, remains the most studied host polymer (Gray, 1991). Likewise, much attention has been focused on lithium and sodium salts, though many salts can be dissolved in PEO. Figure 2.14 schematically shows the ion transport in PEO-based polymer electrolytes.

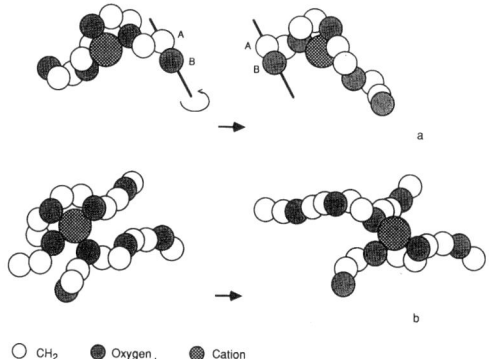

Figure 2.14. Schematic representation of the cation transport mechanism in a PEO-based polymer electrolyte. (a) Lateral displacement of the cation brought about by 180° bond rotation at the C-O bond along the AB line; (b) the transfer of the cation between PEO chains, possibly along with anions as either an ion pair or an ion triplet (After Gray, 1991, copyright 1991 VCH Publishers. Reproduced with permission)

As can be seen in Figure 2.14, the ionic mobility is closely correlated to the relaxation modes of the polymer host, which become effective above the polymer glass transition temperature, T_g (e.g. -60°C for PEO). Indeed, polymer electrolytes have long been considered to transport charge only above T_g, and the relatively slow segmental motion of polymer chains has limited the hopping rate of cations, and hence the conductivity. For this reason, the highest room-temperature conductivity ($ca.10^{-4}$ S/cm) reported for PEO-LiX electrolytes [with X being a negative ion like Cl, ClO_4, AsF_6, CF_3SO_3 and $N(SO_2CF_3)_2$] is still several orders of magnitude less than the corresponding values of most inorganic solid-state ionics. In comparison with inorganic solid-state ionics, however, polymer electrolytes offer many advantages, including the replacement of liquid electrolytes currently used in the lithium-battery technology, the versatility for fabrication of flexible solid-state devices free from seals, and the availability in various geometries (Croce *et al.* 1998).

2.4.2 Transport Properties and Chain Dynamics

As mentioned above, the movement of ions in polymer electrolytes is mediated by the local motion of the polymer chain segments above T_g. The liquid-like motion of polymer segments above T_g causes the local environment at any one point in the polymer sample to change with time. This has a strong influence on the charge transport in polymer electrolytes. In principle, the conductivity, $\sigma(T)$, is given by:

$$\sigma(T) = nq\mu \tag{2.24}$$

where n is the number of charge carriers, q is the charge for a single-charge carrier, and μ is its mobility. For a system with an invariant number of charge carriers and cations of a unity charge, such as PEO-LiX at a constant salt concentration, the conductivity can be related directly to the mobility of the charge species. Therefore, the ion transport in these polymer electrolyte systems can be described solely in terms of "solvent" fluidity. It is known that the temperature dependence of many relaxation and transport processes in liquid-like (amorphous) polymer systems in the vicinity of T_g can be described by the Williams-Landel-Ferry (WLF) equation (Flory, 1953):

$$\log a_T = \frac{-17.4\,(T - T_g)}{51.6 + T - T_g} \tag{2.25}$$

where a_T is a shift factor, which can be expressed in general terms as:

$$a_T = \tau/\tau_0 \tag{2.26}$$

with τ and τ_0 being the relaxation time at temperature T and a reference temperature T_0, respectively. Now since $\mu \propto 1/\tau$, the WLF equation implies that a decrease in T_g could lead to an increase in the ionic conductivity $\sigma(T)$.

Indeed, it was found that the addition of various plasticizers did improve the low-temperature ionic conductivity of PEO-LiX electrolytes due to an enhanced local relaxation and segmental motion of the polymer chains. The typical plasticizers used in the PEO-LiX systems can be divided into two categories: liquid plasticizers, such as low-molecular weight polyethyleneglycols or aprotic organic solvents; and solid-state plasticizers including ceramic powders and nanoparticles (Croce et al., 1998). The gain in conductivity by liquid plasticizers, however, is often accompanied by a deteriorated effect on mechanical properties and a loss of the compatibility with the lithium electrode in lithium secondary batteries. In contrast, the use of ceramic powders in the PEO-LiX electrolytes has been demonstrated to show several distinct advantages: a) increased low-temperature conductivity by promoting localized amorphous regions due to the large surface area of the particle fillers, b) enhanced mechanical stability by the network formation of the dispersed fillers, and c) improved compatibility with the lithium electrode due to the elimination of any liquids or interfacial stabilizing reagents. Figure 2.15 reproduces the temperature dependence of conductivity for nanocomposites of PEO-LiClO$_4$-10 wt% TiO$_2$ and PEO-LiClO$_4$-10 wt% Al$_2$O$_3$, along with that of a pristine PEO-LiClO$_4$ polymer electrolyte (Croce et al., 1998). The enhancement of ionic conductivity for the PEO-LiClO$_4$ electrolytes by adding ceramic nanoparticles is clearly evident.

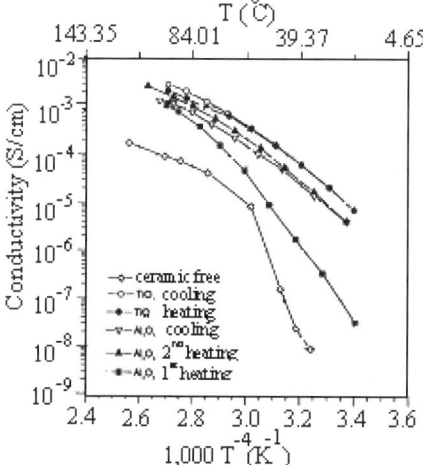

Figure 2.15. Arrhenius plots of the conductivity of ceramic-free PEO-LiClO$_4$, PEO-LiClO$_4$-10 wt% TiO$_2$ and PEO-LiClO$_4$-10 wt% Al$_2$O$_3$ polymer electrolytes (PEO:LiClO$_4$ = 8:1 in all cases) (After Croce *et al.*, 1998, copyright 1998 Macmillan Magazines Limited. Reproduced with permission)

Although the WLF equation does provide some useful information on the temperature dependence of the ionic conductivity in polymer electrolytes, the above transport mechanism is currently being challenged. In contrast to the general belief that ionic conductivity occurs only in *amorphous* polymers above T_g, Gadjourova *et al.* (2001) have recently shown that ionic conductivity in certain *crystalline* polymer electrolytes is even superior to that of their amorphous counterparts. In so doing, these authors prepared crystalline and amorphous PEO$_6$-LiSbF$_6$ and found that ionic conductivities of the crystalline polymer electrolytes are much higher than the corresponding amorphous materials over a range of temperatures (Figure 2.16).

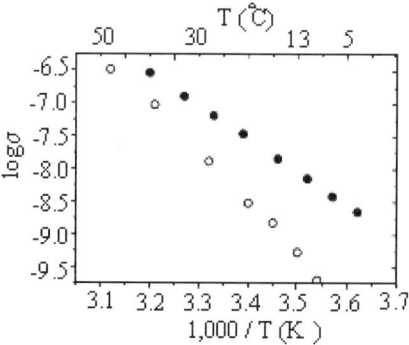

Figure 2.16. Ionic conductivity σ (S/cm) of amorphous (open circles) and crystalline (filled circles) PEO$_6$-LiSbF$_6$ as a function of temperature (After Gadjourova *et al.*, 2001, copyright 2001 Macmillan Magazines Limited. Reproduced with permission)

Detailed structural study shows that the polymer chains fold to form cylindrical tunnels inside of which the lithium ions are coordinated by the ether oxygens, whereas the anions are located outside of the tunnels with no coordinating interaction with the cations. The enhanced ionic conductivity of the crystalline materials, therefore, results from the relatively free movement of cations through the *pre-existing* tunnels.

The favored ion transport in crystalline polymer electrolytes is not inconsistent with the finding that some of the crystalline ceramics (*e.g.* $Li_{0.5}La_{0.5}TiO_3$) showed conductivities up to 1 - 3 orders of magnitude higher than the highest conductivity reported for conventional amorphous polymer electrolytes (Inaguma *et al.*, 1993). Besides, high ionic conductivity has also been recently observed in plastic crystals due to the fast lithium ion motion associated with rotational disorder and/or vacancies in the lattice (MacFarlane *et al.*, 1999). The realization that ordered structures can promote ion transport has opened up new approaches to the design of polymers with high ionic conductivity. This finding is also interesting in the context of electronically conducting conjugated and CT polymers, in which crystallinity favors electron transport.

2.5 Conductively Filled Polymers

An alternative method of inducing electrical conductivity in polymers is to make polymer composite materials with conductive additives or fillers. Typical examples of conductive components used to prepare this type of conducting polymer include conducting solids (carbon-black, carbon fibers, aluminum flake, stainless steel fibers, metal-coated fillers, metal particles, *etc.*) and conjugated conducting polymers. Because the conductivity is introduced through the addition of the conducting components, various polymer materials including both amorphous polymers (polystyrene, PVC, PMMA, polycarbonate, acrylonitrile butadiene styrene (ABS), polyethersulphone, polyetherimides, *etc.*) and crystalline polymers (polyethylene, polypropylene, polyphenylene sulphide, nylons, *etc.*) can be made electrically conducting. Various processing techniques such as hot compression, extrusion, and *in situ* polymerization have been used to prepare the conductively filled polymers for a number of applications ranging from electrostatic discharge protection to electromagnetic interference/radio-frequency interference (EMI/RFI) shielding (Olabisi, 1997).

2.5.1 Polymers Filled with Conductive Solids

As schematically shown in Figure 2.17, dispersion of conducting particles in an insulating polymer matrix can impart the conductivity when the particle volume fraction is greater than a value referred to as the percolation threshold (ϕ_c) (Stauffer, 1985).

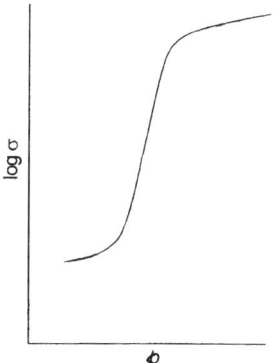

Figure 2.17. Dependence of composite conductivity, σ, on the volume fraction, ϕ, of conductive fillers

The conductivity of the composite is almost the same as that of the insulating polymer below ϕ_c. Since the number of the conductive particles is insufficient to form a continuous conducting path below ϕ_c (Figure 2.18(a)), the conductive domains are insulated from each other by the polymer medium and the electrical conducting behavior could not be observed. In the vicinity of ϕ_c, the isolated conductive particles appear to contact each other forming a continuous network for transportation of electrons (Figure 2.18(b)). Immediately after the percolation threshold (Figure 1.18(c)), a slight increase in the concentration of conductive particles may greatly increase the bridges in the conducting network. The insulating composite is thus transformed into a conducting material in a jumpwise fashion. Further increase in the concentration of the conductive particles, however, may only cause the volume of the conducting domains to increase without any significant increase in the pathways for electrons, leading to a monotonic increase in conductivity.

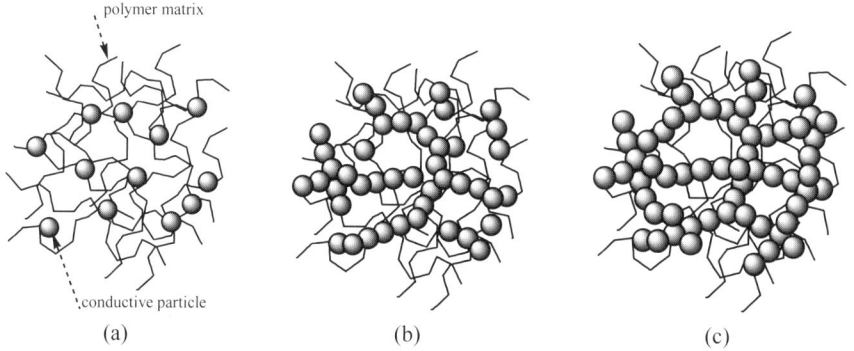

Figure 2.18. Schematic representation of conductive particles dispersed in a polymer matrix at different particle volume fractions: (a) $\phi < \phi_c$; (b) $\phi \approx \phi_c$; (c) $\phi \geq \phi_c$

Obviously, the precise location of the percolation threshold is affected by many factors, including the size, aspect ratio and size/spatial distributions of the conductive particles. To a given particle volume, for example, the surface area available for conductive contact increases with a decrease in the particle size, and hence a lower ϕ_c. Likewise, the loading of metal particles of a high aspect ratio has been demonstrated to be more effective in enhancing conductivity than the corresponding particles of a low aspect ratio (Sichel and Rubner, 1985). For spherical conducting particles dispersed in insulating matrices, the value of ϕ_c was found to be ca.16 vol.%.

Several theories have been developed to explain the percolation behavior observed in polymer composites filled with conducting particles (Balberg and Anderson, 1984; Mikrajuddin et al., 2000; Roldughin and Vysotskii, 2000). Most of them can only predict the general feature of the conductivity change with the filler content. For example, the scaling law of percolation theory (Roldughin and Vysotskii, 2000) gives no information about possible effects of the particle size, particle size distribution and spatial distribution flocculation on the percolation threshold. Nevertheless, it can be used to predict the change in conductivity near the percolation threshold, as seen in Equation (2.27).

$$\sigma \approx \sigma_1 \tau^t, \quad \tau > 0, \qquad \sigma \approx \sigma_1 |\tau|^{-q}, \quad \tau < 0 \qquad (2.27)$$

where $\tau = (\phi - \phi_c)/(1 - \phi_c)$, σ_1 and σ_2 are conductivity of the conductive particle and the polymer matrix, respectively, and t and q are scaling indexes. Numerical calculations have demonstrated that for a two-dimensional space $t = q$, while for a three-dimensional space $t > q$ (Roldughin and Vysotskii, 2000).

On the other hand, Wu (1985, 1988) deduced the correlation between the surface-to-surface interparticle distance, S (Figure 2.19), and the particle volume fraction, ϕ, for polymer blends with homogenously dispersed spherical particles of a diameter D as follows:

$$S = D \, [(\pi/6\phi)^{1/3} - 1] \qquad (2.28)$$

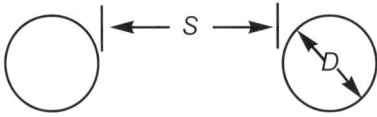

Figure 2.19. Schematic representation of conductive particles in a polymer matrix

By taking the typical electron tunneling distance of 10 nm (Doyen, 1999) as the surface-to-surface interparticle distance, the critical volume fraction (ϕ_c) calculated from Equation (2.28) is plotted against the particle diameter (D) in Figure 2.20 (Jing et al., 2000), which clearly shows that ϕ_c decreases rapidly with decreasing D in the nanometer range.

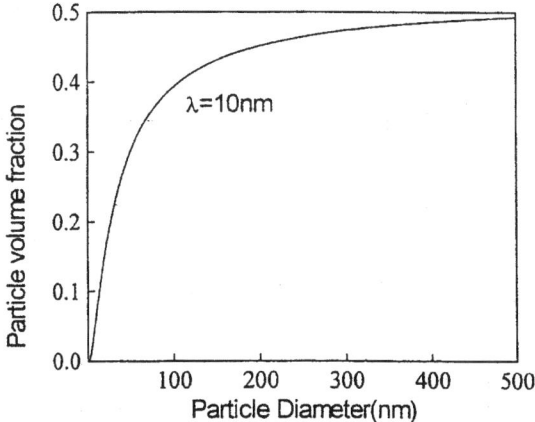

Figure 2.20. The relationship of the particle volume fraction and particle diameter for hexagonal close packed distribution at a surface-to-surface interparticle distance of 10 nm (After Jing *et al.*, 2000, copyright 2000 Kluwer Academic Publishers. Reproduced with permission)

Following the more recent formulations developed by Wu (1988), the effects of the particle size distribution and spatial distribution flocculation on the percolation threshold could also be investigated.

Consistent with Figure 2.20, a percolation threshold as low as 0.75 vol.% has recently been reported for nylon 6/graphite-conducting nanocomposites prepared by intercalation polymerization (Pan *et al.*, 2000). Furthermore, conducting polymer composites of low percolation thresholds have also been developed by using a *binary* host of *immiscible* polymer blends (Cheah *et al.*, 2000). In the binary polymer systems, the conductive particles may preferentially localize at the interface of the polymer blends or within one of the polymer phases (Gubbels *et al.*, 1998). The selective localization of the conductive particles at the interface of polymer blends is particularly interesting not only because percolation thresholds as small as 0.02 vol.% are achievable by forming a continuous conducting path at the interface region between two co-continued polymer phases but also because mechanical properties of the composites could be improved by the *in situ* interfacial modification.

In addition to the particulate conducting fillers, carbon fibers derived from polyacrylonitrile have also been used not only to generate conductivity but also to provide a degree of reinforcement of certain thermoplastic polymers for EMI/RFI shielding applications (Cogswell, 1992). Because of their high resilience, large surface area and good electrical properties, carbon nanotubes (see Chapter 5 for details) could well work better than carbon fibers for making novel conductively filled polymers with enhanced mechanical and electrical properties (Dai and Mau, 2001).

2.5.2 Polymers Filled with Conjugated Conducting Polymers

Composite materials based on conjugated conducting polymers and non-conducting polymers are another major class of conducting polymer composites that often show a low percolation threshold and improved environmental stability with respect to the conjugated polymer. To mention but a few examples, Heeger and co-workers (Reghu *et al.*, 1993; Suzuki *et al.*, 1990) have demonstrated very low percolation thresholds down to *ca.* 0.05 vol.% for spin-cast polymer composites consisting of soluble conjugated conducting polymers (*e.g.* PANI, poly(3-octylthiophene)) and certain non-conducting polymers (*e.g.* poly(methyl methacrylate), poly-(*p*-phenylene terephthalamide), polyethylene). The observed low ϕ_c was attributed to a multiple percolation, with which a continuous network of conducting pathways is formed on pre-existing tenuous interconnected networks of the non-conducting polymer. In addition to solution processing, compounding techniques used for processing of conventional thermoplastics have also been applied to prepare composites of polyacetylene or polypyrrole with certain thermoplastics (Wessling and Volk, 1986). On the other hand, polyacetylene lattices have been prepared by chemically polymerizing acetylene in the presence of poly[(tert-butylstyrene)-*b*-(ethylene oxide)] (Edwards *et al.*, 1983), whereas electropolymerization of pyrrole in a polymeric latex dispersion yielded conductive films after drying (Jasne *et al.*, 1986). In a separate but closely related study, polyacetylene/low-density polyethylene composites containing 1%–18% polyacetylene were prepared by exposing acetylene gas to polyethylene matrices pre-soaked with the Ziegler-Natta catalyst (Galvin and Wnek, 1982). The resultant materials showed conductivities of 5-10 S/cm, after I_2-doping, with as little as 2–4 vol.% polyacetylene. Irregular aggregates of polyacetylene were observed by transmission electron microscopy (TEM) to disperse within the insulating polyethylene matrix, which improved the environmental stability of the conducting component through physical "encapsulation". Later, this *in situ* polyacetylene polymerization method was applied to polybutadiene (Rubner *et al.*, 1983) and polystyrene-polydiene block copolymers (Lee and Jopson, 1983) to yield composite elastomers of conductivities up to *ca.* 570 S/cm. More recently, Frisch and co-workers (Frisch and De Barros, 1992; Frisch and Chen, 1994) have also prepared conducting polymer composites of conductivities up to 10^{-4} S/cm from full and psuedo-interpenetrating polymer networks of poly(carbonate-urethane) and natural rubber by I_2-induced conjugation of the polyisoprene elastomeric component (see Equation (2.12)).

2.6 References

Akamatu, H., Inokutchi, H., Matsunaga, Y. (1954) *Nature* **173**, 168.
Aldissi, M. (1986) *Synth. Met.* **13**, 87.
Allemand, P.-M., Chemani, K.C., Koch, A., Wudl, F., Holczer, K., Donovan, S., Grüner, G., Thompson, J.D. (1991a) *Science* **253**, 301.

Allemand, P.M., Khemani, K.C., Koch, A., Wudl, F., Holczer, K., Donovan, S., Gruner, G., Thompson, J.D. (1991b) *Science* **253**, 301
Armes, S.P. (1996) *Curr. Opin. Colloid Interf. Sci.* **1**, 214, and references cited therein.
Armes, S., Vincent, B., White, J.W. (1986) *J. Chem. Soc., Chem. Commun.* 1525.
Baker, G.L. (1988) In: Bowden, M.J., Turner, S.R. (eds) *Electronic and Photonic Applications of Polymers*. Adv. Chem. Ser. 218, ACS, Washington DC.
Ball, P. (1994) Designing *the Molecular World: Chemistry at the Frontier*, Princeton University Press, New Jersey.
Balberg, I., Anderson, C.H. (1984) *Phys. Rev. B* **30**, 3933.
Bates, F.S., Baker, G.L. (1983) *Macromolecules* **16**, 704.
Becker, J.Y., Bernstein, J., Bittner, S., Sarma, J.A.R.P., Shaik, S.S. (1989) *Chem. Mater.* **1**, 412.
Bolognesi, A., Cattellani, M., Destri, S., Porzio, W., Meille, S., Pedemonte, E. (1986) *Makromol. Chem.* **187**, 1287.
Bolognesi, A., Cattellani, M., Destri, S. (1985) *Mol. Cryst. Liq. Cryst.* **117**, 29.
Borman, S. (1990) *C&E News* **May 7**, 53.
Bradley, D.D.C. (1992) *Adv. Mater.* **4**, 756.
Brédas, J.L., Cornil, J., Heeger, A.J. (1996) *Adv. Mater.* **8**, 447.
Brédas, J. L., Silbey, R. (eds) (1991) *Conjugated Polymers*, Kluwer Academic, Dordrecht.
Brédas, J.L., Street, G.B. (1985) *Acc. Chem. Res.* **18**, 309.
Burn, P.L., Kraft, A., Baigent, D.R., Bradley, D.D.C., Brown, A.R., Friend, R.H., Gymer, R.W., Holmes, A.B., Jackson, R.W. (1993) *J. Am. Chem. Soc.* **115**, 10117.
Calvert, P. (1988) *Nature* **333**, 296.
Cao, Y., Smith, P., Heeger, A.J. (1992a) *Synth. Met.* **48**, 91.
Cao, Y., Treacy, G.M., Smith, P., Heeger, A.J. (1992b) *Appl. Phys. Lett.* **60**, 2711.
Cheah, K., Forsyth, M., Simon, G.P. (2000) *J. Polym. Sci., Polym. Phys.* **38**, 3106, and references cited therein.
Cheng, S.-A., Lin, L.C. (1995) *Macromolecules* **28**, 1239.
Chiang, C.K, Gau, S.C., Fincher, C.R., Jr., Park, Y.W., MacDiarmid, A.G., and Heeger, A.J. (1978) *Appl. Phys. Lett.* **33**, 18.
Chien, J.C.W. (1984) *Polyacetylene, Chemistry, Physics, and Material Science*, Academic Press, London.
Chien, J.C.W., Wenk, G.E., Karasz, F.E., Hirsch, J.A. (1981) *Macromolecules* **14**, 479.
Cogswell, F.N. (1992) *Thermoplastic Aromatic Polymer Composites: A Study of The Structure, Processing, and Properties of Carbon Fibre Reinforced Polyetheretherketone and Related Materials*, Butterworth-Heinemann, Boston.
Coleman, L.B., Cohen, M.J., Sadman, D.J., Yamagishi, F.G., Garito, A.F., Heeger, A.J. (1973) *Solid State Commun.* **12**, 1125.
Croce, F., Appetecchi, G.B., Persi, L., Scrosati, B. (1998) *Nature* **394**, 456.
Dai, L. (1992) *J. Phys. Chem.* **96**, 6469.
Dai, L. (1997) *Macromol. Chem. Phys.* **189**, 1723.
Dai, L. (1999) *J. Macromol. Sci., Rev. Macromol. Chem. Phys. C* **39**, 273.
Dai, L., Huang, S., Lu, J., Mau, A.W.H., Zhang, F. (1998b) *ACS Polym. Prep.* **39**, 171.
Dai, L., Lu, J., Matthews, B., Mau, A.W.H. (1998a) *J. Phys. Chem. B* **102**, 4049, and references cited therein.
Dai, L., Mau, A.W.H. (2001) *Adv. Mater.* **13**, 899.
Dai, L., Mau, A.W.H., Griesser, H.J., Winkler, D.A. (1994) *Macromolecules* **27**, 6728.
Dai, L., Wang, Q., Wan, M. (2000) *J. Mater. Sci. Lett.* **19**, 1645.
Dai, L., White, J.W. (1991) *Polymer* **32**, 2120.
Dai, L., White, J.W. (1997) in: Shi, L., Zhu, D. (eds) *Polymers and Organic Solids*, Science Press, Beijing.
Dai, L., Winkler, B., Dong, L., Tong, L., Mau, A.W.H. (2001) *Adv. Mater.* **13**, 915.

Dai, L., Winkler, B., Huang, S., Mau, A.W.H. (1999) In: Hsieh, B., Galvin, M., Wei, Y. (eds) *Semiconductive Polymers*, American Chemical Society, Washington DC, and references cited therein.
Dandreaux, G.F., Galvin, M.E., Wnek, G.E. (1983) *J. Physique Colloque C3* **44**, 135.
de Meijere, A. (ed.) (1999) *Carbon Rich Compounds II, Macrocyclic Oligoacetylenes and Other Linearly Conjugated Systems*, Springer-Verlag, Berlin.
Deits, W., Cukor, P., Rubner, M., Jopson, H. (1982) *Synth. Met.* **4**, 199.
Doyen, G. (1999) *The Physical Principles of STM and AFM Operation*. Wiley, New York.
Duan, R.G., Miller, L.L., Tomalia, D.A. (1995) *J. Am. Chem. Soc.* **117**, 10783.
Edwards, J.H., Feast, W. J. (1980) *Polymer* **21**, 595.
Edwards, J., Fisher, R., Vincent, B. (1983) *Makromol. Chem. Rapid. Commun.* **4**, 393.
Epstein, A.J., MacDiarmid, A.G. (1995) *Synth. Met.* **69**, 179.
Feast, W.J., Tsibouklis, J., Pouwer, K.L., Groenendaal, L., Meijer, E.W. (1996) *Polymer* **37**, 5017, and references cited therein.
Francois, B., Olinga, T. (1993) *Synth. Met.* **55–57**, 3489.
Frenzel, S., Baumgarten, M., Müllen, K. (2001) *Synth. Met.* **118**, 97.
Frisch, H.L., Chen, Z.J. (1994) *J. Polym. Sci., Polym. Chem.* **32**, 1317.
Frisch, H.L., De Barros, G.G. (1992) *J. Polym. Sci., Polym. Chem.* **30**, 937.
Ferraro, J.R., Williams, J.M. (1987) *Introduction to Synthetic Electrical Conductors*, Academic Press, San Diego.
Flory, P.J. (1953) *Principle of Polymer Chemistry*, Cornell University Press, New York.
Gadjourova, Z., Andreev, Y.G., Tunstall, D.P., and Bruce, P.G. (2001) *Nature* **412**, 520.
Galvin, M.E., Wnek, G.E. (1982) Polymer **23**, 795.
Galvin, M.E., Wnek, G.E. (1985) *Polym. Bull.* **13**, 109.
Garito, A.F., Heeger, AL. (1974) *Acc. Chem. Res.* **7**, 232.
Garnier, F., Hajlaoui, R., Yasser, A., Svrivastava, P. (1994) *Science* **265**, 1684.
Gettinger, C.L., Heeger, A.J., Drake, J.M., Pine, D.J. (1994) *J. Chem. Phys.* **101**, 1673.
Gilch, H.G., Wheelwright, W.L. (1966) *J. Poly. Sci., Polym. Chem.* **4**, 1337.
Gill, R.E., Malliaras, G.G., Wildeman, J., Hadziioannou, G. (1994) *Adv. Mater.* **6**, 132.
Gray, F.M. (1991) *Solid Polymer Electrolytes: Fundamentals and Technological Applications*, VCH Publishers, Weinheim, Germany.
Gubbels, F., Jerome, R., Vanlathem, E., Deltour, R., Blacher, S., Brouers, F. (1998) *Chem. Mater.* **10**, 1227.
Haddon, R.C. (1992) *Acc. Chem. Res.* **25**, 127.
Hammond, G.S., Kuck, V.J. (eds) (1992) *Fullerenes: Synthesis, Properties, and Chemistry of Large Carbon Clusters*, ACS Symposium Series 481, American Chemical Society, Washington DC.
Harrison, W.A. (1979) *Solid State Theory*, Dover Publications, Inc., New York.
Hebard, A.F., Rosseinsky, M.J., Haddon, R.C., Murphy, D.W., Glarum, S.H., Palstra, T.T., Ramirez, A.P., Kortan, A.R. (1991) *Nature*, **350**, 600.
Heeger, A.J., Kivelson, S., Schrieffer, J.R., Su, W.P. (1988) *Rev. Mod. Phys.* **60**, 781, and reference cited therein.
Herrema, J.K., Huten, P.F., Gill, R.E., Wilderman, J., Wieringa, R.H., Haziioannou, G. (1995) *Macromolecules* **28**, 8102.
Hörhold, H.H., Raabe, D., Opfermann, J. (1997) *J. Prakt. Chem.* **139**, 611.
Hsieh, B.R., Yu, Y., Forsythe, E.W., Schaaf, G.M., Feld, W.A. (1998) *J. Am. Chem. Soc.* **120**, 231.
Inaguma, Y., Chen, L.Q., Itoh, M., Nakamura, T., Uchida, T., Ikuta, H., Wakihara, M. (1993) *Solid State Commun.* **86**, 689.
Iqbal, Z., Baughman, R.H., Ramakrishna, B.L., Khare, S., Murthy, N.S., Bornemann, H.J., Morris, D.E. (1986) *Science* **254**, 826.
Ito, T., Shirakawa, H., Ikeda, I.J. (1974) *Polym. Sci., Polym. Chem.* **12**, 11.
Jacoby, M. (2002), *C&E News* **November 4**, 31.

Jandke, M., Strohriegl, P., Berleb, S., Werner, E., Brütting, W. (1998) *Macromolecules* **31**, 6434.
Jasne, S., Chiklis, C.K. (1986) *Synth. Met.* **15**, 175.
Jerome, D., Mazaud, M., Ribault, M., Bechgaard, K. (1980) *J. Phys. Let.* **41**, L95.
Jiang, B., Yang, S.W., Bailey, S.L., Hermans, L.G., Niver, R.A., Bolcar, M.A., Jones, Jr. W.E. (1998) *Coord. Chem. Rev.* **171**, 365.
Jing, X., Zhao, W., Lan, L. (2000) *J. Mater. Sci. Lett.* **19**, 377.
Kanga, R.S., Hogen-Esch, T.E., Randrianalimanana, E., Soum, A., Fontontanille, M. (1990) *Macromolecules* **23**, 4235 and 4241.
Kawakami, S. (1987) JP 01165603.
Kim, Y.H., Webster, O.W. (1992) *Macromolecules* **25**, 5561.
Kminek, I., Trekoval, J. (1984) *Makromol. Chem. Rapid Commun.* **5**, 53.
Knoll, K., Schrock, R.R. (1989) *J. Am. Chem. Soc.* **111**, 7989.
Kuwabara, Y., Ogawa, H., Inada, H., Noma, N., Shirota, Y. (1994) *Adv. Mater.* **6**, 677.
Lee, K.I., Jopson, H. (1983) *Polym. Bull.* **10**, 105.
Leung, L.M., Tan, K.H. (1994) *Polymer* **35**, 1556.
Lin, T., He, Q., Bai, F., Dai, L. (2000) *Thin Solid Films* **363**, 122.
Litt, M.H., Summers, J.W. (1973) *J. Polym. Sci. Polym. Chem. Ed.* **11**, 1339.
Luttinger, L.B. (1962) *J. Org. Chem.* **27**, 1591.
MacDiarmid, A.G., Epstein, A.J. (1994) *Synth. Met.* **65**, 103.
MacDiarmid, A.G., Epstein, A.J. (1995) *Synth. Met.* **69**, 85.
MacFarlane, D.R., Huang, J., Forsyth, M. (1999) *Nature* **402**, 792.
Meier, H., Lehmann, M., Kolb, U. (2000) *Chem. Eur. J.* **6**, 2462.
Mikrajuddin, A., Shi, F.G., Okuyama, K. (2000) *J. Electrochem. Soc.* **147**, 3157.
Morgenroth, F., Reuther, E., Müllen, K. (1997) *Angew. Chem., Int. Ed.* **36**, 631.
Müller, M., Morgenroth, F., Scherf, U., Soczka-Guth, T., Klärner, G., Müllen, K. (1997) *Phil. Trans. R. Soc. Lond. A* **355**, 715.
Olabisi, O. (ed.) (1977) *Handbook of Thermoplastics*. Marcel Dekker, New York.
Pan, Y.X., Yu, Z.Z., Qu, Y.C., Hu, G.H. (2000) *J. Polym. Sci., Polym. Phys.* **38**, 1626.
Patil, A.O., Ikenoue, Y., Wudl, F., Heeger, A.J. (1987) *J. Am. Chem. Soc.* **109**, 1858.
Reghu, M.; Yoon, C.O.; Yang, C.Y.; Moses, D.; Heeger, A.J.; Cao, Y. (1993) *Macromolecules* **26**, 7245, and references cited therein.
Reibel, D., Nuffer, R., Mathis, C. (1992) *Macromolecules* **25**,7090.
Rivera, N.M., Engler, E.M., Schumaker, R.R. (1979) *J. Chem. Soc., Chem. Commun.* 184.
Roldughin, V.I., Vysotskii, V.V. (2000) *Prog. Org. Coat.* **39**, 81.
Rothman, T. (1988) *Sci. Am.* **August**, 12;
Rubner, M.F., Tripathy, S.K., Georger, J., Jr., Cholewa, P. (1983) *Macromolecules* **16**, 870.
Sariciftci, N.S. (1995) *Prog. Quant. Electr,* **19**, 131, and references cited therein.
Schlebusch, C., Kessler, B., Cramm, S., Eberhardt, W. (1996) *Synth. Met.* **77**, 151.
Schön, J.H., Dodabalapur, A., Bao, Z., Kloc, Ch., Schenker, O., Batlogg, B. (2001) *Nature* **410**, 189.
Schrock, R.R. (1990) *Acc. Chem. Res.* **23**, 158.
Sentein, C., Mouanda, B., Rosilio, A., Rosilio, C. (1996) *Synth. Met.* **83**, 27.
Shirakawa, H. (2001) *Angew. Chem. Int. Ed.* **40**, 2574.
Sichel, E.K., Rubner, M.F. (1985) *J. Polym. Sci., Polym. Phys.* **23**, 1629.
Skotheim, T.A., Elsenbaumer, R.L., Reynolds, J.R. (eds) (1998) *Handbook of Conducting Polymers*, Marcel Dekker, New York.
Stauffer, D. (1985) *Introduction to Percolation Theory*, Taylor & Francis, Philadelphia.
Summers, J.W., Litt, M.H. (1973) *J. Polym. Sci. Polym. Chem.* **11**, 1379.
Suzuki, Y.Y., Heeger, A.J., Pincus, P. (1990) *Macromolecules* **23**, 4730.
Swager, T.M., Dougherty, D.A., Grubbs, R. H. (1988) *J. Am. Chem. Soc.* **110**, 2973.
Tamao, K., Kodama, S., Nakajima, I., Kumada, M., Minato, A., Suzuki, S. (1982) *Tetrahedron* **38**, 3347.

Tamura, H., Watanabe, T., Imanishi, K., Sawada, M. (1999) *Synth. Met.* **107**, 19.
Thakur, M. (1988) *Macromolecules* **21**,661.
Tinkham, M. (1975) *Introduction to Superconductivity*, McGraw-Hill, New York.
Tomalia, D.A., Durst, H.D. (1993) *Top. Curr. Chem.* **165**, 193.
Tour, J.M. (1994) *Adv. Mater.* **6**, 190.
Vashishta, P., Mundy, J.N., Shenoy, G.K. (eds) (1979) *Fast Ion Transport in Solids*, North-Holland, Amsterdam.
Wang, Y. (1992) *Nature* **356**, 585.
Wang, F., Rauh, R.D., Rose, T.L. (1997) *J. Am. Chem. Soc.* **119**, 11106.
Wessling, B., Volk, H. (1986) *Synth. Met.* **15**, 183.
Wessling, R.A., Zimmerman, R.G. (1968) U.S. Patent 3401152.
Wessling, R.A., Zimmerman, R.G. (1972) U.S. Patent 3406677.
Winkler, B., Dai, L., Mau, A.W.H. (1999) *Chem. Mater.* **11**, 704.
Winkler, B., Mau, A.W.H., Dai, L. (2000) *Phys. Chem. Chem. Phys.* **2**, 291.
Wright, P.V. (1975) *Br. Polym. J.* **7**, 319.
Wu, S. (1985) *Polymer* **26**, 1855.
Wu, S. (1988) *J. Appl. Polym. Sci.* **35**, 549.
Xia, Y., Wiesinger, J.M., MacDiarmid, A.G., Epstein, A.J. (1995) *Chem. Mater.* **7**, 443.
Xu, Z., Moore, J.S. (1993) *Angew. Chem., Int. Ed.* **32**, 1354.
Yang, Z., Hu, B., Karasz, F.E. (1998) *J. Macromol. Sci., Pure Appl. Chem.* **A35**, 233, and references cited therein.
Yang, Z., Karasz, F.E., Geise, H.J. (1994) *Polymer* **35**, 391.
Yoshino, K., Hirohata, M., Hidayat, R., Tada, K., Sada, T., Teraguchi, M., Masuda, Frolov, S.V., Shkunov, M., Vardeny, Z.V., Hamaguchi, M. (1997) *Synth. Met.* **91**, 283.
Yu, G., Pakbaz, K., Heeger, A.J. (1994) *Appl. Phys. Lett.* **64**, 3422.
Zakhidov, A.A. (1991) *Synth. Met.* **41**, 3393.
Zhong, X.F., Francois, B. (1991) *Makromol. Chem.* **192**, 2277.

Chapter 3

Stimuli-responsive Polymers

3.1 Introduction

A responsive macromolecule is one that changes its conformation and/or properties in a controllable, reproducible, and reversible manner in response to an external stimulus (*e.g.* solvent, pH, or temperature). As we shall see later (Chapters 8–11), these changes in conformation/physicochemical properties of the stimuli-responsive polymers can be utilized to create a large variety of smart devices, such as sensors, actuators and control release systems, for various practical applications. The good processability of most stimuli-responsive polymers facilitates their incorporation into devices and adds additional advantages for the development of smart devices with novel features (*e.g.* all plastic electronic/optical sensors).

In view of the fact that DNA, proteins and polysaccharides constitute the most important components of a living cell, life at its very essence is polymeric. Nature uses these biomacromolecules both as constructive elements and as parts to form complicated cell machinery. The salient feature of biomacromolecules is that they can adopt unique, well-defined conformations according to specific environmental conditions, despite the large number of possible confirmations available (Chapter 1). Although stable, ordered conformational structures can readily be formed by most biomacromolecules, it is equally possible that conformational changes can occur in response to an external stimulus (*e.g.* temperature and solvency). As in the case of most other stimuli-responsive polymers, biomacromolecules can respond to small alterations in an external variable with very dramatic conformational/structural changes, as well exemplified by the temperature-induced helix-coil transition of polypeptides (Cantor and Schimmel, 1980; Flory and Miller, 1966; Lifson and Roig, 1961; Peller, 1959; Poland and Scheraga, 1970).

Along with the well-documented stimuli-responsive biomacromolecules (Cantor and Schimmel, 1980), the study of the synthesis and structure of stimuli-responsive synthetic macromolecules has recently received dramatically increased research attention. This chapter provides an overview of several of the main classes of synthetic responsive polymers (*e.g.* solvent, pH, temperature, electro- and photoresponsive polymers), while their potential device applications will be discussed in subsequent chapters as becomes appropriate.

3.2 Solvent-responsive Polymers

As discussed in Chapter 2, doping with either electron acceptors or electron donors can enhance the conductivity of conjugated polymers. The delocalized electronic structure induced by doping, which is responsible for the conductivity, also causes changes in the π-π^* band gap, accompanied by changes in UV-visible absorption. Based on this delicate combination of electronic and optical properties in conjugated polymers, various electrochromic devices have been developed by reversible doping-dedoping of certain conjugated polymers.

Generally speaking, macromolecular chains usually adopt a coil-like conformation in solution, rather than presenting themselves stretched out as rods (Chapter 1). Certain polymer chains with rigid conjugated sequences, however, may take an extended, more or less rod-like conformation (Aime, 1991). Simple Hückel calculation has indicated that any twisting between adjacent segments in conjugated macromolecular chains will reduce the effective conjugation length (Brèdas et al., 1985; Leclerc and Prud'homme, 1985; Levine, 1974), and hence cause concomitant changes in optical absorption. Therefore, the twisting of substituted conjugated polymer chains (e.g. poly(3-alkylthiophene)s, N-substituted polypyrroles) due to steric effects of the bulky substituents could lead to soluble conjugated polymers with noteworthy electronic and optical properties.

The uncoiling of macromolecular chains may be achieved by a chemical reaction that produces locally rigid sequences extending along the polymer backbone or as a result of increasingly effective rigidity for the polymer chain in response to physical stimuli (e.g. reducing the solvent quality, changing temperature). The coil-to-rod transition of soluble polydiacetylenes and substituted polyacetylenes induced by changes in solvent quality has been widely investigated (Leclerc and Faïd, 1997; Dai and White, 1994). Both the solvent-induced (solvatochromism) and temperature-induced (thermochromism) of poly(3-alkylthiophene)s have been reported. While the thermochromism will be discussed in the next subsection, I am focusing here on the solvatochromism of substituted conjugated polymers by presenting several example studies below.

It has been demonstrated that the protonic-doped polyaniline emeraldine base (e.g. HCSA-doped PANI-EB) is soluble in certain solvents, such as m-cresol, xylene, or formic acid (Cao and Heeger, 1993). As has been mentioned in Chapter 2, the interaction of an HCSA-doped PANI-EB with m-cresol, either in vapor or liquid form, causes a conformational transition of the polymer chain from a "compact coil" to an "expanded coil" through the so-called "secondary doping" process. The change in molecular conformation associated with the secondary doping will not only increase the conductivity but also lead to solvatochromism for the HCSA-doped PANI material (MacDiarmid and Epstein, 1995). Figure 3.1 shows dramatic changes in the UV/vis/NIR spectra when a thin film of the HCSA-doped PANI is exposed to m-cresol vapor, indicating a strong solvatochromism effect. Subsequently, a similar effect was also shown to occur with HCSA-doped PANI secondary-doped by the less toxic carvacrol (Norris *et al.*, 2002) and with certain dendrimers after sorption of some very common solvents (*e.g.* acetone, ethanol, toluene, water) (Miller *et al.*, 1998).

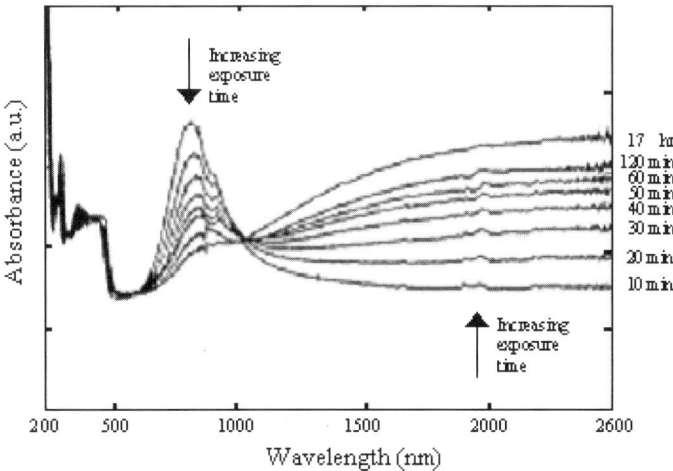

Figure 3.1. Changes in UV/vis/NIR spectra of a thin film of HCSA-doped PANI spun from chloroform solution upon exposure to the vapor of *m*-cresol at room temperature (After MacDiarmid and Epstein, 1995, copyright 1995 Elsevier. Reproduced with permission)

Poly[1-(trimethylsilyl)-1-propyne] ($-[C(Me)=C(SiMe_3)]_n-$, PTMSP) represents a class of substituted polyacetylenes (Chapter 2). Membranes prepared from PTMSP were found to be more permeable to gases such as oxygen and nitrogen than any other polymers (Masuda *et al.*, 1985; Nagai *et al.*, 2001). Although PTMSP contains a polyacetylene-type polymer backbone with alternating single and double bonds, some molecular properties characteristic of conjugated structures are apparently obscured (Masuda *et al.*, 1985). For instance, it, unlike polyacetylene, is electrically insulating, colorless, and soluble in various organic solvents. These unusual properties can be explained on the basis of a highly twisted (non-planar) helical conformation, caused by strong steric interactions between the substituents, as suggested by molecular dynamics calculations (Clough *et al.*, 1991). Such a non-planar conformation would prevent substantial conjugation of double bonds. In spite of the twisted backbone structure, however, dramatic color changes (chromism) characteristic of conjugated polymers (Dai and White, 1994; Inganäs *et al.*, 1988; Rughooputh *et al.*, 1987; Salaneck *et al.*, 1988) were observed for PTMSP both in solution and in the solid state upon interacting with certain electron acceptors, including trifluoroacetic acid and antimony pentachloride. The formation of charge transfer complexes between the polymer segments and those electron acceptors, possibly coupled with conformational rearrangements of chain segments from a helical geometry to a more planar, conjugated structure, is likely to be responsible for the observed color changes.

In solution, color changes from colorless to purple and dark blue were observed following the addition of trifluoroacetic acid into a PTMSP solution in toluene. As shown in Figure 3.2, there was a monotonic increase in optical intensity above 300 nm with the appearance of two weak absorption bands at *ca.* 460 and 660 nm shortly after the addition of trifluoroacetic acid. Further interaction between the trifluoroacetic acid and PTMSP caused a new broad absorption band centered at 560 nm to appear, while the absorption bands at 460 and 660 nm became obscured (Figure 3.2(b–g)). The absorption bands at 460 and 660 nm arise, most probably, from conjugated segments of different conjugation length produced by a conformational change induced by trifluoroacetic acid, which effectively reduces π-conjugation defects in the PTMSP backbone due to the substituent twisting. The appearance of the broad absorption band at 560 nm may indicate the formation of charge transfer complexes between these newly emerged conjugated sequences and trifluoroacetic acid. As expected, the charge transfer band is much stronger than the π-π* polyene bands (Dai and White, 1991; Dai, 1992).

The spectroscopic changes shown in Figure 3.2 are reversible. Similar spectroscopic changes were observed in various other solvents, including cyclohexane and carbon tetrachloride. The induction time for the occurrence of color changes was observed to increase along the series of toluene, cyclohexane, carbon tetrachloride, being closely related to the solubility of PTMSP. The increase in the induction time with decreasing solubility clearly indicates the importance of conformational changes of the PTMSP chains to the solvatochromism.

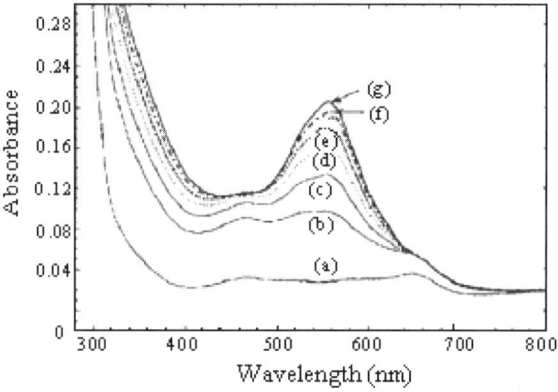

Figure 3.2. Time dependence of UV/vis/NIR spectra of PTMSP in toluene containing trifluoroacetic acid (1%): (a) 2 h; (b) 4 h; (c) 6 h; (d) 8 h; (e) 10 h; (f) 12 h; (g) 14 h (After Bi *et al.*, 1997, copyright 1997 Elsevier. Reproduced with permission)

A similar treatment with a stronger electron acceptor has been shown to generate polyenylic cations, which are also colored species (Carrington *et al.*, 1959; Sorensen, 1965). The evolution in time of the interaction between $SbCl_5$ and the PTMSP in carbon tetrachloride is shown in Figure 3.3. It appears that the optical density increased very rapidly, assuming a shape resembling that observed for I_2-conjugated polyisoprene chains (Dai and White, 1991; Thakur and Elman, 1989). By analogy, the absorption band corresponding to the higher transition energy may be attributed to cation radicals, while the peak at the higher wavelength is believed to be associated with the charge transfer between localized states in the solvent (Dai and White, 1991; Thakur and Elman, 1989).

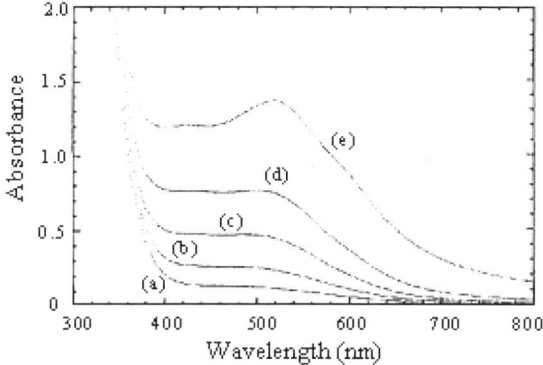

Figure 3.3. Time dependence of UV/vis/NIR spectra of PTMSP in CCl_4 containing $SbCl_5$ (0.4%): (a) upon mixing; (b) 2 min; (c) 4 min; (d) 6 min; (e) 10 min (After Bi *et al.*, 1997, copyright 1997 Elsevier. Reproduced with permission)

In a somewhat related study, Fréchet and co-workers (Gitsov and Fréchet, 1996) have synthesized dendritic copolymers (Chapter 4) that change shape when the polarity of the solvent changes. These stimuli-responsive macromolecules contain a central carbon atom to which four long poly(ethylene glycol), PEG, chains are chemically attached. Each of these flexible hydrophilic PEG "arms" is terminated with a hydrophobic, dendritic 3,5-dihydroxybenzyl alcohol group.

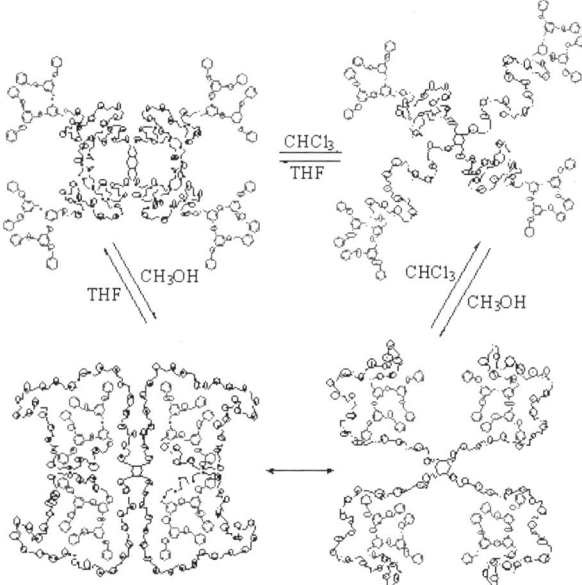

Figure 3.4. Solvent-responsive dendritic copolymers (After Dagani, 1996, reproduced with permission from Fréchet, University of California, Berkeley)

As shown in Figure 3.4, the hybrid star forms a unimolecular micelle with a tightly packed PEG core surrounded by a loose hydrophobic shell of dendritic wedges in THF (*top left*), while in chlorinated solvents such as chloroform, both the PEG core and their dendritic extremities are more extended outward (*top right*). In polar or aqueous solvents (*e.g.* methanol), the hydrophilic PEG arms loop around the hydrophobic dendritic wedges. This rearrangement of the two types of components leads to two possible conformations, namely the one with the dendritic wedges tightly packed toward the middle of the unimolecular micelle (*bottom right*) and the other in which each of the dendritic moieties is individually wrapped in a PEG arm (*bottom left*). Similar conformational changes have also been observed for dendritic polyelectrolytes when the ionic strength changes (Welch and Lescanec, 1996). These stimuli-responsive dendritic macromolecules have potential uses, for example, as rheological modifiers, hosts for catalysts or for controlled release of drugs.

3.3 Temperature-responsive Polymers

3.3.1 Temperature-responsive Polymers in Solution

Optical changes of conjugated polymers can also be induced through conformational changes caused by temperature changes. For example, it has been demonstrated that the order-disorder transitions of their side chains could induce a conformational change of certain polythiophene grafted copolymers (Leclerc and Faïd, 1997). Indeed, neutral poly(3-alkylthiophene)s and poly(3-alkoxy-4-methylthiophene)s exhibit a strong chromic effect upon heating, both in the solid state and in solution (Faïd *et al.*, 1995; Inganäs, 1994; Inganäs *et al.*, 1988; Roux and Leclerc, 1992; Rughooputh *et al.*, 1987).

Figure 3.5 shows the temperature dependence of the optical absorbance of a thin film of poly[3-oligo(oxyethylene)-4-methylthiophene] (Lévesque and Leclerc, 1996). This polymer is highly conjugated with an absorption maximum centered around 550 nm. Heating causes the appearance of a new absorption band around 426 nm and a concomitant decrease in intensity of the absorption band over 550 nm (*i.e.* thermochromism). Upon being heated up to 100 °C, the band at 550 nm eventually disappears, while the newly appearing band reaches its maximum absorption intensity.

The heating-induced blue shift observed above has been attributed to a polymer backbone conformational transition from a highly conjugated coplanar form at low temperatures to a less conjugated non-planar conformation at higher temperatures (Leclerc and Faïd, 1997). It was also found that these optical changes are highly reversible without any further change as a function of the elapsed time at any particular temperature shown in Figure 3.5, indicating that any possible thermal degradation of the polymer chain can be ruled out. The well-defined isosbestic point at ~465 nm suggests the coexistence of two phases in the polymer film, though it is not known whether these two phases do exist on different parts of the same polymer chain or on different ones.

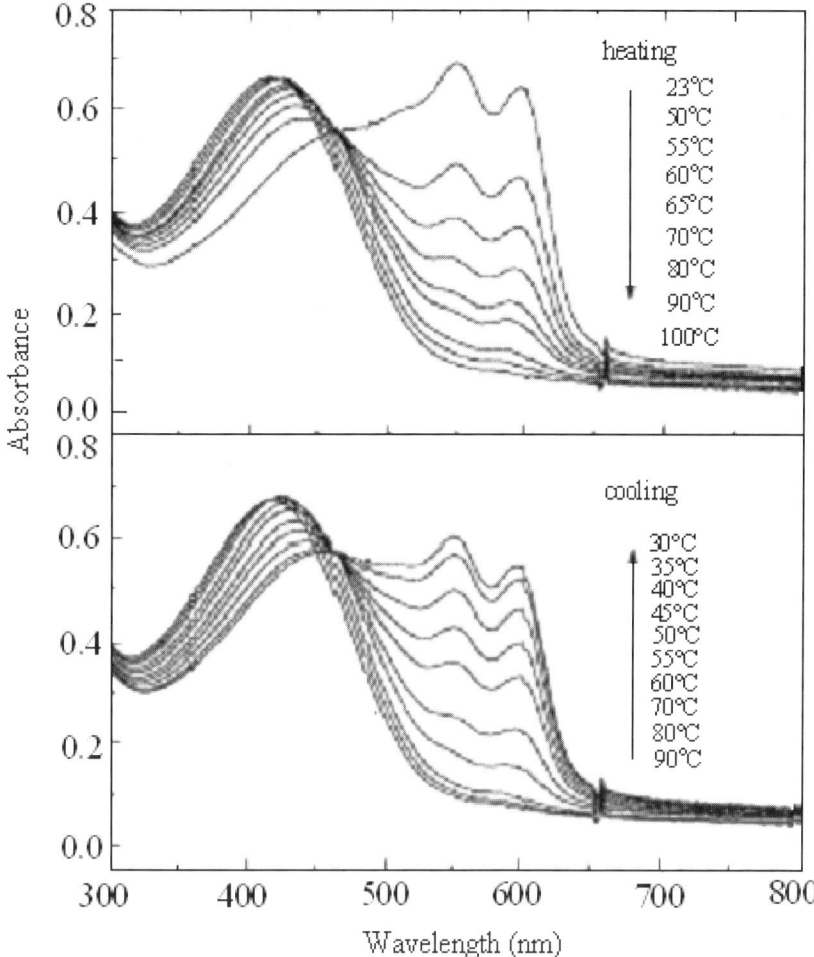

Figure 3.5. Temperature-dependent UV/vis absorption spectra of a highly regioregular poly[3-oligo(oxyethylene)-4-methylthiophene] film (After Lévesque and Leclerc, 1996, copyright 1996 The American Chemical Society. Reproduced with permission)

A similar effect has also been observed for poly[3-oligo(oxyethylene)-4-methylthiophene] chains in a poor solvent. As shown in Figure 3.6, the polymer can be dissolved in poor solvents, such as methanol, to form a violet solution at room temperature. The observed absorption band at 550 nm indicates that the polymer chains are already in a coplanar conformation. Upon heating, the band at 550 nm disappears while a new band at 426 nm increases, as is the case for the solid state.

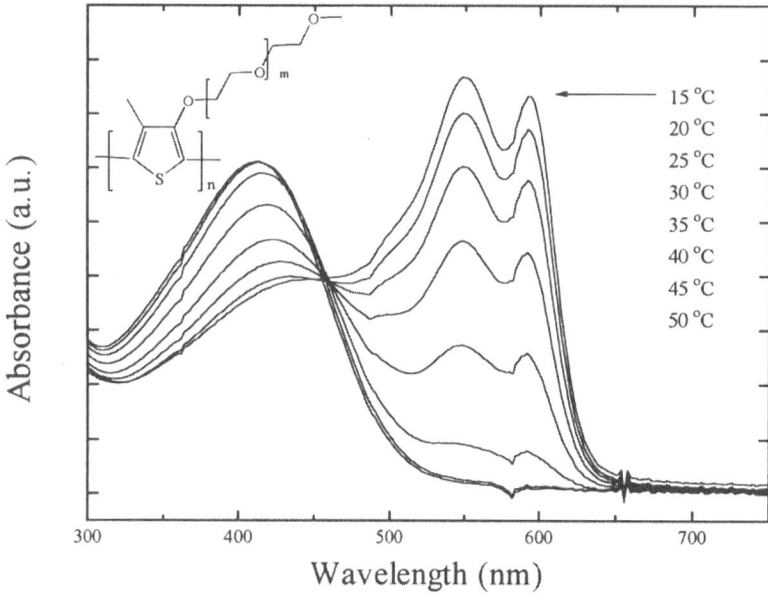

Figure 3.6. Temperature-dependent UV/vis absorption spectra of a highly regioregular poly[3-oligo(oxyethylene)-4-methylthiophene] in a poor solvent (methanol) (After Lévesque and Leclerc, 1996, copyright 1996 The American Chemical Society. Reproduced with permission)

In a good solvent (*e.g.* tetrahydrofuran) at room temperature, however, the maximum absorption of poly[3-oligo(oxyethylene)-4-methylthiophene] is located at 426 nm, indicating the polymer chains are already in a twisted conformation. Cooling induces a reverse conformational transition as evidenced by the color change from yellowish at room temperature to violet at –100 °C. Once again, the transitions observed in both good and poor solvents are found to be fully reversible (Lévesque and Leclerc, 1996).

Poly(*N*-isopropylacrylamide), PNIPAAm, is probably the most widely studied thermo-responsive polymer. PNIPAAm can be prepared by radical polymerization of *N*-isopropylacrylamide via Equation (3.1), using 2,2′-azobisbutyronitrile (AIBN) and 2-aminoethanethiol hydrochloride (AET·HCl) as initiator and chain transfer reagent respectively.

$$NH_2CH_2CH_2SH \cdot HCl + \underset{\substack{\text{N-isopropylacrylamide}\\\text{(NIPAAm)}}}{\text{CH}_2=\text{CH}-\text{C(=O)}-\text{NH}-\text{CH}(CH_3)_2} \xrightarrow[\text{Methanol, 60 °C}]{\text{AIBN}} \xrightarrow{\text{KOH, Methanol}} \underset{\text{Amino-oligoNIPAAm}}{NH_2CH_2CH_2S-[CH_2-CH(C(=O)NH-CH(CH_3)_2)]_n-H}$$

(3.1)

The aqueous solution of PNIPAAm undergoes rod-coil conformational transition at LCST 31°C (Schild, 1992). Polymerization of NIPAAm above LCST permits the preparation of stimuli-responsive PNIPAAm gels with microporous structures. PNIPAAm hydrogels show a thermo-responsive effect, swelling below and shrinking above LCST, which in turn leads also to changes in optical transmittance. Copolymerization of NIPAAm with other appropriate monomers has been shown to change the LCST. In this regard, a series of random copolymers containing PNIPAAm and polyacrylic acid (PAAc) segments were synthesized (Chen G. and Hoffman, 1995a; Dong and Hoffman, 1991). Later, several routes were devised to synthesize various PNIPAm copolymers (Chen G. and Hoffman, 1995b; Yoshioka et al., 1994). Grafted copolymers of poly(N-isopropylacrylamide)-graft-polyacrylic acid (PNIPAm-g-PAAc) can be prepared either by copolymerization of the macromonomer of NIPAAm with AAc [Equation (3.2)] or by coupling end-functionalized oligoNIPAAm onto PAAc [Equation (3.3)].

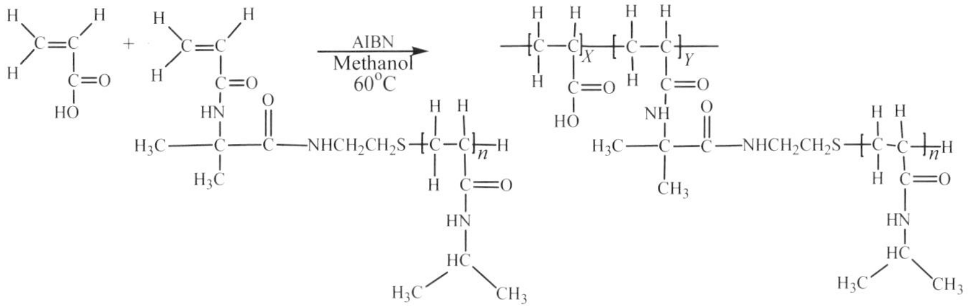

Acrylic Acid (AAc)   Macromonomer of NIPAAm   Graft Copolymer of PNIPAAm-g-PAAc

$$(3.2)$$

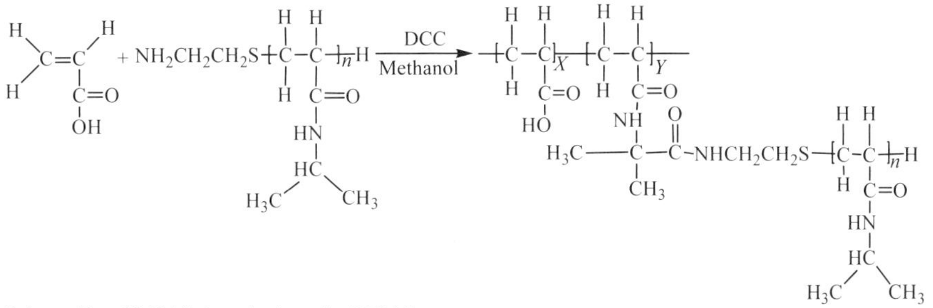

Polyacrylic acid (PAAc)   Amino-oligoNIPAAm   Graft copolymer of PNIPAAm-g-PAAc

$$(3.3)$$

Although the random and graft copolymers have the same basic monomer units, the different distribution of monomer sequences along their polymer backbones enables them to exhibit different temperature-induced phase transition behavior. Figure 3.7 shows the effect of temperature on light transmittance of aqueous solutions of random copolymers.

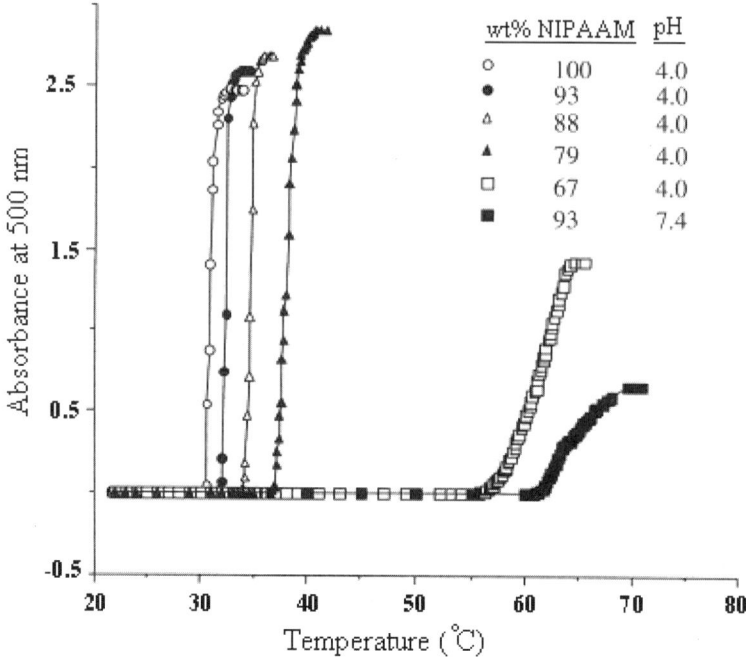

Figure 3.7. Temperature dependence of light transmittance for aqueous solutions of 0.2% random copolymers of NIPAAm and AAc at pH 4.0 and pH 7.4 (After Chen G. *et al.*, 2002, copyright 2002 Taylor & Francis)

As can be seen in Figure 3.7, all of the random copolymers show a higher phase transition temperature than PNIPAAm homopolymers due to the introduction of the more hydrophilic monomer AAc units (Priest *et al.*, 1987). Besides, a much more dramatic shift in LCST is observed for random copolymers with a given content of the PAAc component (*i.e.* 7 wt% AAc) at pH 7.4 than pH 4.0. In the latter case, the –COOH groups of AAc are not ionized. Therefore, copolymers of PNIPAAm could also act as pH-responsive macromolecules (*vide infra*). Even when all of the carboxyl groups in PAAc are ionized at pH 7.4, however, the PAAc graft block was found not to change the phase transition behavior of the oligoNIPAAm grafted chain in PNIPAm-*g*-PAAc.

Furthermore, temperature-responsive PNIPAAm block copolymer hydrogels have also been prepared (Yoshioka *et al.*, 1994; Ebara *et al.*, 2001). Cross-linking polymerization (Kishi and Gehrke, 1992) and radiation techniques (Kishi *et al.*, 1997) have also been used for synthesis of PNIPAAm copolymer hydrogels. Similarly, polymeric nanospheres have been prepared from certain temperature-responsive synthetic polymers (Lee T.B. *et al.*, 2001) and polypeptides (Lee T.A.T. *et al.*, 2000). Polymer nanoparticles have attracted increasing interest as vehicles for the encapsulation and controlled release of drug molecules for medical applications (Chapter 7).

Capsule membranes with surface-grafted poly(*N*-isopropylacrylamide) chains, or the like, may be used as thermoselective and/or pH-selective drug release devices since the end-grafted polymer brush can act as a reversible thermovalve due to the change in its conformation with temperature and/or pH (Figure 3.8).

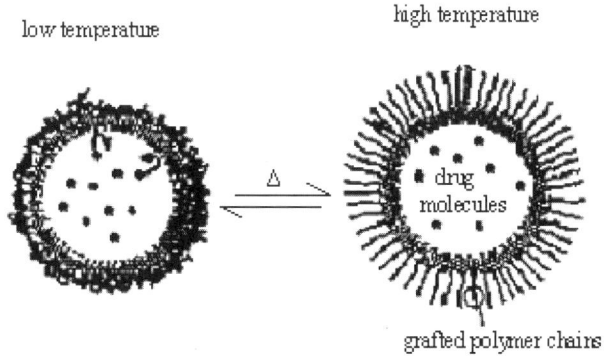

Figure 3.8. Schematic illustration of the polymer-grafted capsule membrane for control release (After Okahata *et al.*, 1986, copyright 1986 The American Chemical Society. Reproduced with permission).

3.3.2 Temperature-responsive Polymers on Surface

As can be seen from the above discussion, the specific balance of hydrophilic and hydrophobic groups is a characteristic feature for many thermo-responsive polymers showing an LCST in aqueous solution. Furthermore, the behavior of thermo-responsive polymers in aqueous solution often contradicts intuition, as they often exhibit an LCST. As such, they precipitate from solution when heated up to a temperature over the so-called cloud point (Chen G. *et al.*, 2002). It should be interesting to find out how polymers that are thermo-responsive in solution will behave when grafted to a surface. In order to address this point, Wischerhoff *et al.* (2000) have recently modified water-soluble poly-*N*-[tris(hydroxymethyl)methyl] acrylamide [P-THMA, Equation (3.4)].

(3.4)

Using these modified P-THMA macromolecules of a controllable thermal transition temperature (LCST), these authors have successfully demonstrated the thermal-responsive effect for the surface-attached P-THMA (Figure 3.9).

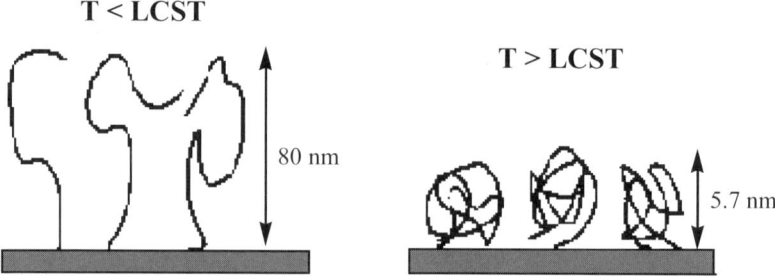

Figure 3.9. Schematic representation of the thermo-responsive effect with an LCST ≈ 26 °C. (After Wischerhoff *et al.*, 2000, copyright 2000 Wiley-VCH Verlag. Reproduced with permission)

More interestingly, Ito *et al.* (1997a) copolymerized *N*-isopropylacrylamide and acrylic acid, followed by coupling with azidoaniline [Equation (3.5)].

(3.5)

The resulting azidoaniline-coupled thermosensitive copolymer (PAzPhPIA) shows a reduced LCST of 21.5 °C with respect to the PNIPAAm homopolymer (32 °C). Furthermore, the incorporation of azidophenyl groups allows these authors to immobilize the copolymer chains onto a polystyrene matrix in a patterned fashion by photolithography, leading to a temperature-responsive surface. For instance, the surface of the immobilized polystyrene plate was shown to be hydrophobic at 37 °C whilst it is hydrophilic at 10 °C. On the other hand, the surface micropattern was clearly observable by means of phase-contrast microscopy at 37 °C, but became invisible at low temperatures due to hydration of the grafted chains (Figure 3.10).

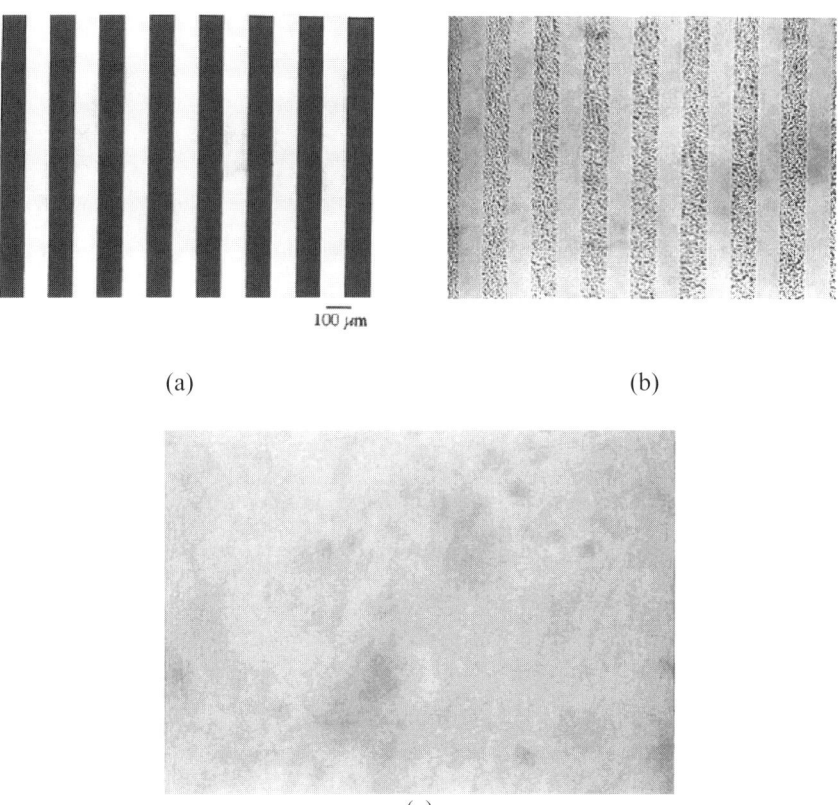

Figure 3.10. Micrograph of the photomask (a) and phase-contrast micrographs of the polystyrene plate micropatterned with PAzPhPIA chains at 37 °C (b) and 10 °C (c) (After Ito *et al.*, 1997a, copyright 1997 The American Chemical Society. Reproduced with permission)

These authors further used the PAzPhPIA-micropatterned surfaces for tissue engineering. They found that mouse fibroblast STO cells grown in the regions covered by the thermo-responsive polymer chains could be selectively detached by lowering the temperature (Figure 3.11).

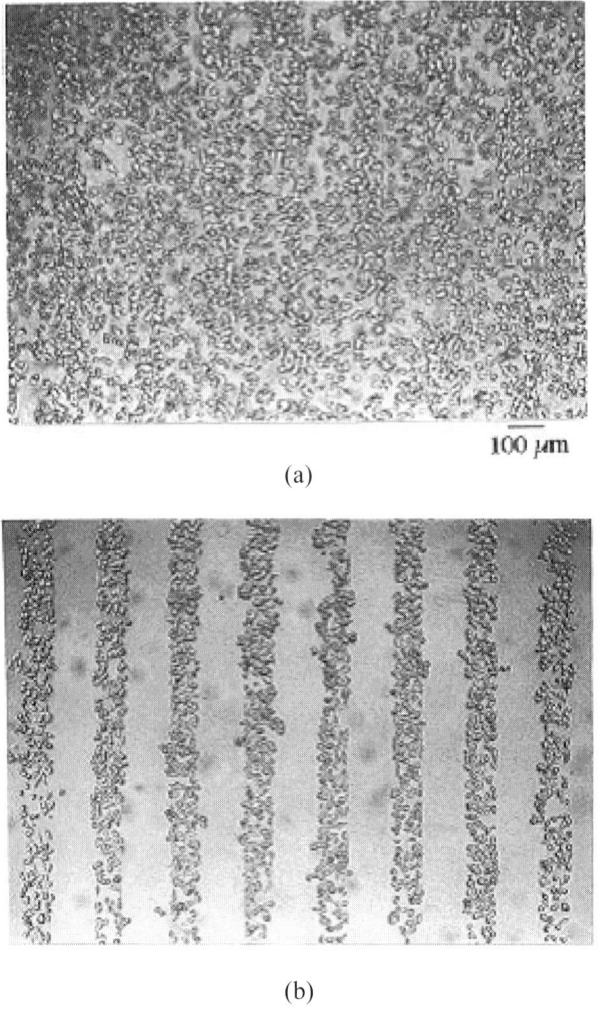

Figure 3.11. Phase-contrast micrographs of the PAzPhPIA-micropatterned surfaces cultured with mouse fibroblast STO cells at 37 °C for 2 h (a) followed by incubation at 10 °C for 30 min (b) (After Ito *et al.*, 1997a, copyright 1997 The American Chemical Society. Reproduced with permission)

Several other groups have also reported surfaces and interfaces that respond in a predictable manner to temperature changes (Bergbreiter *et al.*, 1997; Carey and Ferguson, 1996; Thomas et al., 1995; Jhang *et al.*, 1995; Hosoya *et al.*, 1994; Takei *et al.*, 1994). These responsive surfaces can be used for various smart devices, such as sensors, molecular separation and control release systems (Kost and Langer, 2001).

3.4 pH-responsive Polymers

Certain polymers that contain ionizable functional groups, such as poly(butylmethacrylate), PBMA, poly(acrylic acid), PAA, and poly(N,N-dimethylaminoethyl methacrylate), poly(DMAEMA), can change their conformations in response to a change in pH value of the surrounding medium (Annaka and Tanaka, 1992; Hirokawa and Tanaka, 1984; Brazel and Peppas, 1995; Zhou and Kurth, 2001). At low pH, almost no carboxylic group will be ionized, leading to a greater inter-chain interaction via hydrogen bonding and a reduced LCST temperature. As a result, the polymer chains will be in an unswollen (*i.e.* insoluble) state. In contrast, the polymer chain will take up water, and hence in a swelling (dissolving) state, at high pH due to the ionization of the carboxylic groups. At neutral pH, the incorporation of acrylic acid groups into polymer chains could also increase the LCST due to the hydrophilicity of the charged acrylic acid group (Feil *et al.*, 1993). pH-responsive polymers have been used to coat drugs for controlled release as they can be insoluble in acidic pH and at 37 °C, preventing the drug from gastric degradation, whilst becoming soluble at a neutral pH to release the drug in the intestinal tract where most of the drug absorption takes place (Okahata *et al.*, 1986; Lynn *et al.*, 2001). Recently, pH-dependent polarity changes in dendrimers (Chapter 4) have also been studied (Chen W. *et al.*, 2000). In particular, when the pH of the PNIPAAm solution is sufficiently low, PNIPAm-*g*-PAAc graft copolymers show a shift in the phase transition temperature from 31 °C of pure oligoNIPAAm to *ca.*16 °C (Figure 3.12).

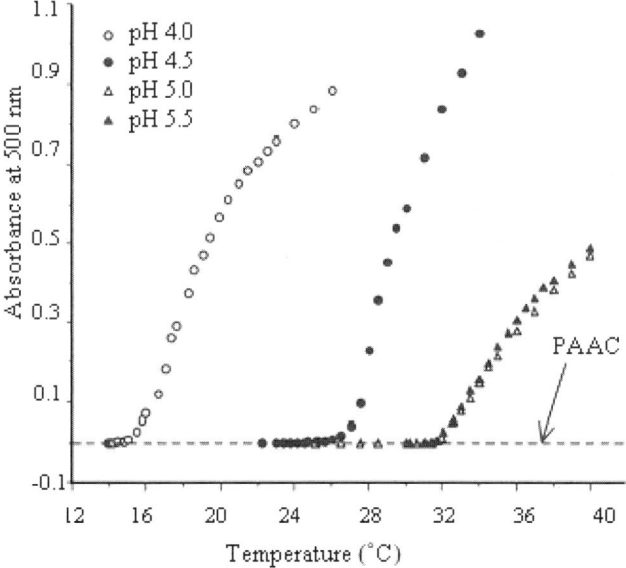

Figure 3.12. Light transmittance measurements showing the pH dependence of LCST for a 0.2% solution of the PNIPAm-*g*-PAAc graft copolymer (containing 50 wt% of NIPAAm) in citric + phosphate buffers at different pH values (After Chen G. *et al.*, 2002, copyright 2002 Taylor & Francis)

Generally speaking, the protonation of polyelectrolytes causes conformational changes. Unusual pH-dependent conformational changes have been observed for both linear and dendritic polyelectrolytes (Rangarajan et al., 1996; Chen W. et al., 2000; Sauer and Meier, 2001). Based on the pH-sensitive effects of certain polypeptides, Ito *et al.* (1997b) graft-polymerized benzyl glutamate *N*-carboxyanhydride onto a porous poly(tetrafluoroethylene) membrane in order to investigate the pH effect on permeation rate. Specifically, they first activated the polymer membrane by ammonia glow-discharged plasma (Chapter 6) and then polymerized benzyl glutamate NCA onto the plasma-induced amine surface groups. As shown in Equation (3.6), the resulting grafted chains were further hydrolyzed to yield poly(glutamic acid).

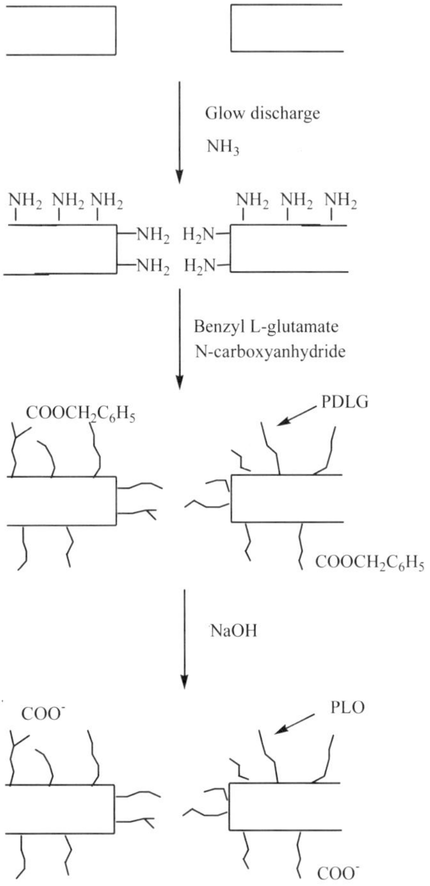

$$\tag{3.6}$$

These authors found that the rate of water permeation through the poly(glutamic acid)-grafted polymer membrane thus prepared was pH-dependent, slow under high-pH conditions and fast under low-pH conditions. This pH-sensitive gating effect has been attributed to a pH-induced conformational transition of the grafted polymer brush between an extended random coil at high pH and a tight helix structure at low pH (Figure 3.13).

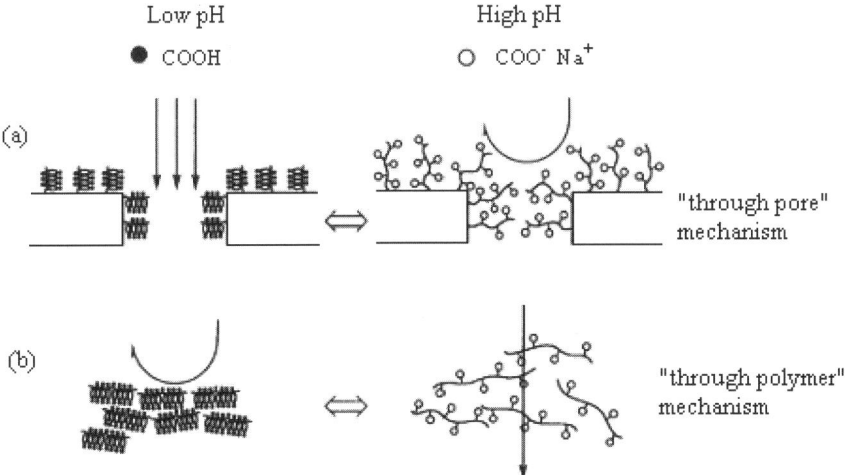

Figure 3.13. Schematic illustration of pH-dependent water permeation through a porous membrane grafted with ionizable polypeptide (After Ito *et al.*, 1997b, copyright 1997 The American Chemical Society. Reproduced with permission)

Similarly, thermo-responsive polymers (*e.g.* PNIPAAm) have also been dispersed in a dense SiO_2 matrix as molecular switches (nanovalves) to selectively control permeability for the hybrid membrane (Rao and Lopez, 2000). Besides, water-soluble polyelectrolyte micro-/nanocapsules that are able to undergo a reversible swelling transition upon a change in pH and/or salt concentration have been prepared using vesicular polymerization (Lynn *et al.*, 2001; Sauer and Meier, 2001). Given that various stimuli-responsive (*e.g.* electroactive, thermal-responsive) polymers are readily available, this approach can be viewed as potentially promising for the use of polymer-carbon nanotube composites in future fluidic devices with nanoscale features.

3.5 Ionically Responsive Polymers

As is well known, ether and crown-ether moieties can form complex structures with alkali metal cations through non-covalent interactions. For a polymer with ether or crown ether substitutes, therefore, its backbone conformation, and hence its optical characteristics, can be modified through interaction with metal ions (*i.e. ionochromism*). Indeed, ionochromic effects have been observed in some polythiophene and poly(*p*-phenylene vinylene) derivatives substituted with ether or crown ether groups (McCullough and Williams, 1993; Marsella and Swager, 1993).

As shown in Figure 3.14, some non-covalent interactions between the ether side chains of the highly regioregular poly[3-oligo(oxyethlene)-4-methylthiophene] and certain ions (K^+, Na^+, NH_4^+, Li^+, *etc.*) induce a cooperative twisting of the main chain. Like the thermochromism (Figure 3.5), the addition of increasing aliquots of KSCN to a polymer solution in methanol led to a two-phase transition characterized by the decrease in the absorption intensity of the planar form over 550 nm and the increase in intensity associated with the new absorption band at 440 nm originated from the non-planar form. The observed ionochromism underpins the design and development of molecular-based sensors using stimuli-responsive macromolecules, as discussed in Chapter 10.

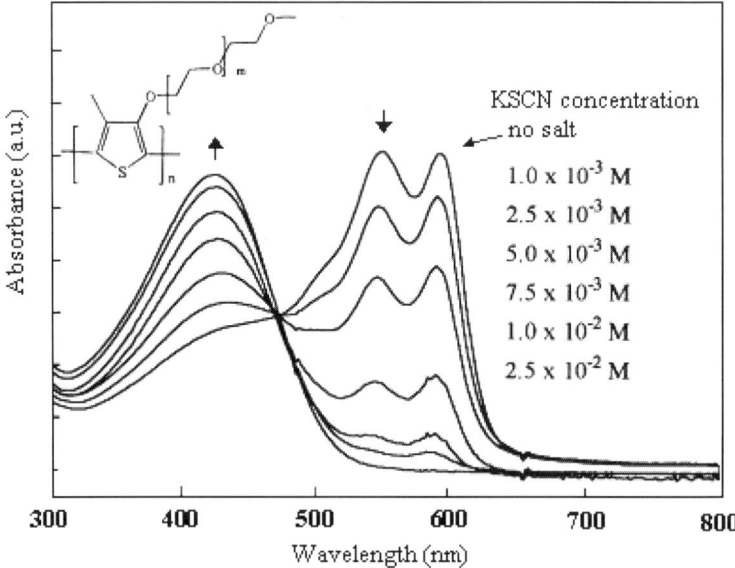

Figure 3.14. UV/vis absorption spectra of poly[3-oligo(oxyethylene)-4-methylthiophene] in methanol with different KSCN concentrations (After Lévesque and Leclerc, 1996, copyright 1996 The American Chemical Society. Reproduced with permission)

3.6 Electrically Responsive Polymers

Apart from the optical detection described above, the polymer backbone conformational changes can also be detected electrochemically. As the highly regioregular planar polymers have smaller band gaps (*e.g.* 1.7 eV for regioregular polythiophene) than their regiorandom (non-planar) analogs (2.0–2.5 eV) (Faïd and Leclerc, 1999), the former should have a lower oxidation potential than the latter. Indeed, the planar (violet) form of poly[3-(2-methyl-1-butoxy)-4-methylthiophene], PMBMT, has an oxidation potential of +0.70 V *vs.* SCE, 0.18 eV lower than that of its non-planar (yellow) counterpart (Figure 3.15).

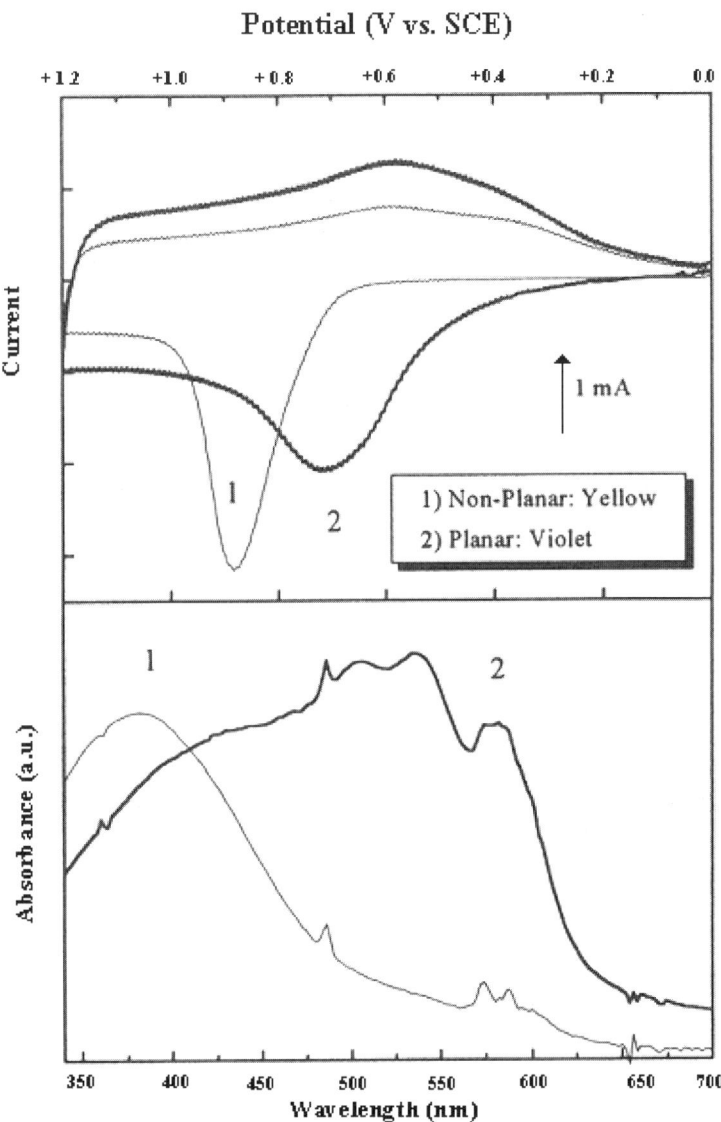

Figure 3.15. Cyclic voltammetry of PMBMT (1) in non-planar and (2) in planar form (After Faïd and Leclerc, 1999, copyright 1999 The American Chemical Society. Reproduced with permission)

Virtually all conducting polymers, including polypyrroles, polyanilines and polythiophenes, show electrochromic effects in a thin film form. Earlier work focused on polypyrrole, polyaniline, polythiophene and their derivatives (Diaz *et al.*,

1981; Diaz and Hall, 1983; Kaneto *et al.*, 1982; Kobayashi *et al.*, 1984; Mohilner *et al.*, 1962; Tourillon and Garnier, 1982). As discussed in Chapter 2, the energy gap between the HOMO and LUMO of conducting polymers determines their intrinsic optical properties. Redox switching (*i.e.* electrochemical doping), accompanied by the transfer of electrons and counter ions, causes changes in the energy gap, giving rise to new optical absorption bands. Chemical modification of the macromolecular structure can significantly change the optical properties. For example, polymer films prepared from 3-methyl-thiophene oligomers show different colors, ranging from pale blue, blue and violet in the oxidized form to purple, yellow, red and orange in the reduced form, depending on the relative positions of the methyl groups on the polymer backbone (Mastragostino, 1993). The driving force for the color change lies in the variation of the effective conjugation length of the polymer backbone caused by redox switching.

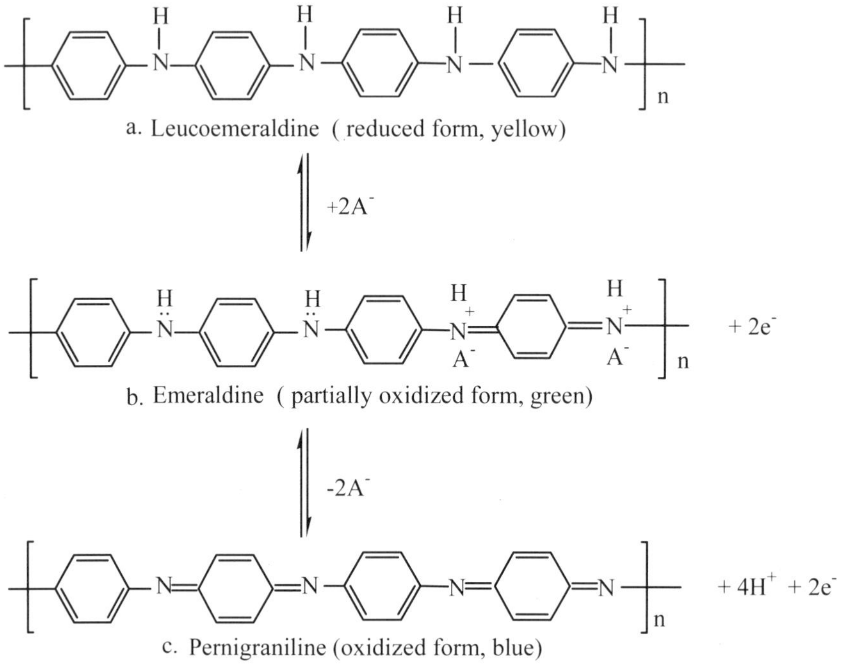

a. Leucoemeraldine (reduced form, yellow)

$+2A^-$

b. Emeraldine (partially oxidized form, green) $+ 2e^-$

$-2A^-$

c. Pernigraniline (oxidized form, blue) $+ 4H^+ + 2e^-$

Figure 3.16. Three predominant intrinsic redox states of polyaniline. (a) leucoemeraldine (fully reduced at –0.2 to +0.14 V *vs.* Ag│AgCl); (b) emeraldine (partially oxidized at +0.14 to +0.45 V *vs.* Ag│AgCl); (c) pernigraniline (fully oxidized at +0.45 to +1.00 V *vs.* Ag│AgCl) (After Massari *et al.*, 2001, copyright 2001 Elsevier)

Figure 3.16 shows three different forms of polyaniline with distinct redox states and colors, which can be regulated by electrochemical oxidation and/or reduction. On this basis, Massari *et al.* (Massari *et al.*, 2001) prepared two-dimensional diffraction gratings by microtransfer molding (Chapter 7) polyaniline thin films in a patterned fashion, followed by electrochemical/chemical oxidation that caused changes in the refractive index of the patterned polyaniline layer. By measuring diffraction efficiency measurements, the periodic change/modulation of the refractive index in one, two, or three dimensions can be further used for sensing applications (Chapter 10).

On the other hand, Reynolds and co-workers (Welsh *et al.*, 1999) electropolymerized a series of poly(3,4-alkylenedioxythiophene)s with different pendant groups. These authors demonstrated that these cathodically coloring polymers could be repeatedly switched between a relatively transmissive light green in the oxidized state to an opaque dark blue in the reduced form. As shown in Figure 3.17, these thin polymer films could be reproducibly switched between their reduced and oxidized forms within seconds with a high contrast of up to 65% transmittance. These results provide a basis for the construction of various organic electrochromic devices (Gazotti *et al.*, 1998), such as "smart windows" (Chapter 9).

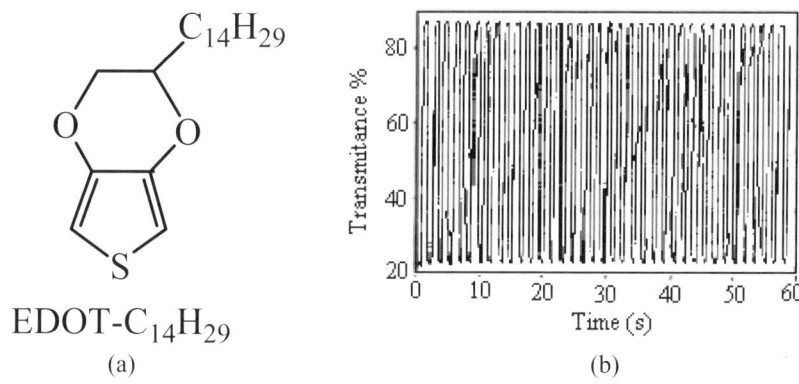

Figure 3.17. (a) Molecular structure of a substituted poly(3,4-ethylenedioxythiophene), PEDTO-$C_{14}H_{29}$, and (b) electrochromic switching the optical absorbance at 590 nm for $C_{14}H_{29}$ (After Welsh *et al.*, 1999, copyright 1999 Elsevier. Reproduced with permission)

Recently, Sutani *et al.* (Sutani *et al.*, 2001) demonstrated electroresponsive synthetic gels for drug release. In so doing, these authors mixed a vinyl monomer (*e.g.* acrylamide or 2-hydroxyethyl methacrylate: HEMA) with a natural polyelectrolyte (hyaluronic acid) in the presence of a cross-linker (*i.e.* polyethyleneglycol dimethacrylate). Hyaluronic acid chains were entrapped into the synthetic polymer gel by polymerizing and cross-linking the vinyl monomers. It was demonstrated that electroresponsive drug release under an "on-off" switching electric field was possible through actuation effect of the polymer gel (Chapter 11).

More recently, Langer and co-workers (Lahann *et al.*, 2003) developed a "smart surface" that can be switched reversibly between the hydrophilic and hydrophobic states upon the application of weak electrical potentials with an opposite bias (Figure 3.18). In particular, these authors prepared a self-assembled monolayer of macromolecules (*i.e.* polymer brush, Chapter 7) with a hydrophobic trunk and a hydrophilic free-end, which was negatively charged (Figure 3.18(b)). When a positive electrical potential was applied, the electrostatic attractive force caused the top to bend down so that the hydrophobic loops were exposed (Figure 3.18(c)). By reversing the electrical potential, the macromolecular chains straighten to their full height to show a hydrophilic surface (Figure 3.18(b)). In order to build in enough space between the molecule chains to allow each to bend down, Lahann *et al.* (2003) added bulky "hats" to each of the constituent chains during the self-assembling process (Figure 3.18(a)), followed by removal of the hats from the tethered monolayer. These smart surfaces with switchable properties should have important implications in materials science, biology and medicine.

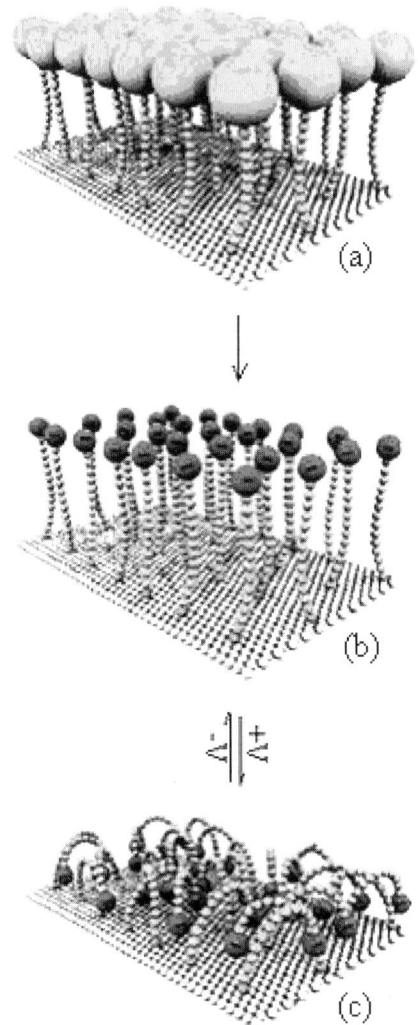

Figure 3.18. A schematic representation of the smart surface that can reversibly switch properties in response to an external electrical potential (After Lahann *et al.*, 2003, copyright 2003 American Association for the Advancement of Science)

3.7 Photoelectrochromism

Photo-induced electron transfer plays important roles in many optoelectronic processes, such as in photovoltaic processes to convert photoenergy to electrical energy (Chapter 9). The combination of the photo-induced electron transfer with electrochromism could enable us to achieve electrochromism by photoillumination, a process which is called *photoelectrochromism* (Kobayashi *et al.*, 1998). In photoelectrochromism, the photo-induced electron transfer is used to oxidize or reduce the reduced or oxidized state of a conducting polymer, respectively (Figure 3.19).

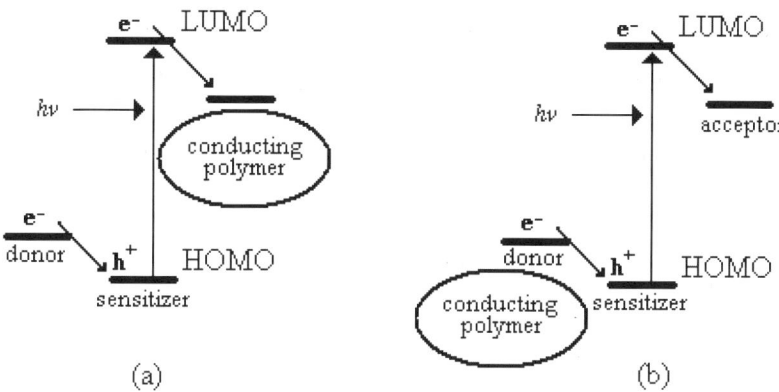

Figure 3.19. Schematic representation of the use of photo-induced electron transfer for (a) reducing and (b) oxidizing conducting polymers (After Kobayashi *et al.*, 1998, copyright 1998 The Royal Society of Chemistry. Reproduced with permission)

Using WO_3 as an electron donor, Yoneyama *et al.* (1992a) have prepared a photoelectrochromic film by electropolymerization of aniline in a solution containing suspended WO_3, followed by light exposure to generate photoexcited electrons in WO_3 for reducing the oxidized polyaniline. By carrying out the light illumination in a patterned fashion with a photomask (Chapter 6), the photoelectrochromism effect can be used for conducting polymer image formation. The reduced polyaniline pattern could be re-oxidized by electrochemical means and the photoreduction and electrooxidation can be recycled (Yoneyama *et al.*, 1992a).

Similarly, Yoneyama *et al.* (1992b) have also incorporated TiO_2 in electropolymerized polyaniline films. In this case, methanol was added to the reaction medium for scavenging holes photogenerated in TiO_2. The methanol molecules are oxidized by the UV irradiation to generate protons, which can induce photoreduction of the deprotonated polyaniline even in a neutral aqueous solution [Equation (3.7)] (Kuwabata *et al.*, 1994). As such, a high-resolution picture image was formed by pattern-illuminating an electrodeposited polyaniline film containing TiO_2 particles in an aqueous neutral solution of methanol (Figure 3.20).

104   Intelligent Macromolecules for Smart Devices

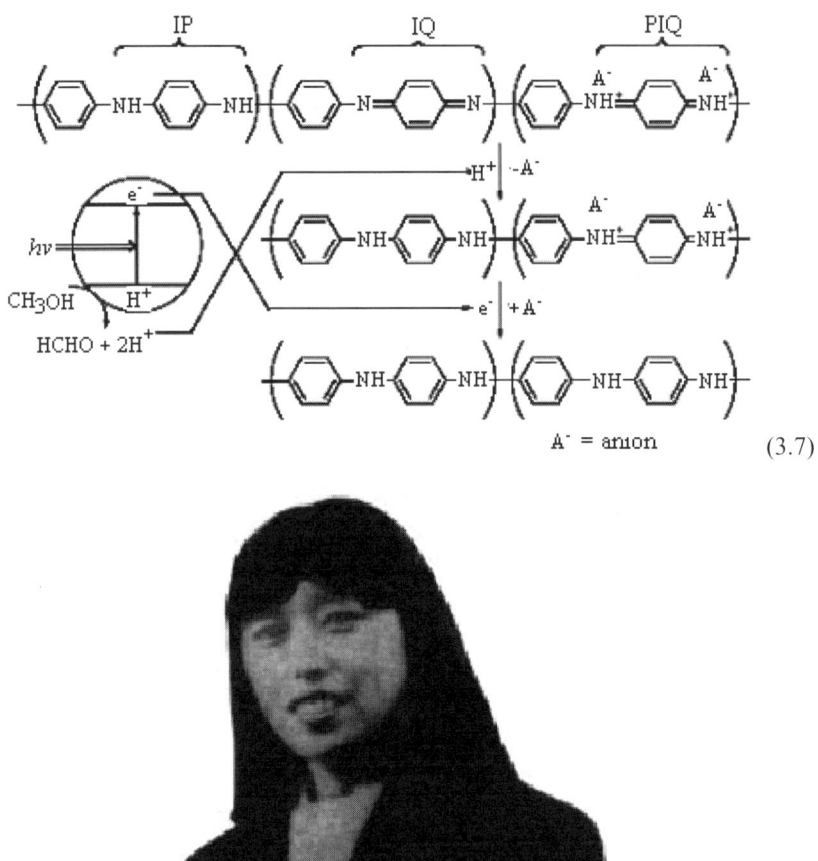

(3.7)

Figure 3.20. Light image formation on polyaniline-TiO_2 film immersed in 0.5 mol/l phosphate buffer (pH 7) containing 20 wt% methanol by projecting the positive image on the polyaniline-TiO_2 film with illumination by 500 W xenon lamp for 1 min (After Yoneyama *et al.*, 1992b, copyright 1992 The Royal Society of Chemistry. Reproduced with permission)

3.8 Photoresponsive Polymers

Photoresponsive polymers are macromolecules that can change their optical and/or electronic properties by being irradiated with light of appropriate wavelengths. These property changes originate from photo-induced electronic structural changes in the macromolecules being irradiated.

Geometric isomerization is a well-known photochemical reaction for alkenes, polyenes or azobenzene (Liu, 2001). By incorporating these photoresponsive chromophores into macromolecular structures, it is possible to make photoresponsive polymers. This concept has stimulated a great deal of research on various photoresponsive polymers with photoresponsive moieties either in the side chains or in the main chains (Kumar and Neckers, 1989; Natansohn, 1999).

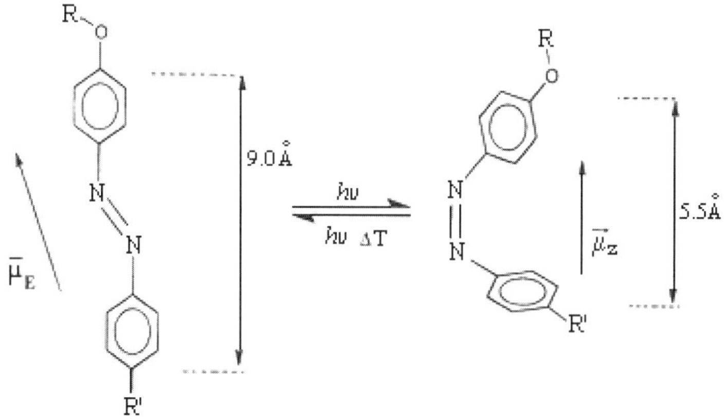

Figure 3.21. Photoisomerization of azobenzene (After Geue et al., 1997, copyright 1997 The American Chemical Society. Reproduced with permission)

Azobenzene is a well-known photoresponsive chromophore that undergoes E-Z photoisomerization upon irradiation (Figure 3.20). As schematically shown in Figure 3.21, photoirradiation causes not only a change in the electronic structure but also in its geometric shape, polarity and transition moment. Using palladium-catalyzed coupling reactions, such as Suzuki coupling and the Heck reaction, Izumi et al. (Izumi et al., 2000; Izumi et al., 2002) have synthesized various conjugated polymers having azobenzenes in the main chain [Equation (3.8)].

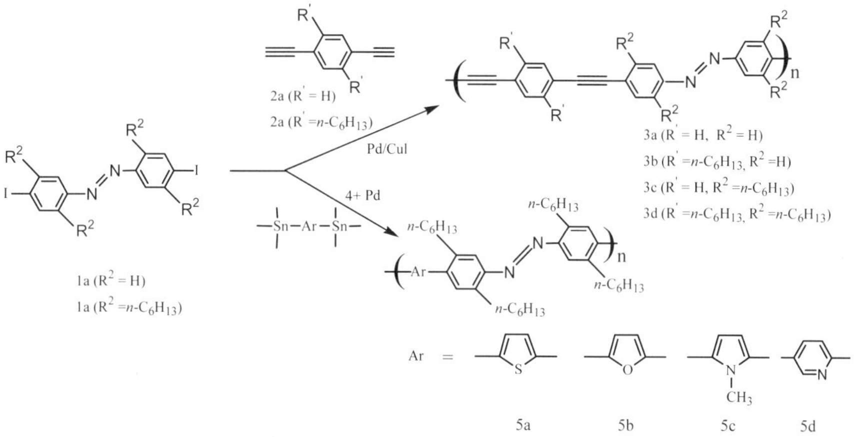

(3.8)

It was demonstrated that polymer **5d** showed a reversible photoisomerization of azobenzene units and a concomitant change in its electrochemical property due to the low degree of main-chain conjugation.

Benzyl aryl ether dendrimers with azobenzene central linkers have also been demonstrated to undergo reversible *cis-/trans-* isomerization upon exposure to ultraviolet light [Equation (3.9)] (Junge and McGrath, 1997).

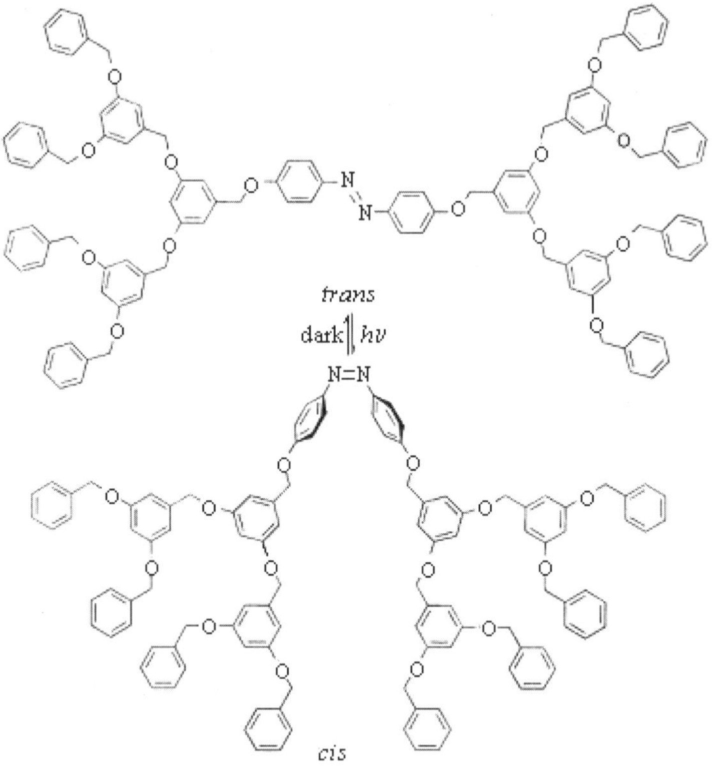

$$\text{(3.9)}$$

On the other hand, the use of azobenzene-based photoresponsive dendritic building blocks has led to the formation of photoswitchable monolayers (Ichimura *et al.*, 2000; Weener and Meijer, 2000) and self-assembled vesicles (Dol *et al.*, 2001), whose macroscopic properties could be completely controlled by light irradiation. By introducing azobenzene moieties into certain Langmuir-Blodgett (LB)-forming macromolecules, Geue *et al.* (1997) have also demonstrated light-induced orientation phenomena in LB multilayers.

In addition to the well-known photoresponsive azobenzene unit, diarylethenes represent another class of unique photochromic molecules as both of the isomers are thermally stable and fatigue-resistant (Irie, 2000). Matsuda and Irie (2000) have recently demonstrated that the exchange interaction between two nitronyl nitroxide radicals at both ends of a diarylethene can be reversibly photoswitched among three different photochromic states: open-open (OO), closed-open (CO), and closed-closed (CC), as exemplified by Equation (3.10).

(3.10)

Using bis(nitronyl nitroxide) molecules with a diarylethene dimer as a photoswitching core, Matsuda and Irie (2001) further demonstrated that the photochromic reaction [Equation (3.10)] proceeded effectively from open-open to closed-closed upon irradiation with 313 nm light (Figure 3.22). Furthermore, it was found that the magnetic interaction in **6(OO)** and **6(CO)** is much smaller than **6(CC)**, indicating potential applications for information storage with **6(OO)** and **6(CO)** as the "off" state and **6(CC)** as the "on" state.

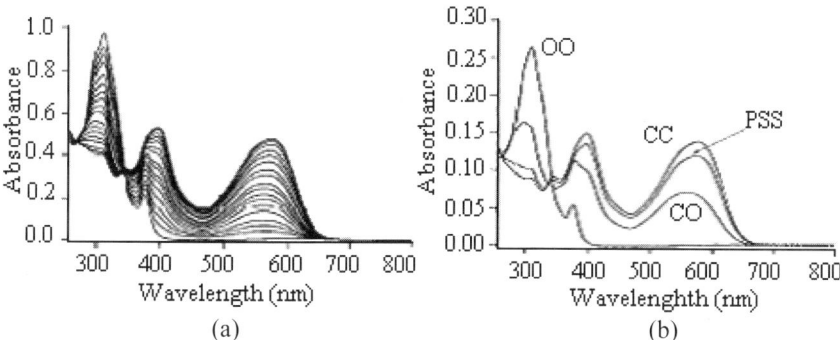

Figure 3.22. (a) Absorption spectral change of **6** under irradiation with 313-nm light for 20 s, 40 s, 60 s, 2 min, 3 min, 4 min, 5 min, 7 min, 9 min, 15 min, 20 min, 30 min, 40 min, 50 min, 70 min, 90 min and 120min; (b) absorption spectra of **6(OO)**, **6(CO)**, **6(CC)** and in the photostationary state under irradiation with 313-nm light (After Matsuda and Irie, 2001, copyright 2001 The American Chemical Society. Reproduced with permission)

Besides, the so-called selective photobleaching, which involves region-selective changes in chemical structure and/or physical properties of photoactive polymer films by irradiation through a photomask, has been successfully used as a very simple approach for certain information storage applications (Kocher et al., 2001). In this context, Weder and co-workers (Kocher et al., 2001) have demonstrated that the photoluminescence of poly(2,5-dialkoxy-p-phenylene ethynylene) and ultra-high molecular weight polyethylene (EHO-OPPE/UHMW-PE) blend films dramatically decreased with photobleaching time under irradiation with a "white" light beam from a xenon lamp (Figure 3.23), due to the photo-induced reduction of the effective conjugation length of EHO-OPPE.

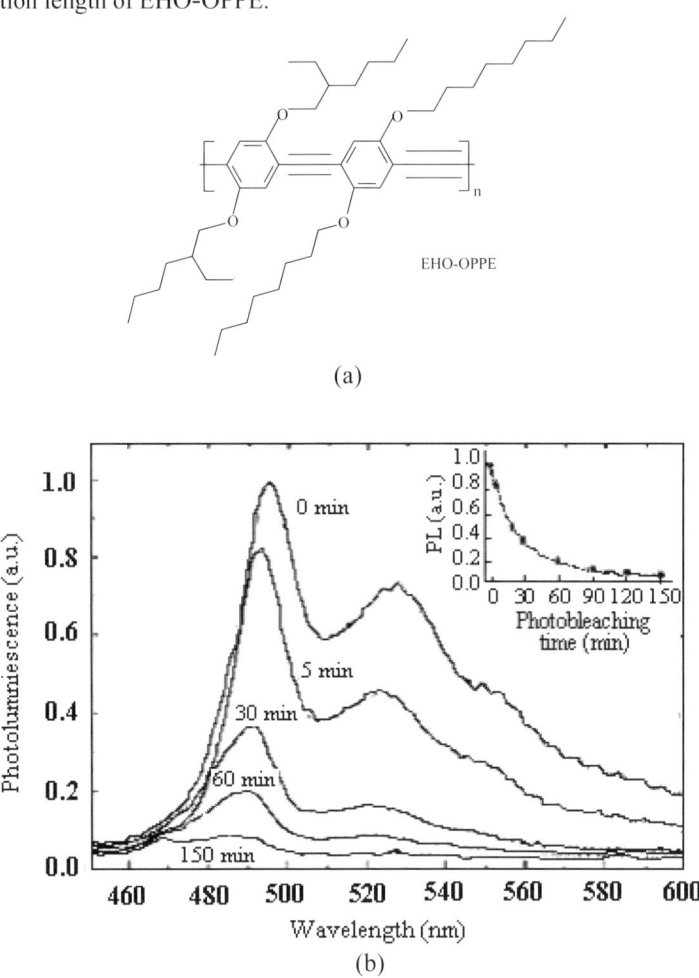

Figure 3.23. (a) Molecular structure of EHO-OPPE; (b) polarized photoluminescence spectra of an oriented EHO-OPPE/UHMW-PE blend film after a different time of photobleaching with white xenon light (After Kocher et al., 2001, copyright 2001 Wiley-VCH Verlag. Reproduced with permission)

Based on the above observation, Weder and co-workers (Kocher *et al.*, 2001) further demonstrated the formation of a photoluminescent image by the photobleaching of an oriented EHO-OPPE/UHMW-PE blend film through a photomask (Figure 3.24), which was obtained by conventional laser printing of a picture of the founder of macromolecular science, Hermann Staudinger (Morawetz, 1985), onto a transparency film. Similar approaches have been previously used for various other micro-/nano-structured conjugated polymers of practical significance (Chapter 6).

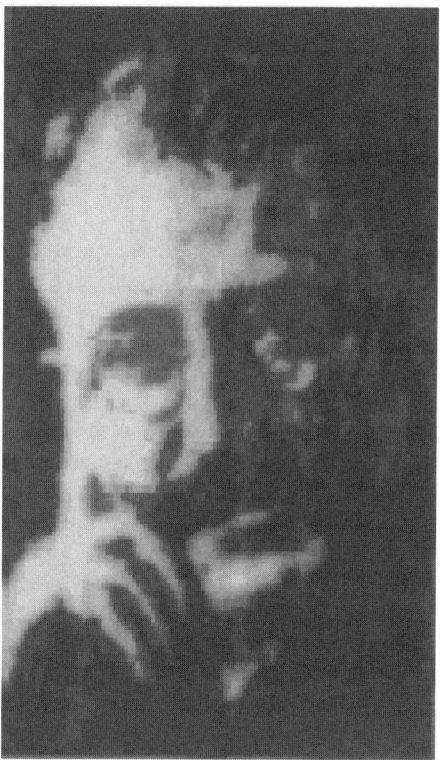

Figure 3.24. Photoluminescent pictures (excitation at 365 nm) of Hermann Staudinger produced by photobleaching an oriented EHO-OPPE/UHMW-PE blend film for 120 min, using a photomask that was fabricated by conventional laser printing of an image onto a transparency film (After Kocher *et al.*, 2001, copyright 2001 Wiley-VCH Verlag. Reproduced with permission)

On the other hand, the photobleaching approach has also been used to develop painless removal bandages (Freemantle, 1998). Removing medical dressings that use adhesives is particularly a problem for patients who have fragile skin, such as the elderly. To overcome this problem, scientists at the Smith & Nephew Group Research Centre in York, England, have developed a light-switchable, pressure-sensitive adhesive that enables a wound dressing to be removed without pain or trauma (Freemantle, 1998). The dressing is backed by two layers laminated together, one opaque and the other transparent to visible light. When the top opaque layer is peeled away, light deactivates the adhesive, allowing the dressing to be removed painlessly. The light-switchable adhesive consists of a copolymer of n-butyl acrylate, 2-ethylhexyl acrylate and acrylic acid. The researchers aimed to add vinyl groups to the copolymer that would cross-link in the presence of a free-radical photoinitiator on irradiation, thus rendering the copolymer non-adhesive [Equation (3.11)].

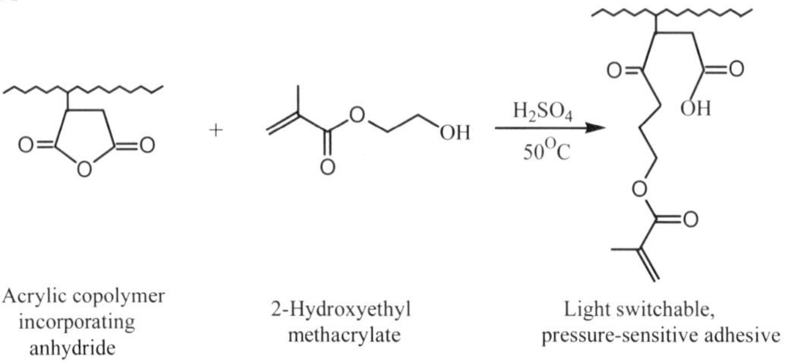

$$\text{Acrylic copolymer incorporating anhydride} + \text{2-Hydroxyethyl methacrylate} \xrightarrow[50^\circ C]{H_2SO_4} \text{Light switchable, pressure-sensitive adhesive} \qquad (3.11)$$

Finally, it is worth pointing out that controlled linear and rotary motions can also be achieved through the construction of molecular devices from photoresponsive elements. Using photoresponsive chiral molecules as optical molecular switches and trigger elements, for example, Feringa (2001) demonstrated the "bottom-up" construction of various light-driven molecular motors. The concept of the photo-switchable molecular rotor can be illustrated in Equation (3.12), though theoretical modeling shows a slightly higher rotation barrier for *trans*-7a.

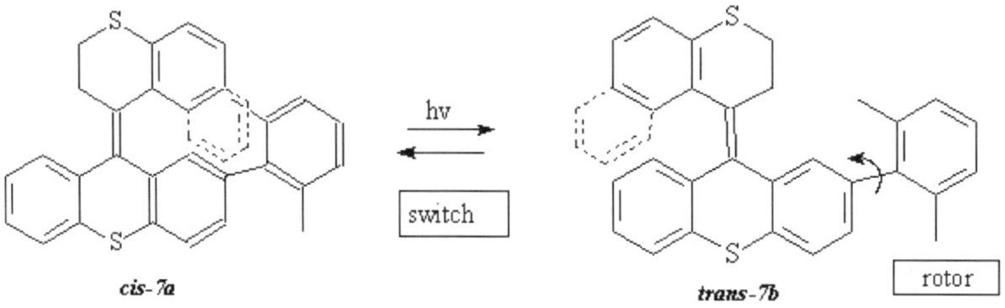

$$cis\text{-}7\mathbf{a} \xrightleftharpoons[]{hv} \text{switch} \quad trans\text{-}7\mathbf{b} \quad \boxed{\text{rotor}} \qquad (3.12)$$

The molecule shown in Equation (3.12) provides both a biaryl-type rotor and a thioxanthene-based switch, much like the molecular brake described earlier by Kelly *et al.* (1994). In this system, the photoisomerization of *cis*-**7a** to *trans*-**7a** would be envisioned to decrease the steric hindrance for biaryl rotation significantly. This is because the xylyl rotor moiety faces the naphthalene unit in *cis*-**7a** whereas in *trans*-**7b** the naphthalene unit should not obstruct the biaryl rotation (Feringa, 2001). Some poly(α-amino acids) conjugated with photochromic side chains have been used as photoresponsive polymers to convert sunlight into mechanical work due to the reversible coil to α-helix (or β-sheet) transitions and/or protein folding (Alonso *et al.* 2000; Cooper *et al.*, 1993; Pieroni *et al.*, 1998).

3.9 Biochromism

The above results indicate that modification of side chains of certain grafted conjugated polymers (*e.g.* poly(3-alkylthiophene)s, poly(3-alkoxy-4-metylthiophene)s) with some targeted chemical or biological moieties for bio-recognition or bio-affinity could lead to biosensors based on the specific complexation reaction between the functional moieties and biomolecular analytes. To demonstrate this possibility, Faïd and Leclerc (1996) have synthesized water-soluble poly(3-alkoxy-4-methylthiophene) copolymers with biotin-containing side chains [Equation (3.13)].

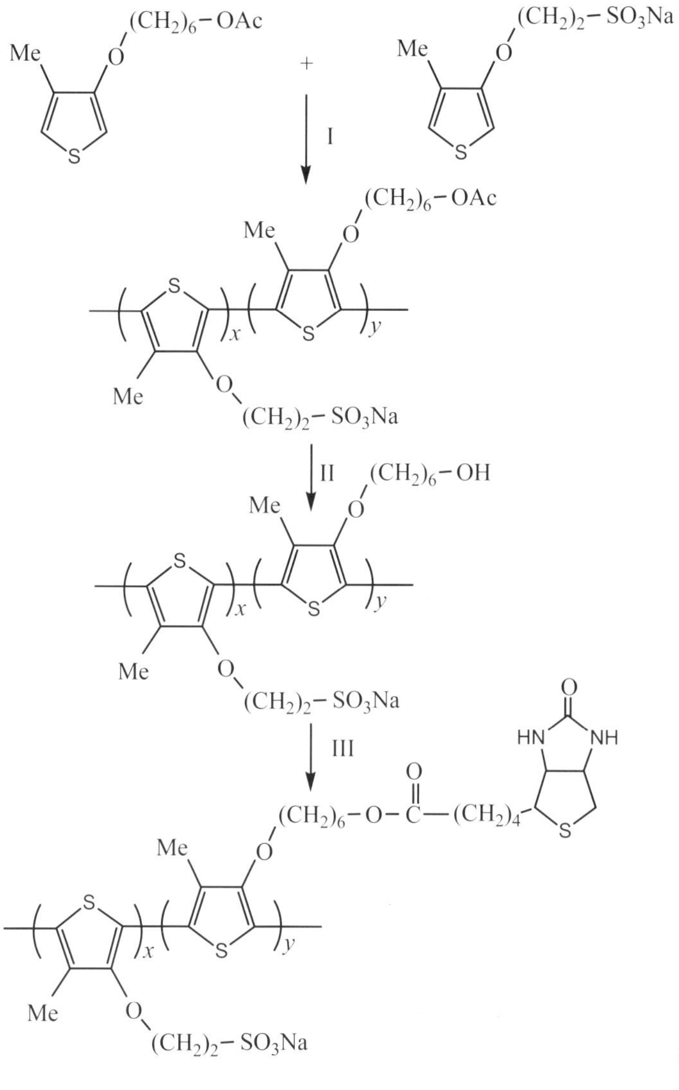

(3.13)

These authors then showed that the specific complexation between avidin and polymer-bound biotin (After Faïd and Leclerc, 1996) caused the side chain conformational changes in the highly regioregular polythiophene derivatives, which in turn altered the main chain conformation. As a consequence, the initially violet copolymer solution, corresponding to the planar conformation for the polymer backbone, changed into yellow in color (non-planar form) upon the addition of avidin into the aqueous solution of the copolymer. The corresponding optical changes shown in Figure 3.25 can therefore be used not only as an optical probe to detect conformational changes of the polymer backbone but also for quantifying the binding event with avidin – a tetrametic protein containing four identical binding sides and forming an essentially irreversible complex with biotin (After Faïd and Leclerc, 1996).

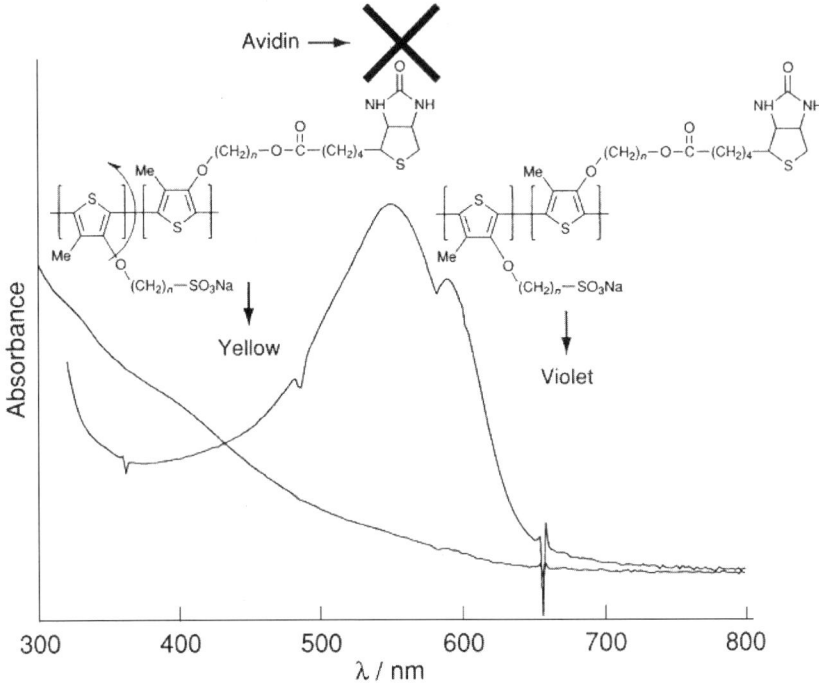

Figure 3.25. UV/vis spectra of polythiophene derivatives of biotin-containing side chains in water at room temperature before and after the addition of avidin (After Faïd and Leclerc, 1996, copyright 1996 The Royal Society of Chemistry. Reproduced with permission)

The results have briefly demonstrated that the ionchromism and biochromism effects can be utilized for chemical- and/or biosensing applications, while various chemical and biosensors based on responsive polymers, dendrimers, and carbon nanotubes will be discussed in detail in Chapter 10.

3.10 Photomodulation of Enzyme Activity

Almost all reactions in the body are mediated by enzymes, which are proteins that act as catalysts to increase the rate of reactions without themselves being changed. Because the structure of the catalytically active protein molecule depends on the ionic character of the amino acid side chains (Chapter 1) and/or the overall conformation, enzymes have been known to show temperature- and pH-dependent activities (Champe and Harvey, 1994). More recently, modulation of the enzyme activity by light has attracted considerable interest as photon absorption by proteins may induce their conformational changes and relaxation dynamics.

Two different approaches have been developed to achieve reversible photoregulation of activities of various proteins (Willner and Rubin, 1993). In one approach, photoisomerizable components (*e.g.* azobenzene or thiophenefulgide units) were covalently attached to the protein chain. The tertiary structure of the proteins is retained and their activity is switched on when the photoisomerizable components adopt one of the photoisomer states, whereas the protein chains are distorted and their activities are switched off in the complementary photoisomer state. The other approach involves immobilization of proteins in photoisomerizable polymers. In this case, the on-off states are triggered by the photostimulated transport of the protein across the polymer matrices.

3.11 References

Aime, J.-P. in (1991) *Conjugated Polymers: The Novel Science and Technology of Highly Conducting and Nonlinear Optically Active Materials*, Brédas, J.L., Silbey, R. (eds), Kluwer Academic Publishers, Dordrecht.
Alonso, M., Reboto, V., Guiscardo, L., Martin, A.S., Rodriguez-Cabello, J.C. (2000) *Macromolecules* **33**, 9480.
Annaka, M., Tanaka, T. (1992) *Nature* **355**, 430.
Bi, J.; Dai, L.; Griesser, H.; Mau, A.W.H. (1997) *Synth. Met.* **86**, 2191.
Brazel, C.S., Peppas, N.A. (1995) *Macromolecules* **28**, 8016.
Bergbreiter, D.E., Ponder, B.C., Aguilar, G., Srinivas, B. (1997) *Chem. Mater.* **9**, 472.
Brèdas, J.L., Street, G.B., Thèmans, B., Andrè, J.M. (1985) *J. Chem. Phys.* **83**, 1323.
Cantor, C.R., Schimmel, P.R. (1980) *Biophysical Chemistry*, W.H. Freeman and Company, New York.
Cao, Y., Heeger, A.J. (1993) *Synth. Met.* **52**, 193
Carey, D.H., Ferguson, G.S. (1996) *J. Am. Chem. Soc.* **118**, 9780.
Carrington, A., Dravnieks, R., Symons, M.C.R. (1959) *J. Chem. Soc.* 847.
Champe, P.C., Harvey, R.A. (1994) *Lippincott's Illustrated Reviews: Biochemistry*, 2nd edn, J.B. Lippincott Company, Philadelphia.
Chen, G.H., Hoffman, A.S. (1995a) *Macromol. Rapid. Commun.* **16**, 175.
Chen, G.H., Hoffman, A.S. (1995b) *Nature* **373**, 49.
Chen, G.H., Hoffman, A.S., Kabra, B. in (2002) *Smart Polymers for Bioseparation and Bioprocessing*, Galev, I., Mattiasson, B. (eds), Taylor & Francis, New York.
Chen, W., Tomalia, D.A., Thomas, J. (2000) *Macromolecules* **33**, 9169.
Clough, S.B., Sun, X.F., Tripathy, S.K. (1991) *Macromolecules* **24**, 4264.
Cooper, T.M., Natarajan, L.V., Crane, R.L. (1993) *Trends Polym. Sci.* **1**, 400.
Dai, L. (1992) *J. Phys. Chem.* **96**, 6469.

Dai, L., White, J.W. (1991) *Polymer* **32**, 2120.
Dai, L., White, J.W. (1994) *Eur. Polym. J.* **30**, 1443.
Dagani, R. (1996) *C&E News* **June 3**, 30.
Diaz, A.F., Castillo, J.I., Logan, J.A., Lee, W.Y. (1981) *J. Electroanal. Chem.* **129**, 115.
Diaz, A.F., Hall, B. (1983) *IBM J. Res. Develop.* **27**, 342.
Dol, G.C., Tsuda, K., Weener, J.-W., Bartels, M.J., Asavei, T., Gensch, T., Hofkens, J., Latterini, L., Schenning, A.P.H.J., Meijer, B.W., De Schryver, F.C. (2001) *Angew. Chem. Int. Ed.* **40**, 1710.
Dong, L.C., Hoffman, A.S. (1991) *J. Controlled Release* **15**, 141.
Ebara, M., Aoyagi, T., Sakai, K., Okano, T. (2001) *J. Polym. Sci. A* **39**, 335.
Faïd, K., Fréchette, M., Ranger, M., Mazerolle, L., Lévesque, I., Leclerc, M. (1995) *Chem. Mater.* **7**, 1390.
Faïd, K., Leclerc, M. (1996) *J. Chem. Soc., Chem. Commun.* 2761.
Faïd, K, Leclerc, M. in (1999) "*Field Responsive Polymer*", Khan, I.M., Harrison, J.S. (Eds.), ACS Symposium Series 726, ACS, Washington DC.
Feil, H., Bae, Y.H., Feijen, J., Kim, S.W. (1993) *Macromolecules* **26**, 2496.
Feringa, B.L. (2001) *Acc. Chem. Res.* **34**, 504.
Flory, P.J., Miller, W.G. (1966) *J. Mol. Biol.* **15**, 284.
Freemantle, M. (1998) *C&E News* **November 30**, 29.
Gazotti, W.A., Casalbore-Miceli, G., Geri, A., De Paoli, M.-A. (1998) *Adv. Mater.* **10**, 60, and references cited therein.
Geue, Th., Ziegler, A., Stumpe, J. (1997) *Macromolecules* **30**, 5729.
Gitsov, I., Fréchet, J.M.J, (1996) *J. Am. Chem. Soc.* **118**, 3785.
Hirokawa, E., Tanaka, T. (1984) *J. Chem. Phys.* **81**, 6379.
Hosoya, K., Sawada, E., Kimata, K., Araki, T., Tanaka, N., Frechet, J.M.J. (1994) *Macromolecules* **27**, 3973.
Ichimura, K., Oh, S.-K., Nakagawa, M. (2000) *Science* **288**, 1624.
Inganäs, O. (1994) *Trends Polym. Sci.* **2**, 189.
Inganäs, O., Salaneck, W.R., Osterholm, J.-E., Laakso, J. (1988) *Synth. Met.* **22**, 395.
Irie, M. (2000) *Chem. Rev.* **100**, 1685.
Ito, Y., Chen, G., Guan, Y., Imanishi, Y. (1997a) *Langmuir* **13**, 2756.
Ito, Y., Ochiai, Y., Park, Y.S., Imanishi, Y. (1997b) *J. Am. Chem. Soc.* **119**, 1619.
Izumi, A., Teraguchi, M., Nomura, R., Masuda, T. (2000) *Macromolecules* **33**, 5347.
Izumi, A., Nomura, R., Masuda, T. (2002) *Macromolecules* **34**, 4342
Jhang, J., Pelton, R., Deng, Y.L. (1995) *Langmuir* **11**, 2301.
Junge, D.M., McGrath, D.V. (1997) *Chem Commun.* 857.
Kaneto, K., Yoshino, K., Inuishi, Y. (1982) *Jpn. J. Appl. Phys.* **21**, L567.
Kelly, T.P., Bowyer, M.C., Bhaskar, K.V., Bebbington, D., Garcia, A., Lang, F., Kim, M.H., Jette, M.P. (1994) *J. Am. Chem. Soc.* **116**, 3657.
Kishi, R., Gehrke, S.H. (1992) *Polym. Commun.* **32**, 322.
Kishi, R., Hirasa, O., Ichijo, H. (1997) *Polym. Gels Networks* **5**, 145.
Kobayashi, N., Teshima, K., Hirohashi, R. (1998) *J. Mater. Chem.* **8**, 497.
Kobayashi, T., Yoneyama, H., Tamura, H. (1984) *J. Electroanal. Chem.* **161**, 419.
Kocher, C., Montali, A., Smith, P., Weder, C. (2001) *Adv. Funct. Mater.* **11**, 31, and references cited therein.
Kost, J., Langer, R. (2001) *Adv. Drug Delivery Rev.* **46**, 125.
Kumar, G.S., Neckers, D.C. (1989) *Chem. Rev.* **89**, 1915.
Kuwabata, S., Kishimoto, A., Yoneyama, H. (1994) *J. Electroanal. Chem.* **377**, 261.
Lahann, J., Mitragotri, S., Tran, T.N., Kaido, H., Sundaram, J., Choi, I.S., Hoffer, S., Somorjai, G.A., Langer, R. (2003) *Science* **299**, 371.
Lee, T.A.T., Cooper, A., Apkarian, R.P., Conticello, V.P. (2000) *Adv. Mater.* **12**, 1105.
Lee, T.B., No, K.T., Cho, S.H., Kim, S.S., Seo, J.K., Lee, J.H., Yuk, S.H. (2001) *J. Poly. Sci. B* **39**, 594.

Leclerc, M., Faïd, K. in (1997) *Handbook of Conducting Polymers*, 2nd edn, Skotheim, T.A., Reynolds, J.R., Elsenbaumer, R.L. (Eds.), Marcel Dekker, New York.
Leclerc, M., Prud'homme, R.E. (1985) *J. Polym. Sci., Polym. Phys. Ed.* **23**, 2021.
Lévesque, I., Leclerc, M. (1996) *Chem. Mater.* **8**, 2843.
Levine, I.N. (1974) *Quantum Chemistry*, Allyn and Bacon, Boston.
Lifson, S., Roig, A. (1961) *J. Chem. Phys.* **34**, 1963.
Liu, R.S.H. (2001) *Acc. Chem. Res.* **34**, 555.
Lynn, D.M., Amiji, M.M., Langer, R. (2001) *Angew. Chem. Int. Ed.* **40**, 1707.
MacDiarmid, A.G., Epstein, A.J. (1995) *Synth. Met.* **69**, 85.
Marsella, M.J., Swager, T. (1993) *J. Am. Chem. Soc.* **115**, 12214.
Massari, A.M., Stevenson, K.J., Hupp, J.T. (2001) *J. Electroanalyt. Chem.* **500**, 185.
Mastragostino, M. in (1993) *Applications of Electroactive Polymers*, Scrosati, B. (ed.), Chapman and Hall, London.
Masuda, T., Isobe, E., Higashimura, T. (1985) *Macromolecules* **18**, 841.
Matsuda, K., Irie, M. (2000) *J. Am. Chem. Soc.* **122**, 8309.
Matsuda, K., Irie, M. (2001) *J. Am. Chem. Soc.* **123**, 9896.
McCullough, R.D., Williams, S.P. (1993) *J. Am. Chem. Soc.* **115**, 11608.
Miller, L.L., Kunugi, Y., Canavesi, A., Rigaut, S., Moorefield, C.N., Newkome, G.R. (1998) *Chem. Mater.* **10**, 1751.
Mohilner, D.M., Adams, R.N., Argersinger, W.J. (1962) *J. Am. Chem. Soc.* **84**, 3612.
Morawetz, H. (1985) *Polymers: The Origins and Growth of a Science*, John Wiley & Sons, Inc., New York.
Nagai, K., Masuda, T., Nakagawa, T., Freeman, B.D., Pinnau, Ingo (2001) *Prog. Polym. Sci.* **26**, 721.
Natansohn, A. (ed.) (1999) *Macromol. Symp.* Vol. 139 (Azobenzene-containing Materials), Wiley-VCH, Weinheim.
Norris, I.D., Kane-Maguire, L.A.P., Wallace, G.G., Dai, L., Zhang, F., Mau, A.W.H. (2002) *Aust. J. Chem.* **55**, 253.
Okahata, Y., Noguchi, H., Seki, T. (1986) *Macromolecues* **19**, 494.
Peller, L. (1959) *J. Phys. Chem.* **63**, 1194.
Pieroni, O., Fissi, A., Popova, G. (1998) *J. Prog. Polym. Sci.* **23**, 81.
Poland, D., Scheraga, H.A. (1970) *Theory of Helic-Coil Transitions in Biopolymers*, Academic Press, New York.
Priest, J.H., Murray, S.L., Nelson, J.R., Hoffman, A.S. (1987) *ACS Symp. Ser.* **350**, 255.
Rangarajan, B., Coons, L.S., Scranton, A.B. (1996) *Biomaterials* **17**, 649.
Rao, G.V.R., Lopez, G. (2000) *Adv. Mater.* **12**, 1692.
Roux, C., Leclerc, M. (1992) *Macromolecules* **25**, 2141.
Rughooputh, S.D.D.V., Hotta, S., Heeger, A.J., Wudl, F. (1987) *J. Polym. Sci., Part B* **25**, 1071.
Salaneck, W.R., Inganas, O., Themans, B., Nilsson, J.O., Sjogren, B., Osterholm, J.-E., Bredas, J.-L., Svensson, S. (1988) *J. Chem. Phys.* **89**, 4613.
Sauer, M., Meier, W. (2001) *J. Chem. Soc., Chem. Commun.* 55.
Schild, H.G. (1992) *Prog. Polym. Sci.* **17**, 163.
Sorensen, T.S. (1965) *J. Am. Chem. Soc.* **87**, 5057.
Sutani, K., Kaetsu, I., Uchida, K. (2001) *Radiat. Phys. Chem.* **61**, 49.
Takei, Y.G., Aoki, T., Sanui, K., Ogata, N., Sakurai, Y. Okano, T. (1994) *Macromolecules* **27**, 6136.
Thakur, M., Elman, B. (1989) *J. Polym. Phys.* **90**, 2042.
Thomas, J.L., You, H., Tirrell, D.A. (1995) *J. Am. Chem. Soc.* **117**, 2949.
Tourillon, G., Garnier, F. (1982) *J. Electroanal. Chem.* **135**, 173.
Weener, J.W., Meijer, E.W. (2000) *Adv. Mater.* **12**, 741.
Welch, P.M., Lescanec, R.L. (1996) *Abstr. Pap. Am. Chem.* **211**, 156.
Welsh, D.M., Kumar, A., Morvant, M.C., Reynolds, J.R. (1999) *Synth. Met.* **102**, 967.

Willner, I., Rubin, S. (1993) *React. Polym.* **21**, 177.
Wischerhoff, E., Zacher, T., Laschewsky, A., Rekaï, E. D. (2000) *Angew. Chem. Int. Ed.* **39**, 4602.
Yoneyama, H., Hirao, S., Kuwabata, S. (1992a) *J. Electrochem. Soc.* **139**, 3141.
Yoneyama, H., Takahashi, N., Kuwabata, S. (1992b) *J. Chem. Soc., Chem. Commun.* 716.
Yoshioka, H., Mikami, M., Mori, Y. (1994) *J. Macromol. Sci., Pure Appl. Chem.* **A31**, 109.
Zhou, W.-J., Kurth, M.J. (2001) *Polymer* **42**, 345.

Chapter 4

Dendrimers and Fullerenes

4.1 Introduction

The quest for the ability to manipulate and control the structure of macromolecules in three dimensions has long fascinated scientists, as it would provide additional freedom for tailoring molecular size, shape and local microenvironments. Unlike most conventional linear macromolecules, three-dimensional macromolecules could possess highly specific molecular architectures and show significantly improved and/or new properties. The best examples of three-dimensional macromolecules include the tree-like molecules known as dendrimers and the soccerball-like fullerenes, such as C_{60}.

Dendrimers are tree-like, three-dimensional macromolecules in which regularly branching repeating units emanate out from a central molecular *core*. In fact, the term "dendrimer" was coined from the Greek word for tree (δενδρον = 'dendron') (Tomalia *et al.*, 1985). As schematically shown in Figure 4.1, all of the repeating units constitute a branch point with the branch cells mathematically amplified in radial, concentric regions called *generations*. These three-dimensional architectural structures will result in an exponential increase in molecular weight for a geometric increase in volume. A noteworthy consequence of this relationship is that the molecular core of dendrimers could be effectively encapsulated by a sterically crowded, closely packed dendritic architecture above a certain critical molecular weight (Hawker *et al.*, 1995; Naylor *et al.*, 1989).

Dendritic macromolecules with their molecular cores sterically shielded from the surrounding environment can be potentially used for various applications, including the attenuation of luminescence quenching of encapsulated dye molecules (Adronov *et al.*, 2000; Kawa and Frechet, 1998), delivering drugs or genes into cells, chemical-/bio-sensors, regulating electron transfer to and from a redox-active core moiety, and molecular antennae for absorbing light energy (Elicker and Evans, 1999; Gorman, 1997; Gorman *et al.*, 1999). While the applications of dendritic macromolecules will be discussed in various chapters as it becomes appropriate, the present chapter focuses on the synthesis and structure of certain cleverly designed dendrimers, as well as on fullerene C_{60}.

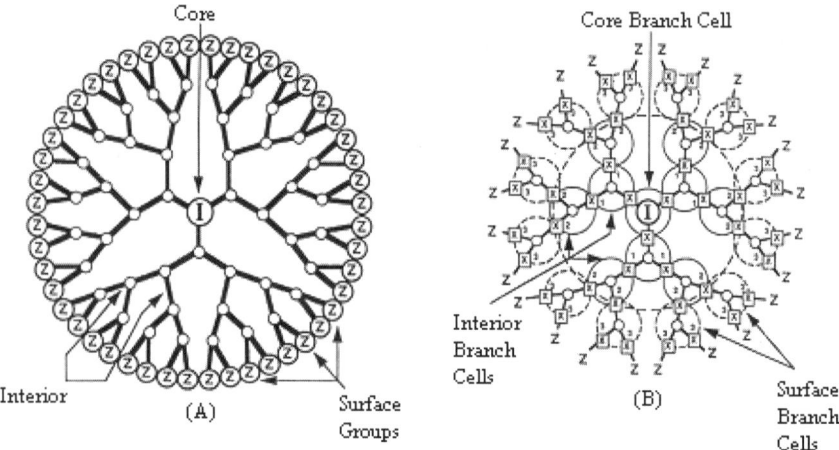

Figure 4.1. Schematic representation of basic structural elements (A) and fundamental repeat units (B) of a dendritic macromolecule. Open circles indicate branching junctures and thick lines indicate connectors. I denotes the core; X denotes junctures; and the numbers refer to generations (After Dvornic and Tomalia, 1996, copyright 1996 Elsevier. Reproduced with permission)

Elemental carbon has long been known to exist in only three forms – amorphous carbon, graphite, and diamond (Marsh, 1989). The recent addition of fullerenes to the carbon family, however, has generated a great deal of renewed interest in carbon physics and chemistry. In 1985, Kroto *et al.* (1985) discovered that fullerene C_{60} has a soccerball-like structure consisting of 12 pentagons and 20 hexagons facing symmetrically (Figure 4.2). This discovery led to an entirely new branch of chemistry (Kroto *et al.*, 1991) and its importance was recognized by the 1996 Nobel Prize in Chemistry (Baum, 1997).

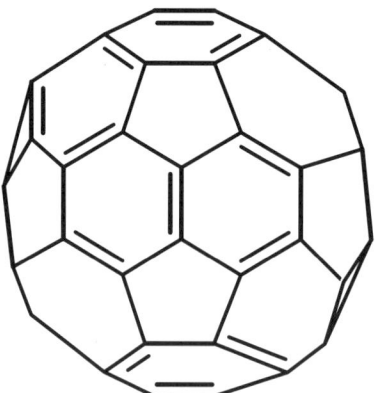

Figure 4.2. The molecular structure of C_{60}

This peculiar hollow, sphere-like structure of fullerenes has made them to exhibit many new intriguing electrical, optical, magnetic and chemical properties in comparison to those more conventional members of the carbon family. For instance, novel photonic, electronic, superconducting, magnetic and biomedical properties have been observed for fullerene C_{60} and its derivatives. These special physicochemical properties, in turn, allow for controlled structural modifications of fullerenes, leading to the formation of various advanced fullerene derivatives with appropriate properties for many potential applications. While the applications of fullerene C_{60} and their derivatives will be discussed in succeeding chapters as becomes appropriate, the important issues related to their syntheses, chemistry, and properties are the subjects of the present chapter.

4.2 Dendrimers

4.2.1 Synthesis

The efforts of early research in the field focused on the synthesis of large dendritic macromolecules by structure amplification towards multi-generations (Newkome, 1994; Tomalia and Durst, 1993). Two main synthetic strategies, termed the *divergent* approach and the *convergent-growth*, have been devised for synthesizing dendrimers. They differ from each other in the direction of growth. The divergent approach pioneered by Newkome *et al.* (1985) and Tomalia *et al.* (1985) builds dendrimers from the central core out to the periphery with rapidly increasing chain ends as the molecule grows. In contrast, the convergent-growth proceeds from the chain ends inwards to produce highly branched molecules with a single functional group at the focal point, termed *dendrons*, which are then attached to a polyfunctional core to form the dendritic macromolecule.

4.2.1.1 Divergent Approach

The first dendritic macromolecule was reported by Vögtle and co-workers (Buhleier *et al.*, 1978). Using the divergent approach, these authors prepared an oligomeric nitrile-terminated polyamine through a series of Michael addition and subsequent reduction steps [Equation (4.1)]. However, the difficulty with which the reduction reactions can be carried out, together with the possible cyclization side reactions, prevented the growth reaction from being beyond the third generation (m.w. 873). Due to the low molecular weight and/or imperfect structure, the resultant oligomeric nitrile-terminated polyamine, like some other early dendritic macromolecules (Aharoni *et al.*, 1982; Denkewalter *et al.*, 1981, 1983), did not display dendrimer-like properties, and hence cannot be considered as real dendrimers. True dendritic macromolecules had not been produced until Tomalia *et al.* (1985) and Newkome *et al.* (1985) reported their approaches in the mid-1980s.

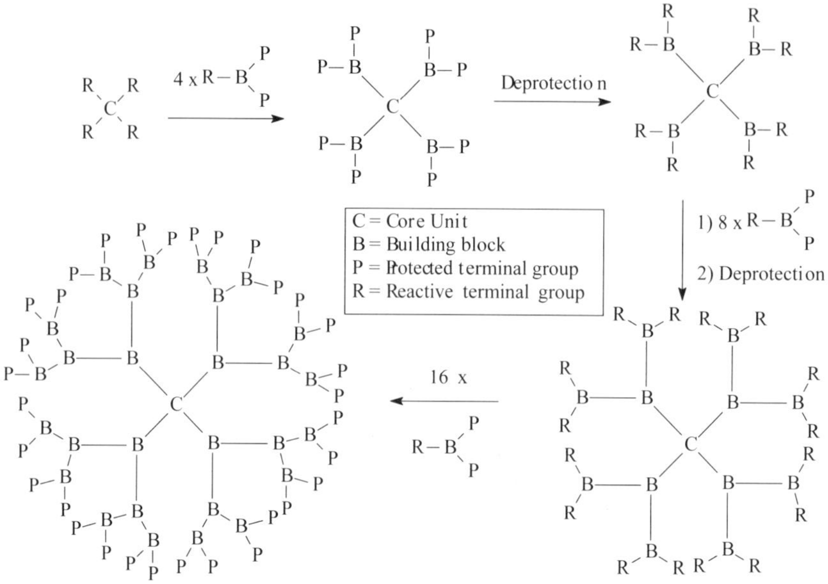

(4.1)

Using a repetitive Michael addition similar to that reported by Vögtle and co-workers (Buhleier *et al.*, 1978), Tomalia *et al.* (1985) carried out reactions of the trifunctional core molecular ammonia with methyl acrylate to form the triester, followed by regeneration of the reactive NH groups through exhaustive amidation in the presence of excess 1,2-diaminoethane [Equation (4.2)].

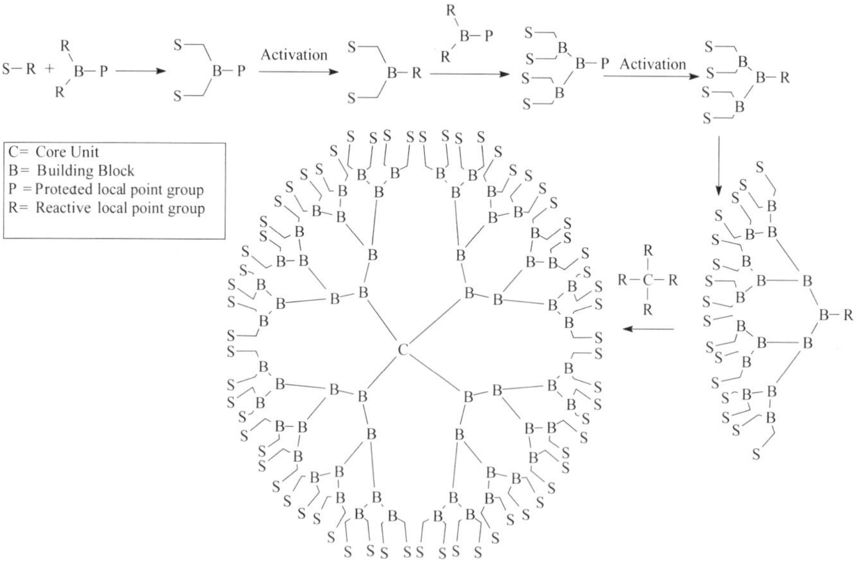

(4.2)

By so doing, these authors circumvented the troublesome reduction step by using exhaustive amidation to regenerate a doubled number of the reactive amino groups. The repetition of the reaction sequences shown in Equation (4.2) with appropriate purification at each step leads to dendritic macromolecules with the number of terminal functional groups being doubled at each growth step, a feature characteristic of the divergent approach.

On the other hand, Newkome et al. (1985) synthesized hydroxymethyl-terminated water-soluble polyetheramides by using a trifunctional core and an AB_3 building block, leading to a speedy growth with three branch points at each junction [Equation (4.3)]:

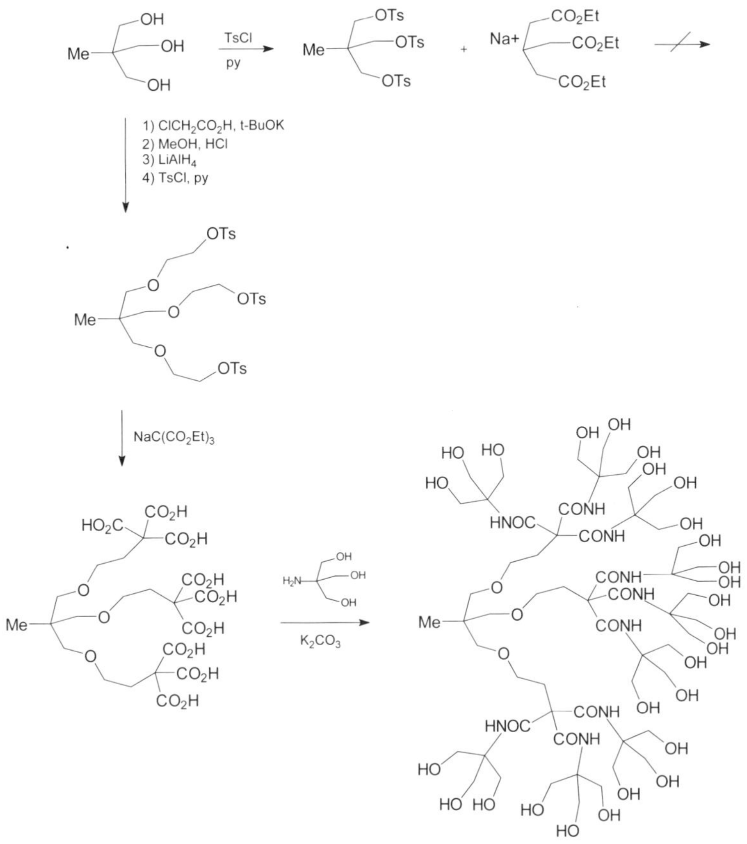

$$(4.3)$$

The success of the above work has led to large-scale production of various dendrimers by the divergent approach (Newkome et al., 1992; Tomalia, 1993).

4.2.1.2 Convergent-growth Approach

As can be seen from the above discussion, the number of reactive groups on the dendrimer periphery, *n*, reacting with monomer units to add a new layer of generation to the dendrimer is doubled or tripled in each of the repeat cycles of the divergent approach, depending on whether the monomer unit's branch multiplicity is 2 or 3 (*i.e.* AB_2 or AB_3 monomer used). Therefore, the number of terminal function groups is large in the divergent approach even at low-generation numbers. When a large number of coupling or condensation reactions have to occur on a congested dendrimer surface, selective functionalization of the end groups becomes impossible and defects begin to accumulate in dendritic macromolecules. For this reason, dendrimers produced by the divergent approach are considered to be statistically inhomogeneous. In order to make more homogeneous dendrimers, the so-called convergent-growth approach has been subsequently developed by Hawker, Fréchet, and co-workers (Hawker and Fréchet, 1990a; Fréchet *et al.*, 1989). To demonstrate this new synthetic strategy, these authors have used 3,5-dihydroxybenzyl alcohol, **1**, as the monomer unit to prepare a series of dendritic polyether macromolecules, **7** [Equation (4.4), see also Hawker and Fréchet, 1990a, b; Fréchet *et al.*, 1989]:

As shown in Equation (4.4), the reaction sequences involve the Williamson reaction between the monomer, **2**, and the surface functional group (*i.e.* **1**) in the presence of potassium carbonate and 18-crown-6. The alkylation of the monomer unit, **1**, occurs regioselectively at the phenolic hydroxy groups, leaving the hydroxymethyl group unchanged. The formation of dendritic fragment, **3**, was then followed by regeneration of the reactive bromomethyl group at the focal point of the growing dendritic fragment, **4**, by activation of the hydroxymethyl group in **3** with carbon tetrabromide and triphenylphosphine. Repetition of the alkylation and bromination leads to monodispersed dendritic macromolecular fragments with a single functional group at the focal point. Finally, these mono-dispersed dendritic macromolecular fragments were attached to a polyfunctional core through the same coupling chemistry (*i.e.* Williamson reaction to form a series of dendritic macromolecules, as exemplified by **7** in Equation (4.4). Subsequently, the convergent growth approach has been proved to be also useful for preparing a wide range of other dendrimers (Miller and Neenan, 1990; Wooley *et al.*, 1991a; Xu Z. and Moore, 1993a, b).

Although both the convergent and divergent approaches can be used to produce the same dendrimers, there are important differences between these two methods. In the convergent-growth approach, for instance, there is always a single functional group at the focal point, which can be exploited to produce a large number of dendrimers with different macromolecular architectures. Therefore, the convergent-growth allows for the production of defect-free dendrimers and provides great control of the focal-point group, the 'interior" building blocks, and the number and nature of the terminal groups. However, the convergent-growth strategy is often limited to dendrimers of low-generation numbers as it normally fails in coupling with very large dendritic fragments to a congested core. In contrast, the divergent approach is better suited to the synthesis of very large dendrimers even with 10 or more generations.

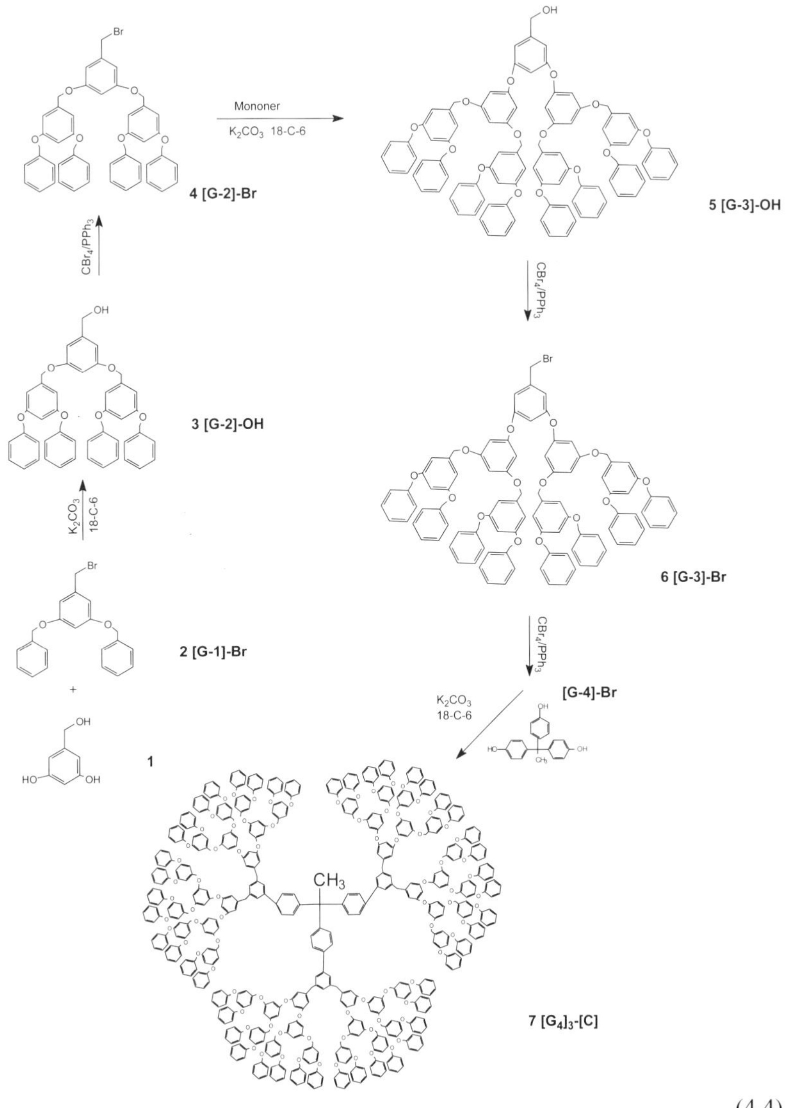

[Scheme 4.4 — structures 1, 2 [G-1]-Br, 3 [G-2]-OH, 4 [G-2]-Br, 5 [G-3]-OH, 6 [G-3]-Br, and 7 [G₄]₃-[C]]

(4.4)

4.2.1.3 Other Miscellaneous Approaches

Double Exponential Dendrimer Growth
As can be seen from the above discussion, the degree of polymerization, dp, in the convergent-growth approach with trifunctional monomers increases with generation number, n, according to Equation (4.5):

$$dp = 2^{(n+1)} - 1 \tag{4.5}$$

To search for synthetic methods for rapidly amplifying dendritic structures to higher molecular weights with fewer reiterative steps, Moore and co-workers (Kawaguchi et al., 1995) developed the double exponential dendrimer growth (DEDG) concept. To demonstrate the DEDG concept, these authors used a single trifunctional monomer of the type $A_p(B_p)$ having orthogonally protected functional groups. As shown in Equation (4.6), the repetitive process involves the selective removal of the protecting groups on A_p in one portion and the protecting groups on B in a second portion, followed by the coupling of the two monoprotected intermediates in a proper stoichiometric ration.

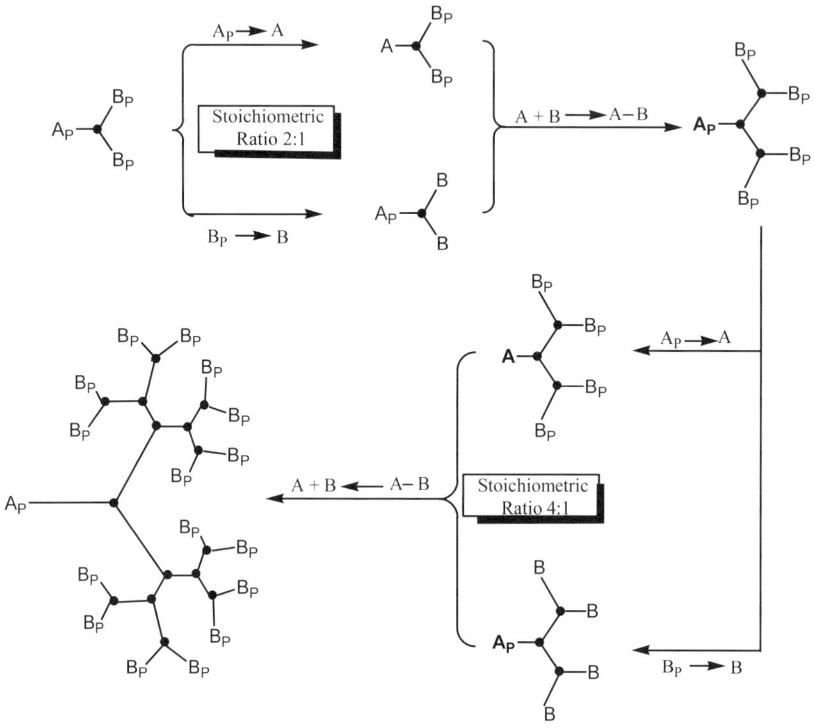

$$\tag{4.6}$$

The newly formed monodendron maintains a single A_p group at its focal point but has $\chi^2 B$ peripheral groups (where χ is the number of peripheral groups in the preceding generation). Therefore, the general form for degree of polymerization dp vs. generation n follows double exponential growth as given by Equation (4.7):

$$dp = 2^n - 1 \tag{4.7}$$

By plotting the degree of polymerization *vs.* the generation number for the conventional convergent approach and the double exponential growth in Figure 4.3, the significance of Eq. (4.7) becomes obvious compared with Equation (4.5). The DEDG approach should, in principle, allow the same-generation dendrimers to be reached after fewer synthetic steps than the classical convergent-growth. Indeed, Moore and co-workers (Kawaguchi *et al.*, 1995) demonstrated the DEDG approach by synthesizing poly(phenylacetylene) with a record-holding molecular weight (40 000 amu) among pure dendritic hydrocarbons just within three generations (nine synthetic steps), whereas the conventional convergent growth would require seven generations (14 synthetic steps) to reach this same size.

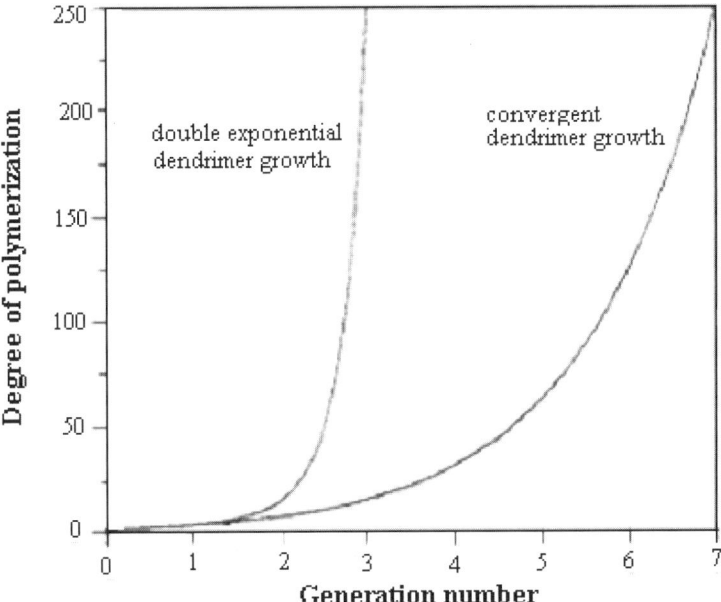

Figure 4.3. Plot showing the ideal degree of polymerisation *vs.* generation number for triconnected monodendrons prepared by the classical convergent approach [*i.e.* Equation (4.5)] and exponential dendrimer growth [*i.e.* Equation (4.7)] (After Kawaguchi *et al.*, 1995, copyright 1995 The American Chemical Society. Reproduced with permission)

One-pot Synthesis
As can been seen from the above discussion, the syntheses of dendrimers often involve multi-step reactions in which protecting groups must be applied or removed, end groups modified or intermediates purified. To circumvent these tedious steps, Rannard and Davis (2000) developed a one-pot synthesis method to prepare dendrimers of several generations by using highly selective chemical reactions. For instance, they have routinely made 100-g batches of third-generation dendrimers shown in Equation (4.8) with more than 95% purity (Stinson, 1997). The success of

the one-pot synthesis approach to dendrimers lies in the reactive selectivity of carbonyldiimidazole (CDI) as a carbonyl-donating reagent to form carbonate linkages between successive branches.

Similar one-pot synthesis methods have also been used to prepare a closely related class of highly branched polymeric materials, termed hyperbranched macromolecules (Mishra and Kobayashi, 1999). In this context, Lin et al. (2000) have synthesized hyperbranched poly(3,5-bisvinylic)benzene through a modified Gilch or Wittig reaction [Equation (4.9)]. The hyperbranched conjugated polymers thus prepared are soluble in most common organic solvents, including chloroform, dichloromethane, THF and ethyl acetate, due, most probably, to the presence of multi-free chain ends and strong steric hindrance between the side chains. The attachment of vinylic bonds at the *m*-positions on each of the constituent benzene rings causes a blue-shift in both the absorption and emission spectra with respect to PPV. The molar absorption coefficient and fluorescence quantum yield were determined to be about 10^4 and 36%, respectively.

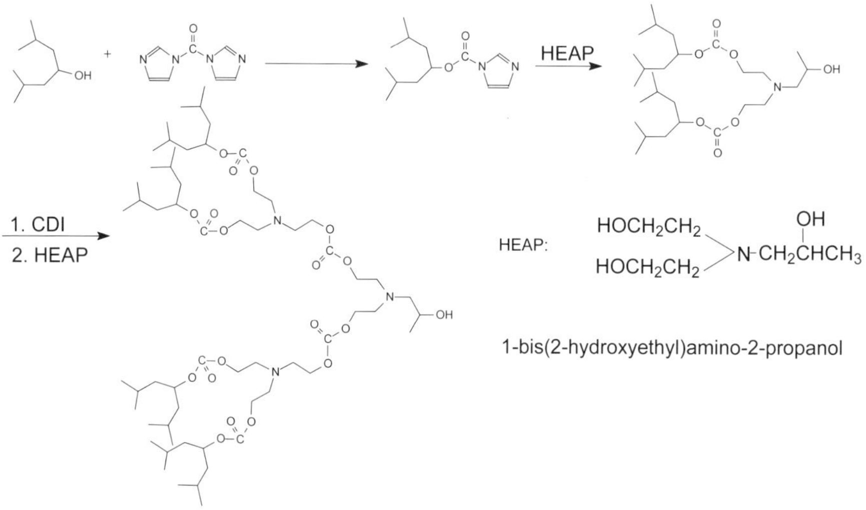

(4.8)

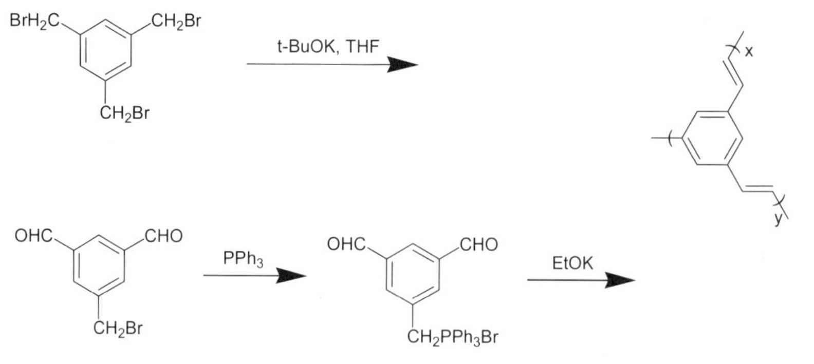

(4.9)

4.2.2 Structure

With so many innovative synthetic routes to dendrimers already discussed above, various dendritic macromolecules with great architectural diversity and hence interesting properties have been produced. In what follows, I will spotlight some cleverly designed dendrimers with unique features for specific applications.

4.2.2.1 Dendrimers with a Metal Core

Dendritic macromolecules with a metal at the central core are of great interest as novel optoelectronic and magnetic materials. In this regard, Constable *et al.* (1996) have developed a convergent method for assembling metallocentric metallodendrimers according to Equation (4.10):

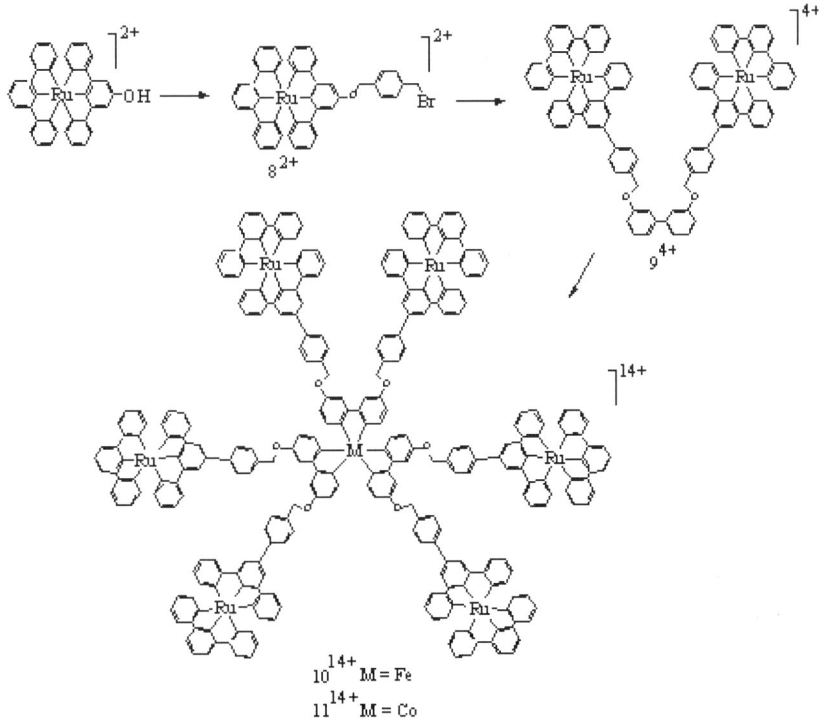

$$\tag{4.10}$$

4.2.2.2 Dendrimers with a Hollow Core

As is well known, all dendrimers have three common structural components: a core unit, surface terminal groups and multiple branching units that span the other two components. The core unit normally links the dendrons. However, Wendland and

Zimmerman (1999) have recently prepared dendrimers that have had their cores chemically removed (Figure 4.4).

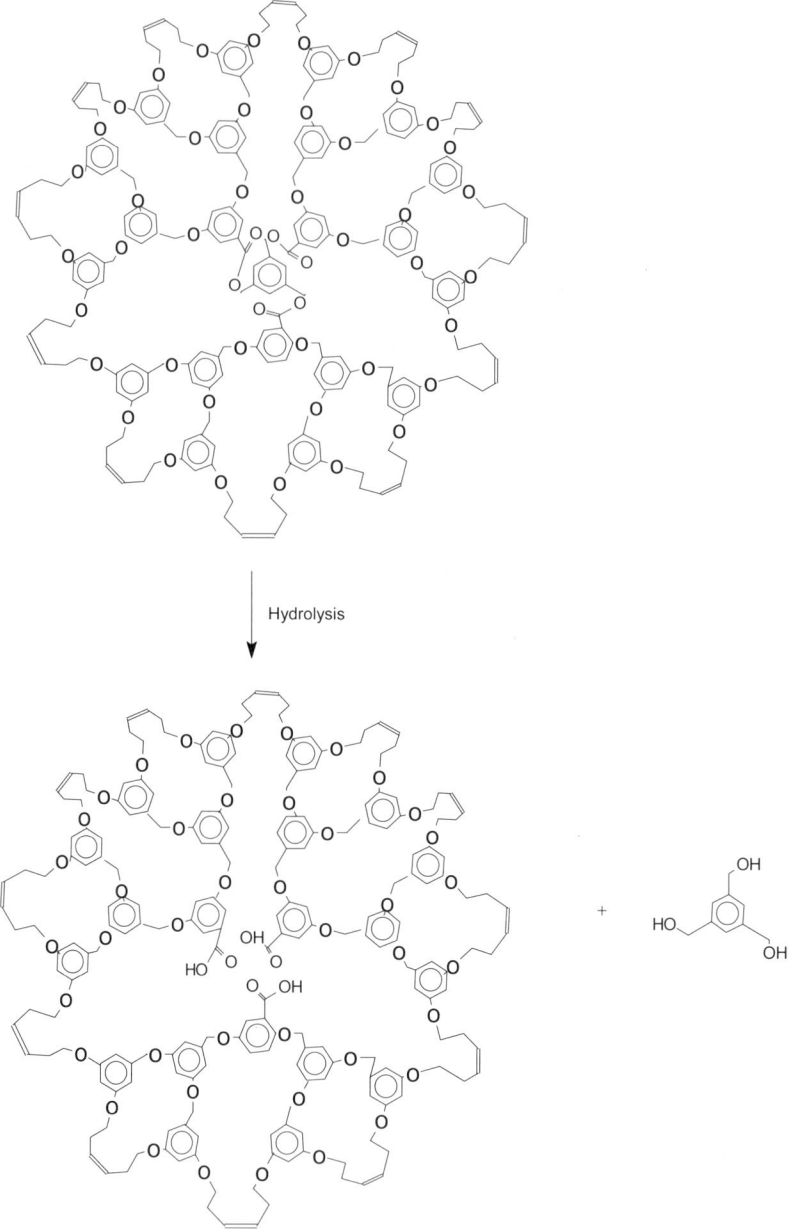

Figure 4.4. Dendrimers with their cores being removed by hydrolysis (After Freemantle, 1999, copyright 1999 The American Chemical Society)

In particular, Zimmerman and co-workers synthesized the hollow dendrimers by covalently attaching dendrons to a molecular core that acts as a template. They then cross-linked the dendrons before cleaving the molecular template, for example by hydrolysis, leading to the whole dendritic macromolecular structure being held together without the core unit. These hollow dendrimers could have an important implication in the development of new dendritic catalysts (Astruc and Chardac, 2001), control release and molecular recognition systems (*vide infra*).

4.2.2.3 Dendrimers with a Hydrophobic Interior and Hydrophilic Exterior Layer

In as early as 1982, Maciejewski (1982) had proposed the possibility of constructing a dendritic "core shell" macromolecule for topologically entrapping small molecules. Since then, various dendrimers with a hydrophobic interior and hydrophilic exterior, or *vice versa*, have been synthesized for this purpose. For instance, Newkome *et al.* (1991) have synthesized demdrimers with a completely hydrophobic (alkane) interior surrounded by polar terminal groups (Figure 4.5).

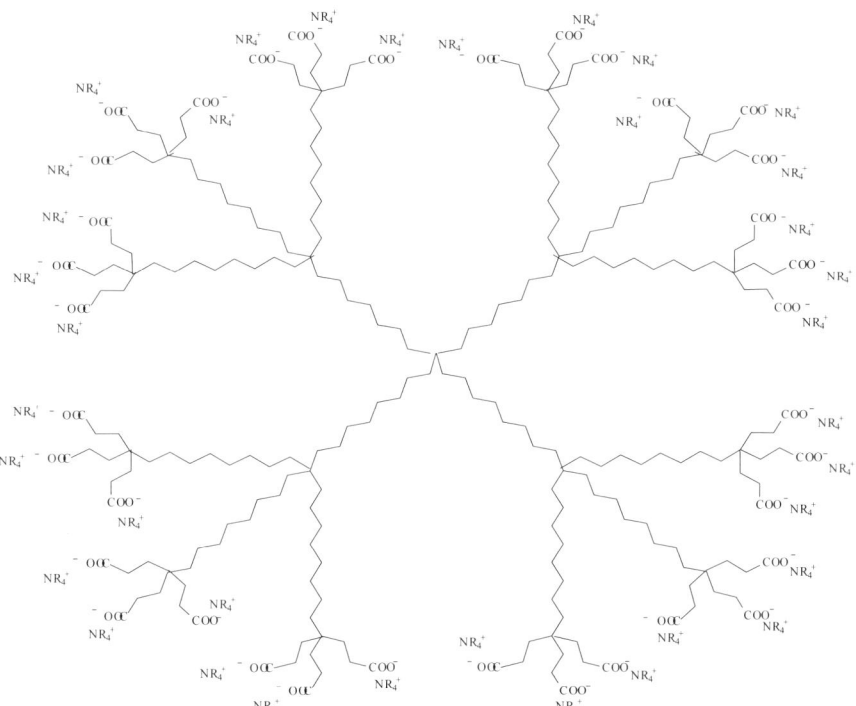

Figure 4.5. A dendritic unimolecular micelle (After Newkome *et al.*, 1991, copyright 1991 Wiley-VCH Verlag. Reproduced with permission)

In contrast to normal micelles that often possess a critical micelle concentration (CMC) below which the micelle dissociates (Dai and White, 1993), these types of dendritic macromolecules, as the one shown in Figure 4.5, could act as a "unimolecular micelle" to exhibit concentration-independent micellar properties. Indeed, dynamic light scattering studies indicated that the dendrimer with a hydrophobic interior and hydrophilic exterior, given in Figure 4.5, was monomolecular in aqueous solution over a wide range of concentrations.

Similarly, Hawker *et al.* (1993) have also prepared a dendritic unimolecular micelle containing a hydrophilic exterior layer and hydrophobic interior (Figure 4.6).

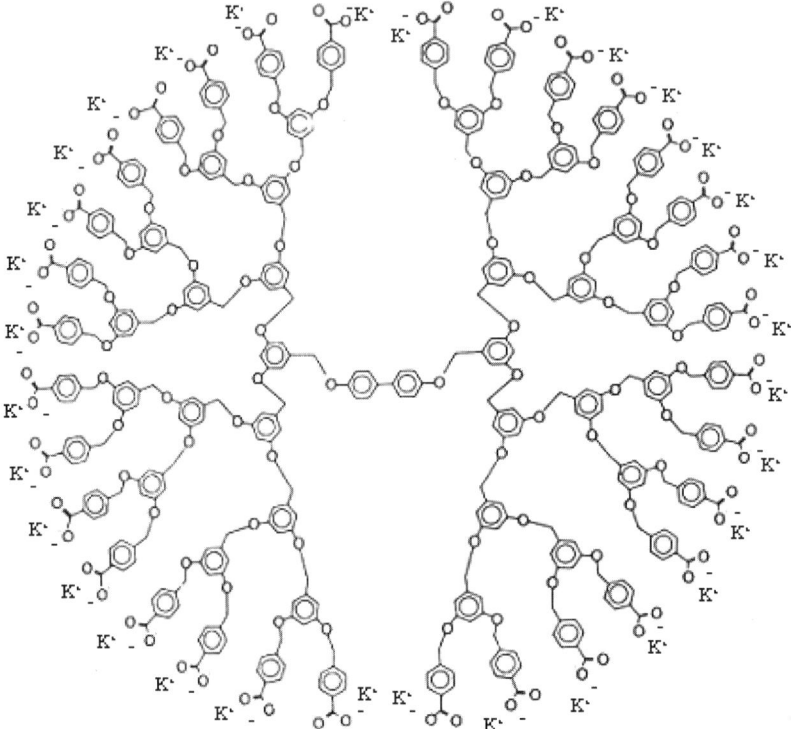

Figure 4.6. Dendritic unimolecular micelle containing a hydrophilic exterior layer and hydrophobic interior (After Hawker *et al.*, 1993, copyright 1993 The Royal Society of Chemistry. Reproduced with permission)

These dendritic unimolecular micelles have been demonstrated to be not only capable of molecular inclusion but also to solubilize hydrophobic molecules in aqueous solutions up to an equal or greater extent than can traditional sodium dodecyl sulphonate (SDS) micelles, though without displaying any critical micelle concentration.

4.2.2.4 Dendrimers with Guest Molecules Trapped in their Cavities

Meijer and co-workers (Jansen *et al.*, 1995a, b) have successfully used the dendritic unimolecular micelles for "unimolecular encapsulation", which is also termed a "dendrimer box". In particular, these authors carried out the surface modification of poly(propylene imine), PPI, dendrimer of five generations and 64 branches on the periphery with 1-phenylalanine or other amino acids (Jansen *et al.*, 1995a, b) to induce dendrimer encapsulation by forming solid-like, densely packed, hydrogen-bonded surface shells. When the addition of the dense shell is carried out in the presence of guest molecules, those molecules can be encapsulated inside the dendrimer box (Figure 4.7).

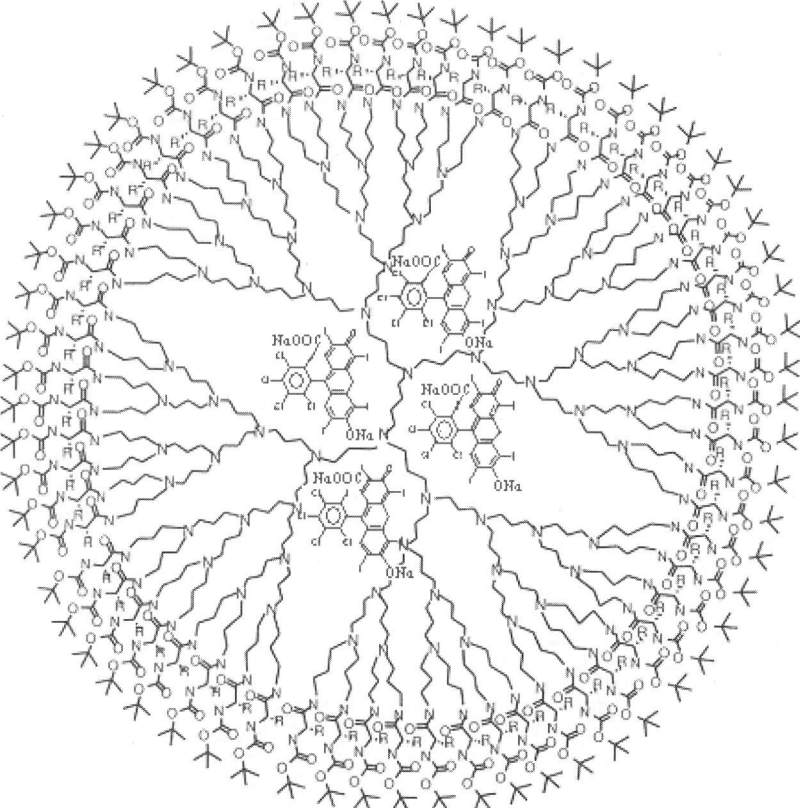

Figure 4.7. A globular dendrimer with a dense amino acid shell holding four molecules of Rose Bengal and many smaller molecules inside the cavities. R = benzyl (After Dagani, 1996. Reproduced with permission from Meijer, Eindhoven University of Technology)

The maximum number of guest molecules which can be encapsulated in a particular dendrimer box depends strongly not only on the shape and the size of the guests but also on the number, shape and size of the available internal dendrimer cavities. For the PPI dendrimer, four large guest molecules (*e.g.* Bengal Rose, Rhodamide B) and 8–10 smaller guest molecules (*e.g.* *p*-nitrobenzoic acid, nitrophenol) could be simultaneously encapsulated within the dendrimer cavities. As schematically shown in Figure 4.7, these dendrimer boxes could be used to release either all or only some of the encapsulated guest molecules by partially or fully hydrolyzing the hydrogen-bonded shell structure. These results suggest the use of dendrimer boxes for many potential applications, including the preparation of nanoscopic markers, targeting and delivery agents, and so on.

4.2.2.5 Dendrimers with Different Terminal Groups – Dendritic Block Copolymers

The block copolymers discussed in the preceding chapters usually consist of different block units chemically linked into a one-dimensional macromolecular chain. The three-dimensional nature of dendritic macromolecules offers additional freedom to make novel hybrid dendritic block copolymers. As schematically shown in Figure 4.8, three different major classes of block copolymers based on dendritic macromolecules, namely dendritic surface block, layer block and segment block copolymers, have been proposed (Hawker and Fréchet, 1995).

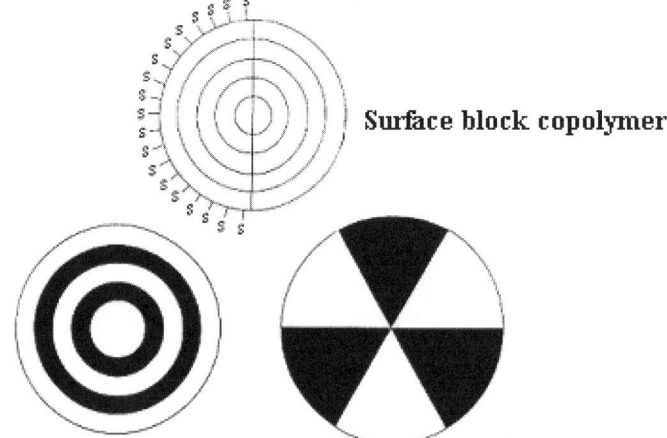

Figure 4.8. Schematic representations of different architectures for various dendritic block copolymers (After Hawker and Fréchet, 1995, copyright 1995 Chapman and Hall Publishers)

Although the divergent approach, convergent-growth approach, and/or their combination have been used to make dendritic block copolymers, the higher degree of structural control has made the convergent-growth approach one of the most common methods for preparing block-type dendrimers (Aoi *et al.*, 1997; Fréchet and Hawker, 1996; Frey *et al.*, 1996; Hawker *et al.*, 1994; Hawker and Fréchet, 1990c, 1992; Newkome *et al.*, 1996; Tomalia and Dvornic, 1996; Tomalia *et al.*, 1993; Wooley *et al.*, 1991b). For instance, Aoi *et al.* (1997) have used a combined divergent and convergent approach with a half-protected initiator core to synthesize surface-block dendrimers as the one shown in Equation (4.11).

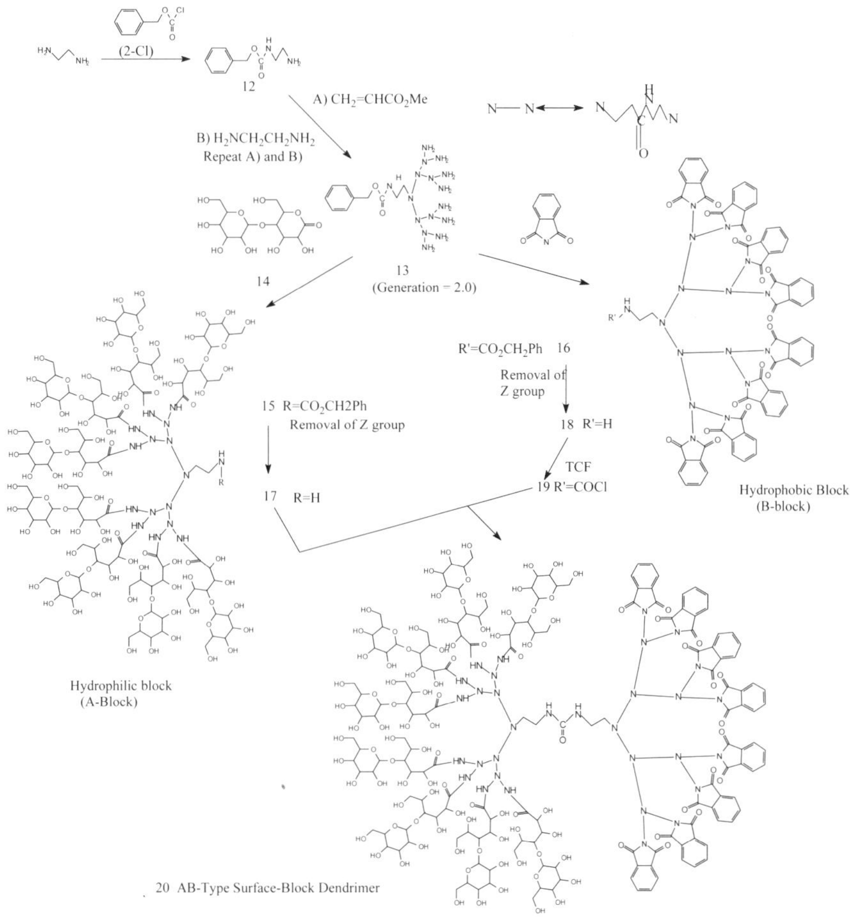

(4.11)

As we shall see later, hybrid combinations of dendritic macromolecules and linear polymers or fullerene C_{60} are also possible.

4.3 Fullerene C_{60}

Fullerenes [also referred to as: buckminsterfullerenes or buckyballs because of the similarity of the structure to the geodesic structures widely credited to R. Buckminster Fuller (Baggott, 1994)] refer to a class of all-carbon molecules, C_n, which have the form of hollow, closed nets composed of 12 pentagons and any number of hexagons, m, such that $m = (C_n - 20)/2$ (*i.e.* Euler's theorem) (Baggott,

1994; Koruga *et al.*, 1993). Fullerene C_{60} ($m = 20$) is one of the most readily available and most studied members of the fullerene family. Shortly after the discovery of fullerene C_{60}, Krätschmer *et al.* (1990) reported a simple way to produce macroscopic amounts of fullerenes in 1990. It was the large-scale production of fullerene C_{60} that enabled scientists to study physicochemical properties of the carbon clusters with all of the conventional spectroscopic methods and to use fullerenes as useful reagents in synthetic chemistry. Although higher fullerenes are still not yet available in larger amounts, they have become more and more accessible now. For instance, pure C_{70} is now available in preparative amounts and some C_{70} chemistry has been reported (Thilgen *et al.*, 1997). Since the fullerene research has been and is still focusing on C_{60}, however, only the chemistry of fullerene C_{60} will be discussed below.

4.3.1 Chemistry of C_{60}

As shown in Figure 4.2, fullerene C_{60} possesses icosahedral symmetry so that all 60 carbon atoms in the buckyball should be equivalent. Due to the presence of both five- and six-membered rings within the structure of C_{60}, however, the chemical bonds in fullerene C_{60} can be distinguished into two groups: i) bonds at the junction between two six-membered rings ([6,6]-bonds, Figure 4.2) that have much double-bond character, and ii) bonds at the junction between a five- and a six-membered ring ([6,5]-bonds, Figure 4.2) with single bond characteristics. X-ray crystal structure determinations on C_{60} and on some of its derivatives have confirmed the existence of short [6,6]-bonds (1.38 Å long) and long [6,5]-bonds (1.45 Å long) (Haddon, 1992). Therefore, fullerene C_{60} has a strongly bond-alternated structure, which governs its reactivity. In contrast to benzene and unlike other superaromatic molecules, double-bond resonant structures in the five-membered rings of buckyballs are not favored (Haddon, 1988) and C_{60} behaves as an electron-deficient olefin in terms of chemical reactivity.

Furthermore, the curvature of the C_{60} molecule causes deviations from planarity for all the double bonds in C_{60} (Haddon, 1992), leading to an excess strain, and hence the enhanced reactivity, for C_{60}. The change of hybridisation from sp^2 to sp^3 associated with most chemical reactions characteristic of C_{60} actually releases the strain (Haddon, 1993). The distribution in energy of those 60 π-electrons in C_{60} has been determined, for example, by Hückel calculations (Haddon, 1992). The theoretically determined low-lying LUMO near to the zero of the Hückel energy axis suggests also a high electron affinity for the C_{60} molecule.

4.3.1.1 Addition Reactions

Owing to its electron deficiency, C_{60} reacts readily with nucleophiles. Most of the reactants attack the electron-rich [6,6]-bonds and insertions into [6,5]-bonds have been reported to occur only as a result of rearrangement from an attached [6,6]-bond. Depending on the nature of the reactants and the special patterns of the attachment on an individual C_{60} molecule, structures of the reaction products can be widely different. However, the addition reactions can be classified into opened additions, cycloadditions and multiple additions (Figure 4.9).

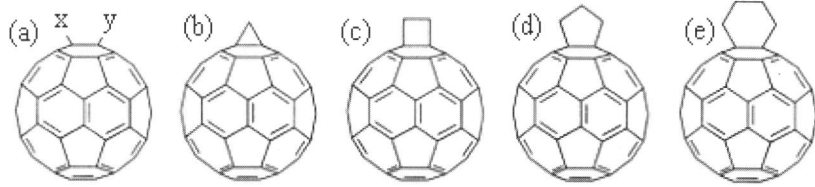

Figure 4.9. Geometrical shapes built onto a 6,6 ring junction of C_{60}: (a) open; (b) three-membered ring; (c) four-membered ring; (d) five-membered ring; (e) six-membered ring (After Prato, 1997, copyright 1997 The Royal Society of Chemistry. Reproduced with permission)

Opened Additions
As shown in Equation (4.12), C_{60} adducts having substituted groups of an opened structure may be obtained through a nucleophilic addition (Nu^-) followed by quenching with an electrophile (E^+) or by hydrogenation (Nu = E = H).

$$(4.12)$$

Examples of the nucleophiles used for Equation (4.12) include Grignard reagents (Hirsch *et al.*, 1992), organolithium derivatives (Hirsch *et al.*, 1992; Fagan *et al.*, 1992), and cyanide ion (Keshavarz-K *et al.*, 1995). Among them, the cyanide addition followed by quenching with toluene-*p*-sulfonyl cyanide produced a dinitrile derivative of C_{60} [*i.e.* the final product in Equation (4.12) with Nu = E = CN], which displayed unique electronic properties allowing fine-tuning of the electron-accepting capacity (Keshavarz-K *et al.*, 1995).

Cycloadditions
In addition to the nucleophilic addition, the electron-withdrawing nature of fullerenes makes C_{60} readily undergo cycloaddition reactions such as the Diels-Alder reaction (Wudl, 1992; Prato *et al.*, 1993a). As shown in Figure 4.10, the cycloaddition products can be further classified into a few broad categories according to the number of atoms within the ring structure. The three-membered ring adducts, including carbon or nitrogen insertion into either a [6,6] or a [6,5] ring junction, represent one of the most interesting and thoroughly studied classes of fullerene derivatives.

The first C_{60} adduct with a three-membered ring substitute was reported by Wudl in 1991 (Suzuki *et al.*, 1991) through the addition of diazomethane derivatives onto C_{60} [Equation (4.13)].

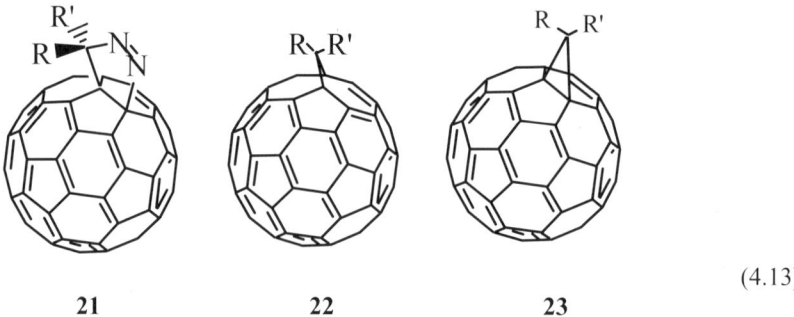

(4.13)

Equation (4.13) gives two different products, commonly called fulleroids **22** and methanofullerenes **23**, and the compound **22** can be thermally (Diederich *et al.*, 1994), electrochemically (Eiermann *et al.*, 1994), or photochemically (Janssen *et al.*, 1995) converted into **23**. Later, it was found that three-membered rings fused on [6,6]-junctions of C_{60} can also be prepared electrochemically (Boulas *et al.*, 1996) or by the addition of nucleophiles, diazirines, carbenes, and sulfonium ylids (Diederich *et al.*, 1994) without the formation of fulleroids.

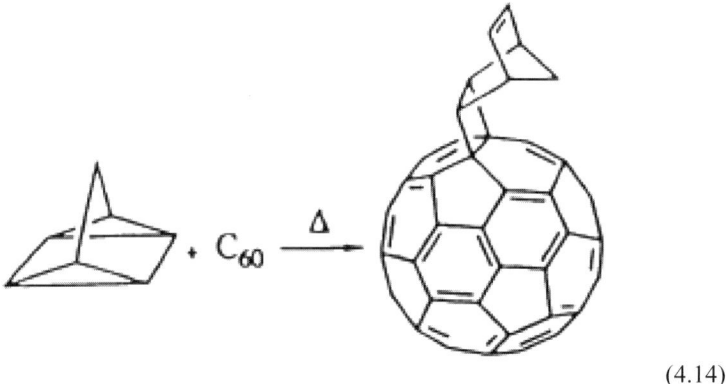

(4.14)

Success has also been achieved with cyclobutanofullerene derivatives by (2 + 2) cycloaddition of benzyne (Hoke II *et al.*, 1992; Tsuda *et al.*, 1992), electron-poor alkenes (Wilson *et al.*, 1993; Liou *et al.*, 1995), electron-rich alkenes and alkynes (Zhang *et al.*, 1996; Tokkuyama *et al.*, 1994) onto C_{60}. Similarly, (2 + 2 + 2) cycloaddition of quadricyclane to C_{60} led to a fullerene-modified norbornene [Equation (4.14), Prato *et al.*, 1993b], which was used for making fullerene-grafted polynorbornenes by the ROMP copolymerisation with pure norbornene [Zhang *et al.*, 1995].

While (3 + 2) cycloadditions of carbocyclic or heterocyclic systems, including cyclopentane derivatives (Prato *et al.*, 1993c), pyrrolidines (Averdung and Mattay,

1996; Ohno et al., 1996a, b), isoxazolines (Meier and Poplawska, 1996), pyrazolines (Muthu et al., 1994), and furans (Jagerovic, 1996), to C_{60} have been reported to produce a five-membered ring fused to a [6,6]-junction of fullerene [Equation (4.15)], the classical (4 + 2) cycloaddition to C_{60} yielded six-membered-ring counterparts (Paquette and Graham, 1995; Ohno et al., 1996a, b; Torres-García and Mattay, 1996) [Equation (4.16)].

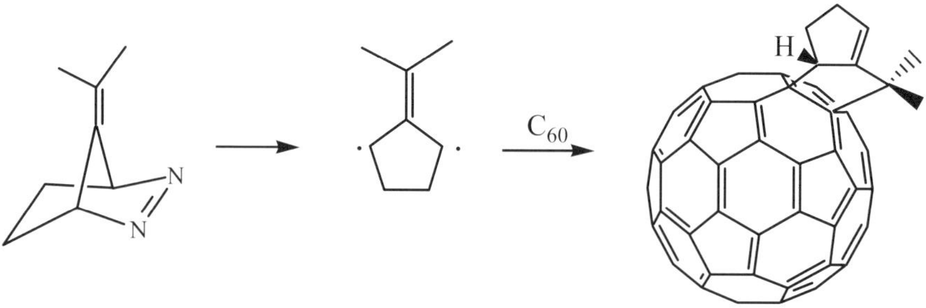

(4.15)

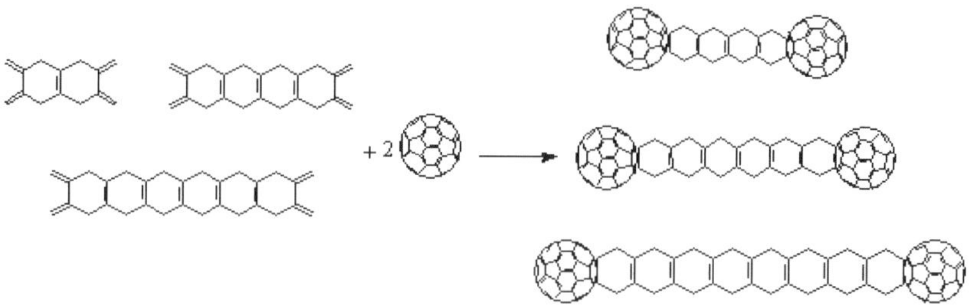

(4.16)

Multiple Additions
The 30 [6,6]-double bonds in the C_{60} molecule all share the same reactivity. As such, the addition of a nucleophile to the [6,6]-bonds of C_{60} often leads to a complex mixture of monoaddition and multiple addition products. As the number of possible isomers increases with the number of additions, the separation of the multiple addition adducts is a very tedious, complicated process. In most cases, therefore, the optimal situation would be to produce a maximum amount of the monoaddition product that can be separated by chromatography, with no attention to the multiple addition entities to be discarded.

Due to the possible use of multi-functional fullerenes as building blocks for the construction of multi-dimensional molecular assemblies, however, the multiple addition chemistry of C_{60} has subsequently attracted considerable interest. For example, several research groups have reported polyadditions of C_{60} using osmylation (Hawhins et al., 1992), hydrogenation (Henderson, 1994; fluorination (Taylor, 1991; Tuinman et al., 1993; Boltalina et al., 1995), silylation (Kusukawa and Ando, 1996), epoxidation (Balch et al., 1995), azide addition (Grösser et al., 1995) and amine addition (Balch et al., 1996). Polyhydroxylated C_{60} derivatives (also called fullerols or fullerenols) have been prepared by several different methods

(Chiang et al., 1994; Darwish et al., 1994). In particular, hydroxy groups can be added to C_{60} either by the reaction with diborane/THF followed by H_2O_2/NaOH (Schneider et al., 1994) or by reacting with nitrating reagents/NaOH (Chiang et al., 1996). The fullerols produced by the nitration reaction [Equation (4.17)] can contain up to 18 –OH groups and fullerols are produced even when sulphuric acid alone is used.

$$C_{60} \xrightarrow[\text{toluene}]{NO_2} C_{60}(NO_2)_n \xrightarrow[H_2O]{NaOH} C_{60}(OH)_n \qquad (4.17)$$

Hydrogensulfated fullerenol derivatives with multiple –OSO$_3$H groups have also been produced from fullerenols by either partially hydrolyzing the polycyclosulfated fullerene derivatives [**24** of Equation (4.18)] (Chiang et al., 1994) or directly sulfating the polyhydroxylated fullerene [**25** of Equation (4.18)] (Lu et al., 1998).

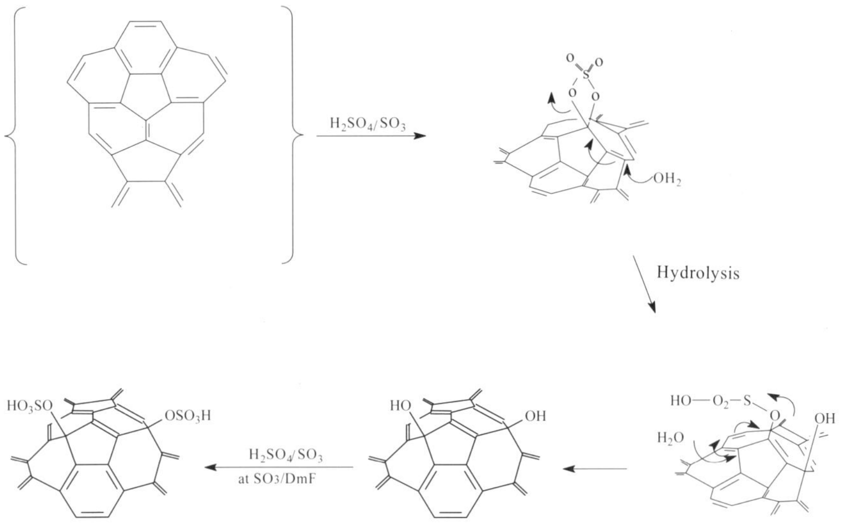

(4.18)

The resultant hydrogensulfated fullerenol derivatives containing multiple –OSO$_3$H groups, $C_{60}(OH)_6(OSO_3H)_6$, have been further used as the protonic acid dopant to make novel highly conducting three-dimensional conducting polyanilines (see Chapter 2).

The polyhydroxylated fullerene thus produced is found to be more soluble in acid than in water due to an acid-induced pinacol rearrangement [Equation (4.19)] from a hemiketal to carbonyl structure (Chiang et al., 1993).

As can be seen in Equation (4.19), the pinacol rearrangement led also to an opened fullerene cage. In fact, opened fullerene cages have also been prepared by several other methods, including light incision (Taliani et al., 1993), regioselective diaddition of azides (Grösser et al., 1995), consecutive pericyclic reactions (Arce et al., 1996), oxidative cleavage (Birkett et al., 1995), and ring opening of

N-methoxyethoxymethyl-substituted [5,6] azafulleroid (Hummelen et al., 1995). The possibility of trapping atoms, molecules or ions inside the fullerene carbon cage has attracted considerable interest from researchers worldwide ever since the discovery of fullerenes in 1985. Although various physical means have been used to trap gases (helium, neon, argon, krypton, xenon, etc.) (Saunders et al., 1996), transition metals (Bethune et al., 1993) and some other species (Smalley, 1992) into fullerenes, the ideal approach to encapsulate things into the fullerene cages is to introduce them into those chemically opened fullerene cages followed by restoring the carbon–carbon bonds.

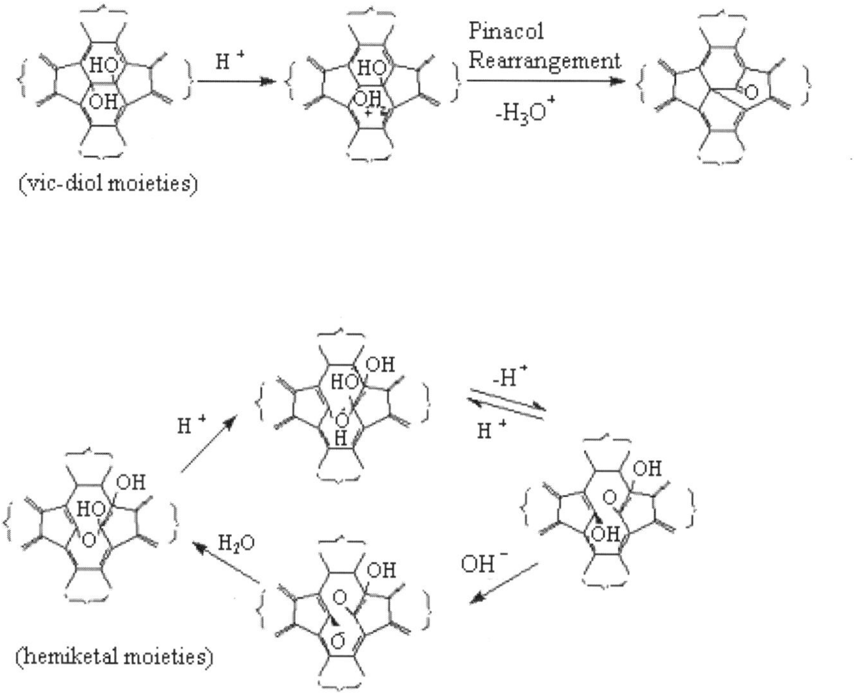

(4.19)

4.3.1.2 Dimerization and Polymerization

In 1993, Rao et al. (1993) discovered that C_{60} molecules arranged in an fcc lattice by a weak van der Waals force could be transformed into a polymerized state upon exposure to visible or ultraviolet light. Supported by the generalized tight-binding molecular dynamic calculations (Menon et al., 1994), a mechanism based on the photochemical (2 + 2) cycloaddition reaction has been proposed (Zhou et al., 1992) that involves the breakdown of the parallel double bonds on adjacent C_{60} molecules (< 4.2 Å apart) and their reforming into a four-membered ring cross-linking the two adjacent C_{60} molecules (Figure 4.10).

140 Intelligent Macromolecules for Smart Devices

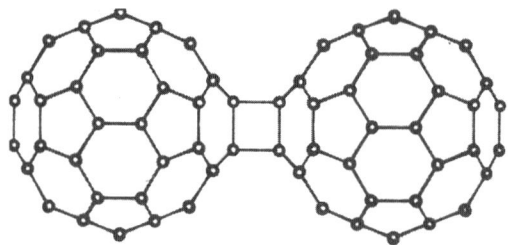

Figure 4.10. The structure of C_{60}-dimer (After Zhou *et al.*, 1992, copyright 1992 American Institute of Physics)

The reaction between two C_{60} molecules to form a C_{60} dimer is the simplest process invoked in C_{60} polymerization. Since the discovery of the photochemical formation of the C_{60} dimer, the study of polymerized fullerite structure has attracted considerable attention, and significant advances in both the theoretical and experimental fronts have been achieved. As the C_{60} molecules in the solid state tend to form a compact face-centered cubic structure, in which the electron-rich double bonds face the electron-depleted pentagons of their nearest neighbors (David *et al.*, 1991), it was found, both experimentally and theoretically, that C_{60} molecules can be chemically bonded into a polymerised lattice under pressure (Duclos *et al.*, 1991), photoirradiation (Rao *et al.*, 1993), plasma (Takahashi *et al.*, 1993), and electron-induced processes (Zhao *et al.*, 1994). In the meantime, a number of possible structures have been reported. Examples of these include one-dimensional polymerized chains (Figure 4.11(a), Núñez-Regueiro *et al.*, 1995), two-dimensional planes (Figure 4.11(b), Xu and Scuseria, 1995) and three-dimensional polymerized structures (Figure 4.11(c), O'Keeffe, 1991).

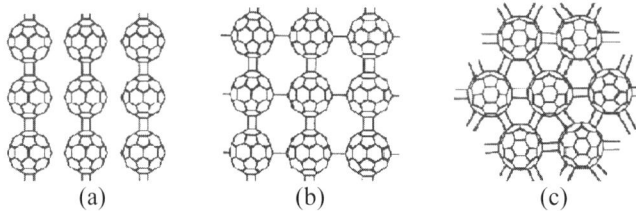

(a) (b) (c)

Figure 4.11. Schematic representation of the polymerised fullerite structures: (a) one-dimensional polymerized C_{60} chains; (b) two-dimensional C_{60} planes; (c) three-dimensional polymerized C_{60} structures (After Núñez-Regueiro *et al.*, 1999, copyright 1999 Spinger-Verlag)

4.3.2 Polymeric Derivatives of C_{60}

As can be seen from the foregoing discussion, fullerene C_{60} and its polymeric composites possess interesting and useful optoelectronic properties for, but not limited to, photovoltaic and optical-limiting applications. However, the poor processability of fullerenes precludes their commercial applications. The combination of the unique molecular characteristics of fullerenes with the good processability of certain polymers through the chemical syntheses of fullerene-containing polymers would open up possibilities for making advanced polymeric materials with exotic physicochemical properties and good processability. Through controlled modification of C_{60}, fullerene-bound polymers with fullerenes either as pendant groups ('charm bracelet' or 'pendant chain', Figure 4.12(a)) (Bergbreiter and Gray, 1993; Hirsh, 1993; Wudl, 1992, 1993), as constituent units of the polymer backbones ('pearl necklace', Figure 4.12(b)) (Taylor, 1993), or as a central core ('flagellenes', Figure 4.12(c)) have been synthesized (Samulski *et al.*, 1992). While several excellent reviews on the chemistry of C_{60}, with or without an emphasis on materials applications, have appeared (Andreoni, 1996; Bergbreiter and Gray, 1993; Diederich *et al.*, 1994; Hirsh, 1993, 1994; Meier, 2000; Taylor, 1995; Wudl, 1992), the synthesis of fullerene-containing polymers is summarized below by including some more recent work.

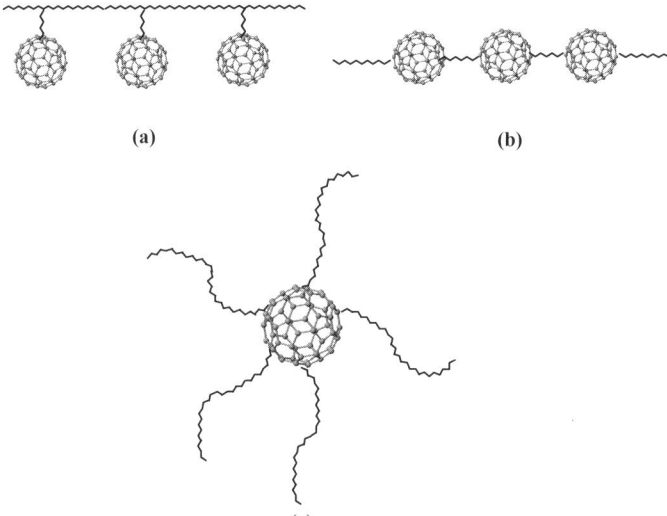

Figure 4.12. Possible structures of C_{60}-containing polymers: (a) "Charm bracelet"; (b) "pearl necklace"; (c) "flagellates"

4.3.2.1 Fullerene Charm Bracelets

The first C_{60}-containing polymer of a highly cross-linked structure was reported by Olah *et al.* (1991). Later, Geckeler and Hirsch (1993) synthesised soluble C_{60}-grafted polymers through the amine addition of aminopolymers to fullerene double bonds (Prato *et al.*, 1993a; Wudl *et al*, 1992). Since then, fullerene C_{60} has also been attached onto aminoethylene propylene terpolymer (EPDM-amine), yielding soluble fullerene-functionalized polymers (Patil and Schriver, 1995). Thermogravimetric analyses (TGA) show that the fullerene-attached aminoethylene propylene terpolymer (EPDM-amine) is thermoxidatively more stable than pure EPDM-amine.

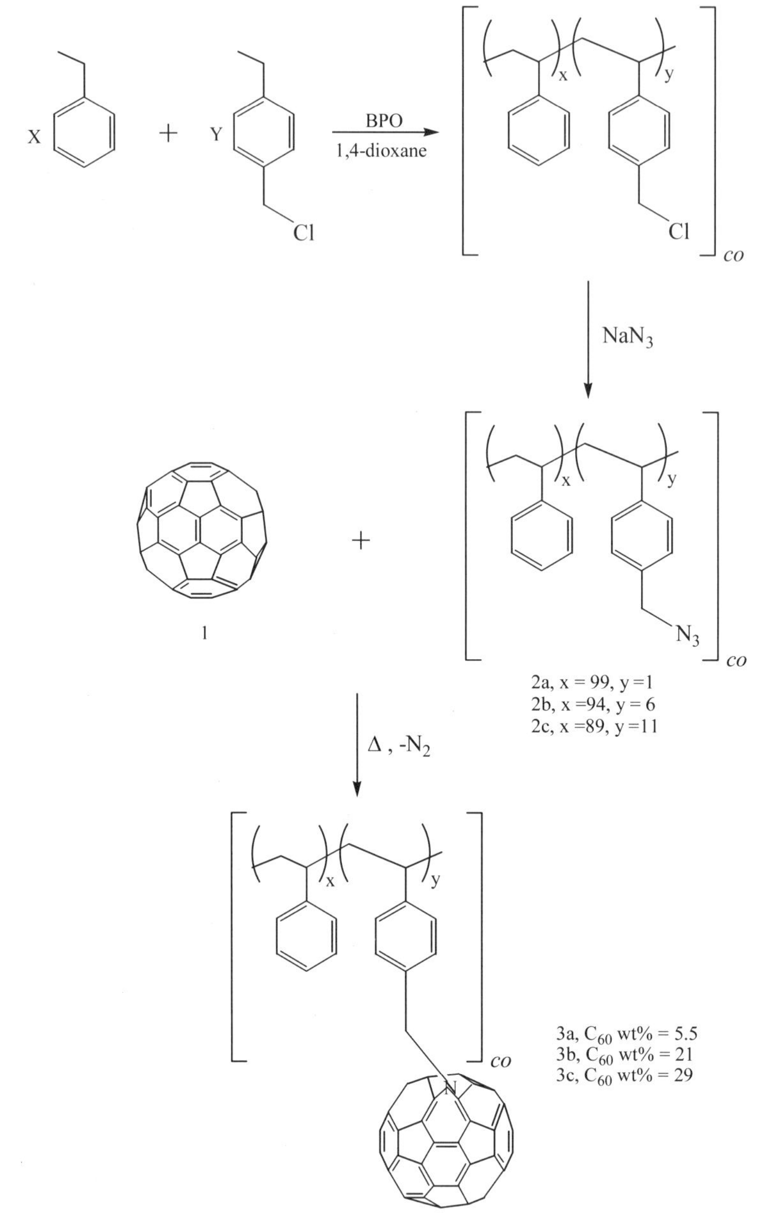

(4.20)

In addition to the amine addition reaction, fullerenes readily undergo cycloaddition reactions, such as the Diels-Alder reaction (Noller, 1996). In this regard, Hawker synthesised soluble C_{60}-grafted polystyrene through the cycloaddition of C_{60} onto azido-substituted polystyrenes [Equation (4.20)] (Hawker, 1994). A single glass transition temperature was observed for the fullerene functionalized polystyrene chains up to 29 wt% C_{60}, indicating homogeneous mixing without phase separation. Although a monoaddition of C_{60} was claimed in this case, it is likely that the double bonds of C_{60} are indiscriminately attacked and that mixtures of isomeric fullerene species are produced, as are the cases for many other fullerene derivatives produced by grafting reactions. In a separate study, Guhr et al. (1994) demonstrated a reversible covalent attachment of C_{60} to cyclopentadiene-functionalized polymers through the Diels-Alder cycloaddition. While the addition of C_{60} to the cyclopentadiene-functionalized polymer occurred readily in decalin at room temperature, heating of the fullerene functionalized polymer in decalin to 180 °C in the presence of maleic anhydride was shown to permanently detach the polymer-bound C_{60}. In the absence of maleic anhydride, however, C_{60} was released from the polymer by heating and re-added to the polymer upon cooling. This process can be repeated many times, indicating the truly reversible nature for the addition and removal of C_{60}.

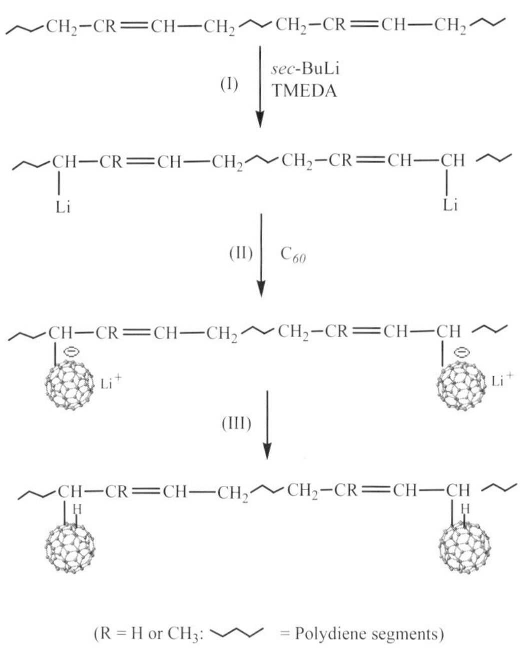

$$\sim\!\!CH_2\!-\!CR\!=\!CH\!-\!CH_2\!\sim\!\!CH_2\!-\!CR\!=\!CH\!-\!CH_2\!\sim$$

(I) $\Big\downarrow$ sec-BuLi / TMEDA

$$\sim\!\!\underset{Li}{CH}\!-\!CR\!=\!CH\!-\!CH_2\!\sim\!\!CH_2\!-\!CR\!=\!CH\!-\!\underset{Li}{CH}\!\sim$$

(II) $\Big\downarrow$ C_{60}

$$\sim\!\!\underset{\text{[C}_{60}\text{]}^{-}\text{Li}^{+}}{CH}\!-\!CR\!=\!CH\!-\!CH_2\!\sim\!\!CH_2\!-\!CR\!=\!CH\!-\!\underset{\text{[C}_{60}\text{]}^{-}\text{Li}^{+}}{CH}\!\sim$$

(III) $\Big\downarrow$

$$\sim\!\!\underset{\text{[C}_{60}\text{H]}}{CH}\!-\!CR\!=\!CH\!-\!CH_2\!\sim\!\!CH_2\!-\!CR\!=\!CH\!-\!\underset{\text{[C}_{60}\text{H]}}{CH}\!\sim$$

(R = H or CH_3: $\sim\!\!\sim$ = Polydiene segments) (4.21)

As C_{60} is also known to readily undergo nucleophilic addition due to its strong electron affinity (Bergbreiter and Gray, 1993; Hirsh, 1993; Wudl, 1992, 1993), C_{60} has been attached as pendant groups along the PVK backbone through nucleophilic addition (Chen et al., 1996). Based also on the nucleophilic addition reaction, Dai et al. (1995, 1998b) have successfully attached C_{60} onto 1,4-polydiene chains with fullerene buckyballs as the pendant groups. As seen in Equation (4.21), the grafting reaction involves lithiation of polydiene chains with sec-BuLi in the presence of tetramethylethylenediamine (TMEDA), followed by the addition of C_{60} onto the lithiated living polydiene chains. The solubility of the resultant C_{60}-functionalized polydiene chains can be varied by changing the molar ratio of [sec-BuLi]/[C_{60}]. For instance, the C_{60}-grafted polydiene materials prepared with [sec-BuLi]/[C_{60}] < 1 are highly soluble in most common organic solvents due largely to the monoaddition of C_{60}, whereas fullerene-cross-linked polymer gels are observed at a high molar ratio of sec-BuLi to C_{60} (i.e. [sec-BuLi]/[C_{60}] > 1) (Dai et al., 1995, 1998b). Differential scanning calorimetric (DSC) analyses on the soluble material indicate that the chain rigidity of the polydienes increases slightly upon grafting with C_{60}, but the final product remains thermally processable.

While the non-conjugated polydiene segments between two adjacent C_{60} pendant groups in the C_{60}-grafted polydiene chains could be converted into a conjugated structure by an I_2-induced conjugation reaction (Dai, 1992; Dai et al., 1994; Dai and White, 1991), various C_{60}-bonded conjugated polymers have recently been reported (Benincori et al., 1996; Cravino et al., 2000; Eckert et al., 2000; Ferraris et al., 1998; Nierengarten et al., 1999). For example, several research groups have synthesized conducting charm bracelet polymers by covalently attaching fullerene moieties onto conjugated polythiophene backbones (Benincori et al., 1996; Cravino et al., 2000), while Hadziioannou and co-workers have prepared fulleropyrrolidine derivatives bearing oligophenylene vinylene substituents by 1,3-dipolar cycloaddition of an azomethine ylide to C_{60} (Eckert et al., 2000; Ferraris et al., 1998; Nierengarten et al.,1999). The resultant C_{60}-bonded polythiophene copolymer materials exhibited electrochemical and optoelectrochemical properties characteristic of both polythiophene and C_{60} (Benincori et al., 1996; Cravino et al., 2000) whereas the fullerene-oligophenylene vinylene hybrids (and some other related donor-linked fullerenes) showed interesting photo-induced energy-transfer behavior useful for photovoltaic cell applications (Eckert et al., 2000; Ferraris et al., 1998; Nierengarten et al., 1999).

4.3.2.2 Fullerene Pearl Necklaces

Cao and Webber (1995), Bunker et al. (1995), and Camp et al. (1995) almost simultaneously, but independently, found that C_{60}-polystyrene random copolymers with fullerenes interdispersed along the polymer backbone can be synthesized by free-radical copolymerization of C_{60} with styrene, initiated either thermally or by using a free radical initiator (e.g. azo-bis-isobutyronitrile, AIBN). In all cases, soluble products (e.g. in toluene, dichloromethane, o-dichlorobenzene) were obtained, albeit with a relatively low yield. Furthermore, an enhancement in fluorescence yields by a factor of 10–20, with respect to pure C_{60} solutions, was observed for the C_{60}-containing copolymer solutions (Cao and Webber, 1995; Bunker et al., 1995). The enhancement in C_{60} fluorescence is presumably due to the reduced symmetry of the fullerenes being incorporated into the copolymer structure. The reduced symmetry could strengthen the $S_1 \rightarrow S_0$ transition so that fluorescence can compete with intersystem crossing (Arbogast and Foote, 1991; Foote, 1994). More recently, Ford and co-workers (Camp et al., 1995; Ford et al., 1997) have also copolymerized C_{60} with methyl methacrylate using an excessive amount of AIBN, yielding C_{60}-poly(methyl methacrylate) random copolymers at a relatively rapid rate with a high degree of C_{60} incorporation.

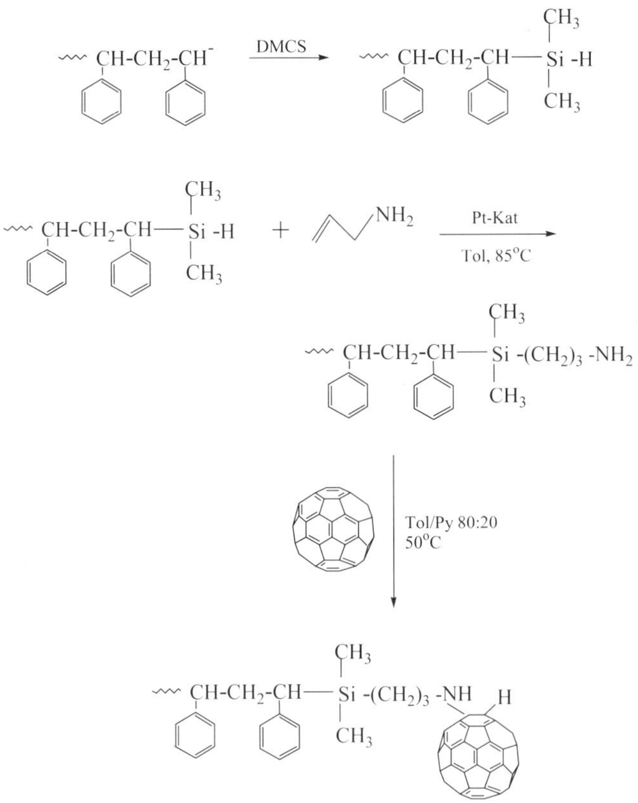

$$\begin{array}{c}
\sim\sim \text{CH-CH}_2\text{-CH}^- \xrightarrow{\text{DMCS}} \sim\sim \text{CH-CH}_2\text{-CH}-\underset{\underset{\text{CH}_3}{|}}{\overset{\overset{\text{CH}_3}{|}}{\text{Si}}}-\text{H}
\end{array}$$

$$\sim\sim \text{CH-CH}_2\text{-CH}-\underset{\underset{\text{CH}_3}{|}}{\overset{\overset{\text{CH}_3}{|}}{\text{Si}}}-\text{H} \; + \; \diagup\!\!\!\diagdown\!\!\text{NH}_2 \xrightarrow[\text{Tol, 85°C}]{\text{Pt-Kat}}$$

$$\sim\sim \text{CH-CH}_2\text{-CH}-\underset{\underset{\text{CH}_3}{|}}{\overset{\overset{\text{CH}_3}{|}}{\text{Si}}}-(\text{CH}_2)_3-\text{NH}_2$$

$$\Big\downarrow \; \text{C}_{60}, \; \text{Tol/Py 80:20, 50°C}$$

$$\sim\sim \text{CH-CH}_2\text{-CH}-\underset{\underset{\text{CH}_3}{|}}{\overset{\overset{\text{CH}_3}{|}}{\text{Si}}}-(\text{CH}_2)_3-\text{NH}-\text{C}_{60}\text{H} \qquad (4.22)$$

Pearl necklace-type C_{60}-polystyrene copolymers with a well-defined structure have also been synthesized through an anionic polymerization of styrene and subsequent nucleophilic addition of C_{60} into the living polymer chains (Okamura *et al.*, 1997). In this case, bis-substitution of C_{60} was identified, by gel permeation chromatography (GPC), to be the main product, and the resulting C_{60}-polystyrene alternating copolymers are soluble in most common organic solvents including tetrahydrofuran, toluene and trichloromethane. The copolymer films exhibit good photoconductivities. Furthermore, C_{60}-end-capped polystyrenes, PS-C_{60}, were prepared by reacting amino-terminated polystyrene chains with C_{60} [Equation (4.22)] (Weis *et al.*, 1995). The final product is soluble and can be spin-cast into homogeneous and transparent films containing up to 27 wt% C_{60}. Preliminary experiments showed some novel electronic properties (*e.g.* higher conductivities than pure polystyrene) and NLO effects (*e.g.* optical-limiting) for the C_{60}-end-capped polystyrene (Kojima *et al.*, 1995). In addition to the third-order NLO effects characteristic of free fullerenes (Blau et al., 1991; Wang and Cheng, 1993; Yang *et al.*, 1994), the unsymmetrically perturbed C_{60} in the fullerene charm bracelets and pearl necklaces could also possess second-order NLO properties (Cao and Webber, 1995), which require the presence of non-centrosymmetrically ordered dipoles (Dai *et al.*, 1999; Prasad and Nigam, 1991). Further studies on the NLO effects of these and other fullerene-containing polymers are clearly needed.

4.3.2.3 Flagellenes

The polymeric fullerene adducts called 'flagellenes' were first synthesized through the addition of living poly(styryl)lithium chains (lithiated at one end only) onto C_{60} to form soluble $C_{60}(PS)x$ with a variable number of polymer arms (x = 1–10) covalently attached onto a single buckyball [Equation (4.23)] (Samulski *et al.*, 1992).

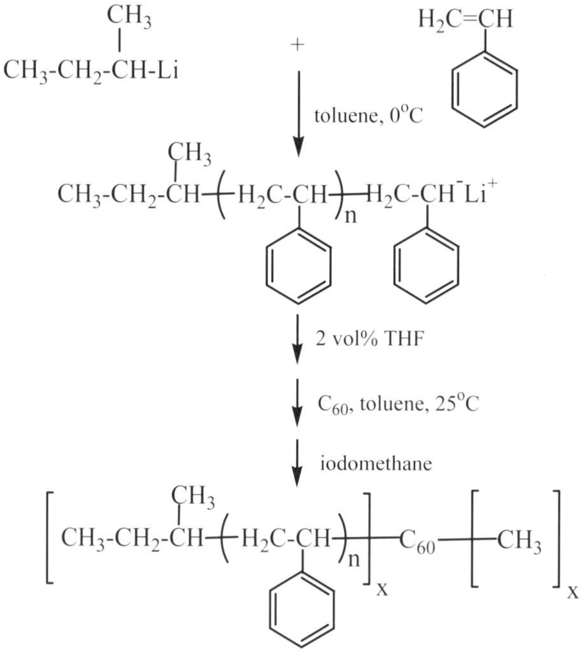

$$CH_3\text{-}CH_2\text{-}\underset{\underset{CH_3}{|}}{CH}\text{-}Li \quad + \quad H_2C\text{=}CH\text{-}C_6H_5$$

$$\downarrow \text{toluene, } 0°C$$

$$CH_3\text{-}CH_2\text{-}\underset{\underset{CH_3}{|}}{CH}\text{-}(H_2C\text{-}CH(C_6H_5))_n\text{-}H_2C\text{-}CH^-(C_6H_5)\,Li^+$$

$$\downarrow \text{2 vol\% THF}$$

$$\downarrow C_{60}, \text{ toluene, } 25°C$$

$$\downarrow \text{iodomethane}$$

$$\left[CH_3\text{-}CH_2\text{-}\underset{\underset{CH_3}{|}}{CH}\text{-}(H_2C\text{-}CH(C_6H_5))_n\right]_x\text{-}C_{60}\text{-}[CH_3]_x$$

(4.23)

The carbanions present on these polymer-grafted C_{60} were shown to be able to initiate anionic polymerization of a vinyl monomer (*e.g.* MMA) leading to the formation of heterostars with, for example, six PS and two PMMA branches (Ederlé and Mathis, 1997). Heterostars with C_{60} as the core and polyacrylonitrile chains / methyl groups as the branches have also been synthesised through the addition of the living polyacrylonitrillithium chains onto C_{60}, followed by a capping reaction with methyl iodine (Chen *et al.*, 1997).

On the other hand, Tajima *et al.* (1997) have used fullerene C_{60} as an effective photo-cross-linking reagent for furan-substituted polymers based on the reaction of C_{60} with furan derivatives (Nie *et al.*, 1995), suggesting that a variety of cross-linking chemistry may be created and explored by using fullerenes and their derivatives. In this regard, Qu *et al.* (2001) have recently demonstrated the photogeneration of polymeric alkyl-C_{60} radical adducts by UV irradiation of low-density polyethylene/benzophenone/C_{60} (LDPE/BP/C_{60}) films in the molten state [413 K, Equation (4.24)]. A well-resolved electron spin resonance (ESR) spectrum characteristic of polymeric alkyl-fullerene (P-C_{60}) radical adducts was recorded. Detailed analyses of hyperfine structures (hfs) and spectroscopic simulation indicated the coexistence of two polymeric radical adducts of C_{60} (*i.e.* the tertiary carbon radical adduct, $-(CH_2)_3C-C_{60}^{\bullet}$, designated as $P_AC_{60}^{\bullet}$, and the secondary carbon radical adduct, $-(CH_2)_2CH-C_{60}^{\bullet}$, designated as $P_BC_{60}^{\bullet}$), which have slightly different g values (g_A = 2.00248 and g_B = 2.00244) and integral intensities I_A/I_B (48.4/51.6). These results clearly indicated the generation of C_{60}-bonded LDPE materials simply by photoirradiation of LDPE/C_{60} with BP as a photoinitiator. The ease with which C_{60}-bound LDPE materials can be effectively produced simply by the photoirradiation of LDPE/BP/C_{60} blends should offer a general approach towards C_{60}-containing polymers and facilitate further chemical derivatization of the otherwise unreactive LDPE chains via various reactions characteristic of C_{60}, including cross-linking through the C_{60} radicals.

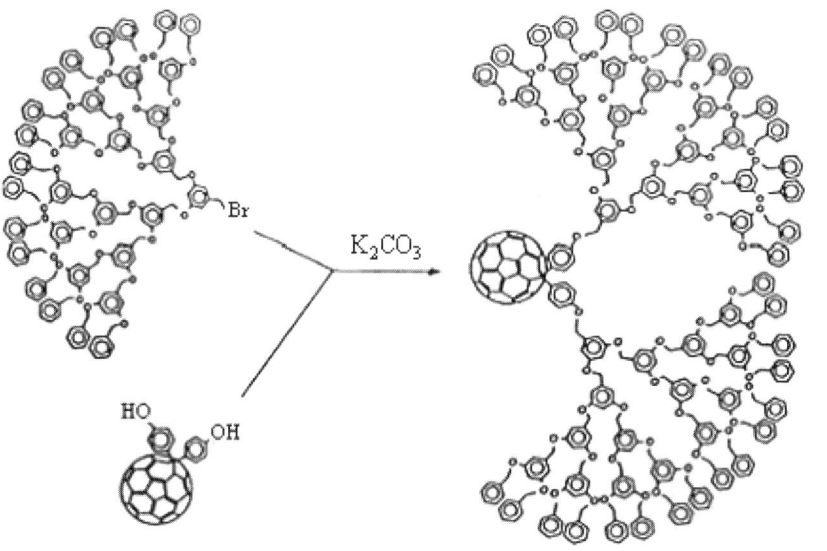

(4.24)

(4.25)

148   Intelligent Macromolecules for Smart Devices

Other interesting studies related to the polymeric C_{60} flagellenes include the syntheses of fullerene-bound dendrimers (Hawker and Fréchet, 1990b; Wooley et al., 1993) through the coupling reaction between a phenol-functionalized fullerene (Shi et al., 1992) and a fourth-generation dendrimer having a single benzylic bromide group at the focal point [Equation (4.25)] or by cycloaddition of dendritic azide onto C_{60} [Equation (4.26)], Hawker et al., 1994). The resulting dendritic fullerenes are extremely soluble in various organic solvents and the latter product has a glass transition temperature of 325 K, 13 K higher than the starting dendrimer, indicating a combination of the properties from both of the constituent components.

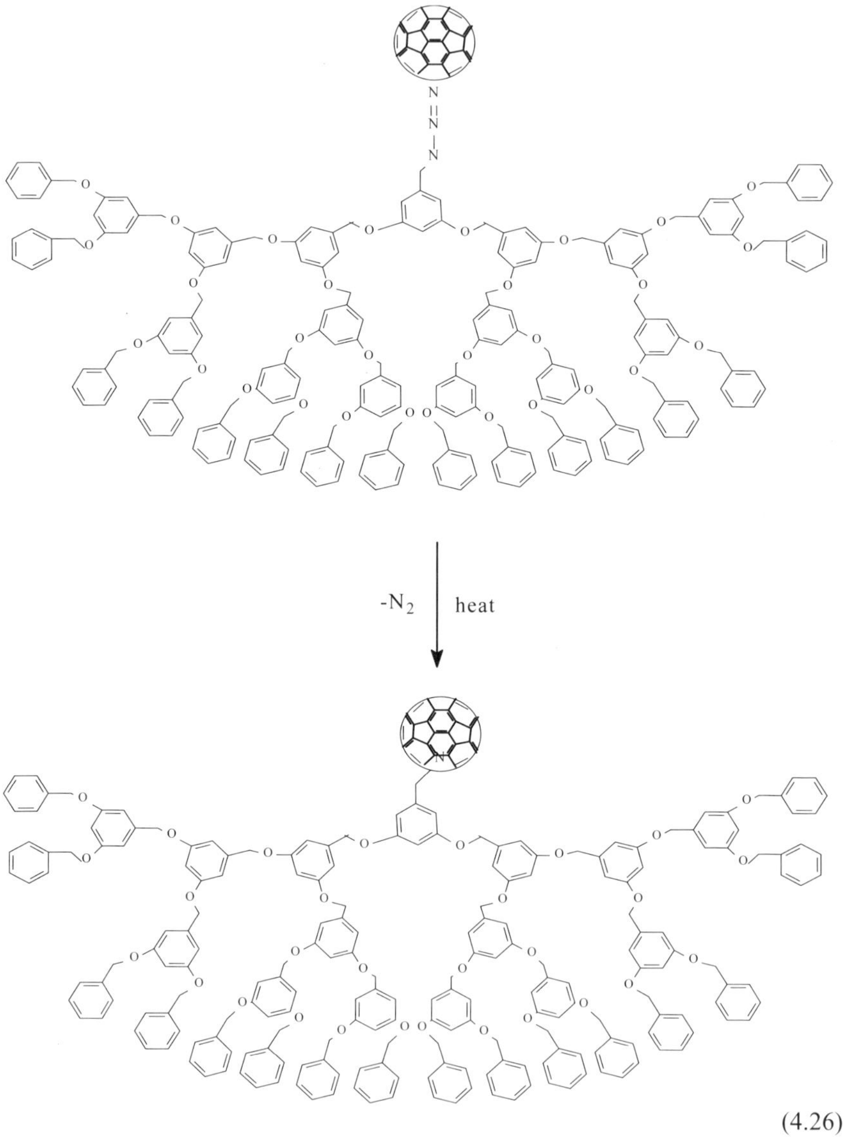

(4.26)

In the context of C_{60}-conjugated polymer composites, Ederlé et al. (1996) have covalently attached polyacetylene chains onto C_{60} buckyballs by thermal conversion of C_{60}-bound poly(phenylvinyl sulfoxide) precursor polymer chains. Using a hydrogensulfated fullerenol derivative containing multiple $-OSO_3H$ groups [designated as $C_{60}(OH)_6(OSO_3H)_6$, i.e. the final product of Equation (4.27)] as the protonic acid dopant, Dai and co-workers (Dai et al., 1998a) have prepared novel three-dimensional conducting polyanilines. Spectroscopic study indicated that, apart from the primary doping, the hydrogensulfated fullerenol can also act as a secondary dopant to "straighten out" the polyaniline emeraldine base chain (PANI-EB) for delocalization of polarons once formed by the protonic acid doping with $-OSO_3H$ groups (Dai et al., 1998a). This is presumably because the electrostatic interaction between the doped polyaniline chains and the "multiple-arms" of the $-OSO_3^-$ counterions in the hydrogensulfated fullerenol molecules could uncoil the doped polymer chains through a "zipping" mechanism (Figure 4.13). The same interaction was also demonstrated to cause self-assembling of the polymer chains into a three-dimensional structure, as indicated by the WAXD study (Lu et al., 1998). The three-dimensional structure thus formed played an important role in regulating the electrical properties of the resultant materials. As expected, the resulting composite material showed metallic characteristics with a limiting room-temperature conductivity of ca.100 S/cm (Dai et al., 1998a), which is about six orders of magnitude higher than the typical value for fullerene-doped conducting polymers (Wei et al., 1993).

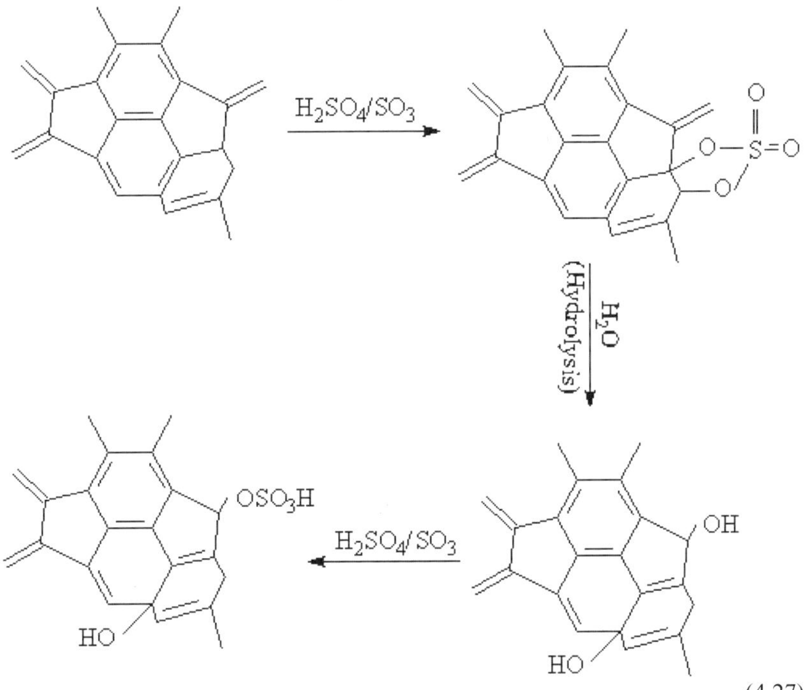

(4.27)

More recently, Qiu et al. (2001) have also synthesized polyaniline (PANI) nanotubes through template-free polymerization using ammonium persulfate ((NH_4)$_2$$S_2$$O_8$) as an oxidant in the presence of $C_{60}(OH)_6(OSO_3H)_6$ or sulfonated dendrimer containing 24 terminal groups of 3,6-disulfo-napthylthiourea (designated as PAMAM4.0 [Naphthyl(SO_3H)$_2$]$_{24}$) as the protonic acid dopant. In this case, water was found to play an important role in regulating the size, and hence the properties, of the final product (Qiu et al., 2001). The PANI–$C_{60}(OH)_6(OSO_3H)_6$ nanotube pellet showed a three-dimensional hopping semiconducting behavior with a room-temperature conductivity of ca. 0.1 S/cm.

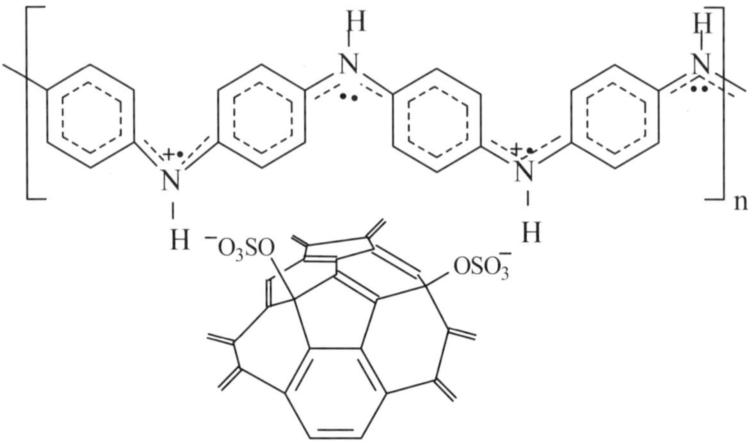

Figure 4.13. Schematic representation of the three-dimensional doping of polyaniline emeraldine base chains with the hydrogensulfonated fullerenol derivatives containing multi –OSO_3H groups. The interchain doping is not shown for reasons of clarity (After Dai et al., 1998a, copyright 2002 The American Chemical Society. Reproduced with permission)

4.4 References

Adronov, A., Gilat, S.L., Frechet, J.M.J., Ohta, K., Neuwahl, F.V.R., Fleming, G.R. (2000) *J. Am. Chem. Soc.* **122**, 1175.
Aharoni, S.M., Crosby, C.R., Walsh, E.K. (1982) *Macromolecules* **15**, 1093.
Andreoni, W. (Ed.), *The Chemical Physics of Fullerenes 10 (and 5) Years Later*, Kluwer Academic Publishers, Netherlands, 1996.
Aoi, K., Itoh, K., Okada, M. (1997) *Macromolecules* **30**, 8072.
Arbogast, J.W., Foote, C.S. (1991) *J. Am. Chem. Soc.*, **113**, 8886.
Arce, M.-J., Viado, A., An, Y.-Z., Khan, S.I., Rubin, Y. (1996) *J. Am. Chem. Soc.* **118**, 3775.
Astruc, D., Chardac, F. (2001) *Chem. Rev.* **101**, 2991.
Averdung, J., Mattay, M. (1996) *Tetrahedron Lett.* **52**, 5407.
Baggott, J. (1994) *Perfect Symmetry, The Accidental Discovery of Buckminsterfullerene*, Oxford University Press, Oxford.
Balch, A.L., Costa, D.A., Noll, B.C., Olmstead, M.M. (1995) *J. Am. Chem. Soc.* **117**, 8926.

Balch, A.L., Ginwalla, A.S., Olmstead, M.M., Herbst-Irmer, R. (1996) *Tetrahedron* **52**, 5021.
Baum, (1997) *C&E News*, **January 6**, 29.
Benincori, T., Brenna, E., Sannicoló, F., Trimarco, L., Zotti, G., Sozzani, P. (1996) *Angew. Chem. Int. Engl.* 35, 648.
Bergbreiter, D.E., Gray, H.N. (1993) *J. Chem. Soc., Chem. Commun.* 645.
Bethune, D.S., Johnson, R.D., Salem, J.R., de Vires, M.S., Yannoni, C.S. (1993) *Nature* **366**, 123.
Birkett, P.R., Avent, A.G., Darwish, A.D., Kroto, H.W., Taylor, R., Walton, D.R.M. (1995) *J. Chem. Soc., Chem. Commun.* 1869.
Blau, W.J., Byrne, H.J., Cardin, D.J., Hare, J.P., Kroto, H.W., Taylor, R., Walton, D.R.M. (1991) *Phys. Rev. Lett.* 67, 1423.
Boltalina, O.V., Abdul-Sada, A.K., Taylor, R. (1995) *J. Chem. Soc., Perkin Trans.* **2**, 981.
Boulas, P.L., Zuo, Y., Echegoyen, L. (1996) *J. Chem. Soc., Chem. Commun.* 1547.
Buhleier, E., Wehner, W., Vögtle, F. (1978) *Synthesis* **2**, 155.
Bunker, C.E., Lawson, G.E., Sun, Y.-P. (1995) *Macromolecules* **28**, 3744.
Camp, A.G., Lary, A., Ford, W.T. (1995) *Macromolecules* **28**, 7959.
Cao, T., Webber, S.E. (1995) *Macromolecule*s 28, 3741.
Chen, Y., Huang, Z.-E., Cai, R.-F. (1996) *J. Polym. Sci., Part B: Polym. Phys.* **34**, 631.
Chen, Y., Huang, W.-S., Huang, Z.-E., Cai, R.-F., Yu, H.-K., Chen, S.M., Yan, X.-M. (1997) *Eur. Polym. J.* **33**, 823.
Chiang, L.Y., Bhonsie, J.B., Wang, L., Shu.F., Chang, T.M., Hwu, J.R. (1996) *Tetrahedron* **52**, 4963.
Chiang, L.Y., Upasani, R.B., Swirczewski, J.W., Sold, S. (1993) *J. Am. Chem. Soc.* **115**, 5453.
Chiang, L.Y., Wang, L.-Y., Swirczewski, J.W., Soled, S., Cameron, S. (1994) *J. Org. Chem.* **59**, 3960, and references cited therein.
Constable, E.C., Harverson, P., Oberholzer, M. (1996) *J. Chem. Soc., Chem. Commun.* 1821.
Cravino, A., Zerza, G., Maggini, M., Bucella, S., Svensson, M., Andersson, M.R., Neugebauer, H., Sariciftci, N.S. (2000) *J. Chem. Soc., Chem. Commun.* 2487.
Dagani, R. (1996) *C&E News* **June 3**, 30.
Dai, L. (1992) *J. Phys. Chem.* **96**, 6469.
Dai, L., Lu, J., Matthews, B., Mau, A.W.H. (1998a) *J. Phys. Chem. B* **102**, 4049, and references cited therein.
Dai, L., Mau, A.W.H., Griesser, H.J., Spurling, T., White, J.W. (1995) *J. Phys. Chem.* **99**, 17302.
Dai, L., Mau, A.W.H., Griesser, H.J., Winkler, D.A. (1994) *Macromolecules* **27**, 6728.
Dai, L., Mau, A.W.H., Zhang, X. (1998b) *J. Mater. Chem.* **8**, 325.
Dai, L., White, J.W. (1991) *Polymer* **32**, 2120.
Dai, L., White, J.W. (1993) *J. Polym. Sci., Part B, Polym. Phys.* **31**, 3, and references cited therein.
Dai, L., Winkler, B., Huang, S., Mau, A.W.H. In *Semiconducting Polymers: Applications, Properties, and Synthesis*, Hsieh, B., Wei, Y. (eds), ACS Symp. Ser. 735, ACS, Washington, DC, 1999.
Darwish, A.D., Kroto, H.W., Taylor, R., Walton, D.R.M. (1994) *J. Chem. Soc., Chem. Commun.* 15.
David, W.I.F., Ibberson, R.M., Matthewman, J.C., Prassides, K., Dennis, T.J., Hare, J.P., Kroto, H., Taylor, R., Walton, D.R.M. (1991) *Nature* **353**, 147.
Denkewalter, R.G., Kolc, J.F., Lukasavage, W.J. (1981) US Patent 4,289,872.
Denkewalter, R.G., Kolc, J.F., Lukasavage, W.J. (1983) US Patent 4,410,688.
Diederich, F., Isaacs, L., Philp, D. (1994) *Chem. Soc. Rev.* **23**, 243, and references cited therein.
Duclos, S.J., Brister, K., Haddon, R.C., Kortan, A.R., Thiel, F.A. (1991) *Nature* **351**, 380, 1991.

Dvornic, P.R., Tomalia, D.A. (1996) *Curr. Opin. Colloid Interf. Sci.* **1**, 221.
Eckert, J.-F., Nicoud, J.-F., Nierengarten, J.-F., Liu, S.-G., Echegoyen, L., Barigelletti, F., Armaroli, N., Ouali, L., Krasnikov, V., Hadziioannou, G. (2000) *J. Am. Chem. Soc.* **122**, 7467.
Ederlé, Y., Mathis, C. (1997) *Macromolecules* **30**, 4262.
Ederlé, Y., Reibel, D., Mathis, C. (1996) *Synth. Met.* **77**, 139.
Eiermann, M., Wudl, F., Prato, M., Maggini, M. (1994) *J. Am. Chem. Soc.* **116**, 8364.
Elicker, T.S., Evans, D.G. (1999) *J. Phys. Chem. A* **103**, 9423.
Fagan, P.J., Krusic, P.J., Evans, D.H., Lerke, S., Johnston, E. (1992) *J. Am. Chem. Soc.* **114**, 9697.
Ferraris, J.P., Yassar, A., Loveday, D.C., Hmyene, M. (1998) *Opt. Mater.* **9**, 34.
Ford, W.T., Graham, T.D., Mourey, T.H. (1997) *Macromolecules* **30**, 6422.
Foote, C.S. (1994) *Top. Curr. Chem.* **169**, 347.
Fréchet, J.M.J., Hawker C.J. (1996) In *Comprehensive Polymer Science*, 2nd Supplement, Allen, G. (Ed.), Pergamon, Elsevier Sci., Oxford.
Fréchet, J.M.J., Jiang, Y., Hawker, C.J., Philippides, A.E. (1989) *Proc. IUPAC Int. Symp. Macromol. Seoul* 19.
Freemantle, M. (1999) *C&E News* **November 1**, 27
Frey, H., Lorenz, K., Mûlhaupt, R. (1996) *Macromol. Symp.* **102**, 19.
Geckeler, K.E., Hirsch, A. (1993) *J. Am. Chem. Soc.* **115**, 3850.
Gorman, C.B. (1997) *Adv. Mater.* **9**, 1117.
Gorman, C.B., Smith, J.C., Hager, M.W., Parkhurst, B.L., Sierzputowska-Gracz, H., Haney, C.A. (1999) *J. Am. Chem. Soc.* **121**, 9958.
Grösser, T., Prato, M., Lucchini, V., Hirsch, A., Wudl, F. (1995) *Angew. Chem. Int. Ed.* **34**, 1343.
Guhr, K.I., Greaves, M.D., Rotello, V.M. (1994) *J. Am. Chem. Soc.* **116**, 5997.
Haddon, R.C. (1988) *Acc. Chem. Res.* **21**, 243.
Haddon, R.C. (1992) *Acc. Chem. Res.* **25**, 127, and other articles in the special issue on buckminsterfullerenes.
Haddon, R.C. (1993) *Science* **261**, 1545.
Hawhins, J.M., Meyer, A., Lewis, T.A., Bunz, U., Nunlist, R., Ball, G.E., Ebbesen, T.W., Tanigaki, K. (1992) *J. Am. Chem. Soc.* **114**, 7954.
Hawker, C.J. (1994) *Macromolecules* **27**, 4836.
Hawker, C.J., Farrington, P.J., Mackay, M.E., Wooley, K.L., Fréchet, J.M.J. (1995) *J. Am. Chem. Soc.* **117**, 4409.
Hawker, C.J., Fréchet, J.M.J. (1990a) *J. Chem. Soc., Chem. Commun.* 1010.
Hawker, C.J., Fréchet, J.M.J. (1990b) *J. Am. Chem. Soc.* **112**, 7638.
Hawker, C.J., Fréchet, J.M.J. (1990c) *Macromolecules* **23**, 4726.
Hawker, C.J., Fréchet, J.M.J. (1992) *J. Am. Chem. Soc.* **114**, 8405.
Hawker, C.J., Frechet, J.M.J. in (1995) *New Methods of Polymer Synthesis, Vol. 2*, Chapman and Hall Publishers, New York.
Hawker, C.J., Wooley, K.L., Fréchet, J.M.J. (1993) *J. Chem. Soc., Perkin Trans 1*, 1287.
Hawker, C.J., Wooley, K.L., Fréchet, J.M.J. (1994) *Macromolecules* **77**, 11.
Henderson, C.C., Rohlfing, C.M., Assink, R.A., Cahill, P.A. (1994), *Angew. Chem. Int. Ed.* **33**, 786.
Hirsh, A. (1993) *Adv. Mater.* **5**, 859.
Hirsch, A. (1994) *The Chemistry of the Fullerenes*, Thieme Verlag, Stuttgart.
Hirsch, A., Soi, A., Karfunkel, H.R. (1992) *Angew. Chem. Int. Ed.* **31**, 766.
Hoke II, S.H., Molstad, J., Dilettato, D. Jay, M.J., Carlson, D., Kahr, B., Cooks, R.G. (1992) *J. Org. Chem.* **57**, 5069.
Hummelen, J.C., Prato, M., Wudl, F. (1995) *J. Am. Chem. Soc.* **117**, 7003.
Jagerovic, N., Elguero, J., Aubagnac, J.L. (1996) *J. Chem. Soc., Perkin Trans. 1*, 499.

Jansen, J.F.G.A., De Brabander-Van Den Berg, E.M.M., Meijer, E.W. (1995a) *J. Am. Chem. Soc.* **117**, 4417.
Jansen, J.F.G.A., Meijer, E.W., De Brabander-Van Den Berg, E.M.M. (1995b) *Science* **266**, 1226.
Janssen, R.A.J., Hummelen, J.C., Wudl, F. (1995) *J. Am. Chem. Soc.* **117**, 544.
Kawa, M., Fréchet, J.M.J. (1998) *Chem. Mater.* **10**, 286.
Kawaguchi, T., Walker, K.L., Wilkins, C.L., Moore, J.S. (1995) *J. Am. Chem. Soc.* **117**, 2159.
Keshavarz-K, M., Knight, B., Srdanov, G., Wudl, F. (1995) *J. Am. Chem. Soc.* **117**, 11371.
Kojima, Y., Matsuoka, T., Takahashi, H., Kurauchi, T. (1995) *Macromolecules*, **28**, 8868.
Koruga, D., Hameroff, S., Withers, J., Loutfy, R., Sundareshan, M. (1993) *Fullerene C_{60}: History, Physics, Nanobiology, Nanotechnology*, Elsevier Science Publishers, Amsterdam.
Krätschmer, W., Lamb, L.D., Fostiropoulos, K., Huffman, D.R. (1990) *Nature* **347**, 354.
Kroto, H.W., Allaf, A.W., Balm, S.P. (1991) *Chem. Rev.* **91**, 1213.; (1993) *C&E News*, **January 4**, 29.
Kroto, H.W., Heath, J.R., O'Brien, S.C., Curl, R.F., Smalley, R.E. (1985) *Nature* **318**, 162.
Kusukawa, T., Ando, W. (1996) *Angew. Chem. Int. Ed.* **35**, 13151.
Lin, T., He, Q., Bai, F., Dai, L. (2000) *Thin Solid Films* **363**, 122.
Liou, K.-F., Cheng, C.-H. (1995) *J. Chem. Soc., Chem. Commun.* 2473.
Lu, L., Dai, L., Mau, A.W.H. (1998) *Acta Polym.* **49**, 371, and references cited therein.
Maciejewski, M. (1982) *J. Macromol. Sci. – Chem.* **A17**, 689.
Marsh, H. (1989) *Introduction to Carbon Science*, Butterworth & Co. (Publishers) Ltd, London.
Meier, M.S. (2000) in *Fullerene Polymers and Fullerene Polymer Composites*, Eklund, P.C., Rao, A.M. (eds), Springer-verlag, Berlin.
Meier, M.S., Poplawska, M. (1996) *Tetrahedron* **52**, 5043.
Menon, M., Subbaswamy, K.R., Sawtarie, M. (1994) *Phys. Rev. B* **49**, 13966.
Miller, T.M., Neenan, T.X. (1990) *Chem. Mater.* **2**, 346.
Mishra, M.K., Kobayashi, S. (eds) (1999) *Star and Hyperbranched Polymers*, Marcel Dekker, New York.
Muthu, S., Maruthamuthu, P., Ragunatha, R., Vasudeva Rao, P.R., Mathews, C.K. (1994) *Tetrahedron Lett.* **35**, 1763.
Naylor A.M., Goddard, W.A. III, Kiefer, G.E., Tomalia, D.A. (1989) *J. Am. Chem. Soc.* **111**, 2339.
Newkome, G.R. (Ed.) (1994) *Advances in Dendritic Macromolecules*, Vol. 1, JAI Press Inc, Connecticut.
Newkome, G.R., Moorefield, C.N., Baker, G.R. (1992) *Aldrichim. Acta* **25**, 31.
Newkome, G.R., Moorefield, C.N., Baker, G.R., Saunders, M.J., Grossman, S.H. (1991) *Angew. Chem. Int. Ed.* **30**, 1178.
Newkome, G.R., Moorefield, C.N., Vögtle, F. (1996) *Dendritic Molecules: Concepts, Synthesis, Perspectives*, VCH, Weinheim.
Newkome, G.R., Yao, Z., Baker, G.R., Gupta, V.K. (1985) *J. Org. Chem.* **50**, 2003.
Nie, B., Hasan, K., Greaves, M.D., Rotello, V.M. (1995) *Tetrahedron Lett.* **36**, 3617.
Nierengarten, J.-F., Eckert, J.-F., Nicoud, J.-F., Ouali, L., Krasnikov, V., Hadziioannou, G. (1999) *Chem. Commun.*, 617.
Noller, C.R. (1966) *Textbook of Organic Chemistry*, W.B. Saunders Company, Philadelphia.
Núñez-Regueiro, M., Marques, L., Hodeau, J.-L., Béthoux, O., Perroux, M. (1995) *Phys. Rev. Lett.* **74**, 278.
Núñez-Regueiro, M., Marques, L., Hodeau, J.L., Xu, C.H., Scuseria, G.E. (1999) In *Fullerene Polymers and Fullerene Polymer Composites*, Eklund, P.C., Rao, A.M. (eds), Springer-verlag, Berlin.
Ohno, M., Azuma, T., Kojima, S., Shirakawa, Y., Eguchi, S. (1996a) *Tetrahedron* **52**, 4983, and references cited therein.

Ohno, M., Yashiro, A., Eguchi, S. (1996b) *J. Chem. Soc., Chem. Commun.* 291.
Okamura, H., Minoda, M., Komatsu, K., Miyamoto, T. (1997) *Macromol. Chem. Phys.* **198**, 777.
O'Keeffe, M. (1991) *Nature* **352**, 674.
Olah, G., Bucsi, J., Lambert, C., Amiszfeld, R., Trivedi, N., Sensharma, D.K., Prakash, G.K.S. (1991) *J. Am. Chem. Soc.* **113**, 9387.
Paquette, L.A., Graham, R.J. (1995) *J. Org. Chem.* **60**, 2958.
Patil, A.O., Schriver, G.W. (1995) *Macromol. Symp.* **91**, 73.
Prasad, P.N., Nigam, J.K. (eds) (1991) *Frontiers of Polymer Research*, Plenum Press, New York,.
Prato, M. (1997) *J. Mater. Chem.* **7**, 1097.
Prato, M., Li, Q.C., Wudl, F., Lucchini, V. (1993a) *J. Am. Chem. Soc.* **115**, 1148.
Prato, M., Maggini, M., Scorrano, G., Lucchini, V. (1993b) *J. Org. Chem.* **58**, 3613.
Prato, M., Suzuki, T., Foroudian, H., Li, Q., Khemani, K., Wudl, F., Leonetti, J. Little, R.D., White, T., Rickborn, B., Yamago, S., Nakamura, E. (1993c) *J. Am. Chem. Soc.* **115**, 1594.
Qu, B., Hawthorn, G., Mau, A.W.H., Dai, L. (2001) *J. Phys. Chem.* **105**, 2129.
Qiu, H., Wan, M., Matthews, B., Dai, L. (2001) *Macromolecules* **34**, 675.
Rannard, S.P., Davis, N.J. (2000) *J. Am. Chem. Soc.* **122**, 11729.
Rao, A.M., Zhou, P., Wang, K.-A., Hager, G.T., Holden, J.M., Wang, Y., Lee, W.-T., Bi, X.-X., Eklund, P.C., Cornett, D.S., Duncan, M.A., Amster, I.J. (1993) *Science* **259**, 955.
Samulski, E.T., DeSimone, J.M., Hunt, M.O. Jr., Menceloglu, Y.Z., Jarnagin, R.C., York, G.A., Labat, K.B., Wang, H. (1992) *Chem. Mater.* **4**, 1153.
Saunders, M., Cross, R.J., Jiménez-Vásquez, Shimshi, R., Khong, A. (1996) *Science* **271**, 1693.
Schneider, N., Darwish, A.D., Kroto, H.W., Taylor, R., Walton, D.R.M. (1994) *J. Chem. Soc., Chem. Commun.* 463.
Shi, S., Khemani, K.C., Li, Q., Wudl, F. (1992) *J. Am. Chem. Soc.* **114**, 10656.
Smalley, R.E. (1992) *Acc. Chem. Res.* **25**, 98.
Stinson, S.C. (1997) *C&E News* **September 22**, 28.
Suzuki, T., Li, Q., Khemani, K.C., Wudl, F., Almarsson, Ö. (1991) *Science* **254**, 1186.
Tajima, Y., Tezuka, Y., Yajima, H., Ishii, T. (1997) *Polymer* **38**, 5255.
Takahashi, N., Dock, H., Matzuzawa, N., Ata, M. (1993) *J. Appl. Phys. Lett.* **74**, 5790.
Taliani, C., Ruani, G., Zamboni, R., Danieli, R., Orlandi, F., Zerbetto, F. (1993) *J. Chem. Soc., Chem. Commun.* 220.
Taylor, R., (Ed.) (1995) *The Chemistry of Fullerenes*, World Scientific, Singapore.
Taylor, R. (1991) *Nature* **355**, 27.
Taylor, R., Walton, D.R.M. (1993) *Nature* **363**, 685. Fischer, J.E. (1994) *Science* **264**, 1548.
Thilgen, C., Herrmann, A., Diederich, F. (1997) *Angew. Chem. Int. Ed.* **36**, 2268.
Tokkuyama, H., Isobe, H., Nakamura, E. (1994) *J. Chem. Soc., Chem. Commun.* 2753.
Tomalia, D.A. (1993) *Aldrichim. Acta* **26**, 91.
Tomalia, D.A., Baker, H., Dewald, J., Hall, M., Kallos, G., Roeck, J., Ryder, J., Smith, P. (1985) *Polym. J.* **17**, 117.
Tomalia, D.A., Dvornic, P.R. in (1996) *Polymeric Materials Encyclopedia*, Salamone, J.C. (Ed), CRC Press, Boca Raton, FL.
Tomalia, D.A., Durst, H.D. (1993) *Top. Curr. Chem.* **165**, 193.
Tomalia, D.A., Swanson, D.R., Klimaash, J.W., Brothers, H.M. (1993) ACS *Polym. Prepr.* **34**, 52.
Torres-García, G., Mattay, J. (1996) *Tetrahedron* **52**, 5421, and references cited therein.
Tsuda, M., Ishida, T., Nogami, T., Kurono, S., Ohashi, M. (1992) *Chem. Lett.* 2333.
Tsuda, M., Ishida, T., Nogami, T., Kurono, S., Ohashi, M. (1993) *Tetrahedron. Lett.* **34**, 6911.

Tuinman, A.A., Gakh, A.A., Adcock, J.L., Compton, R.N. (1993) *J. Am. Chem. Soc.* **115**, 5885.
Wang, Y., Cheng, L.-T. (1992) *J. Phys. Chem.*, **96**, 1530.
Wei, Y., Tian, J., MacDiarmid, A.G., Masters, J.G., Smith, A.L., Li, D. (1993) *J. Chem. Soc., Chem. Commun.* 603, and references cited therein.
Weis, C., Friedrich, C., MÜlhaupt, R., Frey, H. (1995) *Macromolecules* **28**, 403.
Wendland, M.S., Zimmerman, S.C. (1999) *J. Am. Chem. Soc.* **121**, 1389.
Wilson, S.R., Kaprinidis, N.A., Wu, Y., Schuster, D.I. (1993) *J. Am. Chem. Soc.* **115**, 8495.
Wooley, K.L., Hawker, C.J., Fréchet, J.M.J. (1991a) *J. Am. Chem. Soc.* **113**, 4252.
Wooley, K.L., Hawker, C.J., Fréchet, J.M.J. (1991b) *J. Chem. Soc., Perkin Trans 1*, 1059.
Wooley, K.L., Hawker, C.J., Fréchet, J.M.J., Wudl, F., Srdanov, G., Shi, S., Li, C., Kao, M. (1993) *J. Am. Chem. Soc.* **115**, 9836.
Wudl, F. (1992) *Acc. Chem. Res.* **25**, 106.
Wudl, F. in *Buckminsterfullerenes*, Billups, W.E., Ciufolini, M.A. (eds), (1993) VCH, New York.
Wudl, F., Hirsch, A., Khemani, K.C., Suzuki, T., Allemand, P.-M., Koch, A., Eckert, H., *Clusters*, G., Hammond, G.S., Kuck, V.J., (eds), (1992) ACS, Washington, DC.
Xu, Z.F., Moore, J.S. (1993a) *Angew. Chem. Int. Ed.* **32**, 246.
Xu, Z.F., Moore, J.S. (1993b) *Angew. Chem. Int. Ed.* **32**, 1354.
Xu, C.H., Scuseria, G. (1995) *Phys. Rev. Lett.* **74**, 274.
Yang, L., Dorsinville, R., Alfano, R. (1994) *Chem. Phys. Lett.* **226**, 605.
Zhang, X., Fan, A., Foote, C.S. (1996) *J. Org. Chem.* **61**, 5456.
Zhang, N., Schricker, S.R., Wudl, F., Prato, M., Maggini, M., Scorrano, G. (1995) *Chem. Mater.* **7**, 441.
Zhao, Y.B., Poirier, D.M., Pechman, R.J., Weaver, J.H. (1994) *Appl. Phys. Lett.* **64**, 577.
Zhou, P., Rao, A.M., Wang, K.-A., Robertson, J.D., Eloi, C., Meier, M.S., Ren, S.L., Bi, X.X., Eklund, P.C., Dresselhaus, M.S. (1992) *Appl. Phys. Lett.* **60**, 2871.

Chapter 5

Carbon Nanotubes

5.1 Introduction

Just as the discovery of C_{60} has created an entirely new branch of carbon chemistry (Chapter 4), the subsequent discovery of carbon nanotubes by Iijima (1991) has opened up a new era in material science and nanotechnology (Harris, 2001). Carbon nanotubes consist of carbon hexagons arranged in a concentric manner with both ends normally capped by fullerene-like structures containing pentagons. They usually have a diameter ranging from a few Ångstrom to tens of nanometers and a length of up to several centimeters (Harris, 2001; Ajayan *et al.*, 2002). There are two distinct types of carbon nanotubes. The so-called single-walled carbon nanotube (*i.e.* SWNT or graphene tube) is made of one layer of graphene sheet, while the multi-walled carbon nanotube (*i.e.* MWNT or graphitic tube) consists of more than one layer (Figure 5.1).

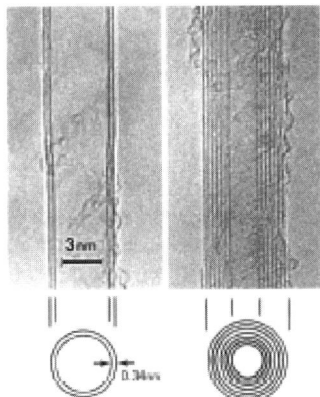

Figure 5.1. High-resolution TEM micrographs of a cross-section of two MWNTs consisting of two graphitic sheets (*left*) and seven graphitic sheets with a diameter of 6.5 nm and a hollow diameter of 2.2 nm (*right*) (After Iijima, 1991, copyright 1991 Macmillan Magazines Limited. Reproduced with permission)

158 Intelligent Macromolecules for Smart Devices

The typical SWNT can be viewed as a fullerene molecule that has been bisected at the equator and the two resulting hemispheres have been joined with a monolayer graphene tube of the same diameter as C_{60}. Because the graphene sheet can be rolled up with varying degrees of twist along its length, SWNTs can have a variety of chiral structures (Harris, 2001). The so-called "armchair" nanotube (Figure 5.2(a)) corresponds to the bisection of the C_{60} molecule along the direction normal to a fivefold axis, whereas a "zigzag" nanotube (Figure 5.2(b)) is formed if the C_{60} molecule is bisected normal to a threefold axis. The so-called "chiral" nanotubes (Figure 5.2(c)) can be produced with other varieties of "hemispherical"-like caps.

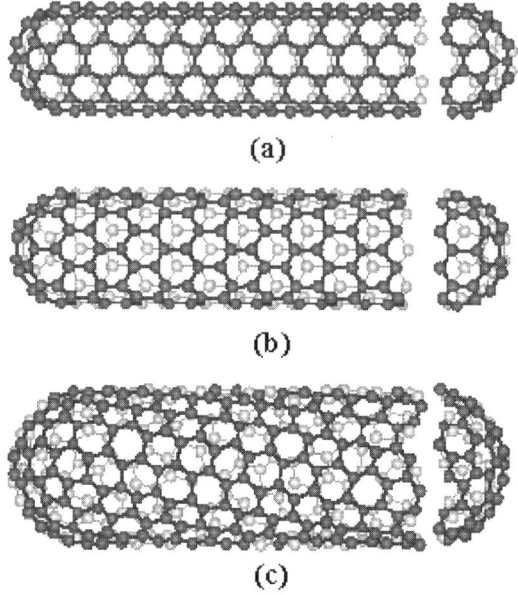

Figure 5.2. Schematic representation of SWNTs: (a) [5,5] armchair nanotube; (b) [9,0] Zigzag nanotube; (c) [10,5] chiral nanotube (Dresselhaus *et al.*, 1996, copyright 1996 Academic Press)

Their peculiar hollow geometry, coupled with a *conjugated* all-carbon structure, has allowed carbon nanotubes to exhibit many new intriguing electrical, mechanical and thermal properties with respect to those more conventional members of the carbon family. In particular, carbon nanotubes can exhibit semiconducting or metallic behavior depending on their diameter and helicity of the arrangement of graphitic rings in the walls. These special physicochemical properties, in turn, facilitate the controlled structural modifications of carbon nanotubes, leading to the formation of various advanced composite materials with, for example, appropriate macromolecules for many potential applications. While the applications of carbon nanotubes and their aligned/micro-patterned derivatives will be discussed in succeeding chapters as becomes appropriate, the important issues concerning their syntheses, chemistry, and properties are the subject of the present chapter.

5.2 Structure

In general, carbon nanotubes can be specified mathematically in terms of the unit cell, the smallest group of atoms that defines its structure (Figure 5.3(a)), through the so-called chiral vector of the nanotube, C_h, and the chiral angle, θ. The C_h is given by (Saito et al., 1992; Ajayan and Ebbesen, 1997; Dresselhaus et al., 1998):

$$C_h = na_1 + ma_2 \qquad (5.1)$$

where, a_1 and a_2 are unit vectors in the two-dimensional hexagonal lattice and n and m are integers. The chiral angle, θ, is then the angle between C_h and a_1. When the graphene sheet is rolled up to form the cylindrical part of the nanotube, the chiral vector forms the circumference of the nanotube's circular cross-section with its ends meeting each other. Different pairs of integers (n, m) define a different way of rolling the graphene sheet to form a carbon nanotube (Figure 5.3(b)). For example, armchair nanotubes are formed when $n = m$ and the chiral angle is 30°, whereas zigzag nanotubes correspond to either n or $m = 0$ and $\theta = 0$. All other chiral nanotubes have the chiral angles intermediate between 0° and 30°. Depending on their diameter and the helicity of the arrangement of graphitic rings, carbon nanotubes can exhibit semiconducting or metallic behavior.

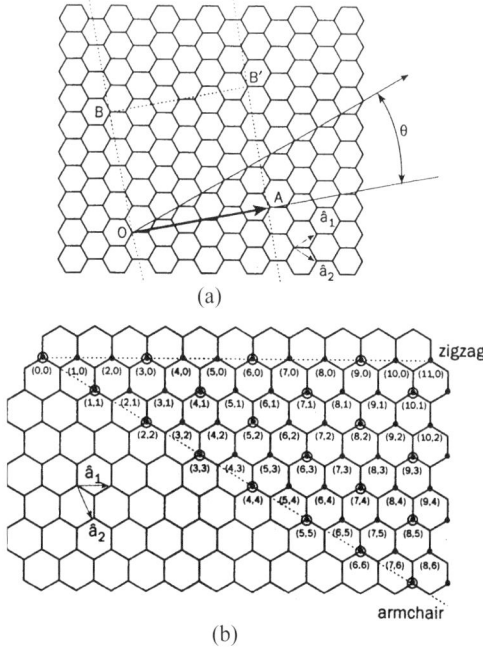

Figure 5.3. (a) Schematic of a two-dimensional graphene sheet illustrating lattice vectors a_1 and a_2, and the roll-up vector $C_h = na_1 + na_2$; (b) possible vectors specified by the pairs of integers [n, m] for general carbon nanotubes, including zigzag, armchair and chiral tubules. The encircled dots denote metallic tubules while the small dots are for semiconducting tubules (After Saito et al., 1992, copyright 1992 American Institute of Physics)

The properties of carbon nanotubes are determined by their diameter, d_t, and chiral angle, θ, which in turn depend on n and m as follows (Figure 5.2) (Odom et al., 2000):

$$d_t = C_h/\pi = 3^{1/2} a_{c-c} (m^2 + mn + n^2)^{1/2}/\pi \qquad (5.2)$$

$$\theta = tan^{-1}[3^{1/2} m/(m + 2n)] \qquad (5.3)$$

where a_{c-c} is the distance between neighboring carbon atoms in the flat sheet (1.421 Å in graphite). Theoretical calculations have indicated that all armchair nanotubes are metallic, as are one third of all possible zigzag nanotubes depending on their diameter and chiral angle. Generally speaking, an (n, m) carbon nanotube will be metallic if $2n + m = 3q$ with q being an integer (Satio R. et al., 1992). It has also been predicted that the presence of a magnetic field will strongly affect the band structure of carbon nanotubes near their Fermi level (Ajiki and Endo, 1993).

5.3 Property

Although theoretical studies have been focusing on SWNTs, almost all of the early experimental work was done on MWNTs. As mentioned earlier, the experimental discovery of SWNTs in 1993 (Bethune et al., 1993; Iijima and Ichihashi, 1993) did not change the situation too much owing to the lack of structural uniformity in these samples. The subsequent breakthrough in the synthesis of SWNTs with high yields and structural uniformity (Thess et al., 1996) facilitates the structure and property measurements and allows a direct link of experimental data to theoretical simulations.

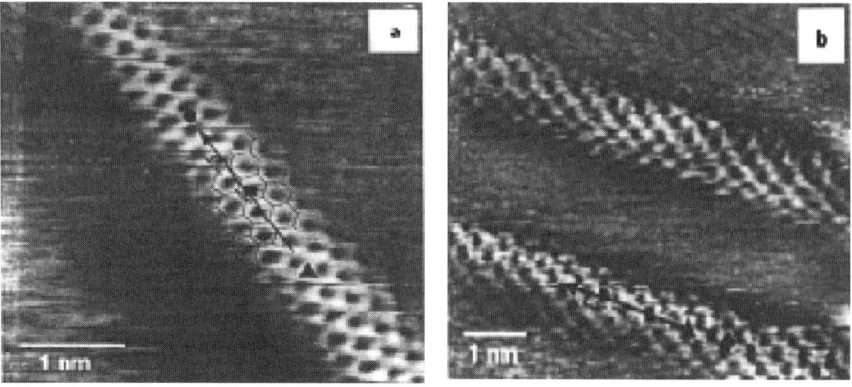

Figure 5.4. STM Images of (a) a metallic SWNT exposed at the surface of a rope; (b) isolated SWNTs on a Au (111) substrate (After Odom et al., 1998, copyright 1998 Macmillan Magazines Limited. Reproduced with permission)

Several groups (Avouris, 2002; Ando et al., 1998; Chen Y. et al., 1996; Ebbesen, 1994; Tans et al., 1997; Odom et al., 1998; Wildöer et al., 1998) have recently measured, by both STM and scanning tunneling spectroscopy (STS), the diameter and helicity of individual SWNTs prepared by Smalley's group (Figure 5.4) and reported experimental conformation for the theoretical predication that carbon nanotubes can be either metallic or semiconducting depending on their diameter and helicity (Satio R. *et al.*, 1992). They have also demonstrated experimentally an inverse linear dependence of the bandgap on the nanotube diameter with a theoretically predicated coefficient of proportionality (Jishi *et al.*, 1994). Other groups have also reported STM/STS studies of carbon nanotubes but not with this level of detail and precision (Ge and Sattler, 1994; Hubler *et al.*, 1998; Lin *et al.*, 1996).

In addition to the above-mentioned STM/STS studies, a few different approaches have been reported for experimentally measuring the electrical properties of individual nanotubes. For example, de Heer and coworkers (Frank *et al.*, 1998) have measured the conductance of MWNTs by replacing the tip of a scanning probe microscope with the nanotube, which can form an electrical circuit with a liquid metal (Chapter 11). Besides, several groups have measured electrical properties of a single nanotube and/or a nanotube bundle by lithographically depositing contact electrodes across them (Figure 5.5) (Bockrath *et al.*, 1997; Ebbesen *et al.*, 1996; Langer *et al.*, 1996; Tans *et al.*, 1997).

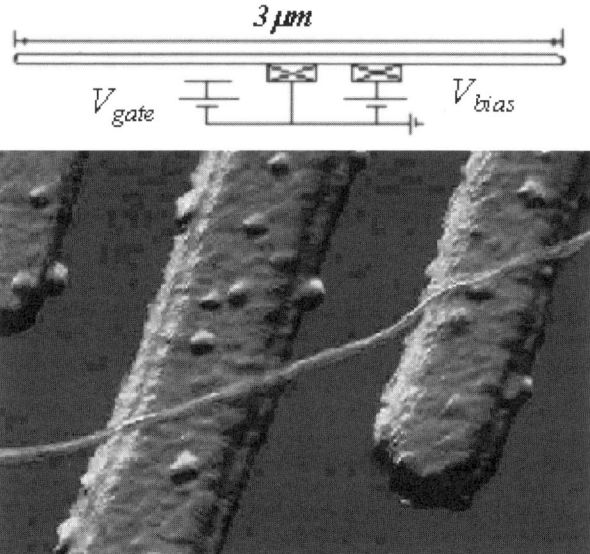

Figure 5.5. AFM tapping-mode image of a carbon nanotube on top of a Si/SiO2 substrate with two 15 nm thick Pt electrodes, and the corresponding circuit diagram (Bockrath *et al.*, 1997; Ebbesen *et al.*, 1996; Langer *et al.*, 1996; Tans *et al.*, 1997, copyright 1997 Macmillan Magazines Limited)

The general application of this method, however, is limited by the uncertainty in size and structure of the individual nanotubes or nanotube bundles in the sample to be measured. A more versatile method has recently been reported by Lieber and coworkers (Dai *et al.*, 1996). These authors started the measurement by depositing a drop of a nanotube suspension on a flat insulating surface, which was subsequently covered with a uniform layer of gold. Then, a pattern of open slots was made in the gold layer by conventional lithographic technique to expose the nanotubes for the measurements (Figure 5.6).

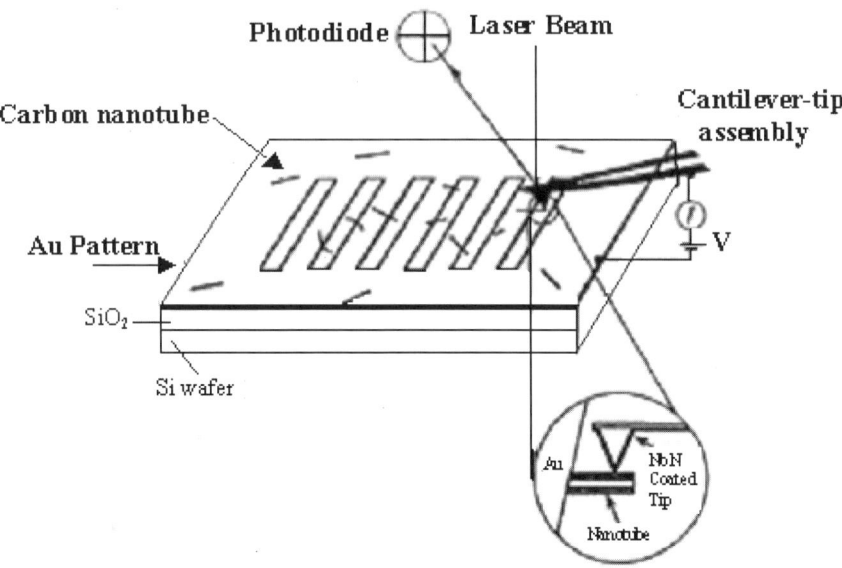

Figure 5.6. Schematic diagram of the measurement of the electrical properties of an individual carbon nanotube (Dai, H. *et al.*, 1996, copyright 1996 American Association for the Advancement of Science)

By so doing, they can measure the axial conduction through a single nanotube to the gold contact using a conducting cantilever-tip of a force microscopy to contact electrically onto the tube end extending into an open slot. The results from this study indicate that structural defects in the nanotube can cause substantial increases in the resistivity. In fact, it has been shown that even a kink in a nanotube may change it from metallic to semiconducting (Chico *et al.*, 1996; Dresselhaus, 1996; Saito *et al.*, 1996; Tamura *et al.*, 1997), suggesting the possibility of building nano-scale electronic devices simply by stretching and bending nanotubes here and there. The recent report on reshaping, moving and cutting individual carbon nanotubes by an atomic force microscopy (AFM) tip may hasten the ongoing trend toward the construction of all-carbon nanoelectronic devices (Hertel *et al.*, 1998; Jacoby, 1998). Similarly, Oekker and coworkers (Venema *et al.*, 1997) have demonstrated the ability to cut individual carbon nanotubes by applying a voltage pulse to the nanotube through an STM tip. These authors found that short nanotubes (30 nm) cut from a longer one showed novel stepwise increases in current with increasing voltage. This was attributable to quantum size effects. In a separate study, Smalley and coworkers (Collins *et al.*, 1997) observed distinct changes in the conductivity as the active length of a nanotube is increased by moving an STM tip along the length of long carbon nanotubes (>2 μm). These results indicate that the on-tube nanodevices could be invaluable building blocks in future nanoelectronic devices.

By extension, Lieber and coworkers (Wong et al., 1997) have been able to measure the mechanical properties of individual carbon nanotubes by AFM using an approach similar to the one described above for the electrical measurements. In this case, they first deposited nanotubes randomly on a flat surface followed by depositing a regular array of square pads to pin down the nanotubes to this surface, and then used the AFM directly to measure the bending force as a function of displacement from the pinning point (Figure 5.7).

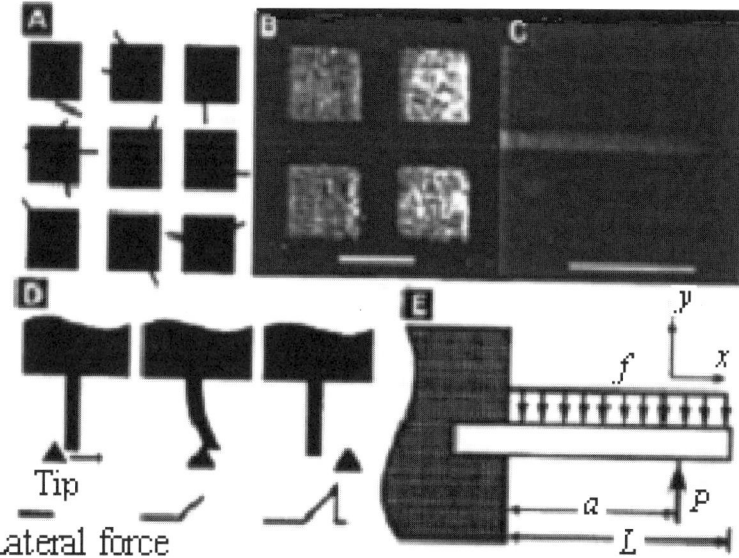

Figure 5.7. Schematic diagram of the measurement of the mechanical properties of an individual carbon nanotube (Wong et al., 1997, copyright 1997 American Association for the Advancement of Science)

This technique provides an unambiguous measure of the stiffness, strength and toughness of nanotubes and/or nanorods. The results thus obtained indicated that the carbon nanotubes exhibited a nearly two times greater Young's modulus than the strongest whisker materials (i.e. SiC nanorods). While continued bending of the SiC nanorods led to fracture, the carbon nanotubes showed an unusal elastic buckling process as predicted by theoretical simulations and confirmed by other workers (Iijima et al., 1996; Yakobson, et al., 1996; Cornwell and Wille, 1997; Falvo et al., 1997). Because of the large decrease in stiffness at the buckling points, the ultimate level of strength of the carbon nanotubes was shown to be less than those of the SiC nanorods. However, the elastically buckling observed for carbon nanotubes at large bending angles enabled them to store or absorb considerable energy. Therefore, carbon nanotubes are of great potential in mechanical applications, especially as energy-absorbing materials. Some of the properties of carbon nanotubes are listed in Table 5.1.

Table 5.1. Properties of carbon nanotubes (After Collins and Avouris, 2000)

Property	SWNTs	By Comparison
 **Size**	0.6–1.8 nm in diameter	Electron beam lithography can create lines 50 nm wide, a few nm thick
 **Density**	1.33–1.40 g/cm^3	Aluminum has a density of 2.7 g/cm^3
 **Tensile Strength**	45 billion pascals	High-strength steel alloys break at about 2 billion pascals
 **Resilience**	Can be bent at large angles and restraightened without damage	Metals and carbon fibers fracture at grain boundaries
 **Current Carrying Capacity**	~ 1 billion amps/cm^2	Copper wires burn out at about 1 million amps/cm^2
 **Field Emission**	Can activate phosphors at 1–3 V if electrodes are spaced 1 μm apart	Molybdenum tips require fields of 50–100 V/μm and have very limited lifetimes
 **Heat Transmission**	Predicted to be as high as 6000 V/m•K at room temperature	Nearly pure diamond transmits 3320 W/m•K
 **Temperature Stability**	Stable up to 2,800 °C in vacuum, 750 °C in air	Metal wires in microchips melt at 600–1000 °C
 **Cost**	$1500/g from Bucky USA in Houston	Gold was selling for *ca.* $10/g in October, 2002

5.4 Synthesis

5.4.1 Multi-wall Carbon Nanotubes (MWNTs)

The large-scale synthesis of MWNTs was first reported by Ebbesen and Ajayan (1992). These authors produced carbon nanotube bundles, along with carbon particles and other disordered carbonaceous material, by operating a carbon arc-discharge generator with a DC current of 50–100 A and voltage of 20–25 V at a discharge temperature above 3000 °C under an inert atmosphere (*e.g.* He). Figure 5.8 gives a typical schematic representation of an arc-discharge generator, which is self-explanatory.

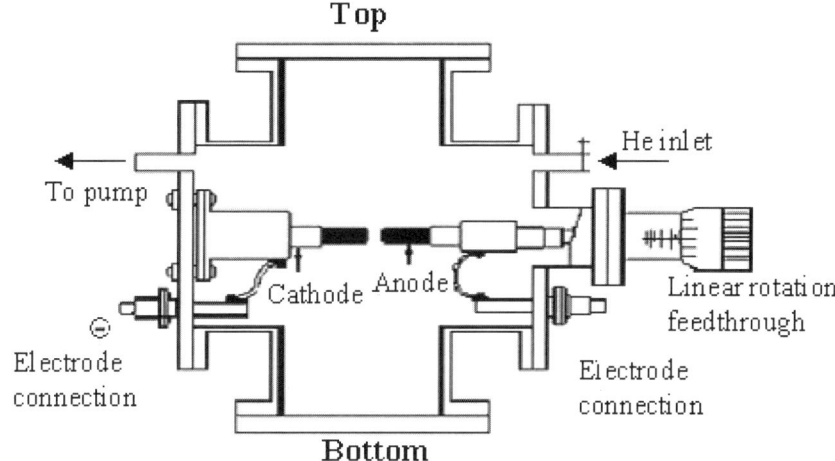

Figure 5.8. Schematic representation of a typical arc-discharge generator (After Satio, Y. *et al.*, 1996, copyright 1996 American Institute of Physics)

Later, it was found that MWNTs could also be produced by the decomposition of hydrocarbons in the presence or absence of metal catalyst(s) (*e.g.* Fe, Co, Ni) at 1100 °C with a much lower gas pressure than that for the preparation of vapor-grown carbon fibers (Endo *et al.*, 1993; Rao and Govindaraj, 2002). In these cases, conventional vapor-grown carbon fibers form at the same time as carbon nanotubes grow. As such, nanotube samples thus prepared often have a poor crystallinity, which, however, can be significantly improved by heating in argon at 2500–3000 °C. The vapor-grown carbon nanotubes exist also in bundles and often show internal bamboo-shaped structures, suggesting the capping-off of the inner layers over the epitaxial growth of the outer layers along the length and tubule diameter (Endo *et al.*, 1993). Compared with the arc-discharge method, the pyrolysis technique has some advantages owing to the simplicity of the process and high yield of carbon nanotubes. Closely related to the pyrolysis method, the preparation of

MWNTs by vaporizing carbon under vacuum with either an electron beam or resistive heating, followed by depositing the carbon vapor on a cold surface, has also been reported (Kosakovskaya *et al.*, 1992; Yamamoto *et al.*, 1996). In addition to the above commonly used methods, electrolysis of molten alkali halide salts using carbon electrodes under an argon atmosphere has recently been reported to yield MWNTs (Hsu *et al.*, 1995; 1996). On the other hand, carbon nanotubes have also been produced by high-temperature graphitizing of the polyacrylonitrile (PAN) nanotubules prepared within the pores of an alumina template membrane or zeolite nanochannels (Parthasarathy *et al.*, 1995; Kyotani *et al.*, 1997a). Furthermore, uniform hollow carbon nanotubes with open ends were generated by the chemical vapor deposition of organic hydrocarbon compounds (*e.g.* ethylene or pyrene) directly within the pores of an alumina template membrane incorporated with or without an Ni catalyst (Che *et al.*, 1998). In these cases, the aluminum oxide template can be dissolved in an aqueous solution of HF at room temperature, leaving pure carbon nanotubes as the product. The most striking feature of the template synthesis is that it allows the production of monodispersed carbon nanotubes with uniform length, diameter and wall thickness (Kyotani *et al.*, 1997b).

5.4.2 Single-wall Carbon Nanotubes (SWNTs)

In 1993, SWNTs were observed for the first time in soot generated by the carbon arc discharge using metal-containing (*e.g.* Fe, Ni, Co or Cu) carbon electrodes (Bethune et al., 1993; Iijima *et al.*, 1993; Saito Y. *et al.*, 1993). To produce the SWNTs, a pure carbon cathode and a carbon anode containing the mixture of transition metal(s) and graphite powder in a hole are normally used. Co-vaporization of carbon and the metal catalyst in the arc generator results in the formation of web-like deposits of SWNTs in the fullerene-containing soot. This allows property characterization for the SWNTs. However, a direct comparison of the experimental results to theoretical simulations for a single nanotube was precluded by the low yields of the SWNTs and the lack of structural uniformity in these samples. It was the efficient synthesis of ordered SWNT bundles by a pulsed laser vaporization of a carbon target containing 1–2% (w/w) Ni/Co in a furnace at 1200 °C, developed by Smalley and co-workers in 1996, that brought the goal of testing the properties of SWNTs against theories within reach (Thess *et al.*, 1996). The cobalt-nickel catalyst used in the pulsed laser vaporization is believed to prevent the ends from being "capped" during the nanotube growth. While a catalyst is not always necessary for MWNT growth, SWNTs can only be grown with a catalyst. Recently, similar arrays of SWNTs have also been synthesized by a refined carbon arc method using Ni/Y (4.2/1 atom %) as the catalyst (Journet *et al.*, 1997).

As an alternative, SWNTs have also been obtained by oxidizing MWNTs in CO_2 at about 850 °C (Tsang *et al.*, 1993) and in air at 700–800 °C (Ajayan *et al.*, 1993). Since the tips of nanotubes consist of reactive five-membered carbon rings and have a higher curvature with higher strain than their cylindrical part, they are more reactive and etched away first, which is followed by a slower layer-by-layer removal of the cylindrical layers until the SWNTs are observed. This method presents both difficult challenges in controlling the experimental conditions and rewarding outcomes, as nanotubes with predetermined layer numbers may be produced at will.

5.5 Purification

The presence of carbon nanoparticles and other impurities in most of the *as-synthesized* carbon nanotubes (typically, 30–70% w/w) (Chen Y. *et al.*, 1996a; Ebbesen and Ajayan, 1992) is a serious impediment to detailed structural and property characterization. Therefore, purification for retaining only the nanotubes becomes essential to the further development of science and technology involving carbon nanotubes. Up to now, the most commonly used purification method is the oxidation of nanotube samples either in a gas (Ando *et al.*, 1998; Chen Y. *et al.*, 1996b; Ebbesen, 1994) or liquid phase (Hiura *et al.*, 1995; Tsang *et al.*, 1994). The nanoparticles, like the tips of nanotubes, are more prone to oxidation and can be first burnt away from the nanotubes in air above 700 °C (Ajayan *et al.*, 1993; Ebbesen *et al.*, 1994; Tsang *et al.*, 1993), and the remaining pure (opened) nanotubes can then be separated from each other by sonication in solvents (*e.g.* ethanol) (Figure 5.9).

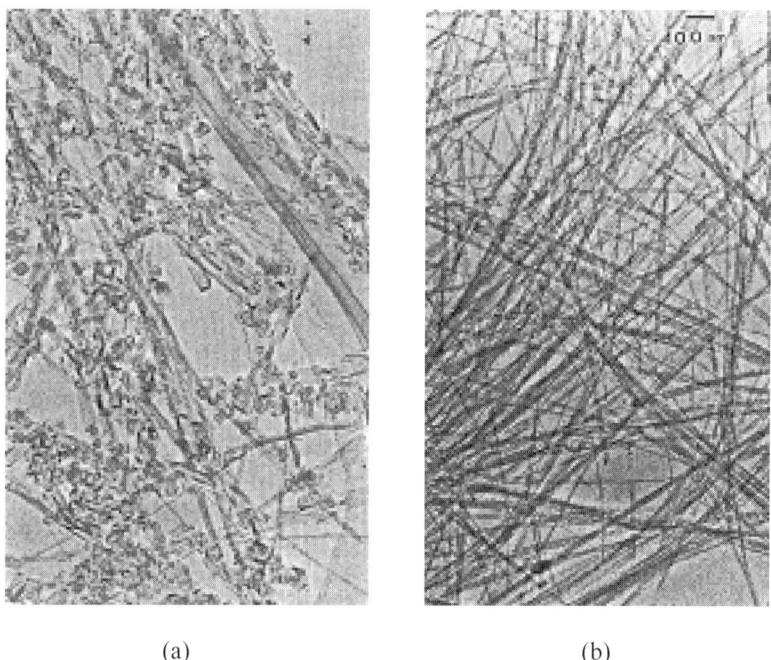

(a)   (b)

Figure 5.9. (a) A typical nanotube sample produced by the carbon-arc method; (b) a purified open carbon nanotube sample (After Ebbesen *et al.*, 1994, copyright 1994 Macmillan Magazines Limited. Reproduced with permission)

However, the oxygen-burning technique may also vaporize nanotubes, albeit at a slightly slower rate than nanoparticles, leading to a very low yield of retaining the nanotubes (*ca.* 1% wt). Much higher yields (up to 40% wt) have been achieved by liquid-phase oxidation, for example, in $KMnO_4/ H_2SO_4$ (Hiura *et al.*, 1995; Kosaka *et al.*, 1995) or liquid bromine (Chen Y. *et al.*, 1996a). It has been demonstrated that carbon nanotubes purified by the oxidation method often contain functional groups on the tube surface, which may cause some undesirable effects on their physical properties (*e.g.* electrical and interfacial properties). They can also be used to functionalize the nanotubes. The tips of nanotubes have been shown, both theoretically and experimentally (Carroll *et al.*, 1997; Tamura *et al.*, 1995), to have different electrical properties from their sidewalls because of the presence of topological defects. In order to maintain the property integrity of carbon nanotubes, therefore, benign purification methods that cause little damage on the tips and sidewalls of carbon nanotubes are highly desirable. In this context, Bonard *et al.* (1997) have recently demonstrated a non-destructive and size-selective purification method based on sonication and controlled flocculation of nanotube samples in a surfactant solution followed by filtration.

The oxidation-purification methods discussed above for MWNTs have been used to purify SWNTs with little success, although the inner SWNTs in certain nanotube bundles were demonstrated to be inert toward oxidation by acids (*e.g.* nitric acid) and have been purified by treatment in boiling nitric acid (Ebbesen *et al.*, 1994; Dujardin *et al.*, 1998). The failure is mainly due to the ease with which the monolayer of graphene sheets in SWNTs may be destroyed and the residual metal particles in the sample can catalyze oxidation reactions. Fortunately, a multistep solution-based purification method that incorporates a hydrothermal treatment for easy removal of metal particles and carbonaceous materials has been developed for purification of the arc-generated SWNTs (Tohji *et al.*, 1996; 1997), while a microfiltration method was devised for the purification of SWNTs produced by the pulsed laser technique (Bandow *et al.*, 1997). Haddon and co-workers have also used a chromatographic purification method to remove non-nanotube materials (Niyogi *et al.*, 2001) and capillary electrophoresis for separating nanotubes by length (Doorn *et al.*, 2002).

5.6 Microfabrication

5.6.1 Opening, Filling and Closing

Most of the *as-synthesized* carbon nanotubes are closed at both ends. As mentioned above, however, the tips of nanotubes are more reactive than their sidewalls, and hence can be preferentially attached by various chemical reagents leading to the opening of the nanotubes (Colbert *et al.*, 1994). Several studies have demonstrated that carbon nanotubes can be either uncapped by oxidation with CO_2 or O_2 at high temperatures (Ajayan *et al.*, 1993; Tsang *et al.*, 1993) or opened through wet-chemical treatments with various oxidants (*e.g.* boiling in HNO_3, H_2SO_4, HF/BF_3) (Hiura *et al.*, 1995; Hwang, 1995; Tsang *et al.*, 1994). While concentrated nitric acid was commonly used to open MWNTs (Satishkumar *et al.*, 1996), milder reagents (such as concentrated hydrochloric acid) were proven to be efficient for opening SWNTs by dissolving the residual metal catalyst (*e.g.* Co) at their tips (Freemantle, 1998; Sloan *et al.*, 1998).

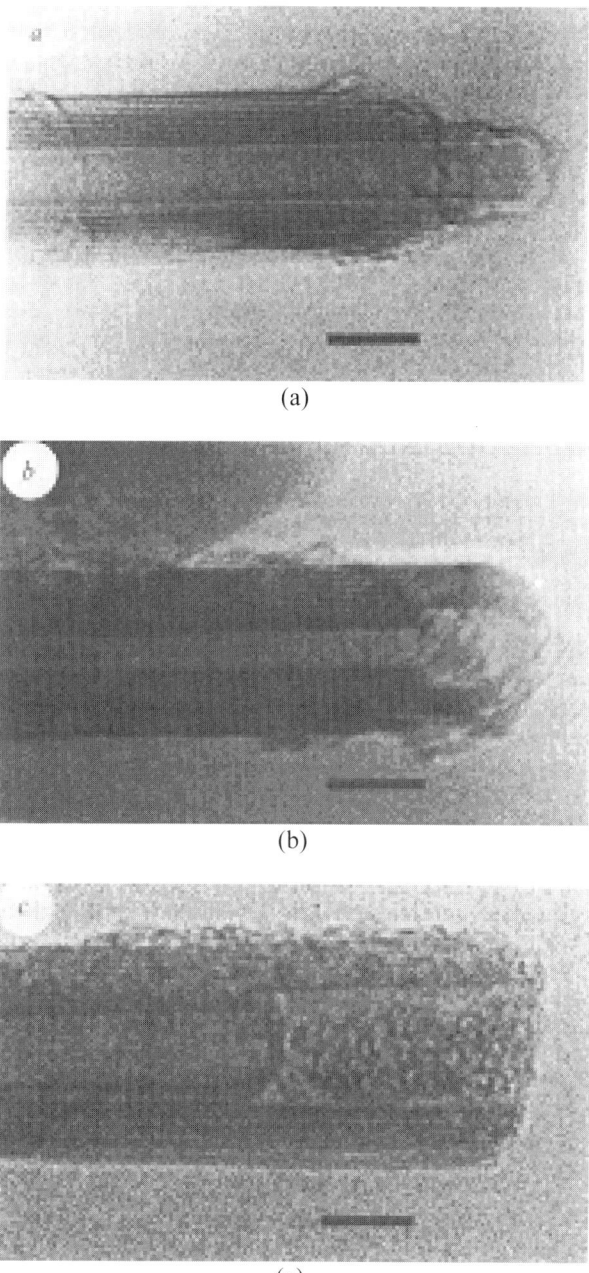

Figure 5.10. Carbon dioxide-treated nanotubes: (a) some outer graphic layers of the cap have been stripped off and a thinner inner tube extends beyond the region; (b) partially opened; (c) completely opened tube with amorphous material inside (After Tsang *et al.,* 1993, copyright 1993 Macmillan Magazines Limited.)

5.6.2 Filling

The possibility of filling the hollow cores of carbon nanotubes with foreign materials is expected to have a large impact on the potential applications of nanotubes (Birkett and Terrones, 1999). Filling can be accomplished either with preformed nanotubes or during the nanotube formation. Both capillarity and surface wettability play important roles in the filling of a preformed nanotube. It was observed that water droplets (surface tension 73 mN/m) spread on the surface of a carbon nanotube could be readily sucked up into the nanotube (Ebbesen, 1994), consistent with the strong nanocapillary action predicated by theory (Pederson and Broughton, 1992). Although liquids with low surface tensions (< 200 mN/m), including most common organic solvents, can wet and fill nanotubes, those with surface tensions higher than 200 mN/m, such as most pure metals, can neither wet carbon nanotubes nor initialize the capillary action (Dujardin et al., 1994). However, lead droplets in the presence of air were found to be sucked up by the capillary effect (Ajayan and Iijima, 1993), presumably because the nanotube was opened by an oxidative interaction of Pb with its end tips at a high temperature (~500 °C) (Ajayan et al., 1993; Ajayan and Iijima, 1993). In general, tube-filling with metals can be catalyzed by the presence of oxygen, as the formation of oxide and/or carbide compounds may enhance the wetting of the carbon nanotube by the guest material (Ajayan et al., 1993). With similar annealing techniques, several other metal oxides with low melting points including Bi_2O_3, MoO_3 and V_2O_5 have also been filled inside carbon nanotubes (Ajayan et al., 1993, 1995; Chen Y. et al., 1996).

A more general method of filling nanotubes with metals is to fill them with corresponding metal-containing inorganic precursor salts [e.g. $Ni(NO_3)_2$] or acids (e.g. H_2PtCl_6, $HAuCl_4$) by sonication in a liquid medium, and then transform them into the metal oxides upon drying/calcination followed by chemical reduction to the pure metal states through heating the metal oxide-containing nanotubes under H_2 flow or reacting them with an aqueous solution of NaB at room temperature. Various metals including Ni, Co, Fe, Ag, Au, Pd and Pt have been successfully filled into carbon nanotubes by this method (Kyotani et al., 1997; Satishkumar et al., 1996; Tsang et al., 1994).

By introducing appropriate metals into the arc electrodes during tubule syntheses, many groups have reported the in situ formation of various pure metal and/or carbide-encapsulated carbon nanotubes (e.g. Co, Cu, Mn, Se, Sb, Ge, Gd, Yt, Mn, Ti, Cr, Fe, Ni, Zn, Mo, Pd, Sn, Ta, W, Dy, Yb, Ru, Rh, Pd, Os, Ir and Pt) (Ajayan et al., 1994; Guerret-Piecourt et al., 1994; Liu and Cowley, 1995; Loiseau and Pascard, 1996; Saito Y. et al., 1993, 1996; Seraphin et al., 1993; Setlur et al., 1996; Tsang et al., 1994). The metals/carbides thus encapsulated exist either as nanoparticles or continuously filled "nanowires" along the internal bore of the nanotubes. The length and quantity of the "nanowires" that form have been shown to increase with increasing number of holes in the incomplete electronic shell of the metal atoms (Guerret-Piecourt et al., 1994). In situ encapsulation has also been achieved during tubule syntheses by the pyrolysis methods. This has been done, for example, by either gas-phase deposition of carbon-containing species on to catalytic metal particles (Kiang et al., 1994; Amelinckx et al., 1994; Bernaerts et al., 1995; Saito Y. et al., 1996) or vapor-phase pyrolysis of metal-containing precursors [e.g. $Fe(CO)_5$] in the presence of another carbon source as CO or C_6H_6 (Sen et al., 1997).

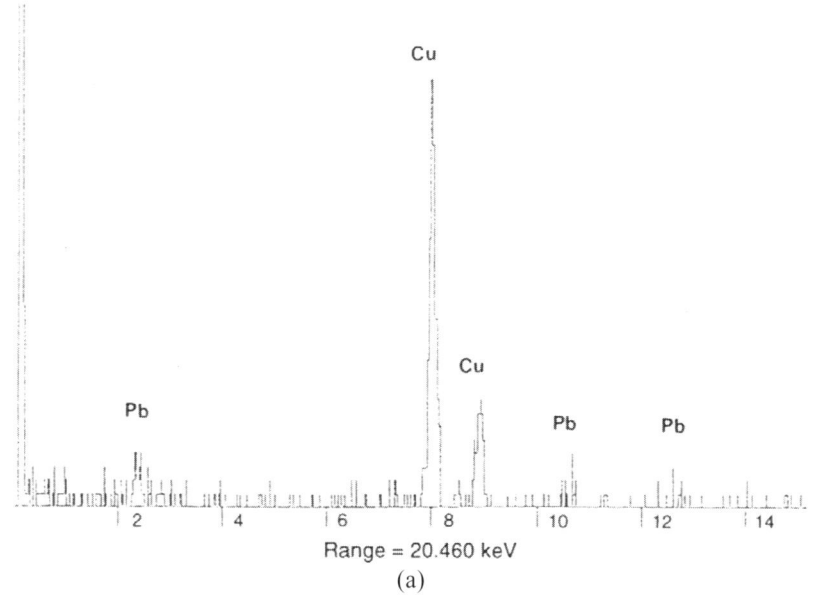

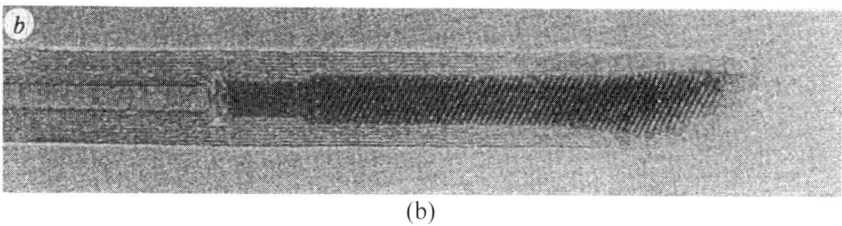

Figure 5.11. (a) Energy-dispersive X-ray data from filled carbon nanotubes, showing the presence of M, La and Lb peaks from the filled lead. The large peaks from copper Ka and Kb are due to the copper grid that supports the specimen; (b) high-resolution TEM image of a crystalline portion of filling material near the open end of a nanotube (After Ajayan and Iijima, 1993, copyright 1993 Macmillan Magazines Limited. Reproduced with permission)

Encapsulation of carbon nanotubes may not only enhance the electrical and magnetic properties of both the encapsulated and encapsulating materials (Ajayan and Ebbesen, 1997; Dresselhaus et al., 1998; Ebbesen, 1994; Hamada et al., 1992; Murakami et al., 1993), but also provide unique one-dimensional systems for studying various physical and/or chemical phenomena. As a consequence, it is now possible to perform the size-limited crystallization at the atomic scale (Sloan et al., 1997), nano-scale-confined chemical reactions (Ugate et al., 1996; Chen Y. et al., 1997), and fabrication of nanocomposite structures (Dai et al., 1995; Han et al., 1997) within the hollow of the nanotubes. In particular, hollow nanotubes of SiO_2, AbO_3, V_2O_5 and MOO_3 have been prepared by coating carbon nanotubes (as templates) with tetraethylorthosilicate, aluminum tisopropoxide, vanadium pentoxide gel, or molybdenum (VI) oxide, followed by calcination and heating at high temperatures *(ca.* 700–1000K) in air to oxidize the carbon (Satishkumar, 1997). Starting with the metal oxide-encapsulated carbon nanotubes, this procedure has also led to the formation of SiO_2 nanotubes filled by transition metal ions, which should be of use as catalysts.

5.6.3 Tip-closing

In comparison with the opening and filling of carbon nanotubes, the closing of opened nanotubes seems to be much less discussed in the literature, although a few approaches have been reported with some success. For example, Rao and co-workers (Satishkumar *et al.*, 1996) have successfully closed the HNO_3-opened nanotubes by the formation of ester and/or ether linkages between the acid-induced –OH/–COOH surface groups *(vide supra)* and appropriate organic hydroxy compounds (e.g. ethylene glycol and propylene glycol). Such closure was shown to be stable against solvents (*e.g.* diethylether, acetone) or thermal treatments up to 670 K. These authors have further demonstrated the possibility of closing opened nanotubes by heating them in a stream of benzene vapor, argon and hydrogen at 1173 K (Satishkumar *et al.*, 1996), as pyrolysis of aromatic hydrocarbons in a reducing atmosphere is known to produce new carbon species including tubular structures.

5.7 Chemical Modification

Carbon nanotubes have been demonstrated to possess excellent electronic, mechanical and thermal properties. As we shall see in succeeding chapters, these useful physicochemical properties make carbon nanotubes very attractive for a wide range of applications, including as electron emitters in field emission displays, reinforcement fillers in nanocomposite materials, scanning probe microscopy tips, actuators and sensors, as well as molecular-scale components in micro-/nano-electronic devices. However, it is difficult to synthesize carbon nanotubes with good surface characteristics as demanded for various specific applications (*e.g.* strong interface with polymers in nanocomposites, good biocompatibility for nanotube sensors). Therefore, surface modification and interfacial engineering are essential in making advanced carbon nanotubes of good bulk and surface properties for the aforementioned, and many other applications. Although the research on chemical modification of carbon nanotubes is still in its infancy, some interesting work on carbon nanotube chemistry has recently been reported in the literature. In what follows, I will summarize carbon nanotube chemistry, covering both the covalent and non-covalent reactions at the tips, outerwall and innerwall of SWNTs and MWNTs.

5.7.1 End-functionalization

5.7.1.1 Oxidation of Carbon Nanotubes

Early work on carbon nanotube chemistry can be traced back to the oxidation of carbon nanotubes at high temperatures in air or O_2 (Ajayan *et al.*, 1993). The oxidation of carbon nanotubes at a temperature above 700 °C in the presence of air for 10 min resulted in the opening of the hemispherical end-caps, indicating that the hemispherical tips are more reactive than the graphite sidewall. This work also led to the prospect of filling foreign materials (*e.g.* metal oxide nanoparticles) into the hollow tubes.

The above-mentioned early work on carbon nanotube chemistry was followed by gas-phase reactions with CO_2, N_2O, NO, NO_2, O_3 and ClO_2 (Niu *et al.*, 2001), and the purification of carbon nanotubes, as amorphous carbon impurities are known to be more susceptible to gas-phase oxidation than carbon nanotubes. Solution chemical oxidation, however, was found to be more efficient for the purification and/or modification of carbon nanotubes. Since Tsang *et al.* (1994) reported the liquid-phase oxidation of carbon nanotubes in HNO_3, various oxidants have been shown to react with carbon nanotubes. Oxygen-containing acids, including HNO_3 (Dillon *et al.*, 1999; Dujardin *et al.*, 1998; Nagasawa *et al.*, 2000), HNO_3 + H_2SO_4 (Yu *et al.*, 1998), $HClO_4$ (Delpeux *et al.*, 1999), H_2SO_4 + $K_2Cr_2O_7$ (Li X. *et al.*, 1999; Yang and Wu, 2001) and H_2SO_4 + $KMnO_4$ (Hiura *et al.*, 1995), remained as the main class of oxidants that have been reported in the literature, though several other oxidants (*e.g.* OsO_4, H_2O_2) have also been used.

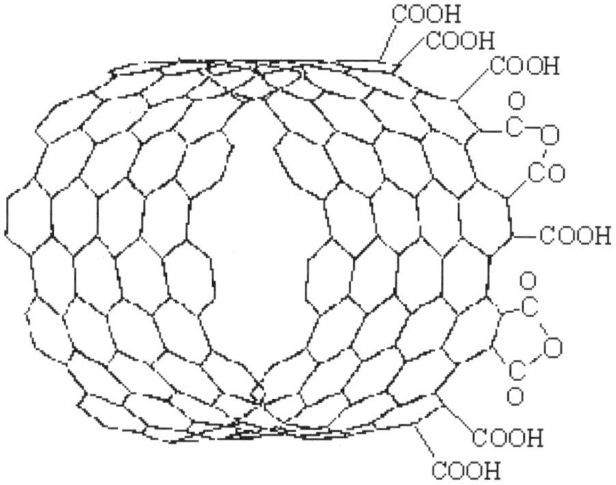

(5.1)

The oxidation reactions discussed above often generate various functional groups (*e.g.* –COOH, –OH, –C=O) at the opened end or defect sites of the carbon nanotube structure [Equation (5.1)]. Other groups may be also introduced due to side reactions. For example, a small amount of sulfuric group may be introduced onto the H_2SO_4/HNO_3 oxidized carbon nanotubes, as shown below (Yu *et al.*, 1998).

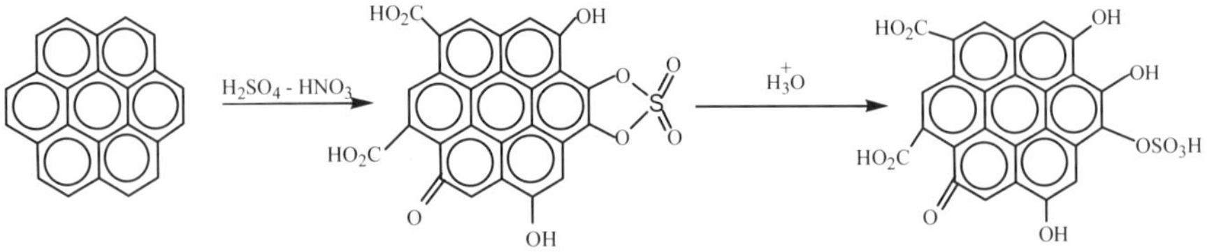

174   Intelligent Macromolecules for Smart Devices

(5.2)

The degree of oxidation depends strongly on the nature of the oxidant and the reaction conditions. More interestingly, SWNTs and MWNTs show different behavior towards the oxidation reactions. As SWNTs are known to exist in bundles, the oxidation reactions take place at the nanotube end-tips and the outer layer of the bundle (Dujardin et al., 1998). Although the oxidation of MWNTs also starts from the nanotube tips and/or the defect sites, the reaction front gradually moves from the outermost graphitic layer toward the nanotube core, leading to a successive removal of the graphene cylinders and thinner nanotubes (Ajayan et al., 1993; Ajayan and Iijima, 1993; Tsang et al., 1993). Due to a relatively high reactivity associated with the opened nanotube tips, the oxidized nanotubes often show a larger internal diameter near the open end than within the tube (Hwang, 1995; Liu et al., 2001). A wide range of techniques, including FTIR, UV/VIS/NIR, Raman (Koshio et al., 2001a; Yu and Brus, 2000), vibration spectroscopic NEXAFS (Kuznetsova et al., 2001), SEM/TEM (Vaccarini et al., 1999), EDS, STM (Biro et al., 2002), XPS (Yu et al., 1998), and X-ray diffractometry (Liu Z.-J. et al., 2001), have been used to characterize the oxidized nanotubes.

5.7.1.2 Covalent Coupling via the Oxidized Nanotube Ends

The carboxylic acid and hydroxyl groups of the oxidized nanotubes can be further used to covalently connect other small and polymeric molecules through reactions characteristic of the –COOH and –OH functionalities. In this context, the reactions of the fluorine molecule, Rhodamine B and *p*-carboxytetraphenylporphine (TPP) with the oxidized nanotubes have been reported (Sun et al., 2000, 2002).

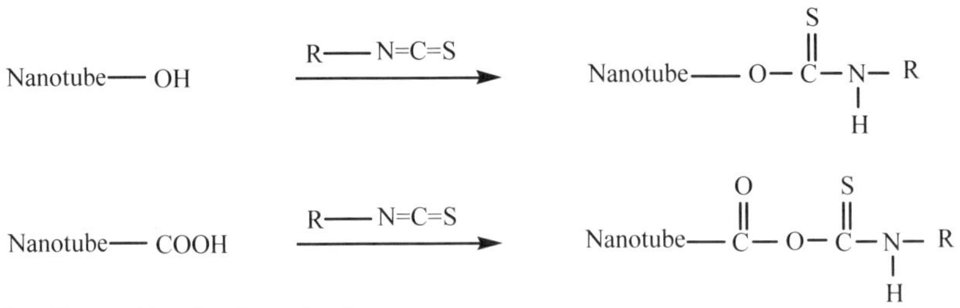

R = Dye small molecules and polymers

(5.3)

Similarly, long alkyl chains and/or polymers have been chemically attached onto the oxidized nanotubes through, for example, the amidation reaction. In this case, the carboxylic acid functionalized nanotubes were first converted into alkyl chloride by reacting with SOCl$_2$; an aryl amine was then reacted with the alkyl chloride to form an amide bond between the nanotube and the aryl group moieties. The attachment of long hydrocarbon chains onto carbon nanotubes improved the nanotube solubility significantly (Chen J. et al., 1998; Hamon et al., 1999), allowing the modified SWNTs to be purified by the conventional chromatographic technique (Niyogi et al., 2001).

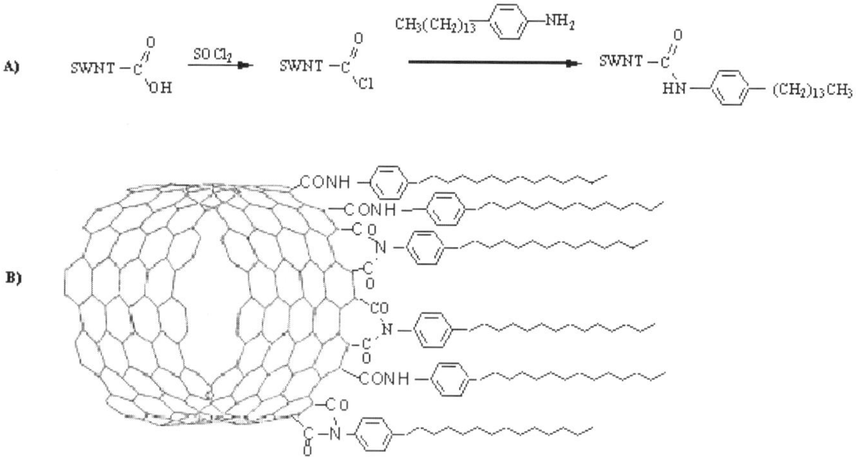

(5.4)

Alkyl chains have also been covalently attached onto the oxidized nanotubes via a more simple reaction between the carboxylic acid and amine groups using a carboxylate-ammonium salt, which also improves the solubility of the nanotubes (Chen J. et al., 2001; Hamon et al., 1999).

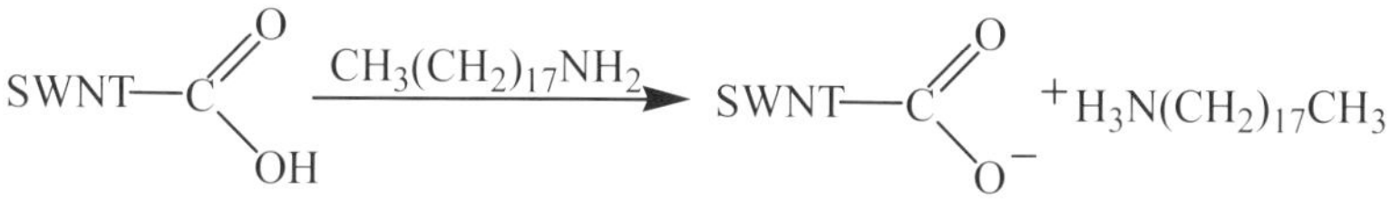

(5.5)

Polymers with amino terminal groups, such as poly(propionylethylenimine-*co*-ethylenimine) (PPEI-EI), have also been grafted onto oxidized nanotubes through the amide formation. The resulting polymer-grafted SWNTs and MWNTs are highly soluble in most common organic solvents and water (Czerw et al., 2001; Riggs et al., 2000a, b).

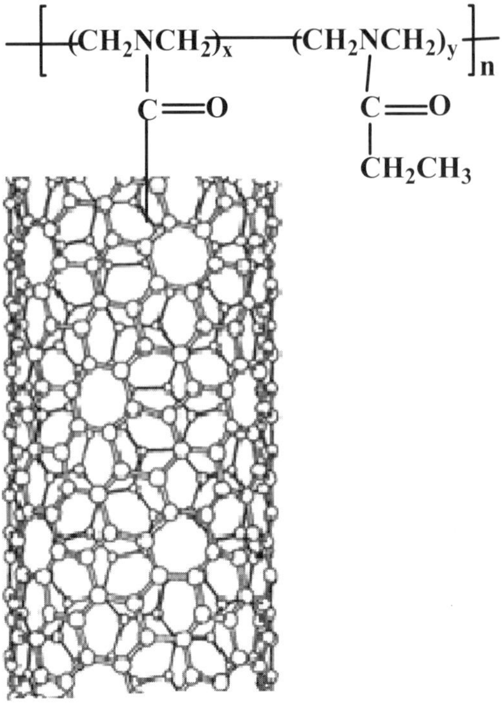

$$\left[-(CH_2NCH_2)_x-(CH_2NCH_2)_y-\right]_n$$
$$\begin{array}{cc} | & | \\ C=O & C=O \\ | \\ & CH_2CH_3 \end{array}$$

(5.6)

Apart from the amide linkage, etherification reaction has also been used to covalently attach alkyl groups and polymers onto oxidized carbon nanotubes. For instance, octadecylalcohol (Hamon et al., 2002) and poly(vinyl acetate-co-vinyl alcohol) (PVA-VA) (Riggs et al., 2000a) have been successfully grafted onto the oxidized SWNTs to show a good solubility.

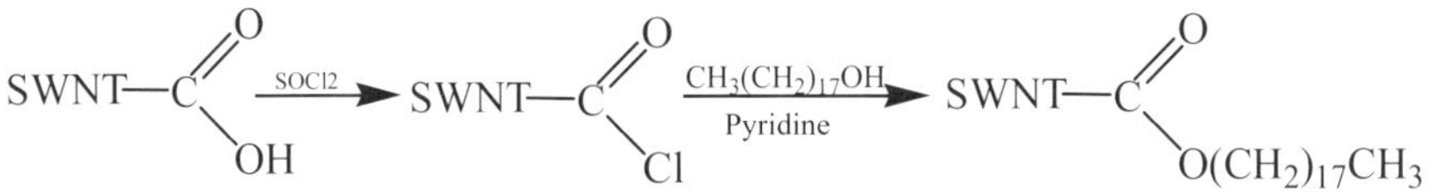

$$SWNT-C\overset{O}{\underset{OH}{\diagup}} \xrightarrow{SOCl2} SWNT-C\overset{O}{\underset{Cl}{\diagup}} \xrightarrow[Pyridine]{CH_3(CH_2)_{17}OH} SWNT-C\overset{O}{\underset{O(CH_2)_{17}CH_3}{\diagup}}$$

(5.7)

Besides, dendrimers or dendrons [e.g. the tenth generation of poly(amidoamine) (PAMAM) starburst dendrimer (G10)] have also been grafted onto the oxidized carbon nanotubes to form nanotube stars (Sano et al., 2001a), as were lipophilic and hydrophilic dendrons that terminated with long alkyl chains and poly(ethylene glycol) oligomers (Fu et al., 2001; Sun Y.-P. et al., 2001).

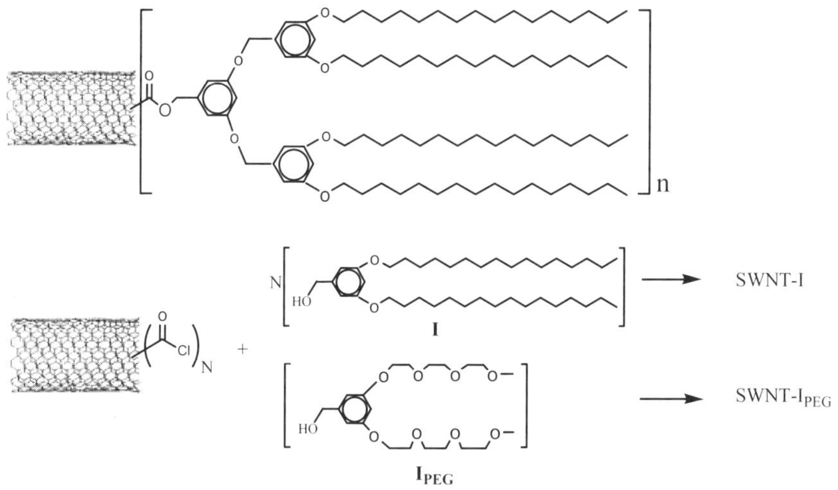

$$\text{(5.8)}$$

Interestingly, Sano *et al.* (2001b) reported the ring formation from the acid-oxidized SWNTs through the esterification between the carboxylic acid and hydroxyl end-groups at the nanotube tips in the presence of a condensation reagent, 1,3-dicyclohexylcarbodiimide [Equation (5.9)]. Carbon nanotube rings with an average diameter of 540 nm and a narrow size distribution were obtained (Figure 5.12).

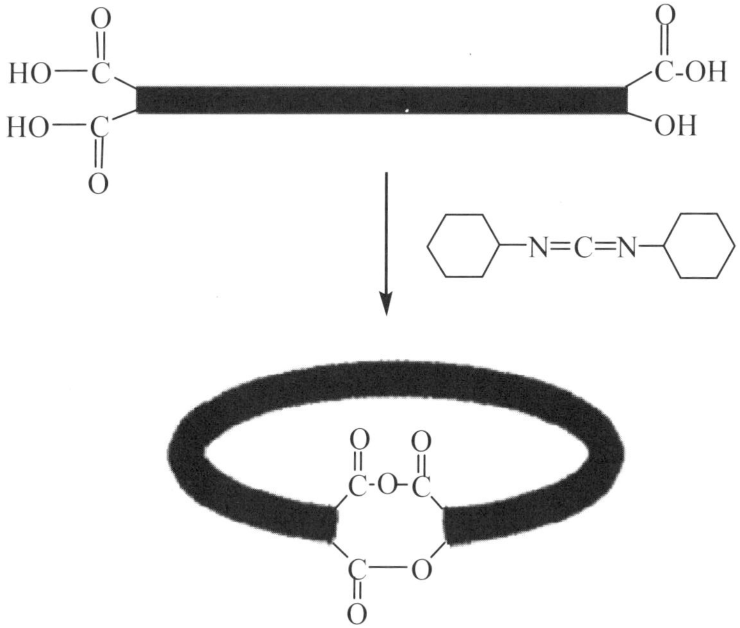

$$\text{(5.9)}$$

178 Intelligent Macromolecules for Smart Devices

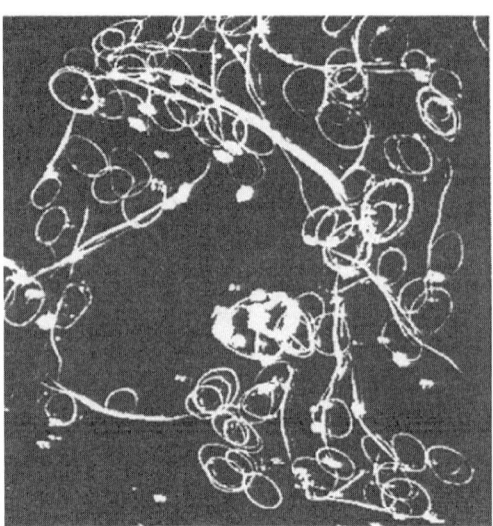

Figure 5.12. A typical AFM image of carbon nanotube rings (After Sano *et al.*, 2001b, copyright 2001 American Association for the Advancement of Science)

More interestingly, nanotube hetero-junctions with two carbon nanotubes joined in either end-to-side or end-to-end configuration have been constructed through a bi-functionalized amine linkage, as schematically shown in Equation (5.10) and Figure 5.13 (Chiu *et al.*, 2002).

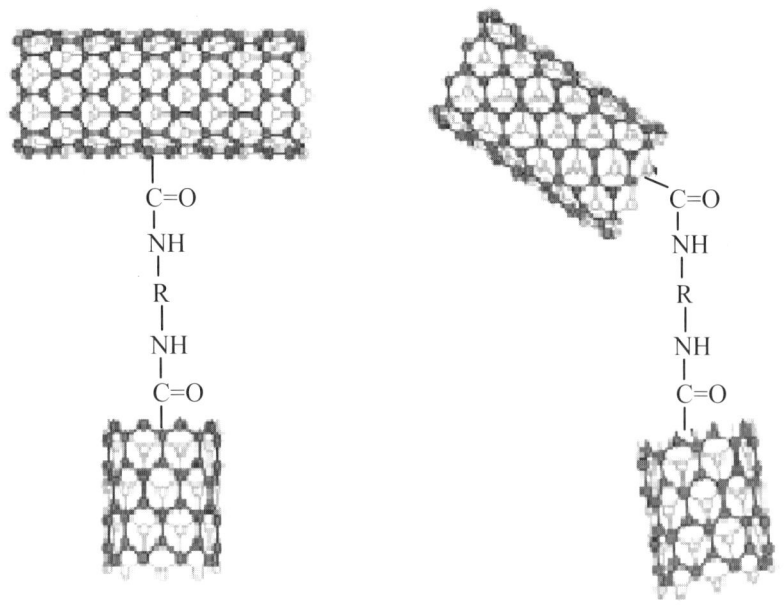

(5.10)

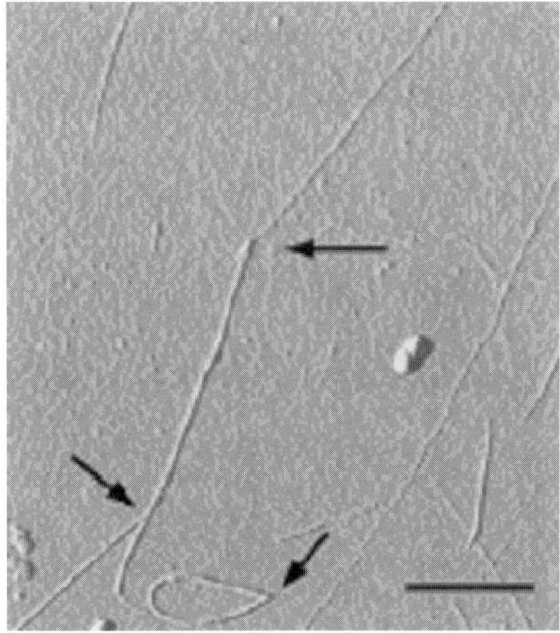

Figure 5.13. SEM image of the interconnected SWNTs (After Chiu *et al.*, 2002, copyright 2002 American Institute of Physics)

5.7.2 Modification of Nanotube Outerwall

5.7.2.1 Sidewall Fluorination of Carbon Nanotubes

Although carbon nanotubes are generally known to be inert towards fluorine at room temperature, chemical fluorination of carbon nanotubes has been achieved at relatively high temperatures. For instance, Nakajima *et al.* (1997) have demonstrated the chemical fluorination of a carbon nanotube wall at temperatures ranging between 250 and 400 °C. Fluorination of SWNTs has also been carried out in the same way as that for the fluorination of graphite (Mickelson *et al.*, 1998), indicating there was considerable room for chemical functionalization of carbon nanotubes under appropriate conditions.

Fluorination at a temperature above 400 °C was found to cause the decomposition of the tube structure. Although fluorination between 250 and 400 °C does not damage the structure to that extent, it reduces the conductivity of the SWNTs due to the partial destruction of the graphitic structure. As a result, the fluorinated SWNTs differ dramatically from their unmodified counterparts in electronic properties. The presence of covalently-bonded fluorine in the fluorinated carbon nanotube sample was confirmed by the appearance of FTIR peaks over 1220~1250 cm^{-1}. The TEM elemental images given in Figure 5.14 further indicate that fluorine atoms are homogeneously attached onto the nanotube wall.

Figure 5.14. (a) TEM image of fluorinated carbon nanotubes; (b) elemental C mapping (in blue) of the image shown in (a); (c) elemental F mapping (in green) of the image shown in (a) (After Hayashi *et al.*, 2002, copyright 2002 The American Chemical Society. Reproduced with permission)

The covalently-bonded fluorine was found to be removable from the fluorinated nanotube surface by reacting with N_2H_4:

$$CF_n + n/4\, N_2H_4 \longrightarrow C + nHF + n/4\, N_2 \qquad (5.11)$$

The above fluorination reaction is of considerable interest because further functionalization of carbon nanotubes can be carried out through the fluorine intermediate. In this context, it was demonstrated that the covalently-bonded fluorine could be replaced by alkyl groups through the alkylation reaction (Boul et al., 1999). For this purpose, alkyl lithium or a Grignard reagent can be used as an alkylation reagent. Interestingly enough, these alkyl groups can also be removed by heating the alkylated nanotubes at ~250 °C, leading to the recovery of the pure SWNTs. Various physicochemical properties of the fluorinated SWNTs, including high-resolution electron energy loss (Hayashi et al., 2002), thermal recovery behavior (Zhao et al., 2002), and solvation in alcohol solvents (Mickelson et al., 1999), have been investigated. Furthermore, Yudanov and co-workers (Yudanov et al., 2002) have recently demonstrated the fluorination of MWNTs at room temperature by using BrF_3 as the fluorination reagent, though only the outside shell of the MWNT was modified.

5.7.2.2 The Attachment of Dichlorocarbene to the Sidewall

As is well known, 1,1-dichlorocarbene can attack the C=C bond connecting two adjacent six-membered carbon rings to produce 1,1-dichlorocyclopropyane. This reaction has been investigated for both SWNTs and MWNTs (Chen J. et al., 1998) using 1,1-dichlorocarbene generated from different sources, including NaOH/chloroform and Hg complex.

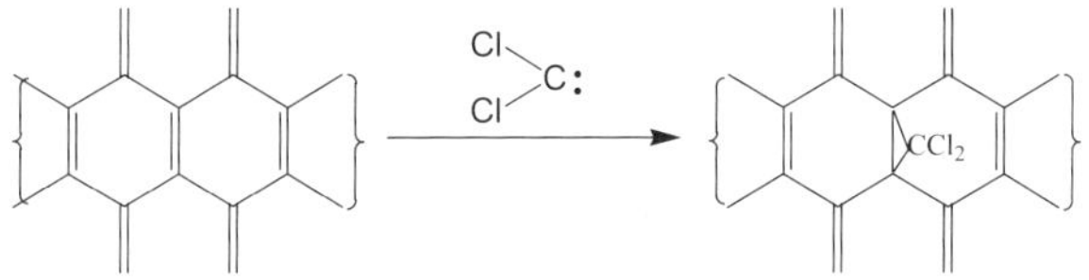

$$(5.12)$$

The surface chemical state of elements C and O in the carbenated nanotubes has been studied by XPS measurements (Lee et al., 2001). A large amount of amorphous carbon was found to attach onto the tubular wall, leading to suspicions as to whether the dichlorocarbene is attached onto the sidewall of the nanotube or just reacts with the amorphous carbon (Chen Y. et al., 1998).

5.7.2.3 Modification via 1,3-Dipolar Cycloaddition of Azomethine Ylides

Like fullerene C_{60}, 1,3-dipolar cycloaddition of azomethine ylides can also take place between the graphite sidewall of SWNTs at the C=C bond connecting two adjacent six-membered rings and an α-amino acid or an aldehyde molecule (Georgakilas et al., 2002). Carbon nanotubes thus modified showed a high solubility in chloroform (up to 50 mg/ml). Compared with pure SWNT bundles, the modified SWNTs also exhibited somewhat poorer electronic properties with a more densely

182   Intelligent Macromolecules for Smart Devices

packed bundle structure, as evidenced by TEM and other electronic spectroscopic measurements.

$$R_1\text{-NHCH}_2\text{COOH} + R_2\text{-CHO} \xrightarrow[\text{DMF, Reflux 120 h}]{\text{NT}}$$

R_1=-CH$_2$CH$_2$OCH$_2$CH$_2$OCH$_2$CH$_2$OCH$_3$, R_2= H
R_1=-CH$_2$(CH$_2$)$_6$CH$_3$, R_2= H
R_1=-CH$_2$CH$_2$OCH$_2$CH$_2$OCH$_2$CH$_2$OCH$_3$, R_2= H$_3$CO-⟨⟩-
R_1=-CH$_2$CH$_2$OCH$_2$CH$_2$OCH$_2$CH$_2$OCH$_3$, R_2= (pyrenyl)

(5.13)

5.7.2.4 The Reaction Between Aniline and Carbon Nanotubes

Recently, it has been found that carbon nanotubes show strong interaction with aniline (Sun Y. *et al.*, 2001). The strong intermolecular interaction leads to a solubility of SWNTs in aniline up to 8 mg/ml. Besides, a new UV-VIS absorption peaked at ~540 nm and a new fluorescence emission peak at ~620 nm were observed (Figure 5.15).

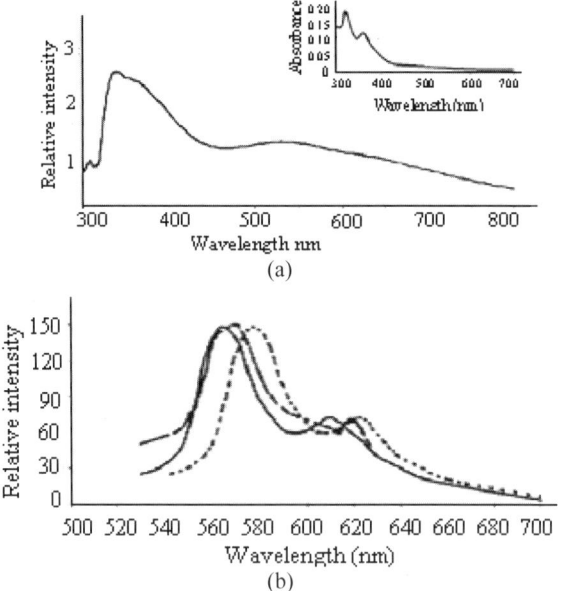

Figure 5.15. (a) UV/vis absorption spectrum of carbon nanotube dissolved in aniline. The inset shows the corresponding UV/vis spectrum for the carbon nanotube/aniline solution diluted with acetone; (b) fluorescence emission spectra of the carbon nanotube/aniline solution diluted in different solvents: acetone (-), toluene (-..-), and methanol (---); all samples were excited at 500 nm (After Sun Y. *et al.*, 2001, copyright 2001 The American Chemical Society. Reproduced with permission)

It is a surprising discovery that the interaction between aniline and carbon nanotubes generates new UV/vis and fluorescence peaks, as aniline normally quenches the fluorescence emission due to its strong electron affinity (Sun Y. *et al.*, 2001). However, a chemical reaction might have taken place between the SWNT and aniline, as schematically shown below:

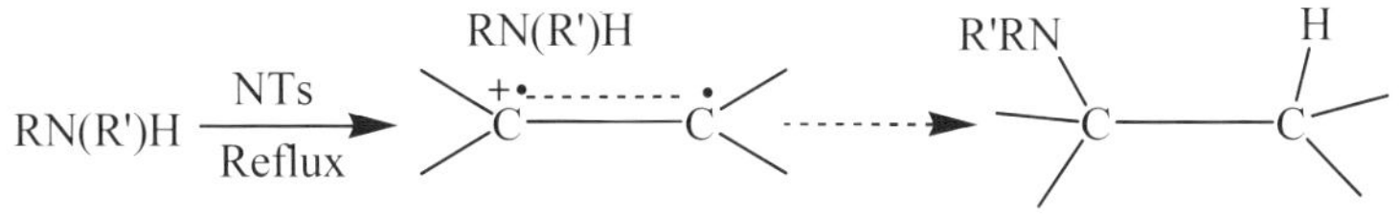

$$RN(R')H \xrightarrow[\text{Reflux}]{\text{NTs}} \quad \text{[intermediate]} \quad \longleftrightarrow \quad \text{[product]} \quad (5.14)$$

Following the modification of the glass-carbon electrode by the electrochemical reduction of aryl diazonium salt, Tour and co-workers (Bahr *et al.*, 2001) applied a similar reaction to the carbon nanotube electrode. This reaction involves the attachment of aryl cations onto the sidewall of SWNTs (Bahr *et al.*, 2001).

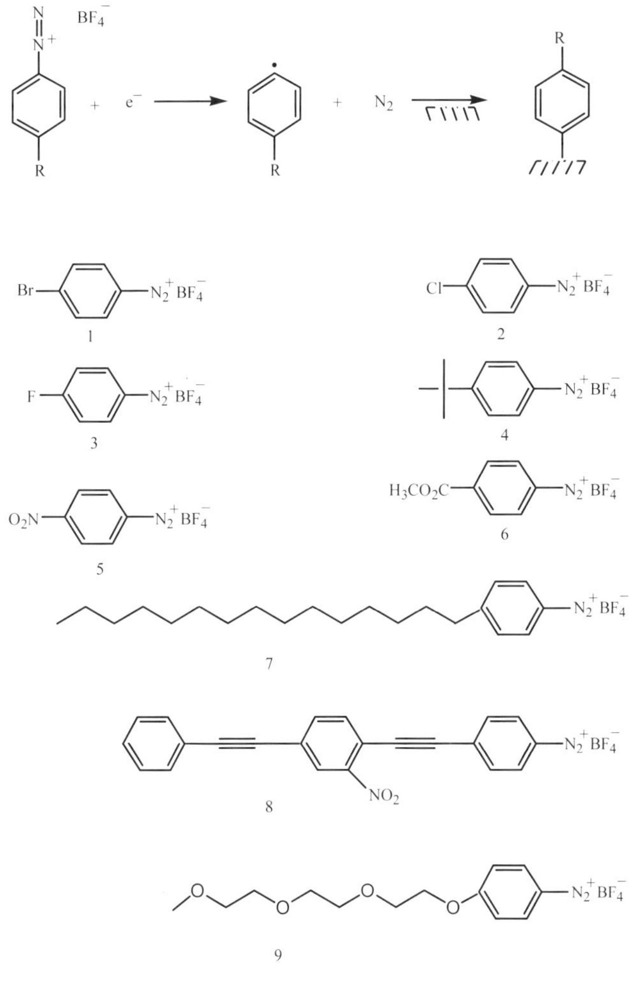

$$(5.15)$$

By extension, a variety of diazonium salts have been attached to the nanotube wall [Equation (5.15)]. The nature of the C-4 substitution of the benzyl group plays an important role in regulating the grafting degree and the solubility of the resulting product. Those aryl cations with long alkyl chains at the *para*-position of the benzyl ring can significantly improve the solubility of carbon nanotubes in organic solvents. The grafted moieties can be removed by simply heating the modified carbon nanotubes in an inert atmosphere at 500 °C.

5.7.3 Functionalization of Carbon Nanotube Innerwall

Compared with the nanotube outerwall modification discussed above, the funcationalization of the nanotube innerwall has been much less discussed in the literature. This is because the innerwall modification often requires the opening of the nanotube tip(s) and the protection of the outerwall; both of these processes are difficult and tedious. In this regard, those carbon nanotubes synthesized by the template technique possess advantages for the innerwall modification (Kyotani *et al.*, 2001).

The template technique involves direct deposition of carbon nanotubes or their precursor polymer nanotubes followed by high-temperature graphitizing within the pores of a nanoporous template (*e.g.* alumina membrane or a mesoporous zeolite). The template synthesis often allows the production of monodispersed carbon nanotubes with opened ends, as well as controllable diameters, lengths and orientation. Therefore, the nanoporous alumina membrane-assisted deposition of aligned carbon nanotubes (Kyotani *et al.*, 1996; Li W. *et al.*, 1996) and the preparation of carbon nanotubes by high-temperature graphitizing of the polyacrylonitrile (PAN) nanotubules synthesized within the pores of an alumina template membrane or zeolite nanochannels are of particular interest (Kyotani *et al.*, 1997a; Parthasarathy *et al.*, 1995). Based on the template method for the nanotube growth, Kyotani *et al.* (2001) have successfully carried out the nitric acid oxidation of nanotube innerwalls within the nanoporous Al_2O_3 pores with the alumina template membrane acting as the protection layer for the nanotube outerwall. Upon the completion of the innerwall oxidation, the template was removed by dissolving it into an aqueous solution of HF to release the innerwall-modified nanotubes (Figure 5.16).

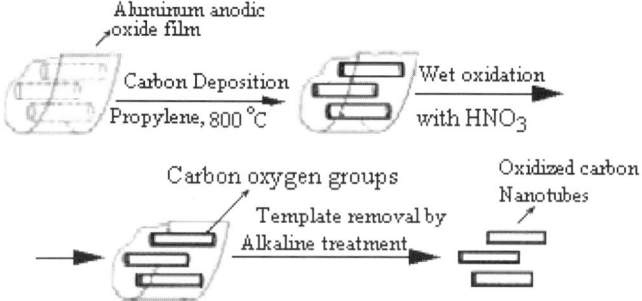

Figure 5.16. Oxidation of the inner surface of template-synthesized carbon nanotubes (After Kyotani *et al.*, 2001, copyright 2001 Elsevier. Reproduced with permission)

5.7.4 Other Physical Chemistries of Carbon Nanotubes

5.7.4.1 Modification of Carbon Nanotubes via Mechanochemical Reactions

Mechanochemistry refers to processes in which mechanical motions/energy control chemical reactions (Drexler, 1992). Specifically, the mechanochemical reaction involves highly reactive centers generated by the mechanical energy (*e.g.* ultrasonication, ball-milling) imparted to the reaction system. Certain organic reactions have been demonstrated to take place efficiently in the solid-state phase (Tanaka and Toda, 2000).

Ultrasonication has been demonstrated to significantly enhance many chemical reactions, including the acid oxidation of carbon nanotubes (Martinez *et al.*, 2002). It was envisioned that the strong vibration arising from the ultrasonication created many defective sites on the SWNT sidewalls (both the inner and outer ones), which could not only facilitate the oxidation reaction but also cut the SWNTs into short segments. Ultrasonication has also been found to induce a chemical reaction between SWNTs and monochlorobenzene in the presence of polymer polymethyl methacrylate (PMMA) (Yudasaka *et al.*, 2000). In particular, SWNTs could be separated from carbonaceous impurities and cut into short segments by ultrasonicating SWNTs and PMMA in monochlorobenzene (Yudasaka *et al.*, 2000). As evidenced by the appearance of new peaks at 1729, 2848 and 2920 cm^{-1}, characteristic of C=C, C=O and C–H bond stretching vibration, respectively, in solid-state diffuse reflectance FTIR spectra, the above polymer-assisted ultrasonication purification process was accompanied by a chemical reaction between the SWNT and monochlorobenzene (Koshio *et al.*, 2001b). PMMA might have enhanced the solubility of SWNTs in monochlorobenzene, and hence facilitated the reaction. High-resolution TEM imaging revealed that a large number of defects formed in the sidewall of SWNTs, which became holes upon heating in oxygen (Zhang *et al.*, 2002).

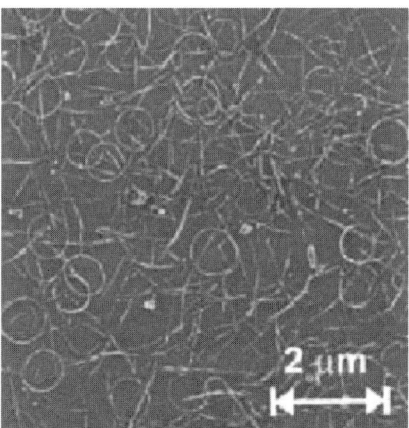 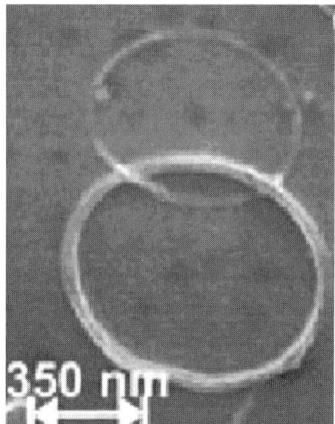

Figure 5.17. Nanotube rings formed by ultrasonication (After Martel *et al.*, 1999, copyright 1999 The American Chemical Society. Reproduced with permission)

186   Intelligent Macromolecules for Smart Devices

Another closely related observation is that certain SWNTs folded back under ultrasonication to form nanotube coils with a narrow distribution of radii (300–400 nm, see Figure 5.17) (Martel et al., 1999). Although the ultrasonication-induced nanotube ring formation shown in Figure 5.17, unlike those nanotube rings formed by the end-to-end ester linkage discussed above [Equation (5.9)] (Sano et al., 2001b), seemed to be largely a physical process, some chemical interactions between the nanotube ends might have been involved as very smooth junctions were observed.

Apart from the ultrasonication, various other physicochemical processes, including microwave (Harutyunyan et al., 2002) and ball-milling (Liu J. et al., 1998), have also been investigated. In particular, ball-milling of carbon nanotubes in the presence of acid was reported to cut the nanotube into 200–300 nm ropes (Liu et al., 1998).

5.7.4.2 Modification of Carbon Nanotubes via Electrochemical Reactions

Owing to the good electronic properties intrinsically associated with carbon nanotubes (Dai and Mau, 2001), electrochemical modification of individual carbon nanotube bundles has been attempted (Kooi et al., 2002). To demonstrate this, Kooi et al. (2002) electrochemically polymerized a thin polymer layer onto the individual SWNT bundle [Equation (5.16)], which was pre-coated on a microelectrode. A thickness of ca. 10 nm for the resultant polymer layer was directly observed by *in situ* AFM imaging (Figure 5.18).

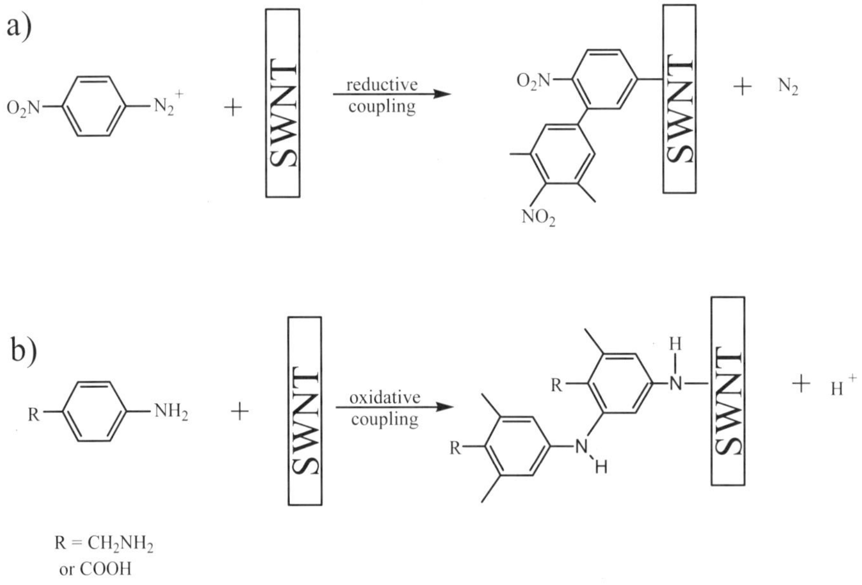

a)

$O_2N-C_6H_4-N_2^+$ + SWNT $\xrightarrow{\text{reductive coupling}}$ [biphenyl product with O_2N and NO_2 substituents]-SWNT + N_2

b)

$R-C_6H_4-NH_2$ + SWNT $\xrightarrow{\text{oxidative coupling}}$ [coupled diamine product with R groups]-SWNT + H^+

R = CH_2NH_2
or COOH

(5.16)

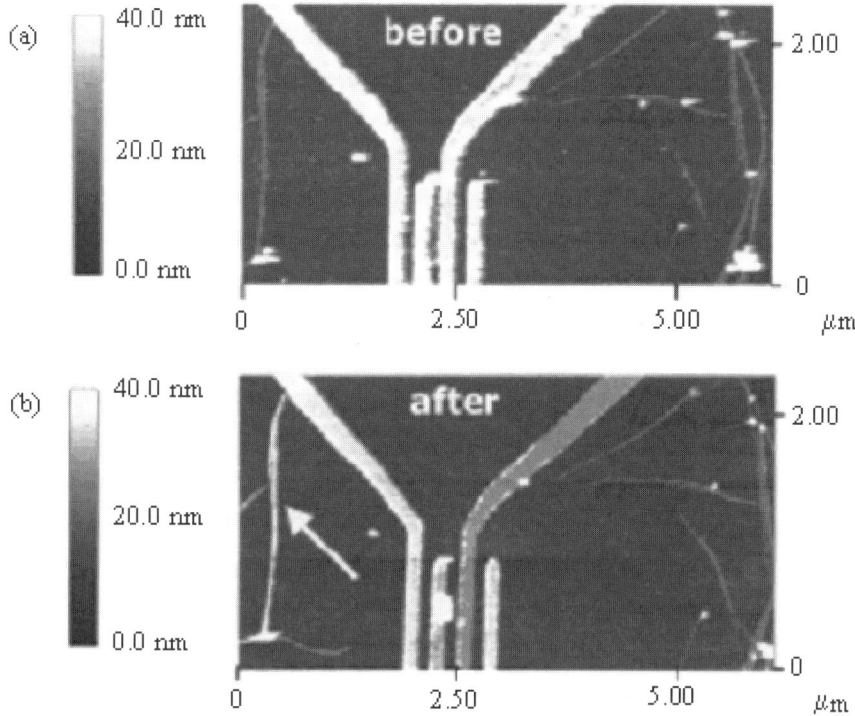

Figure 5.18. AFM image of SWNTs before (a) and after (b) reductive coupling with polymer (After Kooi *et al.*, 2002, copyright 2002 Wiley-VCH Verlag)

5.7.4.3 Modification of Carbon Nanotubes via Photochemical Reactions

In spite of the well-known fact that carbon nanotubes possess unusual optoelectronic properties, the recently reported ignition of some dry and fluffy SWNTs in air by camera flashlight is rather intriguing (Ajayan *et al.*, 2002). In view of the highly efficient absorption of light by carbon nanotubes, the observed ignition may have resulted from a rapid temperature increase within the nanotube sample upon exposure to camera flashlight (it was estimated to be up to at least 1500 °C in a very short time), which is well above the temperature required for ignition when oxygen is present. While the mechanism of ignition by camera flashlight in this system deserves further investigation, such an interesting photochemical reaction may find its application in remote light-triggered combustion or explosion (Ajayan *et al.*, 2002).

Closely related to the above discussion, band gap fluorescence emission from individual SWNTs has also been observed. For instance, O'Connell *et al.* (O'Connell *et al.*, 2002) obtained individual nanotubes through ultrasonication of an aqueous dispersion of raw SWNTs in sodium dodecyl sulfate, followed by centrifuging to remove tube bundles, ropes and residual catalyst. They found that individual semiconducting carbon nanotubes could show fluorescence emission directly across the band gap, whereas aggregation of nanotubes into bundles quenches the fluorescence through interactions with metallic tubes and substantially broadens the absorption spectra. More recently, incandescent light emission has also been observed from certain carbon nanotubes upon the application of a DC voltage (Jiang *et al.*, 2002). Continued research efforts in this embryonic field could give birth to a flourishing area of nanotube optoelectronic technologies.

5.8 Non-covalent Chemistry of Carbon Nanotubes

Non-covalent chemistry involves the self-assembly of molecules or macromolecules to thermodynamically stable structures that are held together by weak non-covalent interactions. These weak non-covalent interactions include hydrogen bonding, π-π stacking, electrostatic forces, van der Waals forces, hydrophobic and hydrophilic interactions.

Because of the fast dynamic and very specific non-covalent interactions involved, self-assembling processes are usually very fast. The resulting supramolecules could undergo spontaneous and continuous de-assembly and re-assembly processes under certain conditions due to the non-covalent nature. Non-covalent supramolecules can, therefore, select their constituents in response to external stimuli or environmental factors and behave as adaptive materials. Although self-assembling (or non-covalent chemistry) of small molecules and certain macromolecules has been an active research area for some time, non-covalent chemistry of carbon nanotubes is a recent development. Non-covalent chemistry, however, offers important advantages as it causes no change to the carbon nanotube structure and hence their electronic properties are largely retained. The literature survey reveals that a few of non-covalent methods, including the self-assembling of planar π systems (Chen R.J. *et al.*, 2001) and wrapping polymer chains onto the nanotube surface (Hirsch, 2002) have been successfully devised for modification of carbon nanotubes, as described below.

5.8.1 Non-covalent Attachment of Small Molecules onto the Nanotube Sidewall

Chen R.J. *et al.* (2001) have recently found that certain aromatic molecules with a planar π moiety (*e.g.* pyrenylene) could strongly interact with the basal plane of graphite on the nanotube sidewall via π stacking. The interaction is so strong that the aromatic molecule is irreversibly adsorbed onto the hydrophobic surface of carbon nanotubes, leading to a high stable self-assembled structure in aqueous solutions. As schematically shown in Equation (5.17), the long alkyl chains end-adsorbed onto the nanotube surface via the aromatic anchors could significantly improve the solubility of the nanotubes.

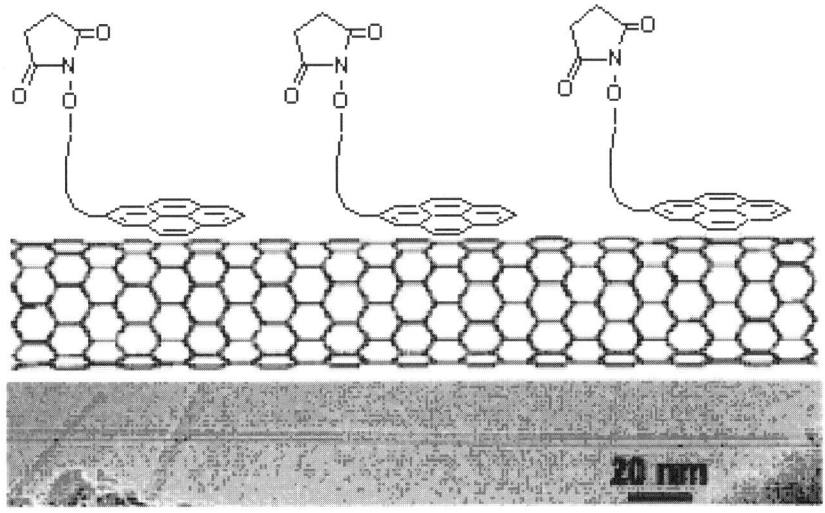

(5.17)

The above approach could be used to attach biochemically active molecules, such as DNA and protein chains, on the side wall of SWNTs through appropriate anchoring molecules. For example, some H_2N-containing bioactive molecules, including ferritin, tavidin and biotinyl-3,6-dioxactanediamine, have been attached onto the sidewall of SWNTs through nucleophilic substitution of the amino functionality by an *N*-hydroxysuccinimide group, which was pre-anchored onto the nanotube surface via a succinimidyl ester linkage by pyrenylene [Equation (5.18) and Figure 5.19] (Chen R.J. *et al.*, 2001).

(5.18)

190   Intelligent Macromolecules for Smart Devices

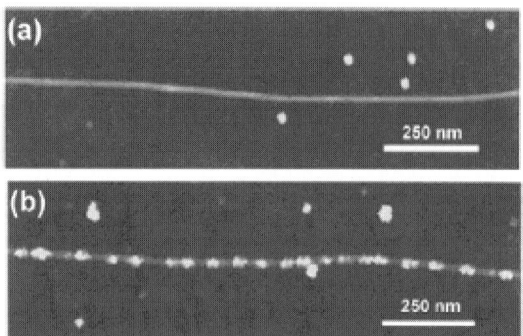

Figure 5.19. (a) An AFM image showing a SWNT bundle (diameter = 4.5 nm) free of absorbed ferritin after incubation in a ferritin solution; (b) an image showing ferritin molecules (apparent heights ~10 nm) adsorbed on a SWNT bundle (diameter = 2.5 nm) functionalized by succinimidyl ester (After Chen R.J. *et al.*, 2001, copyright 2001 The American Chemical Society. Reproduced with permission)

5.8.2 Non-covalent Wrapping of Polymer Chains onto the Nanotube Sidewall

Like the π-conjugated small molecules, many high molecular-weight polymers (especially conjugated polymers) showed the ability to wrap around the surface of carbon nanotubes (Bahr and Tour, 2002; Hirsch, 2002; Niyogi *et al.*, 2002). The wrapping of polymer chains around SWNTs could reduce the intertube van der Waals interaction, and hence enhance the solubility of SWNTs. Strong π-π interactions between carbon nanotubes and conjugated polymers have led to the wrapping of such chains as poly (*m*-phenylene vinylene), PmPV, around the carbon nanotube sidewall, as schematically shown in Figure 5.20.

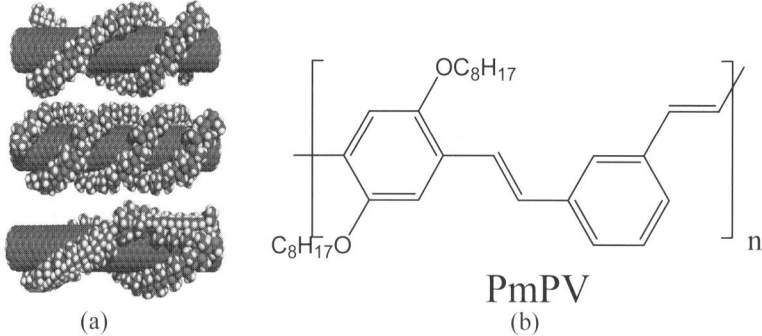

Figure 5.20. (a) Some possible wrapping arrangements of PmPV on a SWNT. A double helix (*top*) and a triple helix (*middle*). Backbone bond rotations can induce switch-backs, allowing multiple parallel wrapping strands to come from the same polymer chain (*bottom*); (b) molecular structure of PmPV (After O'Connell *et al.*, 2001, copyright 2001 Elsevier. Reproduced with permission)

The wrapping process is, at least partially, evidenced by the broadening effect on the UV/vis absorption bands at 329 and 410 nm, characteristic of PmPV, and the appearance of a new broad absorption band over 650–900 nm from the SWNTs/PmPV complexes, as shown in Figure 5.21 (Star *et al.*, 2001).

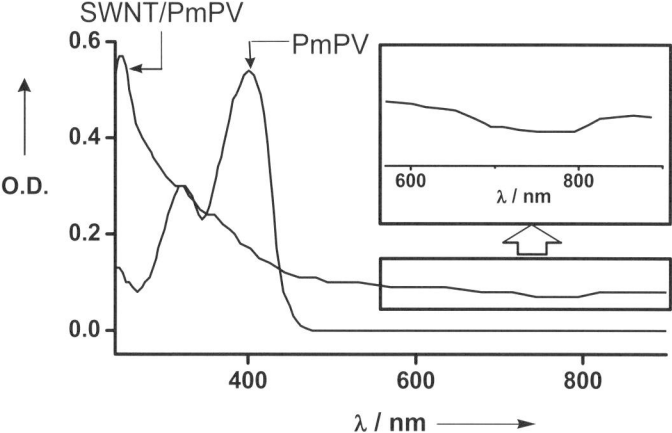

Figure 5.21. UV/vis spectra of PmPV and the SWNT/PmPV complex (After Star *et al.*, 2001, copyright 2001 Wiley-VCH Verlag)

Similarly, conjugated-polymer-wrapped carbon nanotubes have also been prepared by *in situ* polymerization of phenylacetylene, PA, in the presence of shortened SWNTs (Xu and Tang, 1999). Characterization of the resulting material by GPC, NMR, UV, FL SEM/TEM and XRD revealed that the PPA chains thickly wrap nanotubes. The SWNT/PPA composites exhibit a high solubility in organic solvents, good photostability even under harsh laser irradiation, and excellent optical limiting properties.

5.9 Modification of Aligned Carbon Nanotubes

The use of carbon nanotubes, either in aligned or non-aligned form, for many applications, including as the reinforcement agent for making advanced polymer composites and as biosensors/artificial muscle in biomedical devices, inevitably requires the modification of their surface characteristics to meet the specific requirements for particular applications. Due to the rather harsh conditions involved, however, most of the above-mentioned solution reactions would cause detrimental damage to aligned carbon nanotubes. Given that the alignment is an additional advantage for the use of carbon nanotubes in many devices, a particularly attractive option is the surface modification of carbon nanotubes while largely retaining their structural integrity.

5.9.1 Plasma Activation of Aligned Carbon Nanotubes

Dai and co-workers (Chen Q. *et al.*, 2001; Huang and Dai 2002) have developed a novel approach for the chemical modification of aligned carbon nanotubes by carrying out radio-frequency glow-discharge plasma treatment, and subsequent reactions characteristic of the plasma-induced surface groups. For instance, they have successfully immobilized polysaccharide chains onto acetaldehyde-plasma-activated carbon nanotubes through the Schiff-base formation, followed by reductive stabilization of the Schiff-base linkage with sodium cyanoborohydride in water under mild conditions [Equation (5.21)].

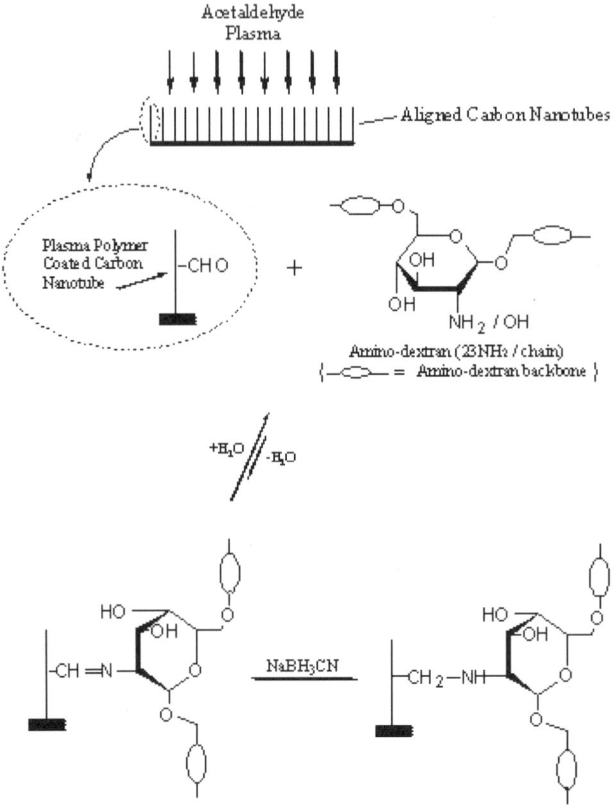

(5.21)

The resulting amino-dextran-grafted nanotube film showed zero air/water contact angles. The (acetaldehyde) plasma-treated carbon nanotube film gave relatively low advancing (90°), sessile (78°) and receding (45°) air/water contact angles compared to the advancing (155°), sessile (146°) and receding (122°) angles for an untreated sheet of aligned carbon nanotubes (Chen Q. *et al.*, 2001). The glucose units within the surface-grafted amino-dextran chains can be further converted into dialdehyde moieties by periodate oxidation (Dai *et al.*, 2000), thereby providing a method for creating multilayer coaxial structures.

These authors have also used the plasma technique for purifying and controlled opening of the aligned MWNTs. As can be seen in Figure 5.22, water plasma etching removed the catalyst tip of aligned carbon nanotubes without damaging the aligned structure (Huang and Dai 2002). Therefore, the plasma-etching method could facilitate the surface modification of the nanotube inner wall.

Figure 5.22. Aligned carbon nanotubes (a) before and (b) after the tip removal by water plasma etching (After Huang and Dai, 2002, copyright 2002 The American Chemical Society. Reproduced with permission)

Similarly, Khare *et al.* (2002) have recently reported the cold hydrogen-plasma treatment for SWNTs. *In situ* FTIR measurements revealed CH_2 and CH_3 bands at 2924, 2955, 2871, 2863 and 2854 cm^{-1} as well as the bending mode of C–H vibrations at 1370 and 1459 cm^{-1} after the plasma irradiation of hydrogen. The plasma modification was further confirmed by the deuterium replacement, as indicated by the C–D stretching bands at 2000 and 2250 cm^{-1}.

5.9.2 Acid Oxidation with Structural Protection

In view of the fact that oxidative treatment of aligned carbon nanotubes in solution for generating carboxylic and/or hydroxyl groups for further chemical modification often leads to total collapse of the aligned nanotubes, Meyyappan and co-workers (Nguyen *et al.*, 2002) developed a very effective method to protect the alignment structure by filling the gaps between the aligned nanotubes with a spin-on glass (SOG) prior to the oxidative reaction (Figure 5.23).

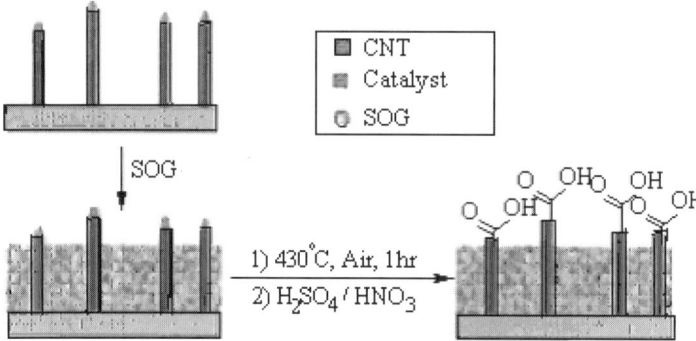

Figure 5.23. Schematic representation of the fabrication and pretreatment of the aligned carbon nanotubes for further functionalization (After Nguyen *et al.*, 2002, copyright 2002 The American Chemical Society. Reproduced with permission)

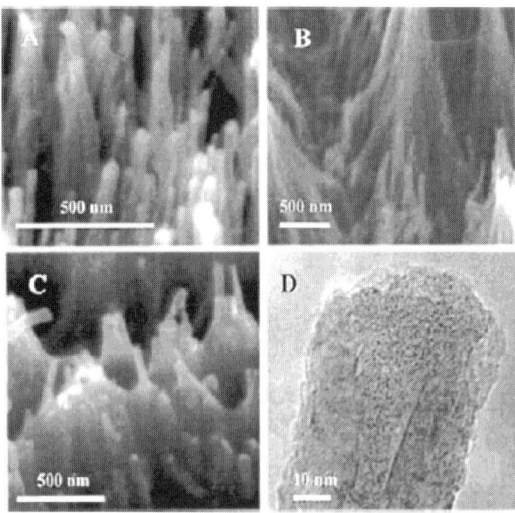

Figure 5.24. SEM images of (A) Vertically aligned carbon nanotube arrays; (B) the collapsed carbon nanotube array without SOG after the oxidative treatment; (C) the aligned carbon nanotube array with SOG after the oxidative treatment; (D) high-resolution TEM image of the opened carbon nanotube (After Nguyen *et al.*, 2002, copyright 2002 The American Chemical Society. Reproduced with permission)

Figure 5.24 shows the oxidation-induced changes in alignment for the aligned carbon nanotubes with and without the SOG protection. Comparing Figure 5.24(C) with Figure 5.24(B) indicates that the SOG film provides structural support to the aligned carbon nanotubes (Nguyen *et al.*, 2002). After having successfully functionalized the tips of the aligned carbon nanotubes with carboxylic groups, these authors demonstrated accomplishments in chemical coupling of nucleic acids to the aligned nanotube array using the standard water-soluble coupling reagents [*i.e.* 1-ethyl-3-(3-dimethylaminopropyl) carbodiimide hydrochloride, EDC, and N-hydroxysulfo-succinimide, sulfo-NHS], as schematically shown in Equation (5.22).

$$1 = H_2N(CH_2)_6 - R_1 - Cy3$$
$$R_1 = AAAAACCCTTTTTTGGAAAAA$$
(5.22)

5.9.3 Electrochemical Modification of Aligned Carbon Nanotubes

Another interesting area closely related to the above chemical modification of the aligned nanotube surface is the synthesis of the coaxial nanowires of polymers and carbon nanotubes. In this regard, Dai and co-workers (Gao *et al.*, 2000) have recently used the aligned carbon nanotubes as nanoelectrodes for making novel conducting coaxial nanowires by electrochemically depositing a concentric layer of an appropriate conducting polymer uniformly onto each of the aligned nanotubes to form the aligned conducting polymer-coated carbon nanotube coaxial nanowires (CP-NT). The SEM image for these CP-NT coaxial nanowires given in Figure 5.25(b) shows the same features as the aligned nanotube array of Figure 5.25(a), but the former has a larger tubular diameter due to the presence of the newly-electropolymerized polypyrrole coating.

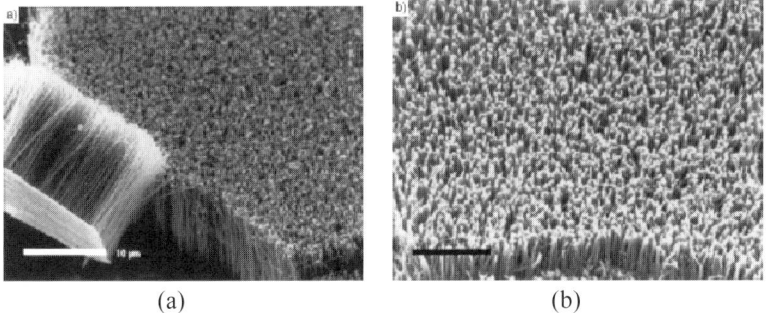

(a)   (b)

Figure 5.25. (a) Aligned nanotubes and (b) CP-NT coaxial nanowires (After Gao *et al.*, 2000, copyright 2000 Wiley-VCH Verlag)

The presence of the conducting polymer layer was also clearly evident in the transmission electron microscopic images (Chapter 7) and by the much stronger redox responses of the conducting polymer layer coated on the nanotube surface than that of the same polymer on a conventional flat conducting electrode under the same condition (Gao et al., 2000). The coaxial structure allows the nanotube framework to provide mechanical stability and efficient thermal and electrical contact with the conducting polymer layer. The large interfacial surface area per unit mass obtained for the nanotube-supported conducting polymer layer is potentially useful in many optoelectronic applications, for example in sensors, organic light-emitting diodes, and photovoltaic cells where the charge injection and separation are strongly limited by the interfacial area available in more conventional devices.

5.10 References

Ajayan, P. M. , Colliex, C., Lambert, J.M., Bernier, P., Barbedette, L., Tence, M., Stephan, O. (1994) *Phys. Rev. Lett.* **72**, 1722.
Ajayan, P.M., Ebbesen, T.W. (1997) *Rep. Prog. Phys.* **60**, 1025.
Ajayan, P.M., Ebbesen, T.W., Ichihashi, T., Iijima, S., Tanigaki, K., Hiura, H. (1993) *Nature* **362**, 522.
Ajayan, P.M., Ichihashi, T., Ijiima, S. (1993) *Chem. Phys. Lett.* **202**, 384.
Ajayan, P.M., Iijima, S. (1993) *Nature* **361**, 333.
Ajayan, P.M., Stephan, O., Redlich, P., Colliex, C. (1995) *Nature* **375**, 564.
Ajayan, P.M., Terrones, M., de la Guardia, A., Huc, V., Grobert, N., Wei, B.Q., Lezec, H., Ramanath, G., Ebbesen, T.W. (2002) *Science* **296**, 705.
Ajiki, H., Ando, T. (1993) *J. Phys. Soc. Jpn.* **62**, 1255.
Amelinckx, S., Zhang, X.B., Bernaerts, D., Zhang, X.F., Ivanov, V., Nagy, J.B. (1994) *Science* **265**, 235.
Ando, Y., Zhao, X., Ohkohchi, M. (1998) *Jpn. J. Appl.* **37**, L61.
Avouris, P. (2002) *Acc. Chem. Res.* **35**, 1026.
Bahr, J.L., Yang, J., Kosynkin, D.V., Bronikowski, M.J., Smalley, R.E., Tour, J.M. (2001) *J. Am. Chem. Soc.* **123**, 6536.
Bahr, J.L., Tour, J.M. (2002) *J. Mater. Chem.* **12**, 1952, and references cited therein.
Bandow, S., Rao, A.M., Williams, K.A., Thess, R., Smalley, E., Eklund, P.C. (1997) *J. Phys. Chem. B* **101**, 8839.
Bethune, D.S., Kiang, C.H., de Vries, M.S., Gorman, G., Savoy, R., Vazquez, J., Beyers, R. (1993) *Nature* **363**, 605.
Bernaerts, D., Zhang, X.B., Zhang, X.F., Amelinckx, S., Vantendeloo, G., Vanlanduyt, J., Ivanov, V., Nagy, J.B. (1995) *Philos. Mag. A* **71**, 605.
Birkett, P.R., Terrones, M. (1999) *Chem. Britain* **May**, 45.
Biro, L.P., Khanh, N.Q., Vertesy, Z., Horvath, Z.E., Osvath, Z., Koos, A., Gyulai, J., Kocsonya, A., Konya, Z., Zhang, X.B., Van Tendeloo, G., Fonseca, A., Nagy, J.B. (2002) *Mater. Sci. Eng. C: Biomi. Supramol. Sys.* **C19**, 9.
Bockrath, M., Cobden, D.H., McEuen, P.L., Chopra, N.G., Zettl, A., Thess, A., Smalley, R.E. (1997) *Science* **275**, 1922.
Bonard, J.-M., Stora, T., Salvetat, J.-P., Maier, F., Stockli, T. , Duschl, C., Forró, L., de Heer, W.A., Chatelain, A. (1997) *Adv. Mater.* **9**, 827.
Boul, P.J., Liu, J., Mickelson, E.T., Huffman, C.B., Ericson, L.M., Chiang, I.W., Smith, K.A., Colbert, D.T., Hauge, R.H., Margrave, J.L., Smalley, R.E. (1999) *Chem. Phys. Lett.* **310**, 367.

Carroll, D.L., Redlich, P., Ajayan, P.M., Charlier, J.C., Blase, X., De Vita, A., Car, R. (1997) *Phys. Rev. Lett.* **78**, 2811.
Che, G., Lakshmi, B.B., Martin, C.R., Fisher, E.R., Ruoff, R.S. (1998) *Chem. Mater.* **10**, 260.
Chen, J., Hamon, M.A., Hu, H., Chen, Y., Rao, A.M., Eklund, P.C., Haddon, R.C. (1998) *Science* **282**, 95.
Chen, J., Rao, A.M., Lyuksyutov, S., Itkis, M.E., Hamon, M.A., Hu, H., Cohn, R.W., Eklund, P.C., Colbert, D.T., Smalley, R.E., Haddon, R.C. (2001) *J. Phys. Chem. B* **105**, 2525.
Chen, Q., Dai, L., Gao, M., Huang, S., Mau, A. (2001) *J. Phys. Chem. B* **105**, 618.
Chen, R.J., Zhang, Y., Wang, D., Dai, H. (2001) *J. Am. Chem. Soc.* **123**, 3838.
Chen, Y., Haddon, R.C., Fang, S., Rao, A.M., Eklund, P.C., Lee, W.H., Dickey, E.C., Grulke, E.A., Pendergrass, J.C., Chavan, A., Haley, B.E., Smalley, R.E. (1998) *J. Mater. Res.* **13**, 2423.
Chen, Y.K., Chu, A., Cook, J., Green, M.L.H., Harris, P.J.F., Heesom, R., Humphries, M., Sloan, J., Tsang, S.C., Turner, J.F.C. (1997) *J. Mater. Chem.* **7**, 545.
Chen, Y.K., Green, M.L.H., Griffin, J.L., Hammer, J., Lago, R.M., Tsang, S.K. (1996) *Adv. Mater.* **8**, 1012.
Chen, Y.K., Green, M.L.H., Tsang, S.C. (1996) *J. Chem. Soc., Chem. Commun.* 2489.
Chico, L., Crespi, V.H., Benedict, L.X., Louie, G., Cohen, L. (1996) *Phys. Rev. Lett.* **76**, 971.
Chiu, P.W., Duesberg, G.S., Dettlaff-Weglikowska, U., Roth, S. (2002) *Appl. Phys. Lett.* **80**, 3811.
Colbert, D.T., Zhang, J., McClure, S.M., Nikolaev, P., Chen, Z, Hafner, J.H., Owens, D.W., Kotula, P.G., Carter, C.B., Weaver, J.H., Rinzler, A.G., Smalley, R.E. (1994) *Science* **266**, 1218.
Collins, P.G., Avouris, P. (2000) Sci. Amer. **December**, 62.
Collins, P.G., Zettl, A., Bando, H., Thess, A., Smalley, R.E. (1997) *Science* **278**, 100.
Cornwell, C.F., Wille, L.T. (1997) *Solid State Commun.* **101**, 555.
Czerw, R. , Guo, Z., Ajayan, P.M., Sun, Y.-P., Carroll, D.L. (2001) *Nano Lett.* **1**, 423.
Dai, L., Mau, A.W.H. (2001) *Adv. Mater.* **13**, 899.
Dai, L., St John, H.A.W., Bi, J., Zientek, P., Chatelier, R.C., Griesser, H.J. (2000) *Surf. Interf. Analysis* **29**, 46.
Dai, H., Wong, E.W., Lieber, C.M. (1996) *Science* **272**, 523.
Dai, H., Wong, E.W., Lu, Y.Z., Fan, S., Lieber, C. (1995) *Nature* **375**, 769.
Delpeux, S., Metenier, K., Benoit, R., Vivet, F., Boufendi, L., Bonnamy, S., Begui, F. (1999) *AIP Confer. Proc.* **486**, 470.
Dillon, A.C., Gennett, T., Jones, K.M., Alleman, J.L.,, Parilla, P.A., Heben, M.J. (1999) *Adv. Mater.* **11**, 1354.
Doorn, S.K., Fields, R.E., III, Hu, H., Hamon, M.A., Haddon, R.C., Selegue, J.P., Majidi, V. (2002) *J. Am. Chem. Soc.* **124**, 3169.
Dresselhaus, M. (1996) *Phys. World* **9**, 18.
Dresselhaus, M., Dresselhaus, G., Eklund, P. (1996) *Science of Fullerenes and Carbon Nanotubes*, Academic Press, San Diego.
Dresselhaus, M., Dresselhaus, G., Eklund, P., Saito, R. (1998) *Phys. World* **January**, 33.
Drexler, K.E. (1992) *Nanosystems: Molecular Machinery, Manufacturing, and Computation*, John Wiley & Sons Inc., New York.
Dujardin, E., Ebbesen, T.W., Hiura, H., Tanigaki, K. (1994) *Science* **265**, 1850.
Dujardin, E., Ebbesen, T.W., Krishnan, A., Treacy, M..M.J. (1998) *Adv. Mater.* **10**, 611.
Ebbesen, T.W. (1994) *Annu. Rev. Mater. Sci.* **24**, 235.
Ebbesen, T.W., Ajayan, P.M. (1992) *Nature* **358**, 220.
Ebbesen, T.W., Ajayan, P.M., Hiura, H., Tanigaki, K. (1994) *Nature* **367**, 519.
Ebbesen, T.W., Lezec, H.J., Hiura, H., Bennett, J.W., Ghaemi, H.F., Thio, T. (1996) *Nature* **382**, 54.
Endo, M., Takeuchi, K., Igarashi,S., Kobori, K., Shiraishi, M., Kroto, H.W. (1993) *J. Phys. Chem. Solids* **54**, 1841.

Falvo, M.R., Clary, G.J., Taylor II, R.M., Chi, V., Brooks, F.P. Jr, Washburn, S., Superfine, R. (1997) *Nature* **389**, 582.
Frank, S., Poncharal, P., Wang, Z.L., de Heer, W.A. (1998) *Science* **280**, 1744.
Freemantle, M. (1998) *C&E News,* **Feb. 16**, 4.
Fu, K., Huang, W., Lin, Y., Riddle, L.A., Carroll, D.L., Sun, Y.-P. (2001) *Nano Lett.* **1**, 439.
Gao, M., Huang, S., Dai, L., Wallace, G., Gao, R., Wang, Z. (2000) *Angew. Chem. Int. Ed.* **39**, 3664.
Ge, M., Sattler, K. (1993) *J. Phys. Chem. Solids* **54**, 1871.
Georgakilas, V., Kordatos, K., Prato, M., Guldi, D.M., Holzinger, M., Hirsch, A. (2002) *J. Am. Chem. Soc.* **124**, 760.
Guerret-Piecourt, C., Bouar, Y.L., Loiseau, A., Pascard, H. (1994) *Nature* **372**, 159.
Hamada, N., Sawada, S., Oshiyama, A. (1992) *Phys. Rev. Lett.* **68**, 1579.
Hamon, M.A., Chen, J., Hu, H., Chen, Y., Itkis, M.E., Rao, A.M., Eklund, P.C., Haddon, R.C. (1999) *Adv. Mater.* **11**, 834.
Hamon, M.A., Hui, H., Bhowmik, P., Itkis, H.M.E., Haddon, R.C. (2002) *Appl. Phys. A: Mater. Sci. Pro.* **74**, 333.
Han, W., Fan, S., Li, Q., Hu, Y. (1997) *Science* **277**, 1287.
Harris, P.J.F. (2001) *Carbon Nanotubes and Related Structures - New Materials for the Twenty-First Century*, Cambridge University Press, Cambridge.
Harutyunyan, A.R., Pradhan, B.K., Chang, J., Chen, G., Eklund, P.C. (2002) *J. Phys. Chem. B* **106**, 8671.
Hayashi, T., Terrones, M., Scheu, C., Kim, Y.A., Ruehle, M., Nakajima, T., Endo, M. (2002) *Nano Lett.* **2**, 491.
Hertel, T., Martel, R., Avouris, P. (1998) *J. Phys. Chem. B* **102**, 910.
Hirsch, A. (2002) *Angew. Chem. Int. Ed.* **41**, 1853, and references cited therein.
Hiura, H., Ebbesen, T.W., Tanigaki, K. (1995) *Adv. Mater.* 7, 275.
Hsu, W.K., Hare, J.P., Terrones, M., Kroto, H.W., Walton, D.R.M., Harris, P.J.F. (1995) *Nature* **337**, 687.
Hsu, W.K., Terrones, M., Hare, J.P., Kroto, H.W., Walton, D.R.M. (1996) *Chem. Phys. Lett.* **262**, 161.
Huang, S., Dai, L. (2002) *J. Phys. Chem. B* **106**, 3543.
Hwang, K.C. (1995) *J. Chem. Soc., Chem. Commun.* 173.
Iijima, S. (1991) *Nature* **354**, 56.
Iijima, S., Brabec, C., Maiti, A., Bernholc, J. (1996) *J. Chem. Phys.* **104**, 2089.
Iijima, S., Ichihashi, T. (1993) *Nature* **363**, 603.
Jacoby, M. (1998) *C & E News* **February 16**, 4.
Jiang, K., Li, Q., Fan, S. (2002) *Nature* **419**, 801.
Jishi, R.A.; Inomata, D., Nakao, K., Dresselhaus, M.S., Dresselhaus, G. (1994) *J. Phys. Soc. Jpn.* **63**, 2252.
Journet, C., Maser, W.K., Bernier, P., Loiseau, A., Lamy de la Chapelle, M., Lefrant, S., Deniard, P., Lee, R., Fischer, J.E. (1997) *Nature* **388**, 756.
Khare, B.N., Meyyappan, M., Cassell, A.M., Nguyen, C.V., Han, J. (2002) *Nano Lett.* **2**, 73.
Kiang, C.H., Goddard, W.A., Beyers, R., Salem, J.R., Bethune, D.S. (1994) *J. Phys. Chem.* **98**, 6612.
Kooi, S.E., Schlecht, U., Burghard, M., Kern, K. (2002) *Angew. Chem. Int. Ed.* **41**, 1353.
Kosaka, M., Ebbesen, T.W., Hiura, H., Tanigaki, K. (1995) *Chem. Phys. Lett.* **233**, 47.
Kosakovskaya, Z.Y., Chernozatonskii, L.A., Fedorov, E.A. (1992) *JEPT Lett.* **56**, 26.
Koshio, A., Yudasaka, M., Iijima, S. (2001a) *Chem. Phys. Lett.* **341**, 461.
Koshio, A., Yudasaka, M., Zhang, M., Iijima, S. (2001b) *Nano Lett.* **1**, 361.
Kuznetsova, A., Popova, I., Yates, J.T. Jr., Bronikowski, M.J., Huffman, C.B., Liu, J., Smalley, R.E., Hwu, H.H., Chen, J.G. (2001) *J. Am. Chem. Soc.* **123**, 10699.
Kyotani, T., Nagai, T., Inoue, S., Tomita, A. (1997a) *Chem. Mater.* **9**, 609.
Kyotani, T., Nakazaki, S., Xu, W.-H., Tomita, A. (2001) *Carbon* **39**, 782.

Kyotani, T., Tsai, L.-F., Tomita, A. (1996) *Chem. Mater.* **8**, 2109.
Kyotani, T., Tsai, L.-F., Tomita, A. (1997b) *J. Chem. Soc., Chem. Commun.* 701.
Langer, L., Stockman, L., Heremans, J.P., Bayot, V., Olk, C.H., van Haesendonck, C., Bruynseraede, Y., Issi, J.-P. (1996) *Phys. Rev. Lett.* **76**, 479.
Lee, W.H., Kim, S.J., Lee, W.J., Lee, J.G., Haddon, R.C., Reucroft, P.J. (2001) *Appl. Surf. Sci.* **181**, 121.
Li, W.Z., Xie, S.S., Qian, L.X., Chang, B.H., Zou, B.S., Zhou, W.Y., Zhao, R.A., Wang, G. (1996) *Science* **274**, 1701.
Li, X., Yang, Z., Chen, Z., Wang, H., Sheng, N., Li, J. (1999) *Xinxing Tan Cailiao* **14**, 32.
Lin, N., Ding, J., Yang, S., Cue, N. (1996) *Carbon* **34**, 1295.
Liu, M., Cowley, J.M. (1995) *Carbon* **33**, 749.
Liu, J., Rinzler, A.G., Dai, H., Hafner, J.H., Bradley, R.K., Boul, P.J., Lu, A., Iverson, T., Shelimov, K., Huffman, C.B., Rodriguez-Macias, F., Shon, Y.-S., Lee, T.R., Colbert, D.T., Smalley, R.E. (1998) *Science* **280**, 1253.
Liu, Z.-J., Yuan, Z.-Y., Zhou, W., Peng, L.-M., Xu, Z. (2001) *Phys. Chem. Chem. Phys.* **3**, 2518.
Loiseau, A., Pascard, H. (1996) *Chem. Phys. Lett.* **256**, 246.
Martel, R., Shea, H.R., Avouris, P. (1999) *J. Phys. Chem. B* **103**, 7551.
Martinez, M.T., Callejas, M.A., Benito, A.M., Maser, W.K., Cochet, M., Andres, J.M., Schreiber, J., Chauvet, O., Fierro, J.L.G. (2002) *J. Chem. Soc., Chem. Comm.* 1000.
Mickelson, E.T., Chiang, I.W., Zimmerman, J.L., Boul, P.J., Lozano, J., Liu, J., Smalley, R.E., Hauge, R.H., Margrave, J.L. (1999) *J. Phys. Chem. B* **103**, 4318.
Mickelson, E.T., Huffman, C.B., Rinzler, A.G., Smalley, R.E., Hauge, R.H., Margrave, J.L. (1998) *Chem. Phys. Lett.* **296**, 188.
Murakami, Y., Shibata, T., Okuyama, K., Arai, T., Suematsu, H., Yoshida, Y. (1993) *J. Phys. Chem. Solids* **54**, 1861.
Nagasawa, S., Yudasaka, M., Hirahara, K., Ichihashi, T., Iijima, S. (2000) *Chem. Phys. Lett.* **328**, 374.
Nakajima, T., Kasamatsu, S., Matsuo, Y. (1997) *Eur. J. Inorg. Chem.* **33**, 831.
Nguyen, C.V., Delzeit, L., Cassell, A.M., Li, J., Han, J., Meyyappan, M. (2002) *Nano Lett.* **2**, 1079.
Niyogi, S., Hamon, M.A., Hu, H., Zhao, B., Bhowmik, P., Sen, R., Itkis, M.E., Haddon, R.C. (2002) *Acc. Chem. Res.* **35**, 1105.
Niyogi, S., Hu, H., Hamon, M.A., Bhomik, P., Zhao, B., Rozenzhak, S.M., Chen, J., Itkis, M.E., Meier, M.S., Haddon, R.C. (2001) *J. Am. Chem. Soc.* **123**, 733.
Niu, C., Moy, D., Chishti, A., Hoch, R. (2001) WO Patent 0107694.
O'Connell, M.J., Bachilo, S. M., Huffman, C.B., Moore, V.C., Strano, M.S., Haroz, E.H., Rialon, K.L., Boul, P.J., Noon, W.H., Kittrell, C., Ma, J., Hauge, R.H., Weisman, R.B., Smalley, R.E. (2002) *Science* **297**, 593.
O'Connell, M.J., Boul, P., Ericson, L.M., Huffman, C., Wang, Y., Haroz, E., Kuper, C., Tour, J., Ausman, K.D., Smalley, R.E. (2001) *Chem. Phys. Lett.* **342**, 265.
Odom, T.W., Huang, J.-L., Kim, P., Lieber, C.M. (2000) *J. Phys. Chem. B* **104**, 2794.
Odom, T.W., Huang, J.L., Kim, P., Lieber, C.M. (1998) *Nature* **291**, 62.
Parthasarathy, R., Phani, K.L.M., Martin, C.R. (1995) *Adv. Mater.* **7**, 896.
Pederson, M.R., Broughton, J.Q. (1992) *Phys. Rev. Lett.* **69**, 2689.
Riggs, J.E., Guo, Z., Carroll, D.L., Sun, Y.-P. (2000a) *J. Am. Chem. Soc.* **122**, 5879.
Riggs, J.E., Walker, D.B., Carroll, D.L., Sun, Y.-P. (2000b) *J. Phys. Chem. B* **104**, 7071.
Satishkumar, B.C., Govindaraj, A., Mofokeng, J., Subbanna, G.N., Rao, C.N.R. (1996) *J. Phys. B: At. Mol.Opt. Phys.* **29**, 4925.
Satishkumar, B.C., Govindaraj, A., Vogl, E.M., Basumallick, L., Rao, C.N.R. (1997) *J.Mater. Res.* **12**, 604.
Saito, R., Dresselhaus, G., Dresselhaus, M.S. (1996) *Phys. Rev. B* **53**, 2044.
Saito, R., Fujita, M., Dresselhaus, G., Dresselhaus, M.S. (1992) *Appl. Phys. Lett.* **60**, 2204.

Saito, Y., Nishikubo, K., Kawabata, K., Matsumoto, T. (1996) *J. Appl. Phys.* **80**, 3062.
Saito, Y., Yoshikawa, T., Okuda, M., Fujimoto, N., Sumiyama, K., Suzuki, K., Kasuya, A., Nishina, Y. (1993) *J. Phys. Chem. Solids* **54**, 1849.
Sano, M., Kamino, A., Shinkai, S. (2001a) *Angew. Chem. Int. Ed.* **40**, 4661.
Sano, M., Kamino, A., Okamura, J., Shinkai, S. (2001b) *Science* **293**, 1299.
Sen, R., Govindaraj, A., Rao, C.N.R. (1997) *Chem. Mater.* **9**, 2078.
Seraphin, S., Zhou, D., Jiao, J., Withers, J.C., Loufty, R. (1993) *Nature* **362**, 503.
Setlur, A.A., Lauerhaas, J.M., Dai, J.Y., Chang, R.P.H. (1996) *Appl. Phys. Lett.* **69**, 345.
Sloan, J., Cook, J., Green, M.L.H., Hutchison, J., Tenne, R. (1997) *J. Mater. Chem.* **7**, 1089.
Sloan, J., Hammer, J., Zwiefka-Sibley, M., Green, M.L.H. (1998) *J. Chem. Soc., Chem. Commun.* 347.
Star, A., Stoddart, J.F., Steuerman, D., Diehl, M., Boukai, A., Wong, E.W., Yang, X., Chung, S.-W., Choi, H., Heath, J.R. (2001) *Angew. Chem. Int. Ed.* **40**, 1721.
Sun, Y.P., Fu, K.F., Lin, Y., Huang, W.J. (2002) *Acc. Chem. Res.* **35**, 1096.
Sun, Y., Wilson, S.R., MacMahon, S., Schuster, D.I. (2000) *Proc. Electrochem.Soc.* 282.
Sun, Y., Wilson, S. R., Schuster, D. I. (2001) *J. Am. Chem. Soc.* **123**, 5348.
Sun, Y.-P., Huang, W., Lin, Y., Fu, K., Kitaygorodskiy, A., Riddle, L.A., Yu, Y.J., Carroll, D.L. (2001) *Chem. Mater.* **13**, 2864.
Tamura, R., Tsukada, M. (1995) *Phys. Rev. B* **52**, 6015.
Tamura, R., Tsukada, M. (1997) *Solid State Commun.* **101**, 601.
Tanaka, K., Toda, F. (2000) *Chem. Rev.* **100**, 1025.
Tans, S.J., Devoret, M.H., Dai, H., Thess, A., Smalley, R.E., Geerligs, L.J., Dekker, C. (1997) *Nature* **386**, 474.
Thess, A., Lee, R., Nikolaev, P., Dai, H., Petit, P., Robert, J., Xu, C., Lee, Y.H., Kim, S.G., Rinzler, A.G., Rinzler, D., Colbert, T., Scuseria, G.E., Tomanek, D., Fischer, J.E., Smalley, R.E. (1996) *Science* **273**, 483.
Tohji, K., Goto, T., Takahashi, H., Shinoda, Y., Shimizu, N., Jeyadevan, B., Matsuoka, I., Saito, Y., Kasuya, A., Ohsuna, T., Hiraga, K., Nishina, Y. (1996) *Nature* **383**, 679.
Tohji, K., Takahashi, H., Shinoda, Y., Shimizu, N., Jeyadevan, B., Matsuoka, I., Saito, Y., Kasuya, A., Ito, S., Nishina, Y. (1997) *J. Phys. Chem. B*, **101**, 1974.
Tsang, S.C., Chen, Y.K., Harris, P.J.F., Green, M.L.H. (1994) *Nature* **372**, 159.
Tsang, S.C., Harris, P.J.F., Green, M.L.H. (1993) *Nature* **362**, 520.
Rao, C.N.R., Govindaraj, A. (2002) *Acc. Chem. Res.* **35**, 998.
Ugate, D., Châtelain, A., de Heer, W.A. (1996) *Science* **274**, 1897.
Vaccarini, L., Goze, C., Aznar, R., Micholet, V., Journet, C., Bernier, P. (1999) *Synth. Met.* **103**, 2492.
Venema, L.C., Wildoer, J.W.G., Temminck Tuinstra, H.L.J., Dekker, C., Rinzler, A.G., Smalley, R.E. (1997) *Appl. Phys. Lett.* **71**, 2629.
Wong, E.W., Sheehan, P.E., Lieber, C.M., (1997) *Science* **277**, 1971.
Xu, H., Tang, B. Z. (1999) *Polym. Mater. Sci. Eng.* **80**, 408.
Yamamoto, K., Koga, Y., Fujiwara, S., Kubota, M. (1996) *Appl. Phys. Lett.* **69**, 4174.
Yakobson, B.I., Brabec, C.J., Bernholc, J. (1996) *Phys. Rev. Lett.* **76**, 2511.
Yang, Z.-H., Wu, H.-Q. (2001) *Gaodeng Xuexiao Huaxue Xuebao* **22**, 446.
Yu, R., Chen, L., Liu, Q., Lin, J., Tan, K.-L., Ng, S.C., Chan, H.S.O., Xu, G.-Q., Hor, T.S.A. (1998) *Chem. Mater.* **10**, 718.
Yu, Z., Brus, L. E. (2000) *J. Phys. Chem. A* **104**, 10995.
Yudanov, N.F., Okotrub, A.V., Shubin, Y.V., Yudanova, L.I., Bulusheva, L.G., Chuvilin, A.L., Bonard, J.-M. (2002) *Chem. Mater.* **14**, 1472.
Yudasaka, M., Zhang, M., Jabs, C., Iijima, S. (2000) *Appl. Phys. A: Mater. Sc. Proc.* **71**, 449.
Zhang, M., Yudasaka, M., Koshio, A., Jabs, C., Ichihashi, T., Iijima, S. (2002) *Appl. Phys. A: Mater. Sci. Proc.* **74**, 7.
Zhao, W., Song, C., Zheng, B., Liu, J., Viswanathan, T. (2002) *J. Phys. Chem. B* **106**, 293.

Part II
From Intelligent Macromolecules to Smart Devices

Chapter 6

Oriented and Patterned Macromolecules

6.1 Introduction

Just as the microchips have revolutionized computers and electronics, nanotechnology has the potential to revolutionize many industrial sectors. Since the discovery of carbon nanotubes by Iijima in 1991, carbon nanotubes have exhibited superior properties for many potential applications (*e.g.* flat panel displays, molecular transistors, scanning probe microscope tips, molecular-filtration membranes and artificial actuators). However, they often need to be aligned or micropatterned, in a similar fashion as silicon or metals, for the fabrication of optoelectronic, and many other, devices. In this chapter, some important work on the development of aligned and micropatterned carbon nanotubes is summarized, along with a brief discussion on some aligned non-carbon nanotubes.

On the other hand, polymers have traditionally been used in many areas of the building and packaging industries for their high mechanical strength and light weight. Due to their excellent electrical insulation properties, polymers have also been widely used as passivation and insulating materials (*e.g.* as protection coatings and intermetal dielectrics) in electronic devices. These applications have had a profound influence on our everyday lives. During the past 20 years or so, however, various conjugated polymers have been synthesized with unusual electrical, magnetic and optical properties owing to the substantial π-electron delocalization along their backbones (Dai, 1999). Having a conjugated, all-carbon structure with unusual molecular symmetries, carbon nanotubes have recently been shown also to possess interesting electronic, photonic, magnetic and mechanical properties (Dai and Mau, 2001; Harris, 2001). These properties make conjugated polymers very attractive for their potential applications in electronic and photonic devices. The construction of oriented and/or patterned conjugated polymers is a key prerequisite for most of these and many other applications, including optoelectronic displays, integrated circuits, field-effect transistors, and optical memory storages. The progress towards advanced microfabrication and the construction of oriented and patterned conjugated polymers and carbon nanotbues is reviewed in this chapter. Although we are mainly concerned here with conjugated polymer and carbon nanotubes due to their significance as intelligent materials, most of the fabrication techniques developed from these particular systems could be applied to a wide range of macromolecules.

6.2 Oriented and Patterned Conjugated Polymers

6.2.1 The Necessity

6.2.1.1 For Electronic Applications

The bulk conductivity of conducting polymers should consist of contributions, at least, from the intra-chain and inter-chain electron transportations (Avlyanov *et al.*, 1995). While the intra-chain diffusion of the charge carriers along the conjugated backbone plays a dominant role in the charge transporting process, the inter-chain transport (via hopping, tunneling, *etc.*) also has a significant influence on the bulk conductivity, since a single chain does not extend throughout the entire length of a sample in most practical applications. The competition between the intra- and inter-chain transports has been discussed based on various theoretical models. Using scaling arguments, for example, de Gennes (de Gennes, 1985) and Heeger and Smith (1991) have independently demonstrated that if the mean lifetime of the charge carrier on the polymer chain (τ_c) is much greater than the time required for a charge carrier to completely explore the polymer chain (τ_i), the conductivity (σ) is limited by inter-chain hopping and is given by:

$$\sigma = (ne^2/kT)a(L/\tau_c) \qquad (6.1)$$

where a is the persistence length of the polymer chain (Dai, 1993), L is the total length of the macromolecule, n is the charge density, e is the charge per carrier, k is the Boltzmann constant and T is the absolute temperature.

On the other hand, if $\tau_i > \tau_c$, the conductivity is independent of molecular weight and σ is given by:

$$\sigma = (ne^2/kT)a(D_i/\tau_c)^{1/2} \qquad (6.2)$$

with $D_i \propto L^2/\tau_i$ being the diffusion coefficient for the charge carriers along the polymer chain.

Pearson *et al.* (1993) derived the same equations through a quantitative analysis and obtained reasonable agreement with some existing experimental data. From Equations (6.1) and (6.2), it can be seen that σ increases with increasing persistence length. This can be caused, for example, by alignment of the polymer chains (*vide infra*).

6.2.1.2 For Non-linear Optical Applications

The necessity for aligning and/or patterning conjugated polymers can also be seen from some recent developments in electroluminescent displays (Bradley *et al.*, 1992; Burroughes *et al.*, 1990; Greenham and Friend, 1995) and non-linear optical (NLO) devices (Messier *et al.*, 1989; Williams, 1983), although in these areas conjugated polymers are used in the non-conducting state. For instance, polymeric light-emitting diodes (LEDs) with polarized light emissions have been made using stretch-oriented (Dyreklev *et al.*, 1995; Lemmer *et al.*, 1996), rubbing-aligned (Hamaguchi and Yoshino, 1995), Langmuir-Blodgett (LB) deposited (Cimrová *et al.*, 1996), and specifically synthesized liquid crystalline (Lüssem *et al.*, 1995, 1996) films of appropriate electroluminescent (EL) polymers, while patterned (multiple-color) emissions have been reported for LEDs based on certain photolithographically patterned (Halliday *et al.*, 1993; Renak *et al.*, 1997; Schmid *et al.*, 1993), and phase-separated conjugated polymers (Berggren *et al.*, 1994). Due to their delocalized π-electron orbitals, some conjugated polymers have also been demonstrated to possess large, ultrafast non-linear optical responses (Arbogast *et al.*, 1991; Dagani, 1996a; Hann and Bloor, 1989; Mukamel *et al.*, 1994; Nalwa, 1993). Non-linear optical processes occur when a medium is polarized by an intense electric field (E) (*e.g.* the one associated with a strong pulse of light), which creates an induced dipole moment (μ_{ind}). μ_{ind} can be expressed as a power series at the molecular level (Hann and Bloor, 1989):

$$\mu_{ind} = \alpha E + \beta EE + \gamma EEE + \cdots \quad (6.3)$$

At the bulk level, an analogous equation is used for the induced polarization (P):

$$P = \chi^{(1)}E + \chi^{(2)}EE + \chi^{(3)}EEE + \cdots \quad (6.4)$$

where $\chi^{(1)}$ is the linear susceptibility which describes the linear response associated with ordinary refraction and absorption; $\chi^{(2)}$ and $\chi^{(3)}$ are the second- and third-order non-linear optical susceptibilities.

As $\chi^{(2)}$ is a third-rank tensor, it is effective only for non-centrosymmetric media. Therefore, the medium used for the second-order non-linear optical effect must have a non-centrosymmetric ordering of the NLO dipoles, possibly by spontaneous ordering or by electric field poling (Prasad, 1991). In contrast, order is not required for third-order non-linear materials (*e.g.* conjugated polymers), but the $\chi^{(3)}$ of a given material is given by (Kajzar *et al.*, 1992):

$$<\chi^{(3)}> = \chi_{xxxx}^{(3)} <\cos^4\theta> \quad (6.5)$$

where $\chi_{xxxx}^{(3)}$ is the third-order susceptibility of an individual macromolecule along the polymer chain direction (x), θ is the angle between the polymer chain and the exciting optical electric field and < > denotes an average over all polymer chain orientations. For certain special cases, the $<\cos^4\theta>$ is given by (Kajzar *et al.*, 1992):

$$<\cos^4\theta> = \begin{cases} 1 & \text{for a uniaxially aligned polymer system} \\ 3/8 & \text{for a bidimensionally disordered system} \\ & \text{(all polymer chains are parallel to a plane} \\ & \text{but randomly disoriented within the plane)} \\ 1/5 & \text{for a three-dimensional disorder} \end{cases} \quad (6.6)$$

Equations (6.5) and (6.6) imply that an increase of a factor of 5 in cubic susceptibility (and a factor of at least 25 in efficiency) can be obtained by aligning a completely disordered system into a uniaxially oriented one (*e.g.* single crystal) (Kajzar *et al.*, 1992).

As can be seen from the above discussion, microconstruction of conjugated polymers is both technologically important and scientifically interesting. As such, a variety of molecularly aligned conjugated polymer systems have been developed, for example, through ordered-matrix-assisted syntheses (Aldissi, 1989; Shirakawa *et al.*, 1988), template syntheses (Cahalane and Labes, 1989; Martin, 1995, 1996), mechanical stretching (Hagler *et al.*, 1991; Joo *et al.*, 1997; Shirakawa *et al.*, 1973), and Langmuir-Blodgett manipulation of the polymer chains (Ando *et al.*, 1989). On the other hand, various techniques including photolithography (Schanze *et al.*, 1996), self-assembly (Rozsnyai and Wrighton, 1994, 1995; Rubner *et al.*, 1991), polymeric phase separation (Berggren *et al.*, 1994) and plasma treatment/polymerization (Dai *et al.*, 1997) have now been used for microlithographic patterning of conjugated polymers.

6.2.2 Oriented Conjugated Polymers

6.2.2.1 Synthesis-induced Orientation

Due to its simple conjugated structure, polyacetylene has served as the prototype for other conducting polymers. The first synthesis of polyacetylene in a film form was reported as early as 1974 (Ito *et al.*, 1974). Initially, the Ziegler-Natta polymerization of acetylene could only produce polyacetylene films with a randomly-oriented fibrillar morphology (Figure 6.1) (Chien, 1984; Ito *et al.*, 1974).

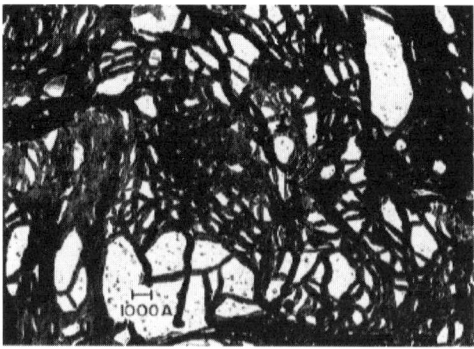

Figure 6.1. A typical SEM micrograph of thin polyacetylene films prepared by the Ziegler-Natta catalyst (After Ito *et al.*, 1974, copyright 1974 John Wiley & Sons, Inc.)

In 1987, however, Naarmann and co-workers (Naarmann and Theophilou, 1987; Theophilou *et al.*, 1987) reported a method to prepare highly oriented polyacetylenes by aging the catalyst mixture of $Ti(OC_4H_9)_4/Al(C_2H_5)_3$ in silicone oil, which was used as a reaction medium. The resulting polyacetylene film showed a copper-like conductivity (up to 10^5 S/cm) after having been stretched and doped with iodine (Schimmel *et al.*, 1989). By replacing the silicone oil with a nematic liquid crystal phase, Shirakawa and co-workers (Akagi *et al.*, 1989; Shirakawa *et al.*, 1988) have also produced partially aligned, highly conducting polyacetylene films. These authors subsequently obtained polyacetylene films with a highly oriented fibrillar morphology (Figure 6.2) by carrying out the polymerization of acetylene on a vertical glass wall of a flask over which the catalyst-containing nematic liquid crystal solution flowed down under the influence of gravity (Araya and Shirakawa, 1986; Shirakawa *et al.*, 1988).

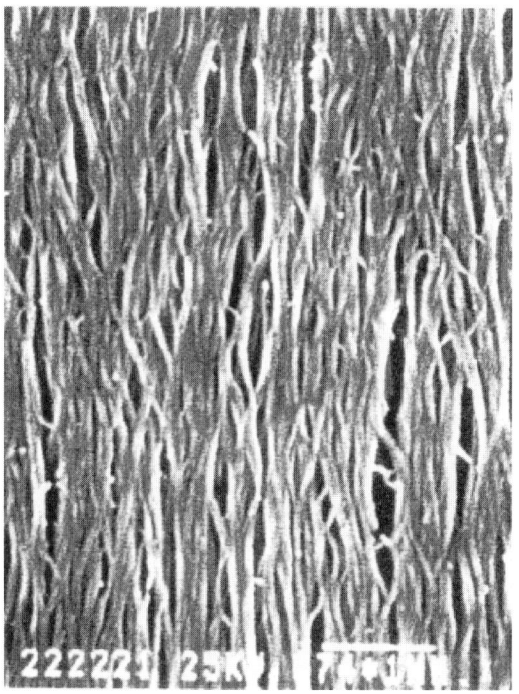

Figure 6.2. SEM micrograph of a polyacetylene film prepared by the flow method. The flow direction is vertical (After Shirakawa *et al.*, 1988, copyright 1988 Marcel Dekker, Inc. Reproduced with permission)

The high infrared (IR) anisotropies ranging from 2.0 to 3.8, together with the high I_2-doping-induced DC conductivities in directions both parallel ($\sigma_{\parallel}$ = 4600 S/cm) and perpendicular ($\sigma_{\perp}$ = 3900 S/cm) to the orientation axis, indicate significant orientation of the polyacetylene chains along the flow direction. Aiming for superior optoelectronic properties, several groups have used a magnetic field (typically, 2–14 kG) to further align the active catalyst sites within the nematic liquid crystal phase (Aldissi, 1985, 1989; Shirakawa *et al.*, 1988, 1989; Montaner *et al.*, 1988, 1989). Although polyacetylene films thus prepared have a morphology quite similar to that shown in Figure 6.2, higher average IR dichroic ratios in the range of 3.3–3.5 and the I_2-doped conductivities as high as ($\sigma_{\parallel}$ = 12 000 S/cm and ($\sigma_{\perp}$ = 4800 S/cm were recorded (Shirakawa *et al.*, 1988).

More recently, Shirakawa and co-workers (Akagi *et al.*, 1998) synthesized *helical* polyacetylene *nanofibers* (Figure 6.3) using *chiral* nematic liquid crystals as solvents for the Ziegler-Natta catalyst. I_2-doping of the helical polyacetylene shows conductivities of *ca.*1500–1800 S/cm. This, together with the peculiar helical nanofiber structure, has implications for these materials in novel electromagnetic and optical applications.

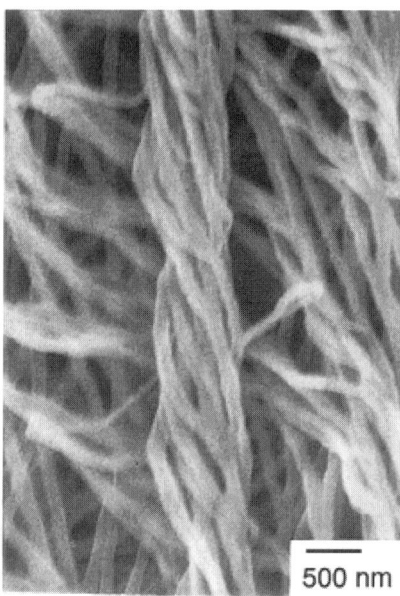

Figure 6.3. A typical SEM micrograph of helical polyacetylene nanofibers (After Akagi *et al.*, 1998, copyright 1998 American Association for the Advancement of Science)

6.2.2.2 Liquid Crystalline Conjugated Polymers

Another important technique for preparing ordered conjugated polymers is to chemically link liquid crystalline (LC) mesogens to conjugated backbones. Interest in liquid crystalline materials has a long history (Collings *et al.*, 1995). However, the synthesis of liquid crystalline conjugated polymers is a very recent development. Since there are many monographs and review articles already available for conventional polymer liquid crystals (Cifferi *et al.*, 1982; Donald and Windle, 1992; Gray and Winsor, 1995; McArdle, 1989; Wiederrecht *et al.*, 1996; Zentel *et al.*,

1997), we will focus our attention on the syntheses and properties of liquid crystalline *conjugated* polymers.

The substitution of conjugated polymers, such as poly(thiophenes), poly(*p*-phenylene) (PPP), PPV, and poly(*p*-phenylene ethynylene), with certain long or bulky side groups combines the order of conjugated backbones with the chain mobility, leading to liquid crystalline behavior (Babudri *et al.*, 1996; Gill *et al.*, 1996; Ringsdorf *et al.*, 1988; Steiger *et al.*, 1997). For instance, a typical Schlieren texture has been observed for poly(3-hexylthiophene) (P3HT) after it has been annealed at 145 °C for 6 h (Qian *et al.*, 1994) and a picture of the nematic morphology has been reported for poly(3-decylthiophenes) (P3DT) with a low molecular weight ($M_w = 5700$) at 70 °C (Bolognesi *et al.*, 1993). Lyotropic liquid crystalline order was also observed in camphorsulfonic acid (HCSA)-doped polyaniline in *m*-cresol (Cao Y. and Smith, 1993). Furthermore, well-organized layered structures with an average interlayer distance of about 30 Å have been produced by thermal doping of polyaniline with dodecylbenzenesulfonic acid (Chien, 1984). The resulting electroactive complexes exhibited a higher maximum conductivity than HCl doped polyaniline by two orders of magnitude, attributable to the well-defined anisotropic order. As mentioned above, the general approach to liquid crystalline conjugated polymers, however, is to graft LC mesogen-containing side chains onto the conjugated backbone. Table 6.1 summarizes some of the side chain liquid crystalline conjugated polymers reported so far.

Table 6.1. Survey of some π-conjugated polymers with liquid crystalline side chains

	Polymer / Ref.	Substituent R
A	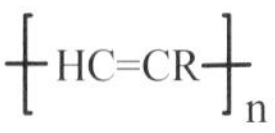 ⁺HC=CR⁺ₙ Akagi *et al.*, 1995, 1997 Goto *et al.*, 1995 Oh *et al.*, 1993a, b Yoshino *et al.*, 1996 Lino *et al.*, 1997 Koltzenburg *et al.*, 1998 Moigne *et al.*, 1991	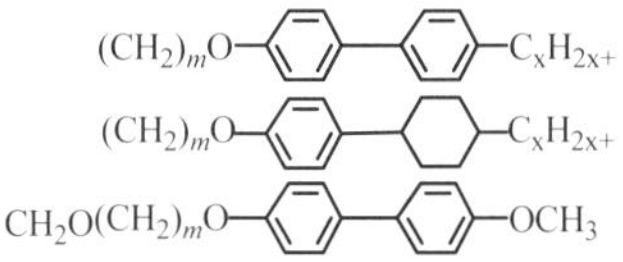 $(CH_2)_mO$—⟨⟩—⟨⟩—C_xH_{2x+1} $m = 3, 4, 6$ $\quad\quad\quad\quad\quad\quad\quad\quad\quad\quad x = 0, 2, 3, 5$-$8,$ $(CH_2)_mO$—⟨⟩—⟨⟩—C_xH_{2x+1} $C_9H_{20}*$ $CH_2O(CH_2)_mO$—⟨⟩—⟨⟩—OCH_3 $m = 4$-8 $(CH_3)_8COO$—Cholesteryl
B	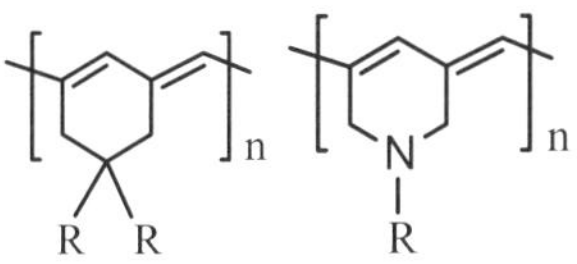 [structures with R R and N-R] Jin *et al.*, 1991, 1993 Choi *et al.*, 1994	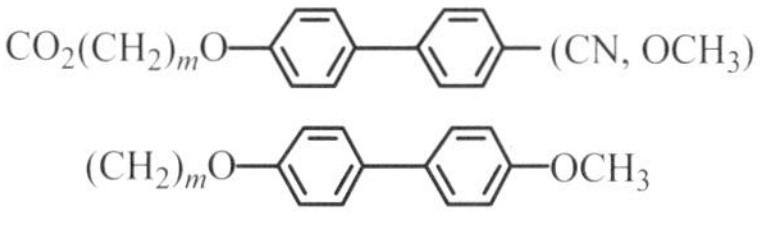 $CO_2(CH_2)_mO$—⟨⟩—⟨⟩—(CN, OCH_3) $(CH_2)_mO$—⟨⟩—⟨⟩—OCH_3

C	⌬—≡—[=CH(CH₂OR)—C≡C—CH(CH₂OR)=]ₙ—≡—⌬	$C_{17}H_{35}CO$, $(C_{12}H_{25}O)_3C_6H_2CO$
	Nierengarten *et al.*, 1997	
D	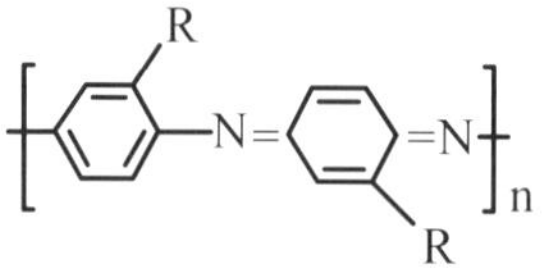 [poly(azo) structure with R groups] Gabaston *et al.*, 1996	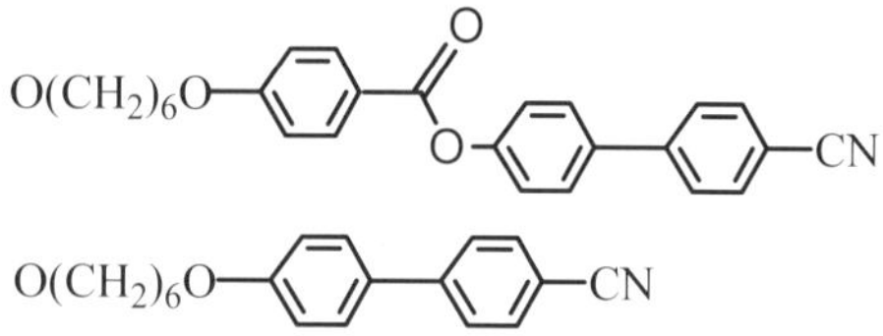 $O(CH_2)_6O$—⌬—COO—⌬—⌬—CN $O(CH_2)_6O$—⌬—⌬—CN
E	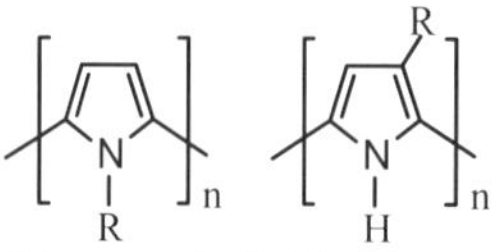 [pyrrole polymer structures] Ibison *et al.*, 1996 Hasegawa *et al.*, 1997 Vicentini *et al.*, 1995	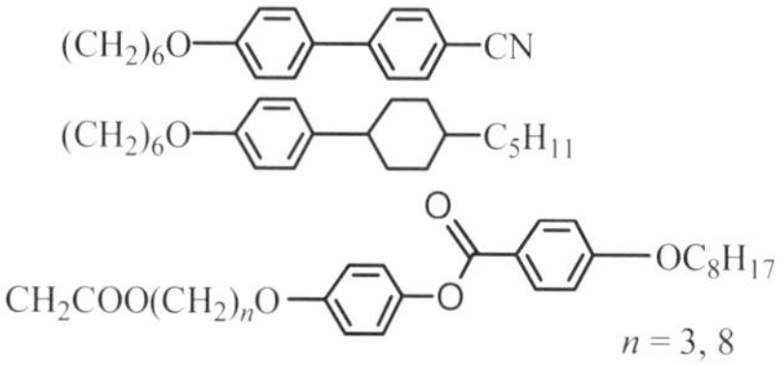 $(CH_2)_6O$—⌬—⌬—CN $(CH_2)_6O$—⌬—⟨H⟩—C_5H_{11} $CH_2COO(CH_2)_nO$—⌬—OOC—⌬—OC_8H_{17} $n = 3, 8$
F	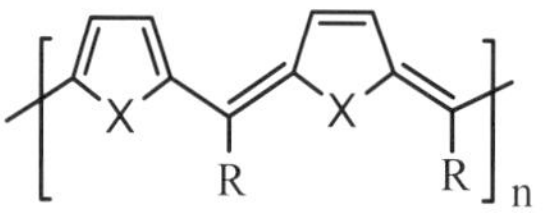 [thiophene/pyrrole polymer X=NH,S] Goto *et al.*, 1997	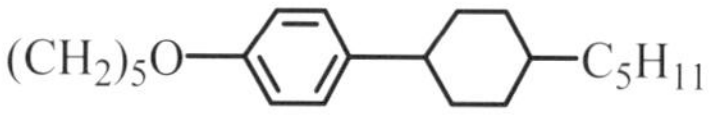 $(CH_2)_5O$—⌬—⟨H⟩—C_5H_{11}
G	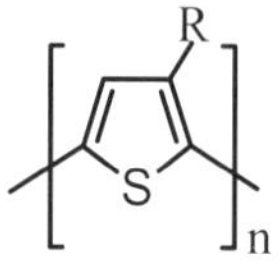 [polythiophene with R] Kijima *et al.*, 1997	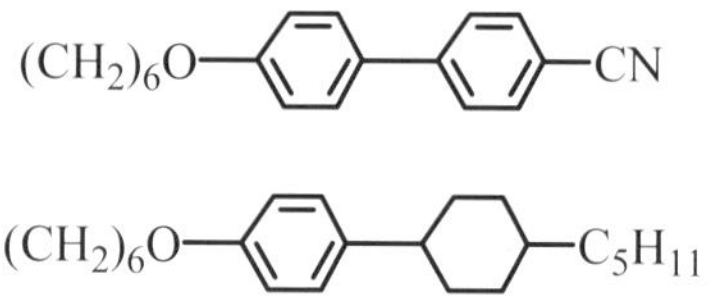 $(CH_2)_6O$—⌬—⌬—CN $(CH_2)_6O$—⌬—⟨H⟩—C_5H_{11}
H	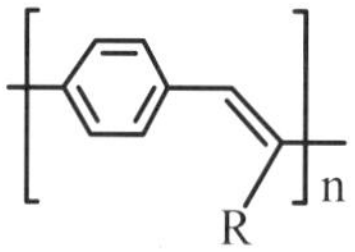 [PPV-type polymer with R] Winkler *et al.*, 1996b	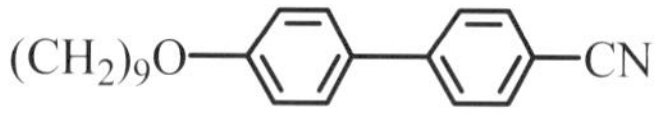  $(CH_2)_9O$—⌬—⌬—CN **H1** $(CH_2)_8COO$—⌬—⌬—CN **H2**

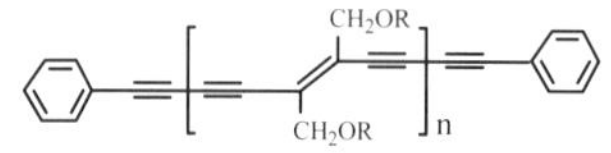

* optically active

Like most other functional polymers, liquid crystalline conjugated polymers can be prepared either by direct polymerization of appropriate functional monomers or by chemical modification of preformed conjugated polymers. Both approaches have been used to synthesize those liquid crystalline conjugated polymers listed in Table 6.1, and the detailed information for each of them can be found in the corresponding reference.

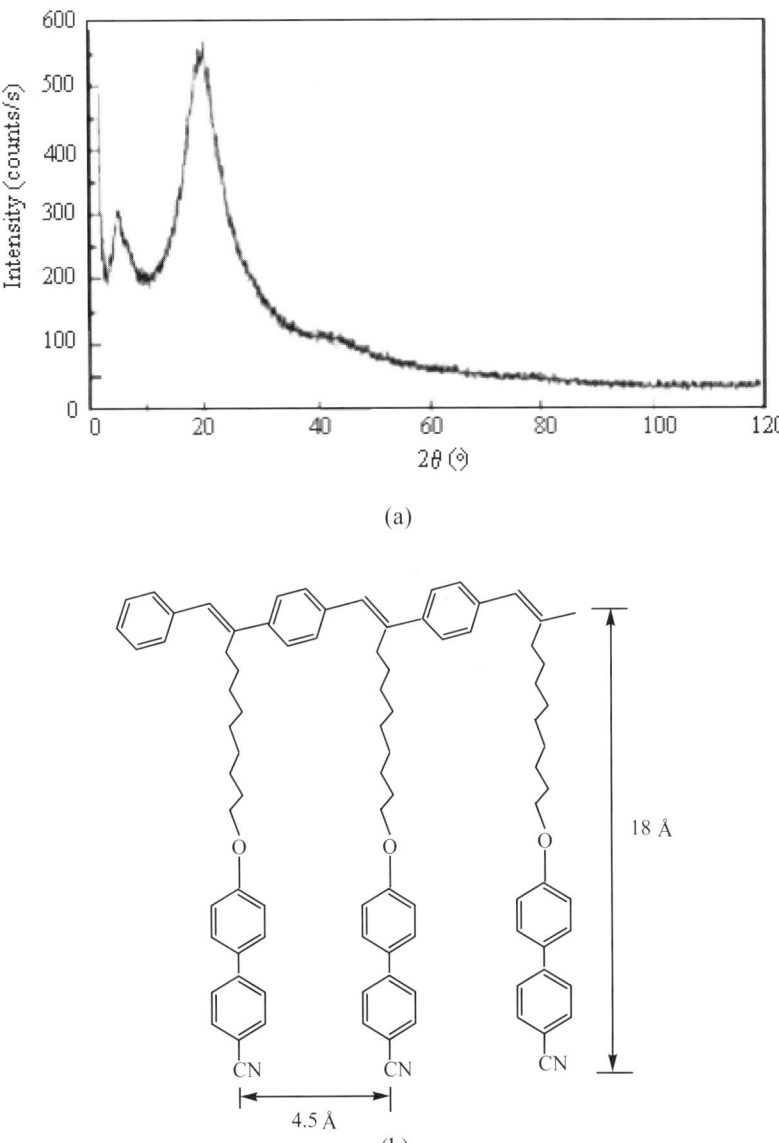

Figure 6.4. (a) X-ray diffraction spectrum of side chain liquid crystalline PPV H1; (b) schematic structure of single-layer of LC-PPV H1 derived from X-ray measurements (After Winkler *et al.*, 1996a, copyright 1996 Wiley-VCH Verlag)

In particular, Winkler and co-workers (Ungerank *et al.*, 1995; Winkler *et al.*, 1996a, 1997) have extended their work on non-conjugated liquid crystalline polynorbornenes and polyoctenylenes to synthesize PPV-type liquid crystalline polymers by chemically grafting them with side chains containing biphenyl-4-carbonitrile mesogens (*i.e.* Polymer **H** in Table 6.1). Its commercial availability coupled with a high NLO hyperpolarizability (Katz *et al.*, 1988) makes biphenyl-4-carbonitrile the mesogen of choice. These yellow-coloured LC-PPVs show the following transition temperatures for phase transitions from the isotropic (i) melt to glass (g) state through the nematic (n) and smectic (s) phases **H1:** i 109 °C n 105 °C s 29 °C g; **H2:** i 165 °C n (s) 28 °C g. The transition from the isotropic melt to the nematic phase for both polymers is very clear, as reflected by corresponding structural evolution from the droplet to Schlieren texture under crossed polarizers. Further evidence for the smectic phase of a quenched **H1** sample is obtained from X-ray diffraction measurements, which show, in good agreement with its molecular structure, a value of *ca.* 4.5 Å for the separation between adjacent mesogenic units and *ca.* 19±2 Å for the distance from the terminal cyano group to the conjugated backbone (Figure 6.4).

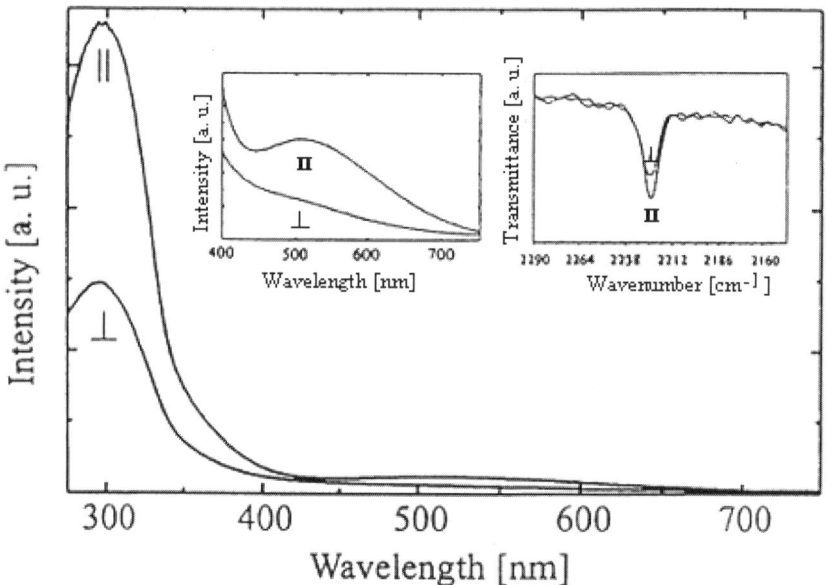

Figure 6.5. Polarized UV/VIS and IR spectra of the oriented liquid crystalline PPV-derivative **H1** in the parallel (II) and perpendicular (⊥) directions to the rubbing axis (After Winkler *et al.*, 1996b, copyright 1996 Elsevier. Reproduced with permission)

The LC-PPVs have similar backbone optical absorption and fluorescence emission as the unsubstituted PPV. However, the liquid crystalline behavior associated with the LC-PPV derivatives facilitates the alignment of the polymer chains by spontaneous ordering or external poling. For instance, Winkler et al. (1997) have obtained highly ordered LC-PPV films by simply casting the polymer solution onto a substrate, which was unidirectionally pre-rubbed with a polytetrafluoroethylene (Teflon) bar (Pooley and Tabor, 1972; Wittmann and Smith, 1991). In contrast, in order to align even a five-ring oligo(*p*-phenylenevinylene) with two (non-LC) *n*-octyl side chains in the *para*- position of the central phenyl ring onto a highly pre-oriented Teflon substrate, vacuum evaporation has to be applied (Gill et al., 1997). Polarized UV/VIS and IR measurements on the aligned LC-PPV films show optical-absorption anisotropies ($A_{\parallel}/A\perp$) ranging from 2 to 3 for both the mesogens and conjugated backbone (Figure 6.5), confirming the significant orientation of the mesogenic groups and the main chain parallel to the rubbing direction (Winkler et al., 1997). Thus, the biphenylene groups in the bulky side chains are energetically preferred to align parallel to the aromatic PPV main chain through strong π-π-stacking, leading to a dense packing for the polymer film. Polarized adsorption and emission have also been reported for some conjugated main-chain LC polymers, as exemplified by rubbing-aligned polyfluorene macromolecules grafted with octyl side chains (Grell et al., 1997).

6.2.2.3 Post-synthesis Orientation

The most straightforward way to stretch-orient conjugated polymers is to mechanically stretch a free-standing conjugated polymer film. In this regard, polyacetylene has been stretch-oriented and subsequently doped into a highly conducting form (Shirakawa et al., 1973; Leising, 1984; Naarmann, 1986), and a stretching-induced crossover from a semiconducting to a metallic state in iodine-doped polyacetylene films was observed (Joo et al., 1997). Highly aligned electrochemically-synthesized polypyrrole films with conductivities along the alignment direction up to 1005 S/cm have also been prepared by stretching them to more than twice their original length (Ogasawara et al., 1986; Satoh et al., 1985). Elongations up to five times the initial length were demonstrated for P3HT (Hotta et al., 1989) and poly(3-octylthiophene) (P3OT) upon stretching (Yoshino et al., 1989). The stretched P3OT gels showed highly anisotropic and reversible size changes with temperature, solvent composition and doping level. Simultaneous heat treatment and the application of stress on films of polyaniline emeraldine base and salt led to the formation of stretch-oriented, partially crystalline polyaniline films with $\sigma_{\parallel}/\sigma\perp$ = 3.0 and 3.5 for the undoped emeraldine base and emeraldine hydrochloride salt, respectively, along with an up to 10-fold increase in $\sigma_{\parallel}$ for the salt sample as compared with the DC conductivity of an unoriented counterpart (Cromack et al., 1991).

Instead of stretching preformed conjugated polymers, orientation can also be achieved by stretching certain non-conjugated precursor polymers, in either a film or fiber form, followed by thermal conversion of the stretched materials into their corresponding conjugated structures (Andreatta et al., 1990; Gagnon et al., 1987; Karasz et al., 1985; Tokito et al., 1991; Townsend et al., 1985). Table 6.2 lists the stretch ratios and the conductivities along the stretching direction for some aligned conjugated conducting polymers thus prepared.

Table 6.2 Typical electrical conductivities of aligned π-conjugated polymers prepared from precursor polymers

Precursor Polymer	Conjugated Polymer	Stretch Ratio	Dopant	$\sigma_{max,\ 25\ °C}$ (S/cm)	Ref.
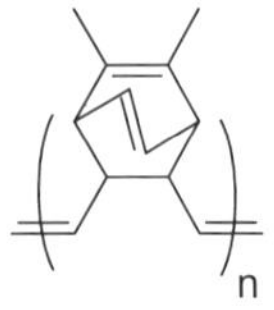		7.5	AsF$_5$	1250	Townsend et al. 1985
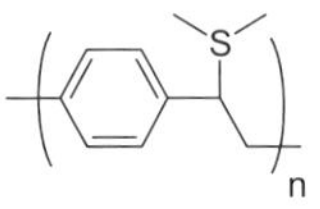	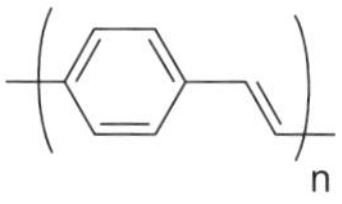	7	AsF$_5$	500	Karasz et al. 1985
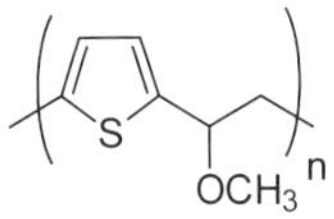	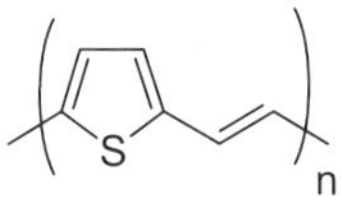	20	I$_2$	2000	Andreatta et al. 1990
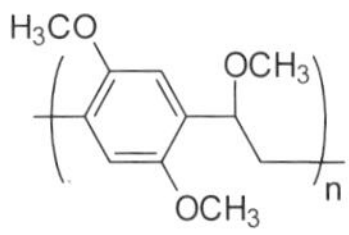	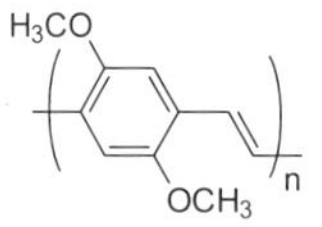	8	I$_2$	1200	Tokito et al. 1991

The ability of certain conjugated polymers to form polymer blends with common non-conjugated polymers or elastomers provides additional possibilities for stretch-alignment of conjugated polymers. For example, Heeger and co-workers (Hagler et al., 1991) have prepared highly oriented polymer blends of poly(2-methoxy-5-(2′-ethyl-hexoxy)-p-phenylene vinylene) (MEH-PPV) and ultrahigh-molecular-weight polyethylene (UHMW-PE) by processing the composite material into a gel intermediate state and followed by tensile drawing. The highly anistropic absorption and photoluminescence emission from the resulting materials indicate a strong orientation of the conjugated polymer guest along the draw axis. Similarly, highly polarized photoluminescent layers based on highly uniaxially stretched blends of a poly(2,5-dialkoxy-p-phenylene ethynylene) derivative and the UHMW-PE have been used as both the polarized light and bright color generator in liquid crystal displays, leading to substantially enhanced device brightness, contrast, efficiency, and viewing angle (Weder et al., 1998). In a separate study, a poly(3-

alkythiophene)/ethylene-propylene elastomer composite has been stretched up to a stretch ratio of 12, causing a blue shift for the absorption peak with a high anisotropy of *ca.* 3 (Yoshino *et al.*, 1988). However, a better oriented poly(3-alkythiophene) was obtained by spinning the polymer solution onto a stretchable substrate (*e.g.* polyethylene) and then stretching the substrate film (Dyreklev *et al.*, 1992). Using polyethylene films as the stretchable substrate, Theophilou and Manohar (1993) have obtained highly aligned polyaniline films with $\sigma_{\parallel}$ as high as 2500 S/cm, while Naarmann and co-workers (Basescu *et al.*, 1987) produced a polyacetylene film with $\sigma_{\parallel}$ = 150 000 S/cm and $\sigma_{\parallel}/\sigma_{\perp}$ = 1000. Furthermore, Inganäs and co-workers (Dagani, 1995; Dyreklev *et al.*, 1995) constructed the first organic LED that could emit polarized light under a voltage as low as 2 V by using a stretch-oriented polythiophene derivative as the electroluminescent material.

Extended conformations can also be obtained for conjugated polymer chains through certain chemical and/or physical interactions (*e.g.* with dopants). While the chain stretching in end-adsorbed polymer brushes has been widely investigated both theoretically and experimentally (Halperin *et al.*, 1992; Dai and Toprakcioglu, 1991, 1992; Dai *et al.*, 1995; Field *et al.*, 1992), Bumm *et al.* (Bumm *et al.*, 1996; Dagani, 1996b) demonstrated the use of STM tip to probe electrical properties of individual conjugated conducting molecules ("molecular wires") dispersed into a self-assembled monolayer film of non-conducting alkanethiolate molecules. Dai and White (1994) have also reported a conformational transition from a random coil to a rod-like structure caused by the I_2-induced conjugation of polyisoprene chains, while MacDiarmid and Epstein (1994, 1995) demonstrated that the interaction of a HCSA-doped polyaniline sample with *m*-cresol could cause a conformational transition of the polymer chain from the so-called "compact coil" to an "expanded coil" with a concomitant increase in conductivity by up to several orders of magnitude. The latter process has been widely known as "secondary doping" (MacDiarmid and Epstein, 1994, 1995). Further to their work on C_{60}-containing polymers (Dai *et al.*, 1995, 1998; Dai and Mau, 1997), Dai and co-workers (Dai *et al.*, 1998; Lu *et al.*, 1998) have recently used hydrogensulfated fullerene derivatives with multiple $-(O)SO_3H$ groups as protonic acid dopants for polyaniline emeraldine bases. It was found that the hydrogensulfated fullerene derivatives can act simultaneously as both primary and secondary dopants, and the resulting material showed metallic characteristics with room-temperature conductivities of up to 100 S/cm (Dai *et al.*, 1998). Besides, Langmuir-Blodgett, LB, (Aoki and Miyashita, 1997; Rubner and Skotheim, 1991) and self-assembling (Stockton and Rubner, 1997) techniques have also been used to control thin film structures at the molecular level. These are discussed at details in other chapters as appropriate.

6.2.3 Patterned Conjugated Polymers

Photolithographic processing has been widely used for the micropattern formation in semiconducting industries (*e.g.* for delineating the circuit elements in todays's large-scale integrated devices) for many years (Bowden and Turner, 1988). Photolithographic patterning of conducting polymers, however, is a recent development (Dai and Mau, 2000).

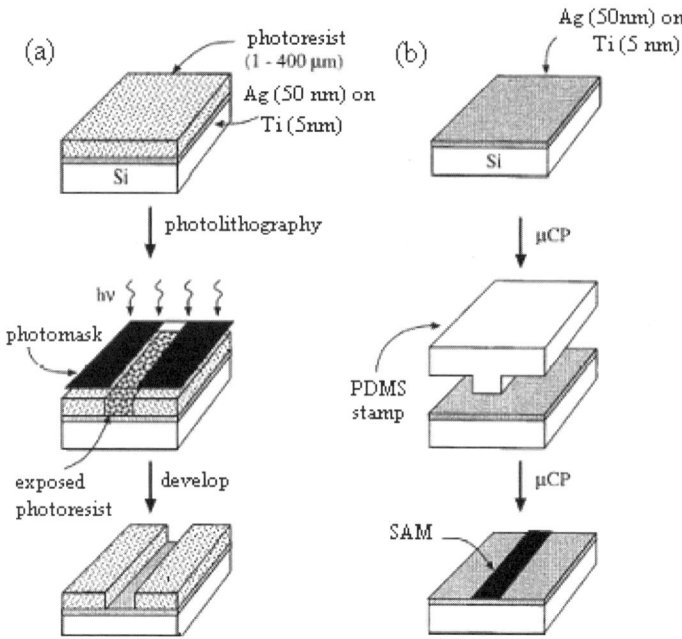

Figure 6.6. Basic steps of the (A) photolithographic and (B) μCP process: (A) a thin layer of photoresist was coated on a substrate (a Ag/Ti-coated silicon wafer in this case). Exposure of the photoresist to UV light is through a patterned photomask. Depending on the chemical nature of the photoresist, the exposed areas may be rendered more soluble in the developer than the unexposed areas, leading to a positive-tone image of the mask. Conversely, the exposed areas may be rendered less soluble, producing a negative-tone image. (B) A polydimethylsiloxane (PDMS) stamp was used to region-selectively transfer a self-assembled monolayer (SAM) of alkanethiols onto the silver surface through the conformal contact, thereby producing patterned areas with different surface properties (After Jackman *et al.*, 1999, copyright 1999 The American Chemical Society. Reproduced with permission)

On the other hand, plasma polymerization has been widely used to generate thin, cohesive, adhering, pinhole-free polymer films for anti-corrosion protection or adhesion promotion apart from the plasma-induced surface modification (Dai *et al.*, 1997; Hollan and Bell, 1974). Although plasma polymerization normally produces electrically insulating organic films, approaches to semiconducting plasma polymer films have been reported and plasma techniques have also been used for micropatterning and microfabrication of electroactive polymers as well as carbon nanotubes. Along with these developments, the so-called soft-lithographic techniques (Jackman and Whitesides, 1999; Xia and Whitesides, 1998), notably micro-contact printing (μCP), has also been used for patterning conjugated polymers and various other materials. In what follows, I will illustrate these developments by discussing several selected examples from the recent literature. The basic steps involved in conventional photolithographic and soft-lithographic processes are given in Figures 6.6(A) and (B), respectively.

6.2.3.1 Photolithographic Patterning

A few approaches for photolithographic patterning of conjugated polymers have been reported in the literature (Schanze *et al.*, 1996). For instance, Clarke *et al.* (Clarke *et al.*, 1981) reported a photochemical doping method using onium salts (*e.g.* triarylsulfonium and diaryliodonium salts) as the photochemical dopants, which allows doping to occur only in the UV illuminated areas of a polyacetylene film. Holdcroft and co-workers (Holdcroft, 2001), Dao *et al.* (1992) and Cai *et al.* (1992) obtained conducting patterns from the soluble form of conjugated polymers [*e.g.* poly(3-alkylthiophenes)] through patterned photo-cross-linking (or cross-linking by electron beam), followed by a solution-based development process. Angelopoulos *et al.* (1992) and Venugopal *et al.* (1995) demonstrated that conducting patterns could also be made from a mixture of the soluble base form of polyaniline and a photoacid generator, which photochemically generated the acidic dopants required for generating conducting patterns of the insoluble polyaniline salts. The solubility of the conducting polymers used in these approaches was acquired by covalently bonding conjugated polymers with soluble side groups or polymeric block chains, which often reduced the conductivity. Besides, the poor solubility of most unfunctionalized conjugated polymers in common organic solvents may limit the general application of the above techniques. Hence, Bargon *et al.* (1991) and Baumann *et al.* (1993) produced conducting patterns through the polymerization of monomers, such as pyrrole, aniline and thiophene, initiated by HCl which was generated photochemically in a patterned fashion from chlorine-containing polymer matrices, including poly(chloro-acrylonitrile), poly(vinylchloride), poly(chloroprene) and poly(chlorostyrene). The conducting regions generated in these cases, however, are composite materials containing newly formed conducting polymer chains within the non-conducting polymer matrices. A particularly attractive option is the formation of conducting patterns from a processable (soluble and/or fusible) insulating polymeric matrix through, for example, a photochemical transition, which can directly convert the microlithographically exposed regions into unsubstituted conjugated sequences.

As discussed in Chapter 2, Thakur (1988) reported that the conductivity of *cis*-1,4-polyisoprene can be increased by about 10 orders of magnitude upon "doping" with iodine. On this basis, he claimed that a conjugated structure is not always necessary for a polymer to be electrically conducting, which generated considerable interest (Borman, 1990; Calvert, 1988; Rothman, 1988). Dai and White (Dai and White, 1991) were the first to demonstrate that "I_2-doping" of 1,4-polydienes produces conjugated sequences of unsaturated double bonds through the polar addition of iodine into the isolated double bonds in the polymer chain, followed by HI elimination [*i.e.* Equation (2.12) of Chapter 2; see also Dai, 1992; Dai *et al.*, 1994]. It was further found that *trans*-1,4-polybutadiene becomes dark in color and conductive when doped with iodine in the solid state. In contrast, *cis*-1,4-polybutadiene does not change color or become conductive on I_2 doping under the same conditions due to an unfavorable combination of electronic and steric interactions that inhibit the HI elimination from the iodinated *cis*-1,4-polybutadiene chains (Dai *et al.*, 1994). This discovery provides means for photolithographic generation of conducting patterns from *cis*-1,4-polybutadiene as the *cis*-isomer can be photoisomerized into its *trans*-counterpart. Using a microlithographic mask, Dai *et al.* (1996) have recently found that *cis*-1,4-polybutadiene films can be photoisomerized into the *trans*- isomer in a patterned fashion without significant lateral diffusion. As a result, only the photoisomerized regions are capable of the generation of conjugated double bonds upon the exposure of the entire polybutadiene film to iodine at room temperature, resulting in the formation of conducting patterns in an insulating matrix of iodinated *cis*-1,4-polybutadiene. The conducting patterns thus formed are colored and show strong fluorescence emission, which enables visualization of the conducting polymer regions.

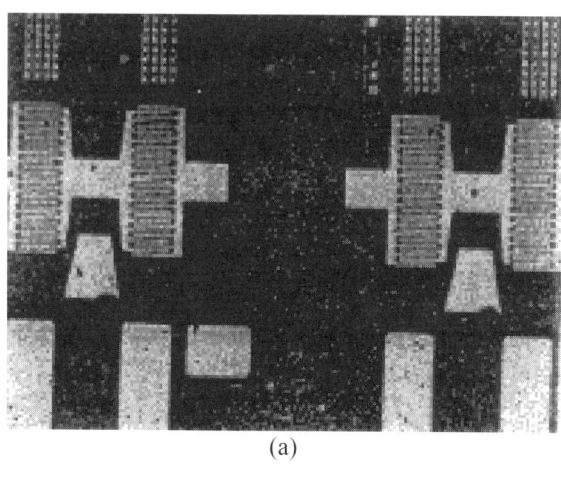
(a)

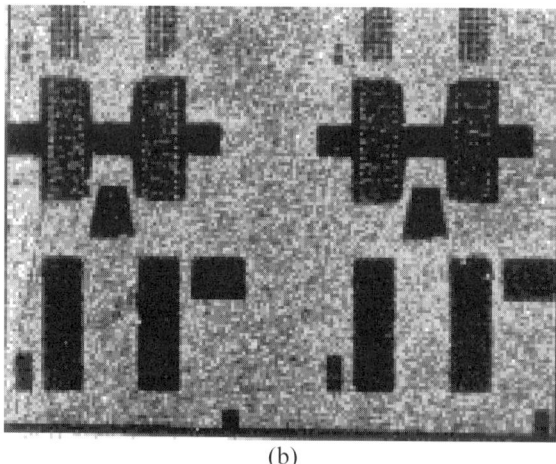
(b)

Figure 6.7 (a) Optical microscopy image of a pattern obtained by "I_2-doping" of the photoisomerized *trans*-1,4-polybutadiene regions in the iodinated *cis*-1,4-polybutadiene matrix. The dark areas are regions of "I_2-doped" polybutadiene, and the width of the white rectangles at the bottom part of the picture is 18 μm; (b) fluorescence micrograph of the conducting pattern (After Dai *et al.*, 1996, copyright 1996 The American Chemical Society. Reproduced with permission)

An example of the conducting patterns thus generated is shown in Figure 6.7(a). It is a close replication of the photomask structure, and conducting wires on a micrometer scale are clearly evident. A corresponding fluorescence microscopic image of the conducting pattern is given in Figure 6.7(b). It shows the same features as the optical micrograph (Figure 6.7(a)), but with inverse intensities in the image. The dark regions characteristic of the "I_2-doped" *trans*-1,4-polybutadiene in Figure 6.7(a) gave rise to a bright fluorescence emission in Figure 6.7(b), consistent with the fluorescence emission originating from the conjugated structures (Skotheim, 1986). The dark regions in Figure 6.7(b) represent non-fluorescent components associated with the *cis-* isomer. While the conducting patterns thus prepared may find applications in certain electronic and photonic devices, the photolithographic method has also been applied successfully to micropattern electroluminescent conjugated polymers (Lidzey *et al.*, 1996) including PPV (Wei P. *et al.*, 1996).

For instance, Wei P. *et al.* (1996) and DeAro *et al.* (1997) have recently demonstrated the use of near-field optical microscopy for writing/reading images on conjugated polymer films including PPV with a resolution of *ca.* 0.1 μm. On the other hand, Cho *et al.* (Cho *et al.*, 1995) have discovered that the methoxy-substituted PPV precursor polymer [*cf.* Equation (2.6)] decomposed at a lower temperature (80–200 °C) after an acid-catalyzed UV irradiation than the non-irradiated precursor polymer (> 220 °C). On this basis, these authors prepared PPV conducting micropatterns (10^{-2}–10^{-3} S/cm) by using triphenylsulfonium salts as both an acid catalyst and photochemical dopant for a selectively photoinitiated thermolysis of the methoxy-substituted PPV precursor polymer at a relatively low temperature. Similarly, micropatterns of PPV with patterned emissions have also been prepared from several copolymers containing PPV precursors through a selective thermolysis, with or without involving a photoinitiator (Halliday *et al.*, 1993; Renak *et al.*, 1997). Besides, PPV has also been microstructured in a non-doped form by UV interferometry (Schmid *et al.*, 1993), as the transformation from the unsubstituted PPV precursor polymer to the conjugated state can occur not only through the conventional thermal conversion but also upon UV irradiation (Schmid *et al.*, 1993). Compared with the conducting PPV micropatterns, the non-doped PPV microstructures thus formed seem to be more suitable for EL applications because of the possible doping-induced fluorescence quenching (Greenham and Friend, 1995). On the other hand, Noach *et al.* (1996) have demonstrated that a PPV copolymer [*i.e.* poly(1,4-phenylenevinylene-*co*-2,6-pyridylene vinylene)]-based light-emitting diode array consisting of micron-sized emitting pixels can be produced through a selective laser ablation process. In addition to the achievement of patterned emissions, these authors also found that emission intensities increased due to the strongly enhanced electric fields at the edges of the conducting (Al/ITO) pixels.

Closely related to the above studies, three-dimensional (3D) polymer nanopatterns have also been prepared by some non-conventional lithographic techniques. Notably, a mask-induced self-assembling, MISA, process has been applied to the nanopattern formation in polymer films (Schäffer *et al.*, 2000; Chou *et al.*, 1999; Chou and Zhuang, 1999; Deshpande *et al.*, 2001). In this process, a mask with protruded patterns is placed at a certain distance above the top surface of a polymer melt (*e.g.* PMMA, PS), in the presence or absence of an external field (*e.g.* thermal or electrical field), allowing attraction of an excess amount of polymer chains to the area below the mask protrusions. Subsequent solidification of the polymer melt thus leads to the formation of polymer patterns with a lateral dimension identical to that of the mask protrusions. Using the MISA patterning technique, 3D polymer patterns with feature size down to nanometer scale can be prepared (Chou *et al.*, 1999). Figure 6.8(A) shows schematically a typical procedure for the MISA patterning. An atomic force microscope, AFM, image of polymer nanopatterns produced by the MISA technique is given in Figure 6.8(B), which shows well-defined 3D nanopatterned structures.

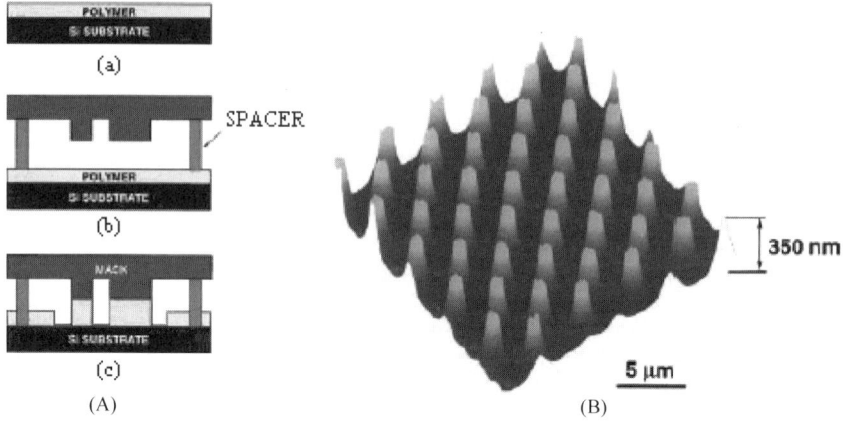

Figure 6.8. (A) Schematic of MISA: (a) a thin polymer film cast on a flat silicon substrate; (b) a mask with protruding patterns placed a distance above the polymer film; (c) the polymer film self-constructs into a mesa under a mask protrusion. The mesa has a lateral dimension identical to that of the mask protrusion, a height equal to the distance between the mask and the substrate. (B) AFM image of the PMMA pillars formed by MISA. Each PMMA pillar is formed under each dot protrusion on the mask (After Chou *et al.*, 1999, copyright 1999 American Institute of Physics)

As can be seen above, the MISA method is a straightforward and feasible technique for fabricating polymer patterns with sub-micrometer/nanometer features. Continued research efforts in this embryonic field could give birth to a flourishing area of nanopatterning technologies.

6.2.3.2 Pattern Formation by Self-assembling

Self-assembled monolayer films (SAMs) form, notably, when materials such as alkanethiols and organosilanes are physically or chemically adsorbed onto gold, silver, aluminium, copper or silicon dioxide surfaces (Terfort *et al.*, 1992; Ulman, 1996). Pattern formation/recognition by the SAMs has recently become a very promising technique for microstructuring organic materials (Akari *et al.*, 1995; Frisbie *et al.*, 1994; Kumar and Whitesides, 1994; Lopez *et al.*, 1993; Schierbaum *et al.*, 1994), and various strategies including photolithographic (or beam-induced) transformation (Dulcey *et al.*, 1991; Wrighton, 1993; Dulcey *et al.*, 1996), mechanical scraping (Ross *et al.*, 1993; Terfort *et al.*, 1992), micro-contact printing (µCP) (Jacoby et al., 1997; Kumar *et al.*, 1994; Kumar and Whitesides, 1993, 1994), scanned-probe based micro-machining (Xu and Liu, 1997), and locally-confined plasma surface modification (Vargo *et al.*, 1992) have been developed for microfabrication and subsequent transformation of SAM patterns.

In particular, the SAM technique has been employed by Rozsnyai and Wrighton (1994, 1995) to pattern conducting polymers of polypyrrole, poly(3-methylthiophene) and polyaniline through selective electropolymerization of the corresponding monomers onto photochemically patterned disulfide SAMs on gold. Meanwhile, Rubner and co-workers (Ferreira *et al.*, 1994; Ferreira and Rubner, 1995) have used a positively charged SAM layer [formed from aminopropyl dimethylethoxy silane or poly(ethyleneimine)] for surface immobilization of poly(thiophene-3-acetic acid). While Garnier and co-worker (Michalitsch *et al.*, 1997) have produced a highly ordered, densely packed, electroactive SAM of thiol-functionalised oligothiophene on platinum surfaces, Kim *et al.* (Kim *et al.*, 1996) prepared polydiacetylene monolayers by first self-assembling diacetylene group-containing n-alkanethiols onto gold surfaces, then photopolymerizing the SAM with UV light. The photopolymerization involved in the latter work suggests that the micropattern formation of the polydiacetylene monolayer should be straightforward by use of a photomask, though this has not yet been reported. Indeed, photofabrication of azobenzene-functionalized polydiacetylenes has led to the formation of regular surface relief gratings with unique NLO properties (Sukwattanasinitt *et al.*, 1998).

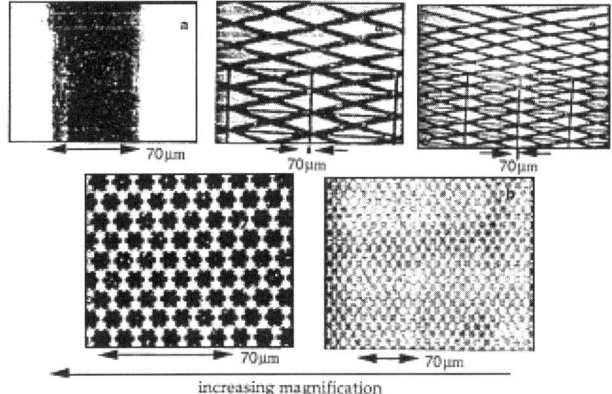

Figure 6.9. Selective deposition of polypyrrole on μCP-patterned hydrophobic surfaces (After MacDiarmid, 1997, copyright 1997 Elsevier. Reproduced with permission)

Microcontact printing (μCP) is a very convenient patterning technique, in which an elastomeric stamp (typically, a polydimethylsiloxane stamp) is used to region-specifically transfer molecular "inks" (*e.g.* alkanethiol, alkylsiloxane, *etc.*) (Kumar and Whitesides, 1993). It has been widely used for generating SAM patterns onto certain metals (*e.g.* Au, Ag, Cu, *etc.*) and Si surfaces with a minimum feature size of *ca.* 100 nm (Xia *et al.*, 1996). The use of μCP has recently been extended to patterning of conducting polymers. For instance, MacDiarmid and co-workers (MacDiarmid, 1997; MacDiarmid *et al.*, 1997) produced hydrophobic patterns of $C_{18}H_{37}SiCl_3$ on hydrophilic glass substrates through a μCP stamp. Then, polypyrrole and polyaniline were adsorbed preferentially onto the hydrophobic regions from dilute aqueous solutions of the polymerizing monomer to form thin film patterns (Figure 6.9). The patterned conducting polymers on the hydrophobic surface were found to take on an extended conformation with an enhanced conductivity (Huang et al., 1997), which have been demonstrated to be useful as electrodes in display devices based on polymer-dispersed liquid crystals (Harlev et al., 1996; Huang et al., 1997). Furthermore, recent reports on the use of certain patterned SAMs for region-specific alignment of liquid crystal molecules (Evans et al., 1997; Gupta and Abbott; 1997) should have an important implication for patterned alignment of the LC conjugated polymers discussed above. More recently, the possibility for multi-dimensional micro-construction by the µCP technique has also been explored (Jeon et al., 1997; Terfort et al., 1997; Zhao et al., 1996), as exemplified by Figure 6.10.

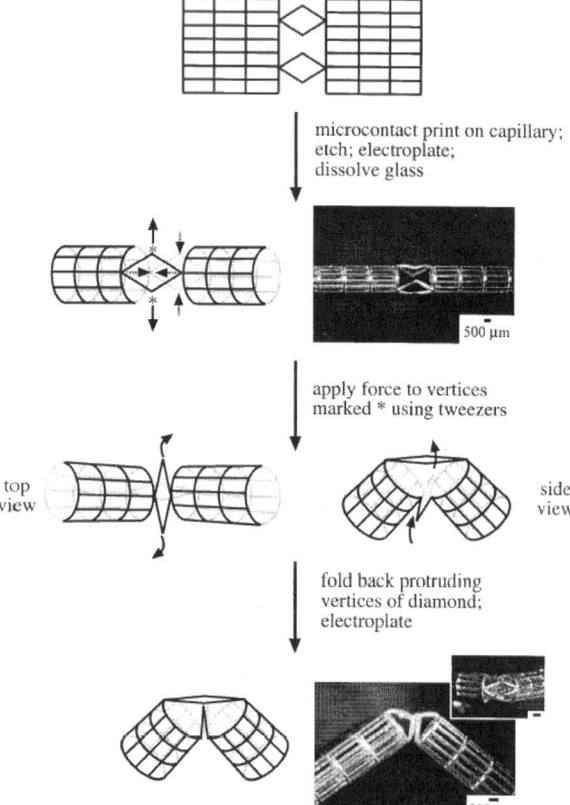

Figure 6.10. 3D cylindrical structures produced by µCP of SAM of hexadecanethiol on silver-coated glass capillaries (~2 mm diameter), followed by wet-chemical etching to remove the silver that remained underivatized after printing, electroplating to increase the rigidity of the structure, and finally dissolution of the glass to produce a free-standing structure. 3D structures can also be produced by applying out-of-plane deformations to arrays of 2D structures, leading to assembly of all components of the array simultaneously (After Jackman et al., 1999, copyright 1999 The American Chemical Society. Reproduced with permission)

6.2.3.3 Pattern Formation by Polymer Phase Separation

Like the SAM technique, polymer phase separation plays an important role in fabrication of microstructures, as it provides the opportunity for nanoscale patterning which otherwise is difficult by lithographic techniques (Fredrickson and Bates, 1996; Park et al., 1997). The conditions necessary for microphase separation in immiscible polymer mixtures depend on their molecular architectures, the nature of the monomers, compositions and molecular weights (de Gennes, 1979). Briefly, for linear homopolymer mixtures of A and B, the free energy of mixing per unit volume is given by (Bates, 1991; de Gennes, 1979):

$$\Delta G/kT = (\phi_A/N_A)\ln\phi_A + (\phi_B/N_B)\ln\phi_B + \chi\phi_A\phi_B \qquad (6.7)$$

where k is the Boltzmann constant, T is the absolute temperature, ϕ_i and N_i are the volume fraction and degree of polymerization for species i, respectively. χ is the Flory-Huggins segmental interaction parameter. $\phi_A + \phi_B = 1$ is a result of incompressible mixing. From Equation (6.7), the conditions for equilibrium, stability and criticality can be derived as follows (Bates, 1991; de Gennes, 1979; Lodge and Muthukumar, 1996):

$$\text{Equilibrium:} \quad \partial\Delta G(\phi_A')/\partial\phi_A = \partial\Delta G(\phi_A'')/\partial\phi_A \qquad (6.8)$$

$$\text{Stability:} \quad \partial^2\Delta G/\partial\phi_A^2 = 0 \qquad (6.9)$$

$$\text{Criticality:} \quad \partial^3\Delta G/\partial\phi_A^3 = 0 \qquad (6.10)$$

where the superscripts ' and " refer to separated phases. Taking the symmetric case of $N_A = N_B = N$ as a simplified example (Bates, 1991; de Gennes, 1979; Lodge and Muthukumar, 1996), it can be derived from the above equations that phase separation occurs only when χ is larger than a threshold value $\chi_c = 2/N$. The large value of N, and hence very small value of χ_c, for macromolecules rationalizes the strong immiscibility often observed in polymer mixtures. For a detailed discussion on the theory of polymer phase separation, interested readers are referred to specialized books and reviews (Bates, 1991; Bates and Fredrickson, 1990; de Gennes, 1979; Flory, 1953; Helfand and Wasserman, 1982; Leibler, 1980; Lodge and Muthukumar, 1996).

Although the above description is ideal and Equation (6.7) may fail to provide a quantitative description of real polymer mixtures (Bates and Fredrickson, 1990), the general trends governed by Equations (6.7–6.10) are even applicable to block copolymer systems, where the phase separation may be reduced to some extent by the covalent linkages between the constituent blocks. Generally speaking, the phase behavior of block copolymers can be classified into two categories according to the χN value (Bates, 1991; Lodge and Muthukumar, 1996): $\chi N \gg 10$, strong segregation region; $\chi N \leq 10$, weak segregation region. In the weak segregation region, entropic factors dominate and diblock copolymers exist in a spatially homogeneous state (Leibler, 1980). When $\chi N \gg 10$, however, well-segregated microdomain structures are formed with relatively sharp interfaces as the number of contacts between the two constituent segments decreases at the cost of chain stretching (Halperin *et al*.,, 1992; Helfand and Wasserman, 1982). In fact, seven well-defined microdomain structures with sizes typically in the range of a few tens of nanometers (Figure 6.11) have been observed in polystyrene-polyisoprene copolymers in the bulk (Bates and Fredrickson, 1990; Lodge and Muthukumar, 1996), which represents a strongly segregated system. Although there is considerable scope for the patterning of polymer chains through polymer phase separation, the fabrication of conjugated polymer patterns by polymer phase separation is still a research field in its infancy. A few reported examples are reviewed below.

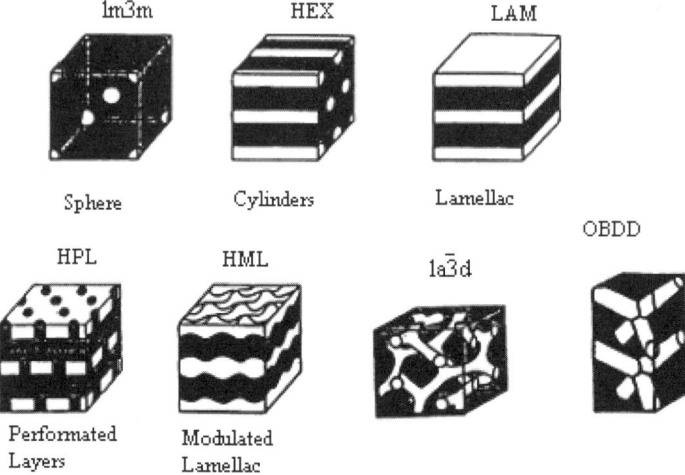

Figure 6.11. Experimentally assigned morphologies for diblock copolymers (After Lodge and Muthukumar, 1996, copyright 1996 The American Chemical Society. Reproduced with permission)

Interestingly, Inganäs *et al.* (Berggren *et al.*, 1994) demonstrated that self-organization (*i.e.* phase separation) in polymer blends consisting of substituted polythiophenes with different band gaps allows the formation of sub-micrometer-sized domains having a range of compositions and emission characteristics. As a result, the emission color can be varied by controlling the operating voltage. The ease with which a voltage-controlled multiple-color emission can be achieved suggests an attractive way for making future polymeric electroluminescent color screens.

As discussed in Chapter 2, polyacetylene could be solubilized by making copolymers with polyisoprene, polystyrene or polybutadiene via a so-called anionic to Ziegler-Natta route. The polyisoprene-polyacetylene (PI-PA) diblock copolymers thus prepared present an unusual situation where the high flexibility and solubility associated with polyisoprene chains and the stiffness and conductivity characteristic of insoluble conjugated polyacetylene chains are combined into one macromolecule.

Due to the large difference in physical properties between these two blocks, the copolymers of PI-PA may be expected to show phase separation in solution and domain structure in the solid state (Dai *et al.,* 1989; Dai and White, 1993, 1997). Indeed, Dai and White (1997) have recently found that I_2-doping of the PI-PA copolymer films leads to the formation of a pseudo-interpenetrating polymer network (PIPN). The aggregation structures (especially the PIPN network) have been demonstrated to play an important role in regulating the electrical properties of the PI-PA copolymers (Hamnett *et al.*, 1993). Therefore, the doping-induced phase separation process may have potential implications for making various multiphase conducting polymers with different ordered structures ranging from the nanometer to micrometer scale.

6.2.3.4 Plasma Patterning of Conjugated Polymers

Vapor deposition techniques, including plasma polymerization and chemical vapor deposition, are solvent-free, low-temperature thin film-forming processes with a clean working environment, which can be used to circumvent problems possibly associated with various solution-processing methods (*e.g.* the deterioration of the mechanical properties of conjugated polymer films by trapped solvent, and difficulty in choosing suitable solvent(s) for preparing pinhole-free single layer or multilayer polymer thin films). In fact, the chemical vapor deposition technique has been widely used for sublimation of low-molecular weight dyes (Tang and VanSlyke, 1987; Väterlein *et al.*, 1996), and recently for forming polymer films in EL devices (Meghdadi *et al.*, 1997; Weaver and Bradley, 1996; Winkler *et al.*, 1998). The plasma polymerization technique, however, seems to be much less discussed for the same purpose. For this reason, I will focus on the plasma technique below.

Radio-frequency glow-discharge is formed when gaseous monomers are exposed to a radio-frequency electric field at low pressure (< 10 Torr) (d'Agostino, 1990; Yasuda, 1978). During the glow-discharge process, energy is transferred from the electric field to free electrons, which inelastically collide with molecules leading to the generation of more electrons, ions and free radicals in the excited state. These plasma species are very reactive towards surfaces, causing surface modification in the case of plasma treatment and polymer deposition in the case of plasma polymerization. Although the electrons activated for generating charged particles may have a very high electron temperature *(e.g.* 2 eV = 23 200 K), they are not at thermodynamic equilibrium with gas molecules in the so-called "cold" plasma process. Therefore, plasma treatment and plasma polymerization can be carried out with the gas molecules being at ambient temperature, and thus without thermal degradation of the sample surface and/or the plasma-polymerized layer. Plasma polymerization has been widely demonstrated to generate thin, cohesive, adhering, pinhole-free polymer thin films, which are often used as protective coatings or adhesion-promoting layers and have potential applications in optical wave guides, sensor technology and electronic/photonic devices (d'Agostino, 1990; Dai *et al.*, 1996&2000; Hollan and Bell, 1974; Liang *et al.*, 1992; Yasuda, 1978).

On the other hand, the preparation of semiconducting plasma polymer films has also attracted a great deal of interest. For example, Bhuiyan and Bhoraskar (1989) produced semiconducting organic thin films through plasma polymerization of acrylonitrile, followed by pyrolysis. The electrical conduction in these plasma-polymerized polymer films was shown to be dominated by the variable-range hopping mechanism (Mott and Daris, 1979). Without involving a plasma precursor polymer, Kawakami (Kawakami, 1987) prepared partially conjugated conducting polymers by the polymerization of halogenated benzene and thiophene, respectively. Semiconductive thin organic polymer films were also prepared by Tanaka *et al.* (1991) through plasma polymerization of 1-benzothiophene. In this case, it was demonstrated that the aromatic skeletons of the monomer molecules are largely retained in the plasma films, which, after doping with iodine, show an increase in their electrical conductivity by eight to nine orders of magnitude (Nishio *et al.*, 1992). Later, these authors and others extended the plasma monomers to include 3,4,9,10-perylenetetracarboxylic dianhydride (PTCDA), 3,4,9,10-perylene-tetracarboxylic diimide (PTCDI), perylene (Tanaka *et al.*, 1993), and aniline (Augestine *et al.*, 1996; Fally *et al.,* 1992; Gong *et al.*, 1998; Hernandez *et al.,* 1984; Kang *et al.*, 1996; Samal *et al.*, 1994). The highest conductivity (0.1 S/cm) reported so far for plasma-polymerized organic films has been obtained with the as-synthesized PTCDA plasma film. In a separate study, Xie *et al.* (1994) prepared conducting organic films by plasma polymerization of a mixture of 7,7,8,8-tetracyanoquinodimethane (TCNQ) and quinoline, and found that the composite plasma polymers have a higher value of conductivity (*ca.* 10^{-5} S/cm) than those of the constituent organic analogues ($<10^{-12}$ S/cm for quinoline plasma polymer and 10^{-9} S/cm for TCNQ plasma polymer) due to a higher conjugation of the π-electrons in the composite film.

Apart from the preparation of plasma polymer films of electronic and photonic properties, plasma techniques have also been applied to pattern formation. For instance, Dai *et al.* (1997) have demonstrated that H_2O-plasma can cause plasma etching of substrate surfaces under certain discharge conditions. On the basis of this finding, these authors have successfully created high resolution surface patterns of hydrophilic regions on, for example, certain hydrophobic polymer films through a patterned plasma treatment using a mask. Furthermore, Dai *et al.* (1997) have also demonstrated that surface patterns of various specific functionalities, including both hydrophilic and hydrophobic groups, on a micrometer scale could be produced by plasma polymerization/treatment in a patterned fashion under appropriate conditions. The steps for the plasma-patterning process are shown in Figure 6.12(a). Figure 6.12(b) shows the surface pattern generated by the patterned plasma polymerization of *n*-hexane on a gold-coated mica surface, in which the dark areas represent *n*-hexane plasma polymer and the bright regions are associated with the plasma-polymer-free gold surface. By extension, these authors developed a versatile method for obtaining patterned conducting polymers by first depositing a thin patterned non-conducting (*e.g.* *n*-hexane) plasma polymer layer onto a metal-sputtered electrode, and then performing electropolymerization of monomers, such as pyrrole and aniline, within the regions not covered by the patterned plasma polymer layer. Figure 6.12(c) represents a typical reflection light microscopic image of a polypyrrole pattern electrochemically polymerized onto platinum-coated mica sheets pre-patterned with the freshly prepared *n*-hexane plasma polymer. It shows the same features as the plasma pattern of Figure 6.12(b), but with inverse intensities. The bright regions characteristic of the uncovered metal surface in Figure 6.12(b) become dark in Figure 6.12(c) due to the presence of a dark layer of the newly electropolymerized polypyrrole film. The bright regions in Figure 6.12(c) represent a more reflective surface associated with the *n*-hexane plasma polymer. The cyclic voltammogram of the polypyrrole pattern shown in Figure 6.12(c) is given in Figure 6.12(d), which shows a quasi-reversible redox process with two reduction peaks in an aqueous solution of sodium perchlorate. The first reduction peak of Figure 6.12(d) is attributable to the polarons in the electrochemically doped polyprrole film (Chapter 2), while the second reduction peak indicates the co-existence of a dicationic species (*i.e.* bipolarons). Therefore, the cyclic voltammogram measurements clearly indicate that the polypyrrole patterns thus prepared are electrochemically active.

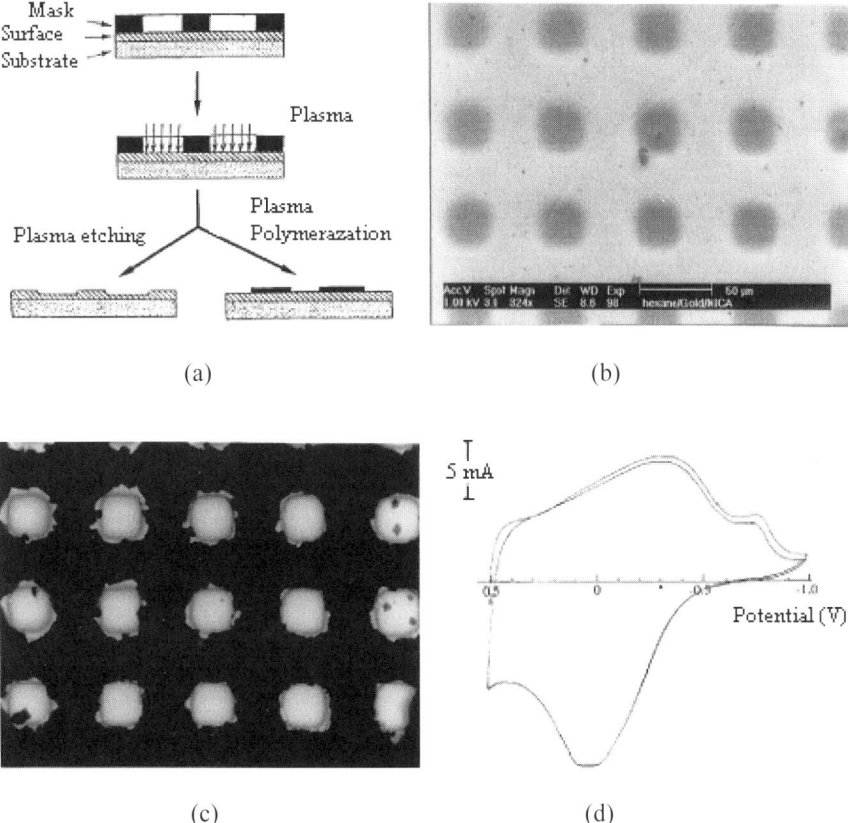

Figure 6.12. (a) Schematic representation of pattern formation by the radio-frequency glow-discharge plasma techniques; (b) a typical SEM micrograph of gold-coated mica sheets patterned by the *n*-hexane plasma polymer with a TEM grid consisting of square windows as the mask; (c) optical microscopy image of a polypyrrole pattern electrochemically polymerized onto a platinum-coated mica surface pre-patterned by the *n*-hexane plasma polymer; (d) typical cyclic voltammogram of the polypyrrole patterns on platinum at 100 mV/s in an aqueous solution containing 0.1 M sodium perchlorate (After Dai *et al.*, 1997, copyright 1997 The American Chemical Society. Reproduced with permission)

Furthermore, some plasma-patterned surfaces can be directly used to deposit conducting polymers under certain conditions. For example, Vargo *et al.* (1995) have region-specifically deposited polycations including polyaniline and polypyrrole onto a H_2/CH_3OH plasma-patterned perfluorinated ethylene-propylene copolymer (FEP) surface. These authors have also demonstrated the possibility for fabrication of both homogeneous and patterned conducting polyaniline multilayer structures with reasonable electrical conductivities. In view of the preparation of semiconducting organic films by plasma polymerization of aromatic monomers discussed above, a patterned plasma polymerization of these monomers should allow for the formation of semiconducting micro-patterns. This approach can be viewed as promising for the fabrication of future electronic and/or photonic devices with region-specific characteristics.

6.3 Aligned and Patterned Carbon Nanotubes

6.3.1 The Necessity

The discovery of carbon nanotubes by Iijima (1991) opened up a new era in material science and nanotechnology. As discussed in Chapter 5, carbon nanotubes can be viewed as a graphite sheet that has been rolled up into a nanoscale tube forming SWNTs, or with additional graphene tubes around the core of an SWNT to form MWNTs (Harris, 2001). These elongated nanotubes consist of carbon hexagons arranged in a concentric manner, with both ends of the tubes normally capped by fullerene-like structures containing pentagons. They usually have a diameter ranging from a few Ångstroms to tens of nanometers and a length of up to several centimeters. Because the graphene sheet can be rolled up with varying degrees of twist along its length, carbon nanotubes have a variety of chiral structures (Harris, 2001). Depending on their diameter and the helicity of the arrangement of graphite rings, carbon nanotubes can exhibit semiconducting or metallic behavior. Dissimilar carbon nanotubes may even be joined together to form molecular wires with unique electronic, magnetic, non-linear optical and mechanical properties (Dresselhaus *et al.*, 1996; Harris, 2001). These properties allow carbon nanotubes to be used as new materials for the development of novel single-molecular transistors (Cleland and Roukes, 1998; Tans *et al.*, 1998), scanning probe microscope tips (Harris, 2001), molecular computing elements (Ouyang *et al.*, 2001; Reed and Tour, 2000; Rueckers *et al.*, 2000), electron field emitting flat panel displays (de Heer *et al.*, 1997; Saito *et al.*, 1997), gas and electrochemical storage (Dillon *et al.*, 1997; Gadd *et al.*, 1997; Niu *et al.*, 1997), molecular-filtration membranes (Che *et al.*, 1998; Jirage *et al.*, 1997; Ren, 2000), artificial muscles (Baughman *et al.*, 1999), and sensors (Dai *et al.*, 2002). For most of the applications, however, an aligned/micropatterned form of carbon nanotubes is highly desirable (Dai and Mau, 2001), as exemplified by those examples to be described in the following sub-sections.

6.3.1.1 Molecular Computing

During the last half century or so, there has been tremendous progress in the field of electronics and computers. The need for cheaper, faster and more accurate calculations has been a driving force in the development of even smaller computing devices. This is where molecular electronics play a vital role. The advances in the field have made it possible to build single-molecule switches and memory elements. They can perform in a way analogous to diodes and transistors, which are the key components of microcircuits. In this context, carbon nanotubes have also been proposed to be used both as active components and molecular wires in molecular-scale electronics. For example, as is to be discussed in Chapter 11, Lieber and co-workers (Rueckers *et al.*, 2000) have recently devised a carbon nanotube-based random access memory (RAM) for molecular computing by using a criss-crossed carbon nanotube relay, which consists of a set of parallel SWNTs on a substrate and a set of perpendicular SWNTs suspended over the parallel nanotube array (Figure 11.22). Owing to their tiny size, as well as their good electrical and mechanical properties intrinsically characteristic of SWNTs, an integration level approaching 10^{12} elements per square centimeter and an element operation frequency in excess of 100 gigahertz are achievable.

6.3.1.2 Electron Emitters

Carbon nanotubes have also been explored for use as new electron field emitters in panel displays (de Heer *et al.*, 1997; Saito *et al.*, 1997). The carbon nanotube electron emitters work on a principle similar to a conventional cathode ray tube, but their small size can lead to a thinner, more flexible and energy-efficient display screen with a higher resolution. As will be mentioned in Chapter 9, Saito *et al.* (Saito *et al.*, 1998) have constructed an electron tube lighting element equipped with MWNT field emitters as a cathode (Figure 9.14(a)). In this study, stable electron emission, bright luminance and long life suitable for various practical applications have been demonstrated. More recently, some prototype carbon-nanotube field-emission displays were reported (Figures 9.14(c) and (d)) (de Heer and Martel, 2000). Although aligned carbon nanotubes are not necessary for these display applications, the use of aligned/micropatterned carbon nanotubes has been shown to offer additional advantages for the development of low field nanotube-based flat panel displays.

6.3.1.3 For Membrane Applications

It has recently been proposed to use the hollow aligned carbon nanotube films as a fast and energy-efficient means for water desalination (Che *et al.*, 1998; Jirage *et al.*, 1997; Ren, 2000). In this case, the vertically aligned carbon nanotube film with opened nanotube tips is electrically charged so that the sodium and chloride ions can be electrostatically adsorbed onto the tube surface when salt water runs through the nanotube hollow core. Subsequently, the electrostatically adsorbed ions can be released into a waste stream by rapidly removing the charge. Due to their high electrical conductivity and large surface area, the aligned carbon nanotube films are far more efficient than ordinary carbon for removing the salt from water.

The above examples clearly illustrate that the aligned and/or micropatterned structures are essential for the use of nanotubes in many applications of practical significance. Therefore, research on the synthesis of aligned nanotubes and their microfabrication has received ever-increasing attention. Recent developments in the field have indicated that the use of various advanced synthetic and micropatterning techniques could lead to a wide range of ordered (*e.g.* both horizontally and perpendicularly aligned) nanotubes of much improved properties and nanodevices of novel features. This section is meant to provide readers with a status summary of recent developments in the synthesis and micropatterning of carbon nanotubes and some of their non-carbon counterparts. In what follows, I will first present an overview of various methods for the preparation of *horizontally* aligned and micropatterned carbon nanotubes. I will then highlight some pyrolytic methods for the growth of *perpendicularly* aligned and micropatterned carbon nanotubes. This is followed by a demonstration of several *self-assembling* approaches towards the vertically aligned carbon nanotubes. The synthesis and microfabrication of aligned non-carbon nanotubes will then be spotlighted with a few examples.

6.3.2 Horizontally Aligned and Micropatterned Carbon Nanotubes

6.3.2.1 Horizontally Aligned Carbon Nanotubes

Horizontally aligned carbon nanotubes were originally prepared either by slicing a nanotube-dispersed polymer composite or by rubbing a nanotube-deposited plastic surface with a thin Teflon sheet or aluminum foil (de Heer *et al.*, 1995). Ajayan (1994) developed a simple technique to produce aligned arrays of carbon nanotubes. He mixed the nanotubes with epoxy-based resin. The nanotube-resin mixture was hardened and then cut into slices ranging in thickness from 50 to 1000 nm with lateral dimensions of a few millimeters. Transmission electron microscopy (TEM) images of the slices show that the nanotubes were preferentially oriented in parallel during the cutting process (Figure 6.13).

On the other hand, de Heer *et al.* (1997) produced thin films of aligned carbon nanotubes by drawing a nanotube suspension through a 0.2-μm-pore ceramic filter, leading to a uniform black deposit on the filter. The deposited nanotubes can be aligned either parallel or perpendicular to the surface of the films. The horizontally aligned surfaces are birefringent, reflecting differences in the dielectric function along and normal to the tubes. Recently, Kroto and co-workers (Terrones *et al.*, 1997) reported a method to align carbon nanotubes by pyrolysis of 2-amino-4,6-dichloro-*s*-triazine on a silica substrate pre-patterned with cobalt catalyst via laser ablation.

Figure 6.13. TEM image of aligned carbon nanotube arrays in a thin film of polymer expoxy (100 nm) obtained by cutting the nanotube/polymer composite (After Ajayan, 1995, copyright 1995 Wiley-VCH Verlag)

More recently, Pan *et al.* (2003) developed a new approach to aligned carbon nanotube films by self-assembling of modified carbon nanotubes with multiple hydroxyl groups (designated as: carbon nanotubols). In particular, these authors prepared the nanotubols via a simple solid-phase mechanochemical reaction (ball milling) with potassium hydroxide (KOH) at room temperature. The carbon nanotubols thus produced are highly soluble in water and can readily self-assemble into aligned arrays upon drying.

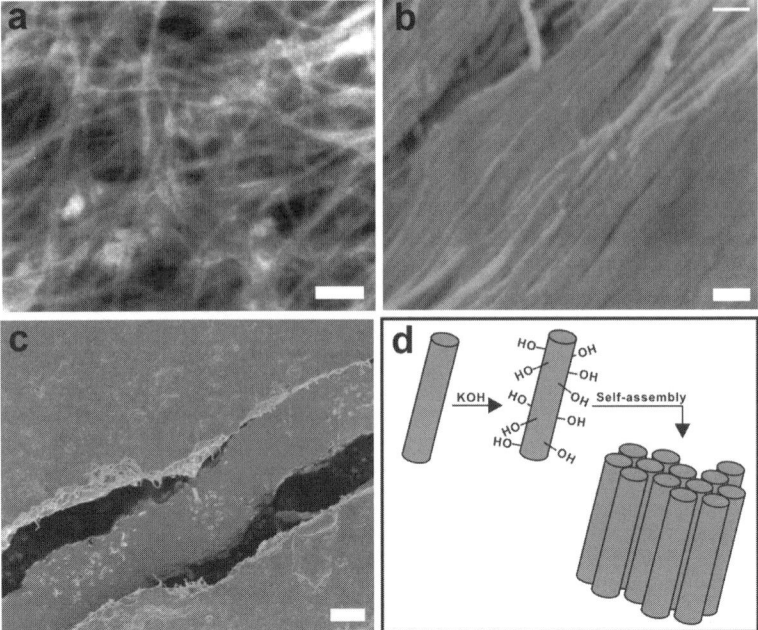

Figure 6.14. (a) SEM images of the starting materials (Tube@Rice), scale bar, 100 nm; (b) a higher magnification view of the aligned tubes, scale bar, 100 nm; (c) top view of the aligned carbon nanotubols, scale bar, 1 µm; (d) a scheme of the self-assembly process (After Pan *et al.*, 2003, copyright 2003 The American Chemical Society. Reproduced with permission)

As shown in Figure 6.14, the scanning tunneling electron microscopy (SEM) image for the starting HiPco SWNTs (Figure 6.14(a)) shows a randomly entangled morphology. In contrast, the cross-section SEM image of a centrifuged nanotubol film prepared by the ball milling of a piece of buckypaper in the presence of KOH reveals a well-aligned self-assembled structure (Figure 6.14(b)). Figure 6.14(c) shows a large-area flat surface for the nanotubol film, indicating the occurrence of a large-scale self-assembling process. As schematically illustrated in Figure 6.14(d), the strong hydrogen bonding interaction between the nanotubols is believed to be the driving force for the formation of the highly oriented, self-assembled nanotubol arrays.

6.3.2.2 Micropatterns of Horizontally Aligned Carbon Nanotubes

Burghard *et al.* (1998) have recently reported the region-specific deposition of carbon nanotubes grafted with surfactants containing negatively charged groups (*e.g.* sodium dodecylsulfate) onto substrate surfaces pre-patterned with positively charged functionalities (*e.g.* ammonium groups on silanized silica). More recently, Liu J. *et al.* (1999) demonstrated that individual single-wall carbon nanotubes purified by refluxing in HNO_3 (2.6 M) can be region-specifically deposited onto surfaces pre-patterned with a self-assembled monolayer (SAM) of $-NH_2$

functionalities. In a separate study, Chen and Dai have developed a versatile method for making patterns of non-aligned carbon nanotubes (Chen and Dai, 2000). In particular, they first generated plasma-induced (either by non-depositing plasma treatment or by plasma polymerization) surface patterns of $-NH_2$ groups onto a substrate (*e.g.* quartz glass plate, mica sheet, polymer film), and then performed region-specific adsorption of the COOH-containing carbon nanotubes from an aqueous medium, through the polar-polar interaction between the COOH groups and the plasma-patterned $-NH_2$ groups. The COOH-containing carbon nanotubes were prepared by acid treatment (HNO_3) of the FePc-generated nanotubes (Tsang *et al.,* 1994). Figure 6.15 reproduces a SEM image of the COOH-containing carbon nanotubes region-selectively adsorbed (*ca.* 2.5 mg/10 ml H_2O) onto a mica sheet pre-patterned with the heptylamine-plasma polymer (200 kHz, 10 W, and a monomer pressure of 0.13 Torr for 30 s). The adsorbed carbon nanotubes are clearly evident by inspection of the plasma-patterned areas of Figure 6.15 under higher magnification (inset of Figure 6.15). The corresponding high magnification SEM image for the plasma-polymer-free areas reveals an almost featureless smooth surface characteristic of mica sheet. No adsorption of the carbon nanotubes was observed in a controlled experiment when a pure mica sheet was used as the substrate.

Figure 6.15. SEM image of adsorbed COOH-containing carbon nanotubes (within the squared areas) on a heptylamine-plasma patterned mica sheet. Inset gives a higher magnification image of the plasma covered areas, showing the individual adsorbed carbon nanotubes.

More recently, Zhou and co-workers (Shimoda *et al.,* 2002) have demonstrated the preparation of ordered/micropatterned carbon nanotubes through self-assembling of pre-formed carbon nanotubes on glass by vertically immersing the substrate into an aqueous solution of shortened (acid-oxidized) SWNTs (Tsang *et al.,* 1994). Figure

6.16(a) shows schematically the self-assembling process. No carbon nanotube deposition was observed for a hydrophobic substrate (*e.g.* a polystyrene spin-coated glass slide) under the same conditions. This selectivity enabled the fabrication of carbon nanotube patterns by using glass substrates with pre-patterned hydrophobic and hydrophilic regions. Figure 6.16(b) shows optical microscopic images for some of the SWNT micropatterns thus prepared.

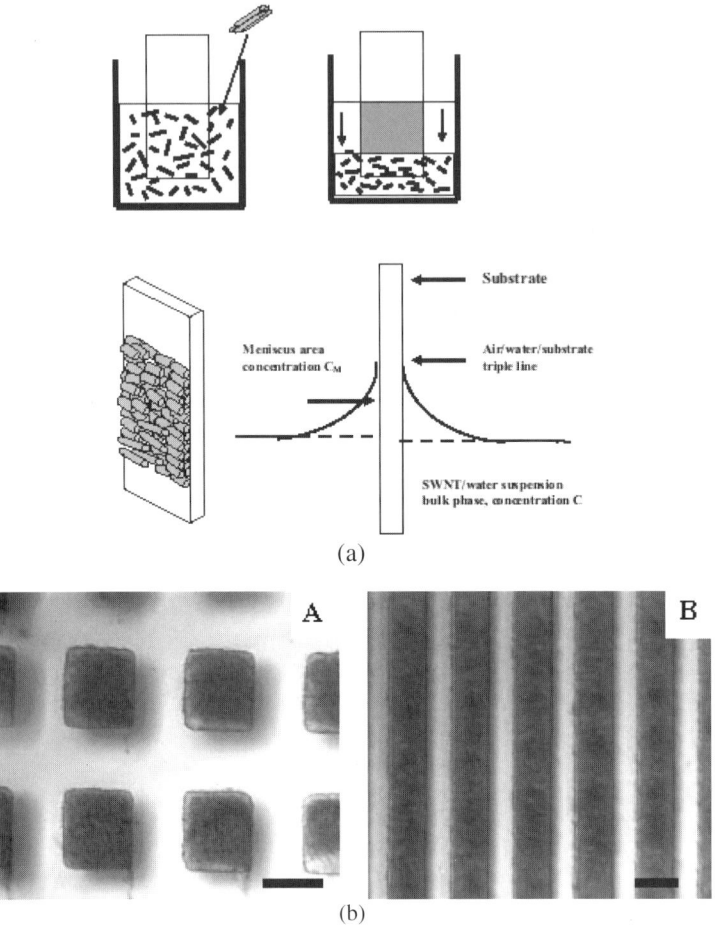

Figure 6.16. (a) A schematic representation of the self-assembling process. A hydrophilic glass slide was vertically immersed into an aqueous solution of acid-oxidized SWNTs, followed by the gradual evaporation of the water at room temperature. The SWNT bundles self-assembled on the glass substrate around the air/water/substrate interface. As the interface progressed downwards, a continuous SWNT film was formed on the glass substrate; (b) patterned SWNT structures formed by the self-assembling process on hydrophobic substrates pre-patterned with periodic hydrophilic regions: (A) the squares are 100 μm × 100 μm and (B) the strips are 100 mm in width. The shadows are due to reflections from the surface on which the samples were placed. The scale bars in both (A) and (B) are 100 μm (After Shimoda *et al.*, 2002, copyright 2002 Wiley-VCH Verlag. Reproduced with permission)

In order to realize potential applications for the aligned and/or micropatterned carbon nanotubes discussed above, considerable efforts are required to make proper connections from these nanoscale entities to the macroscale world. As far as the published work is concerned, it would be unfair to comment that not much has yet been achieved in this area. For instance, Dai and co-workers (Kong *et al.*, 1998; Cassell *et al.*, 1999) at Stanford have demonstrated the ability to grow SWNT wires between controlled surface sites by catalyst patterning, leading to a variety of interconnecting SWNT architectures, including a suspended SWNT power line and a square of suspended SWNT bridges (Figure 6.17).

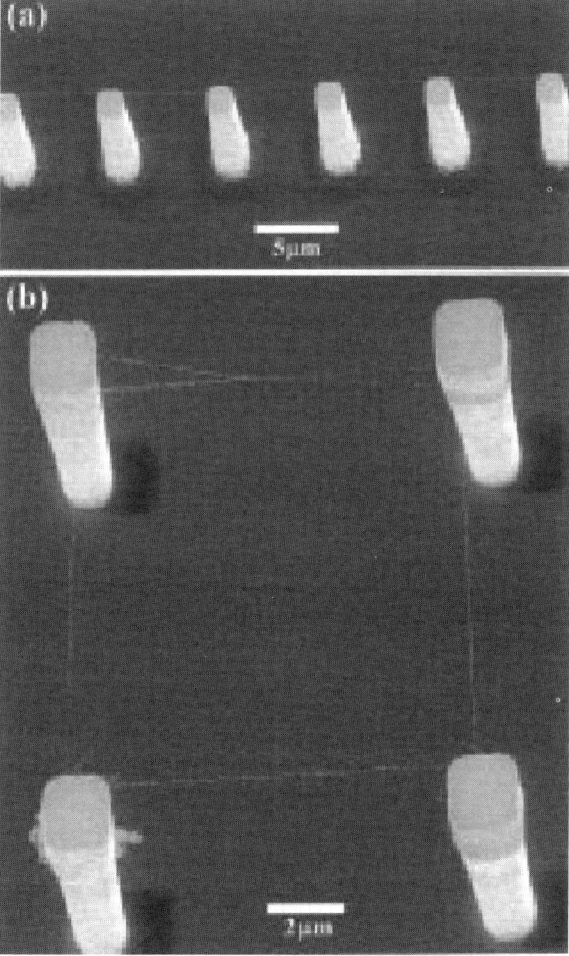

Figure 6.17. (a) SEM images of (a) a suspended SWNT power line and (b) a square of suspended SWNT bridges (After Cassell *et al.*, 1999, copyright 1999 The American Chemical Society. Reproduced with permission)

In particular, these authors developed a liquid-phase catalyst precursor, which was patterned onto silicon towers using the contact printing (μCP) technique. Calcination of the silica-tower-supported catalyst, followed by CVD growth of SWNTs, resulted in the formation of a SWNT network between adjacent towers. The SWNT network thus formed can be utilized as power lines to address nanoscale components in nanodevices. By controlling the arrangement of silicon towers, the growth of the suspended SWNTs can be directed to create a desired architecture.

Recently, the interconnected growth of MWNTs has also been reported by Ajayan and co-workers (Wei B. *et al.,* 2001). These authors have micro-patterned magnesium oxide cubes, over which nanotubes can be grown into network structures, by burning Mg ribbons in ambient air and depositing the smoke on clean, micropatterned Si wafers. The silicon wafers micropatterned with MgO cubes were then used as a substrate for the CVD growth of the MWNT network between the MgO cubes.

6.3.3 Perpendicularly Aligned and Micropatterned Carbon Nanotubes

6.3.3.1 Perpendicularly Aligned Carbon Nanotubes

As can be seen from the above discussion, carbon nanotubes perpendicularly aligned to substrates offer additional advantages for many applications, including their use as electron emitters and molecular membranes. In order to construct a nanotube field emitter, for example, de Heer *et al.* (1997) first made an ethanol dispersion of arc-produced carbon nanotubes. These authors then passed the nanotube dispersion through an aluminum oxide micropore filter to align nanotubes perpendicularly onto the filter surface, which can be transferred onto the cathode substrate in the field-emitting device.

The so-called template synthesis technique has also been used to prepare aligned carbon nanotubes within a porous membrane (*e.g.* mesoporous silica, alumina nanoholes), (Iwasaki *et al.*, 1999; Li J. *et al.*, 1999a). In particular, Xie and co-workers (Li W. *et al.*, 1996) prepared large-scale aligned carbon nanotubes through CVD deposition on iron nanoparticles embedded in mesoporous silica. The growth direction of the nanotubes could be controlled by the orientation of the pores from which nanotubes grow. Similarly, aligned carbon nanotubes have also been prepared by pyrolysis of hydrocarbons (*e.g.* ethylene, propylene or pyrene) within the pores of porous silicons in the presence or absence of metal catalysts (*e.g.* Ni, Co) (Li J. *et al.*, 1996; Li Z. *et al.*, 1999). The pores of porous silicons were produced by electrochemical etching in an aqueous HF solution, in which the anode is the crystalline silicon wafer and the cathode is Pt wire.

Using branched nanochannel alumina templates, Li J. *et al.* (1999a, b) have also produced carbon nanotube Y-junctions (Figure 6.18). Subsequently, Y-shaped carbon nanotubes were prepared by the pyrolysis of methane over cobalt supported on magnesium oxide (Li W. *et al.*, 2001) and direct nano-bridging of carbon nanotubes between photolithographically micro-sized islands (Lee *et al.*, 2002).

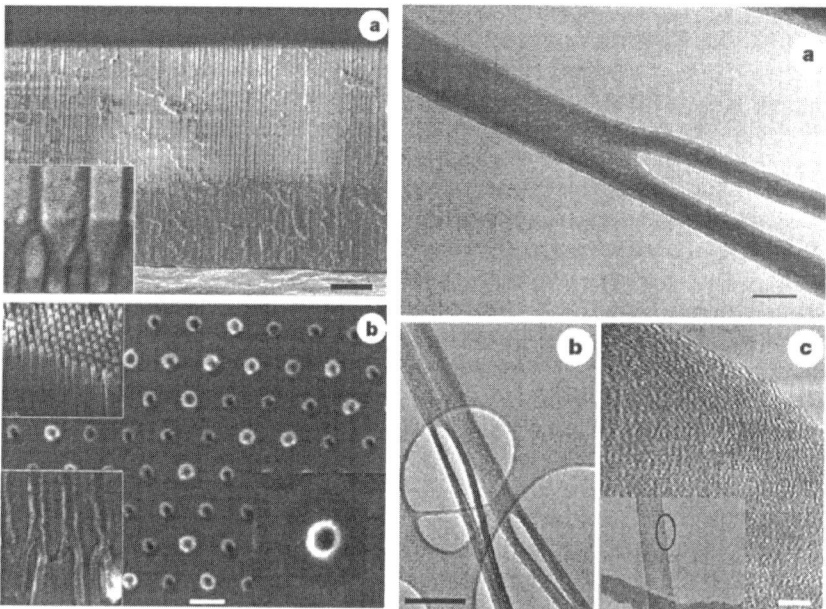

Figure 6.18. *Left:* (a) SEM image of Y-branched nanochannel template (scale bar, 1 μm). The template was formed by first anodizing a sheet of high-purity aluminum in 0.3 M oxalic acid at 10 °C under a constant voltage of 50 V for 15 h, resulting in a hexagonal arry of pores near the aluminium surface. After chemically removing the original film, a second anodization was performed under the same conditions, typically for 30 min. The anodization voltage was then reduced to about 35 V. Because the pore cell diameter is proportional to the anodization voltage, reducing the voltage by a factor of $1/\sqrt{2}$ results in twice as many as pores appearing in order to maintain the original total area of the template, and nearly all pores branching into two smaller-diameter pores. The resulting template consists of parallel Y-branched pores with stems 40 nm in diameter and branches 28 nm in diamater. The arrow shows where Y branches start to grow (see inset for close-up); (b) top-view SEM image of carbon nanotubes aligned in the template after ion-milling of amorphous carbon on the surface (scale bar, 100 nm). The nanotube diameter is larger than the original pore owing to thermal expansion of the template during growth. Top inset, stem part of the Y-junction tubes. Bottom left inset, close-up of the region between stem and branch portions still embedded in the template. Bottom right inset, close-up of the top of the nanotube in its hexagonal cell. *Right:* TEM image of (a) the Y-junction tube (scale bar, 50 nm) with stem *ca.* 90 nm and branches 50 nm in diameter; (b) Y-junction formed by using higher anodization voltages, resulting in stems *ca.*100 nm and branches 60 nm in diameter (scale bar, 200 nm); (c) high-resolution TEM image of typical Y-junction nanotube wall showing graphitic multiwall structure (scale bar, 5 nm). Inset shows the part of the tube that was imaged (After Li J. *et al.,* 1999b, copyright 1999 Macmillan Magazines Limited. Reproduced with permission)

Without using the template pores, Rao *et al.* (1998) have prepared aligned carbon nanotube arrays by high-temperature (*ca.* 900 °C) pyrolysis of ferrocene, which contains both the metal and carbon source required for the nanotube growth, while Kamalakaran *et al.* (2000) reported the formation of aligned carbon nanotube arrays by pyrolysis of a jet solution of ferrocene and benzene in an argon atmosphere at relative low temperatures (850 °C). In a separate but somewhat related study, Ren *et al.* (1998) synthesized large arrays of well-aligned carbon nanotubes by radio-frequency sputter-coating a thin nickel layer onto display glass, followed by plasma-enhanced hot filament chemical vapor deposition of acetylene in the presence of ammonia gas below 666 °C. These authors have also prepared highly aligned MWNTs on polished polycrystalline and single-crystal nickel substrates by plasma-enhanced hot filament chemical vapor deposition at temperatures below 666 °C (Huang Z.P. *et al.*, 1998). The plasma intensity, acetylene to ammonia gas ratio and their flow rates were found to play important roles in regulating the diameter and uniformity of the resulting aligned carbon nanotubes. Tanemura *et al.* (2001) have also produced aligned carbon nanotubes by DC plasma-enhanced chemical vapor deposition of acetylene and ammonia mixtures on Co- or Ni-covered W wires. These authors found that the nature of the catalyst support material, the catalyst material itself and the positive ions in the plasma strongly affect the growth mechanism of aligned carbon nanotubes, and hence their structure and properties. It was further demonstrated that aligned carbon nanotubes deposited on Co-covered W wires formed via the so-called tip growth mode.

Along with the aforementioned and many other plasma-enhanced CVD techniques (Hippler, 2001), a microwave plasma-enhanced CVD method has also been reported for the synthesis of aligned carbon nanotubes. In this regard, Tsai *et al.* (2002) have fabricated aligned carbon nanotubes with open ends on silicon wafer by using a microwave plasma-enhanced chemical vapor deposition system (MPE-CVD) with a mixture of methane and hydrogen as precursors. High-concentration hydrogen plasma and a high-negative bias voltage to the substrate induced anisotropic etching of carbon nanotubes and could effectively reduce the randomly oriented carbon nanotubes. Young and Lee (2002) also reported that the vertically aligned carbon nanotubes were synthesized by microwave plasma-enhanced CVD on Ni-coated Si substrate. The diameter, growth rate and density of carbon nanotubes can be controlled by the grain size and morphology of the Ni thin films, which can be altered by changing the rf power density during rf magnetron sputtering. They found that the diameter of the nanotubes decreased by decreasing the grain size of Ni thin films, whereas the growth rate and density increased. By growing aligned carbon nanotubes using high-frequency, microwave plasma-enhanced chemical vapor deposition, Bower *et al.* (2001) observed that the alignment was primarily induced by the electrical self-bias imposed on the substrate surface from the microwave plasma, and that nanotubes were grown always perpendicular to the local substrate surface regardless of the orientation and shape of the surface from which the nanotubes grow. Unlike the plasma-enhanced growth of perpendicularly aligned carbon nanotubes, Avigal and Kalish (2001) reported a new method for the growth of aligned carbon nanotubes by applying an electric field to a Co-covered substrate in a regular cold-wall chemical vapor deposition reactor (with no plasma applied) containing a flowing mixture of methane and argon at 800 °C. They found that perpendicularly aligned carbon nanotubes formed when a positive bias is applied to the substrate.

Dai and co-workers (Huang S. *et al.*, 1999) have recently prepared large-scale aligned carbon nanotubes perpendicular to the substrate surface by the pyrolysis of iron (II) phthalocyanine ($FeC_{32}N_8H_{16}$, designated as FePc) under Ar/H_2 at 800–

1100 °C. Figure 6.19 shows that the constituent carbon nanotubes have a fairly uniform length and diameter. The constituent straight nanotubes were reported to have a well-graphitized multiwall structure with an outer diameter in the range of 35–55 nm (Li D.-C. *et al.*, 2000).

Figure 6.19. A typical SEM image of the aligned carbon nanotube film prepared by pyrolysis of FePc. The misalignment seen for some of the nanotubes at the edge was caused by the peeling action for the SEM sample preparation (After Huang S. *et al.*, 1999, copyright 1999 The American Chemical Society. Reproduced with permission)

6.3.3.2 Micropatterns of Perpendicularly Aligned Carbon Nanotubes

Dai and co-workers (Huang S. and Dai, 2002), among others (Fan *et al.*, 1999; Li J. *et al.*, 1999a; Ren *et al.*, 1999; Suh *et al.*, 1999; Wei B. *et al.*, 2002), have also prepared micropatterns of *aligned* carbon nanotubes normal to the substrate surface suitable for field emission. This method not only allows for the preparation of micropatterns and substrate-free films of the perpendicularly aligned nanotubes, but also their transfer onto various substrates, including those which would otherwise not be suitable for nanotube growth at high temperatures (*e.g.* polymer films) (Huang S. *et al.*, 1999). On further investigation of the aligned carbon nanotubes produced by the pyrolysis of FePc, Dai and co-workers (Yang *et al.*, 1999) have developed a novel method for photolithographic generation of the perpendicularly aligned carbon nanotube arrays with resolutions down to a micrometer scale.

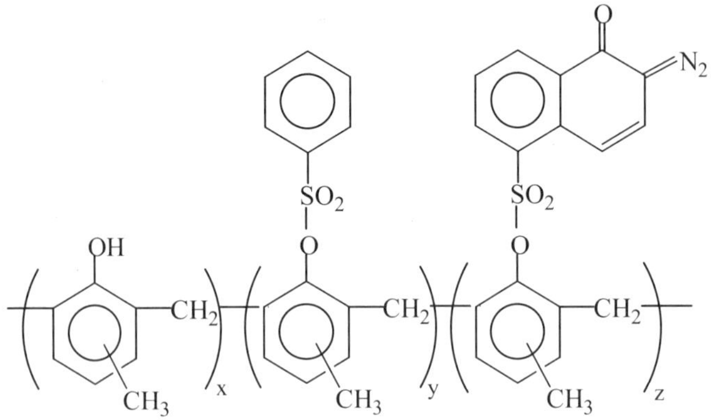

(a)

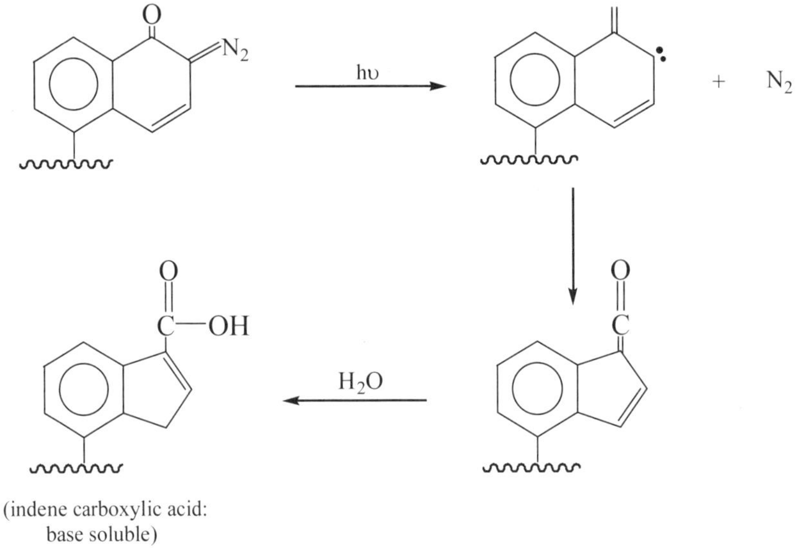

(b)　　　　　　　　　　　　　　　　　　　　　　　　　　(6.11)

(indene carboxylic acid: base soluble)

Figure 6.20(a) shows the steps of the photolithographic process. In practice, Yang *et al.* (1999) first photolithographically patterned a positive photoresist film of diazonaphthoquinone (DNQ)-modified cresol novolak [Equation (6.11)] onto a quartz substrate. Upon UV irradiation through a photomask, the DNQ-novolak photoresist film in the exposed regions was rendered soluble in an aqueous solution of sodium hydroxide due to photogeneration of the hydrophilic indene carboxylic acid groups from the hydrophobic DNQ via a photochemical Wolff rearrangement (March, 1992) [Equation (6.11)]. These authors then carried out the pyrolysis of FePc, leading to region-specific growth of the aligned carbon nanotubes in the UV-exposed regions (Figure 6.20(b)).

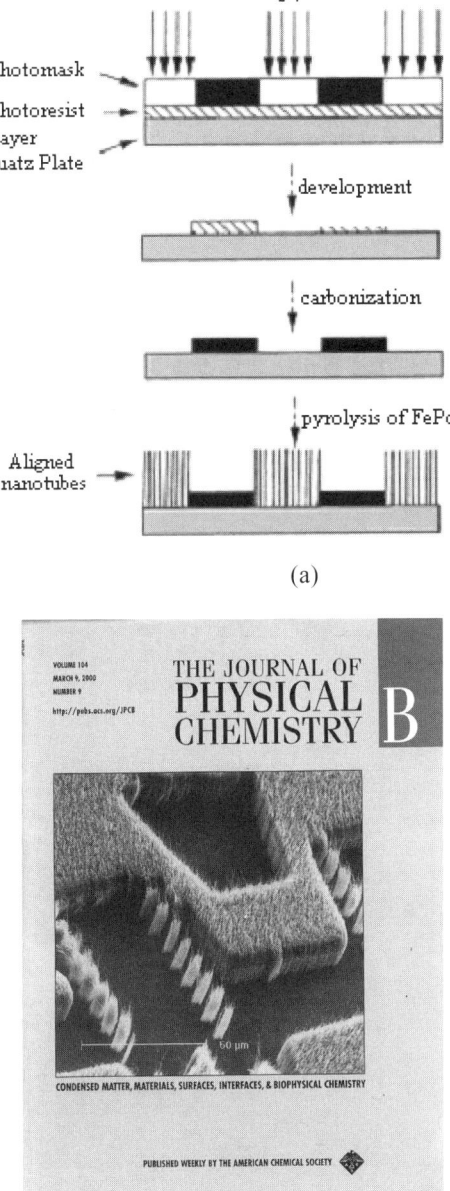

Figure 6.20. (a) Schematic representation of the micropattern formation of aligned carbon nanotubes by photolithographic process; (b) typical SEM micrographs of patterned films of aligned nanotubes prepared by the pyrolysis of FePc onto a photolithographically pre-patterned quartz substrate published as a cover page in *The Journal of Physical Chemistry* (After Yang *et al.*, 1999, copyright 1999 The American Chemical Society. Reproduced with permission)

In this case, the photolithographically patterned photoresist film, after an appropriate carbonization process, acts as a shadow mask for the patterned growth of the aligned nanotubes. This method is fully compatible with existing photolithographic processes (Wallraff and Hinsberg, 1999).

Using a modified photolithographic method for patterned pyrolysis of FePc, Chen and Dai (2001) have recently prepared three-dimensional (3D) micropatterns of *aligned* carbon nanotubes normal to the substrate surface with region-specific tubular lengths and packing densities. The photoresist system used in this approach consists of novolac/hexamethoxymethyl melamine (HMMM) as the film former, phenothiazine (RH) as the photosensitizer, and diphenyliodonium hexafluorophosphate ($Ph_2I^+X^-$) as a photoacid generator that can photochemically generate the acid, through a photomask, required for the region-specific cross-linking of the photoresist film (Havard *et al.*, 1999). Figure 6.21 represents the steps of the photolithographic process with the associated photochemical reactions shown in Equations (6.12)-(6.14).

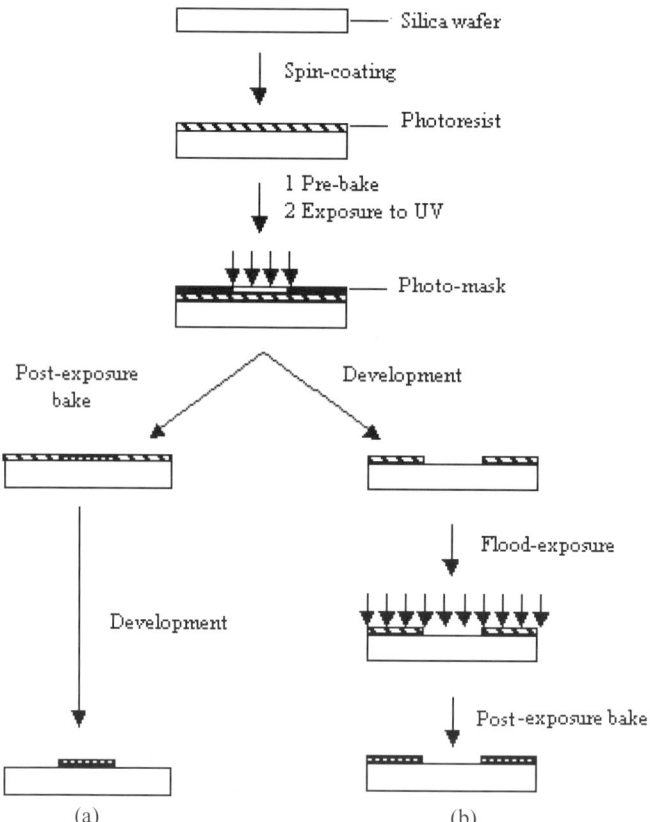

Figure 6.21. Schematic of the procedures for photolithographic patterning of the chemically amplified photoresist into a negative (a) and positive (b) pattern (After Chen and Dai, 2001, copyright 2001 American Scientific Publishers)

$$Ph_2I^+X^- \xrightarrow[2RH]{h\nu} PhI + PhH + H^+X^- + 2R^{\bullet} \qquad (6.12)$$

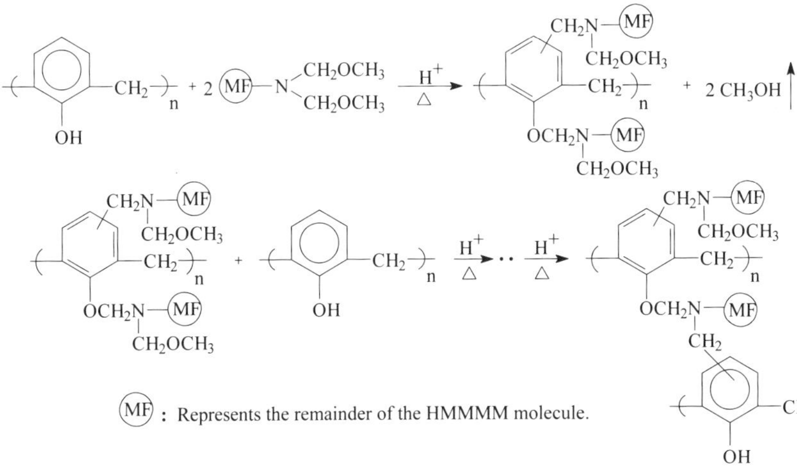

(MF) : Represents the remainder of the HMMMM molecule.

UV irradiation through a photomask creates a latent acid pattern formed by the photolithographic generation of acid from $Ph_2I^+X^-$ [Equation (6.12)]. Post-exposure baking at 110 °C for 10 min (Figure 6.21(a)) caused an acid-induced cross-linking of the novolac resin and HMMM [Equations (6.13) and (6.14)], rendering the photoresist film in the UV-exposed regions insoluble in an aqueous solution of sodium hydroxide (3 wt%) and ethanol (10 wt%). In contrast, the photoresist film in the regions unexposed to the UV light was removed simply by immersing in the developer solution for 10–20 s, leading to the formation of a negative polymer pattern on the substrate. The cross-linking reactions between HMMM and novolac resin were further completed by immersing the photoresist patterned substrate into an aqueous solution of *p*-toluenesulfonic acid (10 wt%) for 30 min and further baking at 150 °C for 30 min (Havard *et al.*, 1999).

The photoresist pre-patterned silica wafer or quartz plate was then directly used as the substrate for the region-specific growing of aligned carbon nanotubes without carbonization. Figure 6.22 represents a typical scanning electron microscope image of the three-dimensional-aligned carbon nanotube micropatterns thus prepared, which clearly shows a region-specific packing density tubular length. Owing to the highly cross-linked structure between the novolac resin and HMMM within the photoresist patterns, the integrity of the photoresist layer was maintained, even without the carbonization, at the high temperatures required for the aligned nanotube growth from FePc (Yang *et al.*, 1999). In contrast to the DNQ-novolak photoresist film, the resultant HMMM-cross-linked novolac photoresist film supported the nanotube growth, most probably due to its delicate surface characteristics that allowed the Fe nanoparticles to deposit onto this particular photoresist layer at the initial stage of the FePc pyrolysis (Li D.-C. *et al.*, 2000). The three-dimensional micropatterns of aligned carbon nanotubes thus prepared offer the possibility for construction of advanced microdevices with multi-dimensional features.

244   Intelligent Macromolecules for Smart Devices

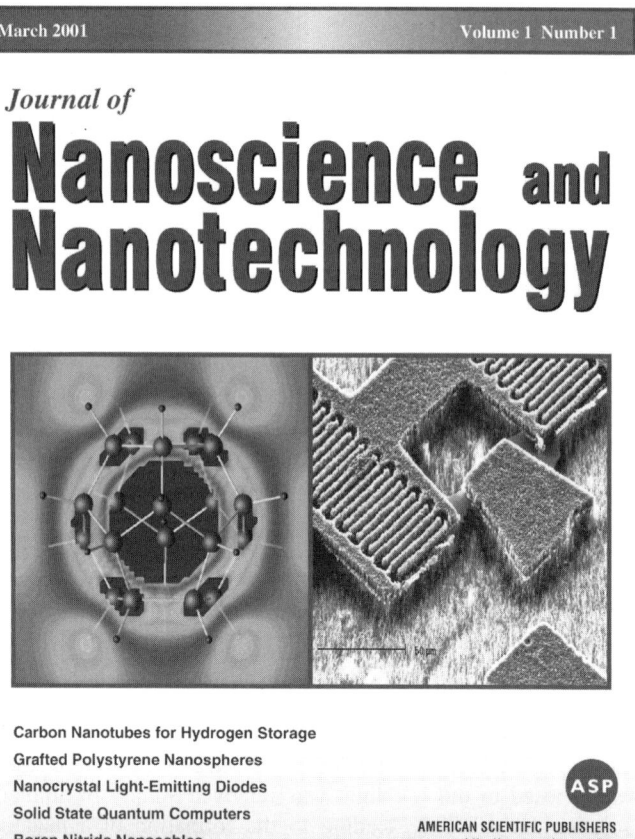

Figure 6.22. Typical SEM micrographs of the 3D-aligned carbon nanotube micropattern (*right*) published as a cover page for the first issue of the *Journal of Nanoscience and Nanotechnology* (After Chen and Dai, 2001, copyright 2001 American Scientific Publishers).

Recently, Dai and co-workers (Huang S. *et al.*, 2000) used the microcontact printing (μCP) and micro-molding techniques to prepare micro-patterns of carbon nanotubes aligned in a direction normal to the substrate surface. While the μCP process involves the region-specific transfer of self-assembling monolayers (SAMs) of alkylsiloxane onto a quartz substrate and subsequent adsorption of polymer chains in the SAM-free regions (Figure 6.23(a)), the micro-molding method (Kim *et al.*, 1997; Rogers *et al.*, 1998) allows the formation of polymer patterns through solvent evaporation from a precoated thin layer of polymer solution confined between a quartz plate and a polydimethylsiloxane (PDMS) elastomer mold (Figure 6.23(b)).

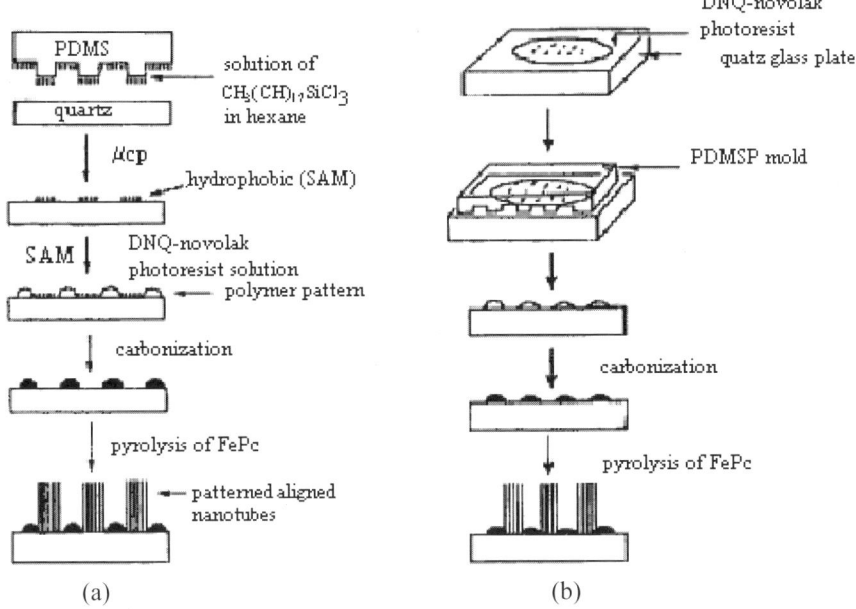

Figure 6.23. Schemtic illustration of the procedure for fabricating patterns of aligned carbon nanotubes by (a) microcontact printing; (b) solvent-assisted micro-molding (After Huang *et al.*, 2000, copyright 2000 The American Chemical Society. Reproduced with permission)

The DNQ-novolak photoresist patterns formed in both cases were then carbonized into carbon black for region-specific growth of the aligned nanotubes in the polymer-free regions by pyrolysis of iron(II) phthalocyanine (FePc) under an Ar/H_2 atmosphere at 800–1100 °C, as is the case for the above-mentioned photolithographic patterning. The spatial resolution is limited by the resolution of the mask used. Micropatterns of aligned nanotubes thus prepared have resolutions down to 0.8 μm, suitable for fabrication of various electronic and photonic devices (Figure 6.24). Kind *et al.* (1999) have also used the μCP technique to pattern catalysts for region-specific growth of aligned carbon nanotubes.

The ease with which micro-/nano-patterns of organic materials can be made on curved surfaces by the soft lithographic techniques (Jackman and Whitesides, 1999; Xia and Whitesides, 1998) should provide additional benefits to this approach with respect to the photolithographic method, especially for the construction of flexible devices.

Figure 6.24. A typical SEM image of an aligned nanotube pattern prepared by the pyrolysis of FePc onto the photoresist prepatterned quartz via micromolding technique, appeared in a cover page of *The Journal of Physical Chemistry* (After Huang S. *et al.*, 2000, copyright 2000 The American Chemical Society. Reproduced with permission)

As can be seen from the above discussion, both the photolithographic and soft-lithographic patterning methods involve a tedious carbonization process prior to the aligned nanotube growth. In order to eliminate the carbonization process, Chen and Dai (2001) prepared high-resolution carbon nanotube arrays aligned in a direction normal to the substrate surface by radio-frequency glow-discharge plasma polymerization of a thin polymer pattern onto a quartz substrate, followed by region-specific growth of the aligned carbon nanotubes in the plasma-polymer-free regions by pyrolysis of FePc (Figure 6.25). The highly-cross-linked structure of plasma-polymer films (Dai, 2001; Hollan and Bell, 1974) could ensure the integrity of the plasma polymer layer, even without carbonization, at the high temperatures necessary for the nanotube growth from FePc (Huang S. *et al.*, 1999). Therefore, the carbonization process involved in the early work on photolithographic (Yang *et al.*, 1999) and soft-lithographic (Huang S. *et al.*, 2000) patterning of the aligned carbon nanotubes can be completely eliminated in the plasma patterning process. Owing to the generic nature characteristic of the plasma polymerization, many other organic vapors can also be used efficiently to generate plasma polymer patterns for the patterned growth of the aligned carbon nanotubes.

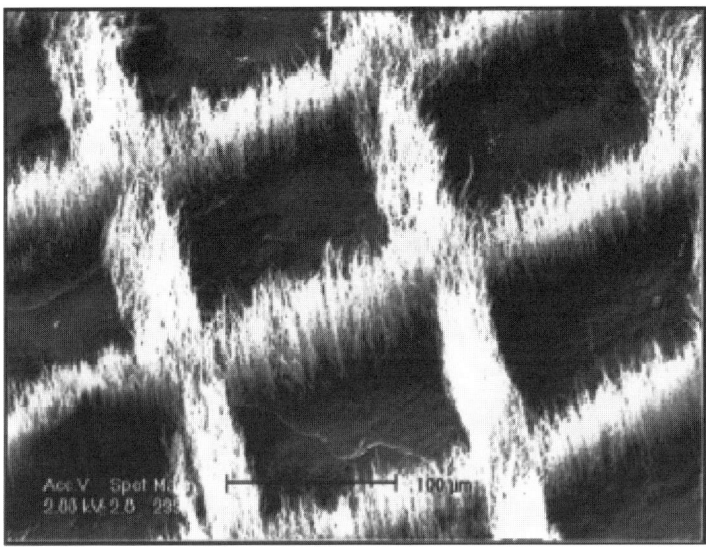

Figure 6.25. SEM images of aligned nanotube arrays growing out from the plasma-polymer-free regions (After Cheng and Dai, 2000, copyright 2000 American Institute of Physics)

Many other groups have also devoted a great deal of effort on micropatteerning perpendicularly aligned carbon nanotubes. In what follows, some of the important papers are spotlighted. For instance, Fan *et al.* (1999) reported the synthesis of massive arrays of monodispersed carbon nanotubes on patterned porous silicon substrate with Fe films by electron beam evaporation through shadow masks. Sohn *et al.* (Sohn *et al.*, 2001; Wen *et al.,* 2001) also used a similar method to produce aligned carbon nanotubes on Fe nanoparticles deposited by a pulsed laser on a porous Si substrate. Furthermore, Ren *et al.* (1999) reported the growth of a freestanding multiwall carbon nanotube onto a grid of patterned sub-micron nickel dot(s) by plasm-enhanced-hot-filament-chemical-vapor deposition (PE-HF-CVD) using acetylene gas as the carbon source and ammonia as the catalyst and dilution gas. The single-carbon nanotubes were observed to grow on the grid and separated well. A thin film nickel grid was fabricated on a silicon wafer by electron beam lithography and metal evaporation. Using this method, the device requiring freestanding vertical carbon nanotubes, including scanning probe microscopy and field emission flat panel display, were prepared (Ren *et al.*, 1999; Teo *et al.*, 2001). Teo *et al.* (2001) also fabricated a uniform array of nanotubes or single freestanding aligned carbon nanotubes by PE-CVD technology with either photolithographically or electron-beam lithographically patterned nickel catalyst substrates (Figure 6.26) (Teo *et al.*, 2001).

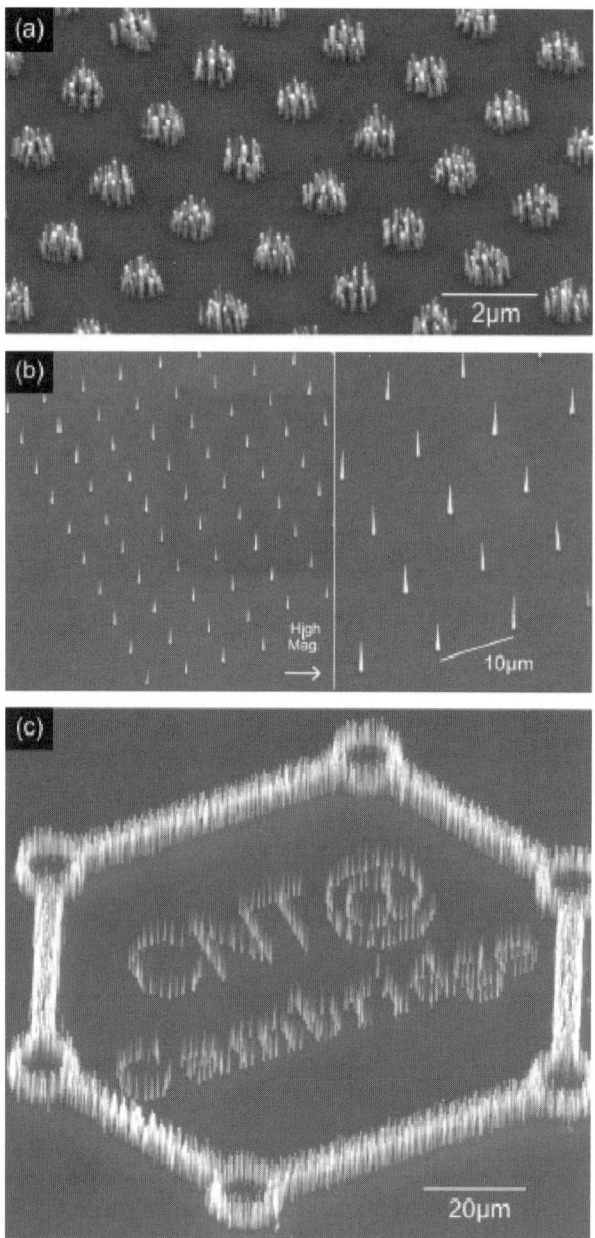

Figure 6.26. (a) Bunches of aligned carbon nanotubes (*ca.*100 nm) are deposited on 1 μm nickel dots by breaking up the nickel catalyst film into multiple nanoparticles; (b) single aligned nanotubes are deposited when the nickel dot size is reduced to 100 nm as only a single nickel nanoparticle is formed from the dot; (c) the selective growth of high yield and uniform aligned nanotubes with different densities (After Teo *et al.*, 2001, copyright 2001 American Institute of Physics)

Apart from the application of the lithographic techniques, the nanopores of porous alumina membranes have also been used to produce micro-patterned aligned carbon nanotubes. Li J. *et al.* (1999b) fabricated the large arrays of parallel carbon nanotubes with a level of periodicity and uniformity by pyrolysis of acetylene on cobalt within a hexagonal closed-packed nanochannel aluminum template at 650 °C. Using this method, ordered nanotubes with diameters from 10 nm to several hundred nm and lengths up to 100 μm can be produced. In practice, the pores of the anodic alumina film can self-organize into a highly ordered hexagonal array under appropriate anodizing conditions. The high level of ordering and uniformity in these arrays is useful for applications in data storage, field emission displays and sensors, and offers the prospect of deriving computational functions from the collective behavior of symmetrically coupled nanotubes. Suh *et al.* (1999) also fabricated the highly ordered carbon nanotubes by using porous anodic aluminum oxide templates prepared by a two-step anodization process. In the first-step of the anidization process employed in this work, the thick aluminum oxide film was obtained after long anodization, a porous thin alumina film with highly ordered pores was obtained by a subsequent reanodization. This technique offered a potential application for cold-cathode flat panel display.

Although a micro-pore template, such as mesoporous silicon and aluminum oxide film, provided a simple way to synthesize aligned carbon nanotubes, the characteristic of the template has played a key role in production aligned carbon nanotubes. Recently, Zhang *et al.* (2000) have reported the highly substrate-dependent site-selective growth of aligned carbon nanotubes by CVD on patterned SiO_2/Si substrates using the conventional lithography. SEM results indicated that carbon nanotubes have grown on the SiO_2 substrate, with no observable growth on the Si substrate. Using a similar method, Cao *et al.* (2001) synthesized regular nanotube networks using the different growth rates of nanotubes on both Au and quartz substrates by catalytic pyrolysis of ferrocene and xylene. The way to synthesize large area aligned carbon nanotube networks sounds simple and controllable, but the resolution for the resulting nanotube patterns is rather limited on the order of micrometer magnitudes.

6.3.3.3 Perpendicularly Aligned and Micropatterned Carbon Nanotubes by Self-assembly

Just as the chemically and non-covalently modified carbon nanotubes have significantly facilitated the characterization of their physicochemical properties at the molecular level and their efficient incorporation into devices, an aligned/micropatterned form of carbon nanotubes has been demonstrated to be highly desirable for device applications (Dai and Mau, 2001). Compared with the large number of publications on the alignment and micro-patterning of *unmodified* carbon nanotubes (Dai and Mau, 2001), the ordering of the chemically and/or non-covalently modified carbon nanotubes is still an early development. So far, ordered structures have been constructed from a few of the end-functionalized carbon nanotubes on certain pre-treated substrates, though most of them deserve further investigation. For instance, Liu *et al.* (2000) have prepared aligned SWNTs by self-assembling of nanotubes end-functionalized with thiol groups on a gold substrate [Equation (6.15)].

250 Intelligent Macromolecules for Smart Devices

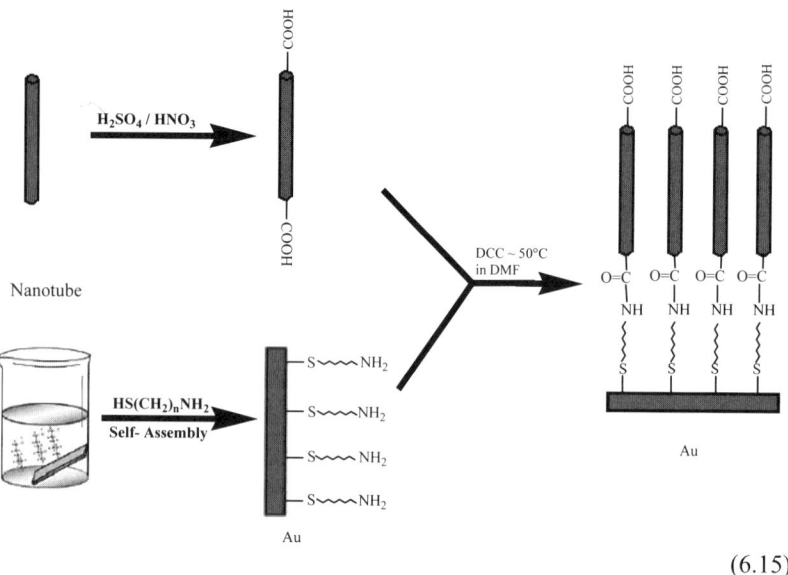

(6.15)

As seen in Equation (6.15), the carboxyl-terminated short SWNTs prepared by acid oxidation were used as the starting material for further functionalization with thiol-containing alkyl amines through the amide linkage. Then, the self-assembly of aligned SWNTs was obtained by dipping a gold(111) ball into the thiol-functionalized SWNT suspension in ethanol, followed by ultrasonication and drying in high-purity nitrogen.

The resulting self-assembled aligned nanotube film is so stable that ultrasonication cannot remove it from the gold substrate. The packing density of the self-assembled aligned carbon nanotubes was found to strongly depend on the incubation time. As seen in the AFM images given in Figure 6.27, a shorter nanotube assembly with a lower packing density was formed with a shorter adsorption time while longer adsorption time resulted in the formation of an aligned nanotube array with a longer length and a higher packing density.

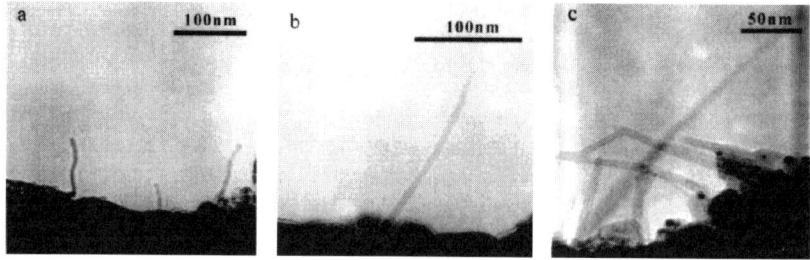

Figure 6.27. (a) A shorter nanotube assembly with a lower packing density was formed with a shorter adsorption time while (b) and (c) longer adsorption time resulted in the formation of aligned nanotube array with a longer length and a higher packing density (After Wu *et al.*, 2001, copyright 2001 The American Chemical Society. Reproduced with permission)

Given that carboxylic acids could be deprotonated by various metal oxides (*e.g.* Ag, Al, Cu), Liu and co-workers (Wu *et al.*, 2001) have investigated the formation of aligned nanotube array by self-assembling COOH-terminated carbon nanotubes onto certain metal oxide substrates (*e.g.* Ag). The deprotonation between the carboxylic groups on nanotubes and the metal surface effectively led to the anchoring of aligned carbon nanotubes on the metal substrate via their carboxylated anion head groups by Coulombic force. Raman spectroscopic measurements confirmed the process shown in Equation (6.16) below (Wu *et al.*, 2001).

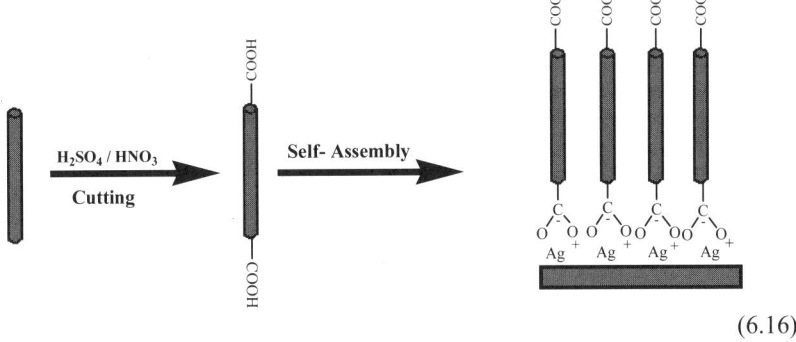

(6.16)

More recently, Nan *et al.* (2002) have demonstrated that individual acid oxidized SWNTs can be deposited onto a gold substrate pre-treated with $NH_2(CH_2)_{11}SH$ to form an aligned nanotube array on gold through the amide linkage. In this case, the microcontact printing (μCP) technique was used to pre-pattern the thiol self-assembled monolayer on the gold surface for region-specific deposition of the aligned carbon nanotubes. As shown in the optical microscope image (Figure 6.28(a)), the dark dot regions are covered by carbon nanotubes. The corresponding AFM image given in Figure 6.28(b) clearly shows the aligned structure.

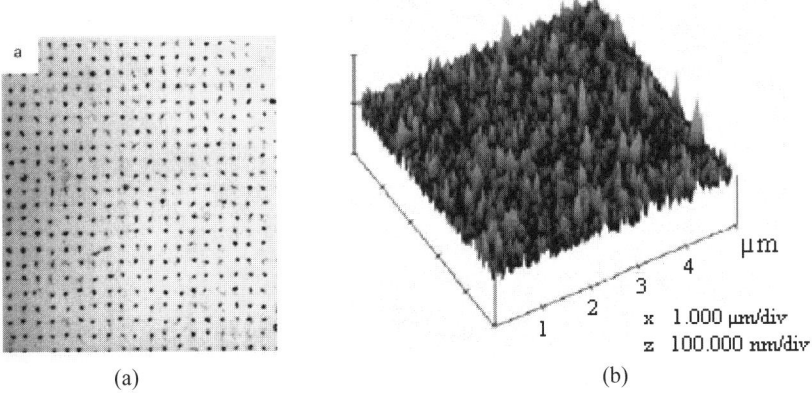

Figure 6.28. Aligned carbon nanotubes formed by self-assembling of COOH-terminated nanotubes onto a gold surface pre-patterned with thiol molecules. (a) Optical microscope image; (b) AFM image (After Nan *et al.*, 2002, copyright 2002 Elsevier. Reproduced with permission)

6.4 Aligned Non-carbon Nanotubes

6.4.1 Aligned B:C:N Nanotubes

Unlike the pure carbon nanotubes, Zettl reported the synthesis of highly aligned $B_xC_yN_z$ nanotubes by using ferrocene, melamine, boron oxide and ammonia as starting materials (Han *et al.*, 2001). The sizes of the aligned $B_xC_yN_z$ nanotubes are 10–30 μm in length and 20–140 nm in diameter. The x/z ratio of $B_xC_yN_z$ nanotubes is about 1:1. The x/y ratio of $B_xC_yN_z$ nanotubes is up to 0.6.

6.4.2 Aligned Inorganic Nanotubes

Gong *et al.* (2001) fabricated well-aligned titanium oxide nanotubes by anodic oxidation of a pure titanium sheet in an aqueous solution containing 0.5 to 3.5 wt% hydrofluoric acid. The aligned tubes can organize into high-density uniform arrays, the tops of the tubes are open and the bottoms of the tubes are closed, forming a barrier layer structure. The average tube diameter is in the range of 25–65 nm, which was found to increase with the increasing anodizing voltage.

Bao *et al.* (2001) prepared highly ordered magnetic Ni nanotubules of up to 35 μm in length and 160 nm in outer diameter by electrodeposition in the pores of an alumina membrane modified with an organoamine as a porewall modifying agent. The coercivity of the nanotubules is enhanced compared with that of the bulk nickel. The authors believed that such metal nanotubules could have a variety of promising applications, such as porous electrodes to be filled with ferromagnetic and non-magnetic metals for the fabrication of magnetic multilayer nanostructures, or filling with other materials to prepare novel nanocomposite materials with special magnetic, optical or electrical properties. They also obtained arrays of polyaniline nanotubules with the pores of an alumina template membrane. Using this array, they have prepared iron, nickel, and cobalt nanowires within the nanotubules of polyaniline.

Michailowski *et al.* (2001) prepared compact, continuous and uniform anatase nanotubules. They impregnated the porous anodic alumina pores with titanium isopropoxide then oxidatively decomposed the reagent at 500 °C. The uniform anatase nanotubes were obtained with diameters in the range 50–70 nm. Cleaning the surface of the template and repeating the process several times produced titania nanotubules with a wall thickness of ~3 nm per impregnation. They also used chemical vapor deposition to create alternating metallic (carbon) and insulating (boron nitride) layers within the template pores in anodic aluminum oxide templates (Shelimov *et al.*, 2000). The composite metal/insulator/metal nanotubules can be converted to an array of nanocapacitors connected in parallel, attaining specific capacitances of up to 2.5 $\mu F/cm^2$ for a 50-μm-thick template.

6.4.3 Aligned Polymer Nanotubes

Fu *et al.* (2001) reported the synthesis of aligned polythiophene micro- and nanotubules. They took microporous alumina membranes as templates, and obtained flexible polythiophene micro and nanotubules, and aligned tubule/gold bilayer films with diameters of 20 nm and 200 nm by direct oxidation of thiophene in freshly distilled boron trifluoride diethyl etherate solutions. These kinds of micro- or nanotubules/gold bilayer film show broad and strong redox responses and its charge/discharge capacity is about 30 times greater than that of the usual polythiophene film. A more general method for the formation of polymer nanotube in a membrane template will be discussed in Chapter 7. Without using any templates, Wan and co-workers (Qiu *et al.*, 2001) have also synthesized polyaniline (PANI) *nanotubes* (Figure 6.29) via a solution chemistry using $(NH_4)_2S_2O_8$ as an oxidant in the presence of a self-assembled $C_{60}(OH)_6(OSO_3H)$. More recently, these authors have developed an electrochemical technique to grow perpendicularly aligned conducting polymer nanotubes (Huang Z. *et al.*, 2003).

Figure 6.29. SEM image of PANI-$C_{60}(OH)_6(OSO_3H)_6$ nanotubes (After Qiu *et al.*, 2001, copyright 2001 The American Chemical Society. Reproduced with permission)

6.4.4 Aligned Peptide Nanotubes

Nanotubes constructed with peptides offer a new, very powerful approach to treating bacterial infections (Fernandez-Lopez *et al.*, 2001; Rouhi, 2001). These nanotubes are formed by the self-assembly of cyclic peptides with an even number of alternating D- and L-amino acids. The cyclic peptides are stable in solution when they are in an open, flat conformation with all the side chains pointing outward. They stack by intermolecular hydrogen bonding through the amide backbone, forming a tube that resembles protein-sheets (Figure 6.30). The nanotubes are membrane-active. They readily insert themselves into cell membranes and align perpendicularly to the membranes. Cells die instantly if their membranes become permeable and leaky. Ghadiri's cyclic peptides assemble to a higher order structure, the nanotubes, and it's that structure which kills the bacteria, not the single molecule. This is an entirely new concept in pharmaceutical research and development. The researchers have successfully designed the right peptides that form nanotubes in bacterial cell membranes only and exclusively to kill bacterial cells.

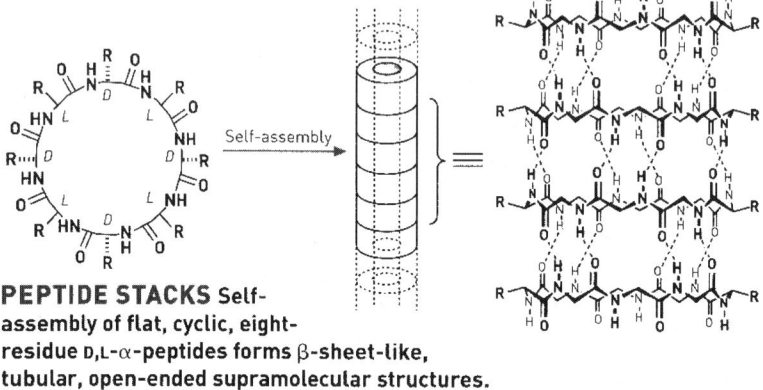

PEPTIDE STACKS Self-assembly of flat, cyclic, eight-residue D,L-α-peptides forms β-sheet-like, tubular, open-ended supramolecular structures.

Figure 6.30. A schematic representation of aligned peptide nanotubes by self-assembling (After Rouhi, 2001; Fernandez-Lopez *et al.*, 2001, copyright 2001 Macmillan Magazines Limited.)

6.5 References

Ajayan, P.M. (1995) *Adv. Mater.* **7**, 489, and references cited therein.
Aldissi, M. (1985) *J. Polym. Sci. B, Polym. Lett.* **23**, 167.
Aldissi, M. (1989) *J. Polym. Sci., Part C: Polym. Lett.* **27**, 105.
Akagi, K., Goto, H., Kadokura, Y., Shirakawa, H., Oh, S.-Y., Araya, K. (1995) *Synth. Met.* **69**, 13.
Akagi, K., Goto, H., Shirakawa, H. (1997) *Synth. Met.* **84**, 313.
Akagi, K., Katayama, S., Ito, M., Shirakawa, H., Araya, K. (1989) *Synth. Met.* **28**, D51.
Akagi, K., Piao, G., Kaneko, S., Sakamaki, K., Shirakawa, H., Kyotani, M. (1998) *Science* **282**, 1683.
Akari, S., Horn, D., Keller, H., Schrepp, W. (1995) *Adv. Mater.* **7**, 549.
Andreatta, A., Tokito, S., Smith. P., Heeger, A.J. (1990) *Mol. Cryst. Liq. Cryst.* **189**, 169.
Ando, M., Watanabe, Y., Iyoda, T., Honda, K., Shimidzu, T. (1989) *Thin Solid Films* **179**, 225.
Angelopoulos, M., Shaw, J. M., Lee, K.-L., Huang, W.-S., Lecorre, M.-A., Tissier, M. (1992) *Polym. Eng. Sci.* **32**, 1535.

Aoki, A., Miyashita, T. (1997) *Adv. Mater.* **9**, 361.
Araya, K., Shirakawa, H. (1986) *Synth. Met.* **14**, 199.
Arbogast, J.W., Darmanyan, A.P., Foote, C.S., Rubin, Y., Diederich, F.N., Alvarez, M..M., Anz, S.J., Whetten, R.L. (1991) *J. Phys. Chem.* **95**, 11.
Augestine, M., Sebastian, M., Samkumar, M.K.A., Joseph, M. J., Joseph, T., Sakthikumar, D., Jayalakshmi, S., Rasheed, T.M.A., Anantharaman, M.R. (1996) *Indian J. Pure Appl. Phys.* **34**, 966.
Avigal, Y., Kalish, R. (2001) *App. Phys. Lett.* **78**, 16.
Avlyanov, J.K., Min, Y., MacDiarmid, A.G., Epstein, A. (1995) *J. Phys. Rev. B* **72**, 65.
Babudri, F., Cicco, S.R., Farinola, G.M., Naso, F., Bolognesi, A., Porzio, W. (1996) *Macromol. Rapid. Commun.* **17**, 905.
Bao, J., Tie, C., Xu, Z., Zhou, Q. Shen, D., Ma, Q. (2001) *Adv. Mater.* **13**, 1631.
Bargon, J., Behnck, W., Weidenbruck, T., Ueno, T. (1991) *Synth. Met.* **41–43**, 1111.
Basescu, N., Liu, Z.-X., Moses, D., Heeger, A.J., Naarmann, H., Theophilou, N. (1987) *Nature* **327**, 403.
Bates, F.S. (1991) *Science* **251**, 898.
Bates, F.S., Fredrickson, G.H. (1990) *Annu. Rev. Phys. Chem.* **41**, 525.
Baughman, R.H., Cui, C., Zakhidov, A.A., Iqbal, Z., Barisci, J.N., Spinks, G.M., Wallace, G.G., Mazzoldi, A., De Rossi, D., Rinzler, A.G., Jaschinski, O., Roth, S., Kertesz, M. (1999) *Science* **284**, 1340.
Baumann, R., Lennarz, K., Bargon, J. (1993) *Synth. Met.* **54**, 243.
Berggren, M., Inganäs, O., Gustafsson, G., Rasmusson, J., Andersson, M.R., Hjertberg, T., Wennerström, O. (1994) *Nature* **372**, 444.
Bhuiyan, A.H., Bhoraskar, S.V. (1989) *J. Mater. Sci.* **24**, 3091.
Bolognesi, A., Porzio, W., Provasoli, F., Ezquerra, T. (1993) *Makromol. Chem.* **194**, 817.
Borman, S. (1990) *C&E News* **May 7th**, 53.
Böltau, M., Walheim, S., Mlynek, J., Krausch, G., Steiner, U. (1998) *Nature* **391**, 877.
Bowden, M.J., Turner, S.R., (eds) (1998) *Electronic and Photonic Applications of Polymers*, ACS, Washington.
Bower, C.A., Jin, S., Zhu, W. (2001) *Eur. Pat. Appl.*18.
Bradley, D.D.C., Brown, A.R., Burn, P.L., Friend, R.H., Holmes, A.B., Kraft, A. (1992) In *Electronic Properties of Polymers*, Kuzmany, H., Mehring, M. (eds), Springer Ser. Solid-State Sci. 107, Springer-Verlag, Heidelberg.
Bumm, L.A., Arnold, J.J., Cygan, M.T., Dunbar, T.D., Burgin, T.P., Jones II, L., Allara, D.L., Tour, J.M., Weiss, P.S. (1996) *Science* **271**, 1705.
Burghard, M., Duesberg, G., Philipp, G., Muster, J., Roth, S. (1998) *Adv. Mater.* **10**, 584.
Burroughes, J.H., Bradley, D.C.C.; Brown, A.R., Marks, R.N., Mackay, M.K., Friend, R.H., Burn, P.L., Holmes, A.B. (1990) *Nature* **347**, 539.
Cahalane, W., Labes, M.M. (1989) *Chem. Mater.* **1**, 519.
Cai, S.X., Kanskar, M., Nabity, J., Keana, J.F.W., Wybourne, M.N. (1992) *J. Vac. Sci. Technol. B* **10**, 2589.
Calvert, P. (1988) *Nature* **333**, 296.
Cao, Y., Smith, P. (1993) *Polymer* **34**, 3139.
Cao, A., Zhang, X., Xu, C., Liang, J., Wu, D., Wei, B. (2001) *Appl. Surf. Sci.* **181**, 234.
Cassell, A.M., Franklin, N.R., Tombler, T.W., Chan, E.M., Han, J., Dai, H., (1999) *J. Am. Chem. Soc.* **121**, 7975.
Che, G., Lakshmi, B.B., Fisher, E.R., Martin, C.R. (1998) *Nature* **393**, 346.
Chen, Q., Dai, L. (2000) *Appl. Phys. Lett.* **76**, 2719.
Chen, Q., Dai. L. (2001) *J. Nanosci. Nanotechnol.* **1**, 43.
Chien, J.C.W. (1984) *Polyacetylene: Chemistry, Physics and Materials Science*, Academic Press, New York.
Cho, I., Kim, J., Lee, S. (1995) *Macromol. Rapid Commun.* **16**, 851.
Choi, S.-J., Cho, H.-N., Choi, S.-K. (1994) *Polym. Bull. (Berlin)* **32**, 179.

Chou, S.Y., Zhuang, V. (1999) *J. Vac. Sci. Technol.* B **17**, 3197.
Chou, S.Y., Zhuang, L., Guo, L. (1999) *Appl. Phys. Lett.* **75**, 1004.
Cifferi, A, Kriegbaum, W.R., Meyer, R.B. (1982) *Polymer Liquid Crystals*, Academic Press, New York.
Cimrová, V., Remmers, M., Neher, D., Wegner, G. (1996) *Adv. Mater.* **8**, 146.
Clarke, T.C., Krounbi, M.T., Lee, V.Y., Street, G.B. (1981) *J. Chem. Soc., Chem. Commun.* **384**.
Cleland, A.N., Roukes, M.L. (1998) *Nature* **392**, 160.
Collings, P.J. (1995) In *Kirk-Othmer Encyclopedia of Chemical Technology*, Vol. 15, 4th edn, Wiley, New York.
Cromack, K.R., Józefowicz, M.E., Ginder, J.M., Epstein, A.J., McCall, R.P., Du, G., Leng, J.M., Kim, K., Li, C., Wang, Z.H. (1991) *Macromoleculs* **24**, 4157.
Dagani, R. (1995) *C&E News* **January 23th**, 28.
Dagani, R. (1996a) *C&E News*, **March 4th**, 22.
Dagani, R. (1996b) *C&E News*, **March 25th**, 7.
d'Agostino, R. (1990) *Plasma Deposition, Treatment, and Etching of Polymers*, Academic Press, San Diego.
Dai, L. (1992) *J. Phys. Chem.* **96**, 6469.
Dai, L. (1993) *Eur. Polym. J.*, **29**, 645.
Dai, L. (1999) *J. Macromol. Sci., Rev. Macromol. Chem. Phys.* **C39(2)**, 273.
Dai, L. (2001) *Rad. Phys. Chem.* **62**, 55.
Dai, L., Griesser, H.J., Hong, X., Mau, A.W.H., Spurling, T.H., Yang, Y., White, J.W. (1996) *Macromolecules* **29**, 282.
Dai, L., Griesser, H. J., Mau, A.W.H. (1997) *J. Phys. Chem. B* **101**, 9548.
Dai, L., Lu, J., Matthews, B., Mau, A.W.H. (1998) *J. Phys. Chem.* **102**, 4049; and references cited therein.
Dai, L., Mau, A.W.H. (1997) *Synth. Met.* **86**, 2277.
Dai, L., Mau, A.W.H. (2000) *J. Phys. Chem. B* **104**, 1891.
Dai, L., Mau, A.W.H. (2001) *Adv. Mater.* **13**, 899, and references cited therein.
Dai, L., Mau, A., Griesser, H.J., Spurling, T., White, J.W. (1995) *J. Phys. Chem.* **99**, 17302.
Dai, L., Mau, A.W.H., Griesser, H.J., Winkler, D.A. (1994) *Macromolecules* **27**, 6728.
Dai, L., Mau, A.W.H., Zhang, X. (1998) *J. Mater. Chem.* **8**, 325.
Dai, L., Soundarrajan, P., Kim, T. (2002) *Pure Appl. Chem.* **74**, 1753, and references cited therein.
Dai, L., St John, H., Bi, J., Zientek, P., Chatelier, R., Griesser, H. (2000) *Surf. Interf. Anal.* **29**, 46
Dai, L., Toprakcioglu, C. (1991) *Europhys. Lett.* **16**, 331.
Dai, L., Toprakcioglu, C. (1992) *Macromolecules* **25**, 6000.
Dai, L., Toprakcioglu, C., Hadziioannou, G. (1995) *Macromolecules* **28**, 5512.
Dai, L., White, J.W. (1991) *Polymer* **32**, 2120.
Dai, L., White, J.W. (1993) *J. Polym. Sci., Part B, Polym. Phys.* **31**, 3.
Dai, L., White, J.W. (1994) *Eur. Polym. J.* **30**, 1443.
Dai, L., White, J.W. (1997) *Polymer* **38**, 775.
Dai, L., White, J.W., Kerr, J., Thomas, R.K., Penfold, J., Aldissi, M. (1989) *Synth. Met.* **28**, 69.
Dai, L., Zientek, P., St John, H., Pasic, P., Chatelier, R., Griesser, H.J. (1996) In *Surface Modification of Polymeric Biomaterials*, Ratner, B.D., Castner, D. (eds), Plenum Press, New York.
Dao, L.H., Nguyen, M.T., Do, T.N. (1992) *Polym. Prepr.* **33**, 408.
DeAro, J.A., Weston, K.D., Buratto, S.K., Lemmer, U. (1997) *Chem. Lett.* **277**, 532.
de Gennes, P.G. (1979) *Scaling Concepts in Polymer Physics*, Cornell University Press, Ithaca, New York.
de Gennes, P.G. (1985) *C. R. Acad. Sci. Paris Ser. 2*, **302**, 1.

de Heer, W.A., Bonard, J.-M., Fauth, K., Châtelain, A., Forró, L., Ugarte, D. (1997) *Adv. Mater.* **9**, 87, and references cited therein.
de Heer, W.A., Châtelain, A., Ugarte, D. (1995) *Science* **270**, 1179.
de Heer, W.A., Martel, R. (2000) *Phys. World*, **June**, 49.
Deshpande, P., Sun, X., Chou, S.Y. (2001) *Appl. Phys. Lett.* **79**, 1688.
Dillon, A.C., Jones, K.M., Bekkedahl, T.A., Kiang, C.H., Bethune, D.S., Heben, M.J. (1997) *Nature* **386**, 377.
Donald, M., Windle, A.H. (1992) *Liquid Crystalline Polymers*, Cambridge University Press, Cambridge.
Dresselhaus, M.S., Dresselhaus, G., Eklund, P.C. (1996) *Science of Fullerenes and Carbon Nanotubes*, Academic Press, New York.
Dulcey, C.S., Georger, J.H., Chen, M.-S., McElvany, S.W., O'Ferrall, C.E., Benezra, V.I., Calvert, J.M. (1996) *Langmuir* **12**, 1638.
Dulcey, C.S., Georger, J.H., Kleinfeld, V., Stenger, D.A., Fare, T.L., Calvert, J.M. (1991) *Science* **252**, 551.
Dyreklev, P., Berggren, M., Inganäs, O., Andersson, M.R., Wennerström, O., Hjertberg, T. (1995) *Adv. Mater.* **7**, 43.
Dyreklev, P., Gustafsson, G., Inganäs, O., Stubb, H. (1992) *Solid State Commun.* **82**, 317.
Evans, S.D., Allinson, H., Boden, N., Flynn, T.M., Henderson, J.R. (1997) *J. Phys. Chem. B* **101**, 2143.
Fally, F., Riga, J., Verbist, J.J. (1992) In *Polymer-Solid Interface*, Pireaux, J.J., Bertrand P., Bredas, J.L. (eds), IOP Publ. Ltd, London.
Fan, S., Chapline, M.G., Franklin, N.R., Tomber, T.W., Cassell A.M., Dai, H. (1999) *Science* **283**, 512.
Fernandez-Lopez, S., Kim, S.-H., Choi, E.C. Delgado, M., Granja, J.R., Khasanov, A., Kraehenbuehl, K., Long, G., Weinberger, D.A., Wilcoxen, K.M., Ghadiri, M.R. (2001) *Nature* **412**, 452.
Ferreira, M., Cheung, J.H., Rubner, M.F. (1994) *Thin Solid Films* **244**, 806.
Ferreira, M., Rubner, M.F. (1995) *Macromolecules* **28**, 7107.
Field, J.B., Toprakcioglu, C., Stanley, H.B., Dai, L., Barford, W., Penfold, J., Smith, G., Hamilton, W. (1992) *Macromolecules* **25**, 434.
Flory, P.J. (1953) *Principles of Polymer Chemistry*, Cornell University Press, Ithaca, New York.
Fredrickson, G.H., Bates, F.S. (1996) *Annu. Rev. Mater Sci.* **26**, 503.
Frisbie, C.D., Rozsnyai, L.F., Noy, A., Wrighton, M.S., Lieber, C.M. (1994) *Science* **265**, 2071.
Fu, M., Zhu, Y., Tan, R., Shi, G. (2001) *Adv. Mater.* **13**, 1874.
Gabaston, L.I., Foot, P.J.S., Brown, J.W. (1996) *J. Chem. Soc., Chem. Commun.* 429.
Gadd, G.E., Blackford, M., Moricca, S., Webb, N., Evans, P.J., Smith, A.M., Jacobsen, G., Leung, S., Day, A., Hua, Q. (1997) *Science* **277**, 933.
Gagnon, D., Karasz, F.E., Thomas, E.L., Lenz, R.W. (1987) *Synth. Met.* **20**, 85.
Galvin, M.E., Wnek, G.E. (1985) *Polym. Bull. (Berlin)* **13**, 109.
Gill, R.E., Hadziioannou, G., Lang, P., Garnier, F., Wittmann, J.C. (1997) *Adv. Mater.* **9**, 331.
Gill, R.E., Meetsma, A., Hadziioannou, G. (1996) *Adv. Mater.* **8**, 212.
Gong, X., Dai, L., Mau, A.W.H., Griesser, H.J. (1998) *J. Polym. Sci., Part A, Polym. Chem.* **36**, 633.
Gong, D., Grimes, C.A., Varghese, O.K., Hu, W., Singh, R.S., Chen, Z., Dickey, E.C. (2001) *J. Mater. Res.* **16**, 3331.
Goto, H., Akagi, K., Shirakawa, H. (1997) *Synth. Met.* **84**, 385.
Goto, H., Akagi, K., Shirakawa, H., Oh, S.-Y., Araya, K. (1995) *Synth. Met.* **71**, 1899.
Gray, G.W., Winsor, P.A. (eds.) (1974) *Liquid Crystals and Plastic Crystals*, Ellis Horwood, Chichester.
Greenham, N.C., Friend, R.H. (1995) *Solid State Phys.* **49**, 1.

Grell, M., Bradley, D.D.C., Inbasekaran, M., Woo, Ed P. (1997) *Adv. Mater.* **9**, 798.
Gupta, V.K., Abbott, N.L. (1997) *Science* **276**, 1533.
Hagler, T.W., Pakbaz, K., Voss, K.F., Heeger, A.J. (1991) *Phys. Rev. B* **44**, 8652.
Halliday, D.A., Burn, P.L., Bradley, D.D.C., Friend, R.H., Gelsen, O.M., Holmes, A.B., Kraft, A., Martens, J.H.F., Pichler, K. (1993) *Adv. Mater.* **5**, 40.
Halperin, A., Tirrell, M., Lodge, T.P. (1992) *Adv. Polym. Sci.* **100**, 31.
Hamaguchi, M., Yoshino, K. (1995) *Appl. Phys. Lett.* **67**, 23.
Hamnett, A., Kerr, J.C.H., White, J.W., Dai, L (1993). *J. Chem. Soc., Faraday Trans.* **89**, 277.
Han, W.-Q., Cumings, J., Huang, X., Bradley, K., Zettl, A. (2001) *Chem. Phys. Lett.* **346**, 368.
Hann, R.A., Bloor, D. (eds) (1989) *Organic Materials for Nonlinear Optics*, Royal Society of Chemistry, London.
Harlev, E., Gulakhmedova, T., Rubinovich, I., Aizenshtein, G. (1996) *Adv. Mater.* **8**, 994.
Harris, P.J.F. (2001) *Carbon Nanotubes and Related Structures - New Materials for the Twenty-First Century*, Cambridge University Press, Cambridge.
Hasegawa, H., Kijima, M., Shirakawa, H. (1997) *Synth. Met.* **84**, 177.
Havard, J.M., Shim, S.-Y., Fréchet, J.M.J., Lin, Q., Medeiros, D.R., Willson, C.G., Byers, J.D. (1999) *Chem. Mater.* **11**, 719.
Heeger, A.J., Smith, P. (1991) In *Conjugated Polymers*, Bredas, J.L., Silbey, R. (eds), Kluwer Academic Publishers, Dordrecht.
Helfand, E., Wasserman, Z.R. (1982) In *Developments in Block and Graft Copolymers – 1*, Goodman, L. (ed.), Applied Science, New York.
Hernandez, H., Diaz, A. F., Waltman, R., Bargon, J. (1984) *J. Phys. Chem.* **88**, 3333.
Hippler, R. (ed.) (2001) *Low Temperature Plasma Physics: Fundamental Aspects and Applications*, Wiley-VCH, Berlin.
Holdcroft, S. (2001) *Adv. Mater.* **13**, 1753.
Hollan, J.R., Bell, A.T. (eds) (1974) *Techniques & Applications of Plasma Chemistry*, Wiley, New York.
Hotta, S., Soga, M., Sonoda, N. (1989) *J. Phys. Chem.* **93**, 4994.
Huang, S., Dai L. (2002) *J. Nanopartticles Res.* **4**, 145.
Huang, S., Dai, L., Mau, A.W.H. (1999) *J. Phys. Chem. B* **103**, 4223.
Huang, S., Mau, A.W.H., Turney, T.W., White, P.A., Dai, L. (2000) *J. Phys. Chem. B* **104**, 2193.
Huang, Z., Shi, G., Zhang, J., Hong, X. (2003) *Chin. Sci. Bull.* **48**, 434.
Huang, Z., Wang, P-C., MacDiarmid, A. G. (1997) *Langmuir* **13**, 6480.
Huang, Z.P., Xu, J.W., Ren, Z.F., Wang, J.H., Siegal, M.P., Provencio, P.N. (1998) *Appl. Phys. Lett.* **73**, 3845.
Ibison, P., Foot, P. J. S., Brown, J. W. (1996) *Synth. Met.* **76**, 297.
Iijima, S. (1991) *Nature* **354**, 56.
Ito, T., Shirakawa, H., Ikeda, S. (1974) *J. Polym. Sci. Polym. Chem.* **12**, 11.
Iwasaki, T., Motoi, T., Den, T. (1999) *Appl. Phys. Lett.* **75**, 2044.
Jackman, R.J., Brittain, S.T., Adams, A., Wu, H., Prentiss, M.G., Whitesides, S., Whitesides, G.M. (1999b) *Langumuir* **15**, 826.
Jackman, R.J., Whitesides, G.M. (1999a) *CHEMTECH* **May**, 18.
Jacoby, M. (1997) *C&E News* **October 6**, 34.
Jeon, N.L., Clem, P., Jung, D.Y., Lin, W., Girolami, G.S., Payne, D.A., Nuzzo, R.G. (1997) *Adv. Mater.* **9**, 891.
Jin, S.-H., Choi, S.-J., Ahn, W., Cho, H.-N., Choi, S.-K. (1993) *Macromolecules* **26**, 1487.
Jin, S.-H., Kim, S.-H., Cho, H.-N., Choi, S.-K. (1991) *Macromolecules* **24**, 6050.
Jirage, K.B.; Hulteen, J.C.; Martin, C.R. (1997) *Science* **278**, 655.
Joo, J., Du, G., Tsukamoto, J., Epstein, A. (1997) *J. Synth. Met.* **88**, 1.

Kajzar, F., Le Moigne, J., Thierry, A. (1992) In *Electronic Properties of Polymers*, Kuzmany, H., Mehring, M. (eds), Springer Ser. Solid-State Sci., Springer-Verlag, Heidelberg.
Kamalakaran, R, Terrones, M., Seeger, T., Kohler-Redlich, Ph., Ruhle, M., Kim, Y.A., Hayashi, T., Endo, M., (2000) *Appl. Phys. Lett.* **77**, 21.
Kang, E.T., Kato, K., Uyama, Y., Ikada, Y. (1996) *J. Mater. Res.* **11**, 1570.
Karasz, F.E., Capistran, J.D., Gagnon, D.R. (1985) *Mol. Cryst. Liq. Cryst.* **118**, 327.
Katz, H.E., Dirk, C.W., Singer, K.D., Sohn, J.E. (1988) *Mol. Cryst. Liq. Cryst.* **157**, 525.
Kawakami, S. (1987) *JP 01165603*.
Kijima, M., Akagi, K., Shirakawa, H. (1997) *Synth. Met.* **84**, 237.
Kim, E., Xia, Y., Zhao, X.-M., Whitesides, G.M. (1997) *Adv. Mater.* **9**, 651.
Kim, T., Ye, Q., Sun, L., Chan, K.C., Crooks, R.M. (1996) *Langmuir* **12**, 6065.
Kind, H., Bonard, J.-M., Emmenegger, C., Nilsson, L.-O., Hernadi, K., Maillard-Schaller, E., Schlapbach, L., Forro, L., Kern, K. (1999) *Adv. Mater.* **11**, 1285.
Koltzenburg, S., Winkler, B., Stelzer, F., Nuyken, O. (1998) *Designed Monomers and Polymers* **1**, 207.
Kong, J., Soh, H.T., Cassell, A.M., Quate, C.F., Dai, H., (1998) *Nature* **395**, 878
Kumar, A., Biebuyck, H.A., Whitesides, G.M. (1994) *Langmuir* **10**, 1498.
Kumar, A., Whitesides, G.M. (1993) *Appl. Phys. Lett.* **63**, 2002.
Kumar, A., Whitesides, G.M. (1994) *Science*, **263**, 60.
Lee, Y.-H., Jang, Y.-T., Choi, C.-H., Kim, E.-K., Ju, B.-K., Kim, D.-H., Lee, C.-W., Yoon, S.-S. (2002) *J. Appl. Phys.* **91**, 6044.
Leibler, L. (1980) *Macromolecules* **13**, 1602.
Leising, G. (1984) *Polym. Commun.* **25**, 201.
Lemmer, U., Vacar, D., Moses, D., Heeger, A.J., Ohnishi, T., Noguchi, T. (1996) *Appl. Phys. Lett.* **68**, 21.
Li, D.-C., Dai, L., Huang, S., Mau, A.W.H., Wang, Z.L. (2000) *Chem. Phys. Lett.* **316**, 349.
Li, Z., Qu, S., Rafailovich, M.H., Sokolov, J., Tolan, M., Turner, M.S., Wang, J., Schwarz, S. A., Lorenz, H., Kotthaus, J.P. (1997) *Macromolecules* **30**, 8410.
Li, J., Papadopoulos, C., Xu, J.M. (1999a) *Appl. Phys. Lett.* **75**, 367.
Li, J., Papadopoulos, C., Xu, J.M. (1999b) *Nature* **402**, 253.
Li, W.Z., Wen, J.G., Ren, Z.F. (2001) *Appl. Phys. Lett.* **79**, 1879.
Li, W.Z., Xie, S.S., Qian, L.X., Chang, B.H., Zou, B.S., Zhou, W.Y., Zhao, R.A., Wang, G. (1996) *Science* **274**, 1701.
Li, Z., Zhao, W., Liu, Y., Rafailovich, M.H., Sokolov, J. (1996) *J. Am. Chem. Soc.* **118**, 10892.
Liang, W.B., Masse, M.A., Karasz, F.E. (1992) *Polymer* **35**, 3101.
Lidzey, D.G., Pate, M.A., Weaver, M.S., Fisher, T.A., Bradley, D.D.C. (1996) *Synth. Met.* **82**, 141.
Lino, K., Goto, H., Akagi, K., Shirakawa, H., Kawaguchi, A. (1997) *Synth. Met.* **84**, 967.
Liu, J., Casavant, M.J., Cox, M., Walters, D.A., Boul, P., Lu, W., Rimberg, A.J., Smith, K.A., Colbert, D.T., Smalley, R.E. (1999) *Chem. Phys. Lett.* **303**, 125.
Liu, Z., Shen, Z., Zhu, T., Hou, S., Ying, L., Shi, Z., Gu, Z. (2000) *Langmuir* **16**, 3569.
Lodge, T.P., Muthukumar, M. (1996) *J. Phys. Chem.* **100**, 13275.
Lopez, G.P., Biebuyck, H.A., Whitesides, G.M. (1993) *Langmuir* **9**, 1513.
Lu, J., Dai, L., Mau, W.H.A. (1998) *Acta Polym.* **49**, 371.
Lüssem, G., Festag, R., Greiner, A., Schmidt, C., Unterlechner, C., Heitz, W., Wendorff, J. H., Hopmeier, M., Feldmann, J. (1995) *Adv. Mater.* **7**, 923.
Lüssem, G., Geffarth, F., Greiner, A., Heitz, W., Hopmeier, M., Oberski, M., Unterlechner, C., Wendorff, J.H. (1996) *Liq. Cryst.* **21**, 903.
MacDiarmid, A.G. (1997) *Synth. Met.* **84**, 27.
MacDiarmid, A.G., Epstein, A.J. (1994) *Synth. Met.* **65**, 103.
MacDiarmid, A.G., Epstein, A.J. (1995) *Synth. Met.* **69**, 85.

MacDiarmid, A.G., Epstein, A.J. (1997) In *Photonic and Optoelectronic Polymers*, Jenekhe, S.A., Wynne, K.J. (eds), ACS Symp. Ser. 672, ACS, Washington.
March, J. (1992) *Advanced Organic Chemistry*, 4th edn, John Wiley & Sons, Inc., New York.
Martin, C.R. (1995) *Acc. Chem. Res.* **259**, 957.
Martin, C.R. (1996) *Chem. Mater.* **8**, 1739.
McArdle, C.B. (1989) *Side Chain Liquid Crystalline Polymers*, Blackie, Glasgow-London.
Meghdadi, F., Tasch, S., Winkler, B., Fischer, W., Stelzer, F., Leising, G. (1997) *Synth. Met.* **85**, 1441.
Messier, J., Kajzar, F., Prasad, P., Ulrich, D. (eds.) (1989) *Nonlinear Optical Effects in Organic Polymers*, Kluwer Academic, Boston.
Michailowski, A., Almawlawi, D., Cheng, G., Moskovits, M. (2001) *Chem. Phys. Lett.* **349**, 1.
Michalitsch, R., Lang, P., Yassar, A., Nauer, G., Garnier, F. (1997) *Adv. Mater.* **9**, 321.
Moigne, J.L., Hilberer, A., Kajzar, F. (1991) *Makromol. Chem.* **192**, 515.
Montaner, A., Rolland, M., Sauvajol, J.L., Meynadier, L., Almairac, R., Ribet, J.L., Galtier, M., Gril, C. (1989) *Synth. Met.* **28**, D19.
Montaner, A., Rolland, M., Sauvajol, J.L., Galtier, M., Almairac, R., Ribet, J.L. (1988) *Polymer* **29**, 1101.
Mott, N.F., Daris, E.A. (1997) *Electronic Process in Non-Crystalline Materials*, Clarendon Press, Oxford.
Mukamel, S., Takahashi, A., Wang, H.X., Chen, G. (1994) *Science* **266**, 250.
Nalwa, H. S. (1993) *Adv. Mater.* **5**, 341.
Naarmann, H. (1986) *Springer Ser. Solid State Sci.* **76**, 12.
Naarmann, H., Theophilou, N. (1987) *Synth. Met.* **22**, 1.
Nan, X., Gu, Z., Liu, Z. (2002) *J. Col. Inter. Sci.* **245**, 311.
Nierengarten, J.-F., Guillon, D., Heinrich, B., Nicoud, J.-F. (1997) *J. Chem. Soc., Chem. Commun.* 1233.
Niu, C., Sichel, E.K., Hoch, R., Moy, D., Tennent, H. (1997) *Appl. Phys. Lett.* **70**, 1480.
Nishio, S., Takeuchi, T., Matsuura, Y., Yoshizawa, K., Tanaka, K., Yamabe, T. (1992) *Synth. Met.* **46**, 243.
Noach, S., Faraggi, E.Z., Cohen, G., Avny, Y., Neumann, R., Davidov, D., Lewis, A. (1996) *Appl. Phys. Lett.* **69**, 3650.
Ogasawara, M., Funahashi, K., Demura, T., Hagiwara, T., Iwata, K. (1986) *Synth. Met.* **14**, 61.
Oh, S.-Y., Akagi, K., Shirakawa, H., Araya, K. (1993a) *Macromolecules* **26**, 6203.
Oh, S.-Y., Ezaki, R., Akagi, K., Shirakawa, H. (1993b) *J. Polym. Sci., Polym. Chem.* **31**, 2977.
Ouyang, M., Huang, J., Cheung, C.L., Lieber, C.M. (2001) *Science* **292**, 702.
Pan, H., Liu, L., Guo, Z.-X., Dai, L., Zhang, F., Zhu, D., Czerw, R., Carroll, D.L. (2003) *Nano. Lett.* **3**, 29.
Park, M., Harrison, C., Chaikin, P.M., Register, R.A., Adamson, D.H. (1997) *Science* **276**, 1401.
Pearson, D.S., Pincus, P.A., Heffner, G.W., Dahman, S.J. (1993) *Macromolecules* **26**, 1570.
Pooley, C.M., Tabor, D. (1972) *Proc. R. Soc. Lond.* **A329**, 251.
Prasad, P.N. (1991) In *Frontiers of Polymer Research*, Prasad, P.N., Nigam, J.K. (eds), Plenum Press, New York.
Qian, R., Chen, S., Song, W., Bi, X. (1994) *Macromol. Rapid Commun.* **15**, 1.
Qiu, H., Wan, M., Matthews, B., Dai, L. (2001) *Macromolecules* **34**, 675.
Rao, C.N.R., Sen, R., Satishkumar, B.C., Govindaraj, A. (1998) *J. Chem. Soc., Chem. Commun.* **1525**.
Reed, M.A., Tour, J.M. (2000) *Sci. Am.* **June**, 86.
Ren, Z.F. (2000) *Technol. Rev.* **January/February**, 25.

Ren, Z.F., Huang, Z.P., Wang, D.Z., Wen, J.G., Xu, J.W., Wang, J.H., Calvet, L.E., Chen, J., Klemic, J.F., Reed, M.A., (1999) *Appl. Phys. Lett.* **75**, 1086.
Ren, Z.F., Huang, Z.P., Xu, J.H., Wang, P.B., Siegal, M.P., Provencio, P.N. (1998) *Science* **282**, 1105.
Renak, M. L., Bazan, G. C., Roitman, D. (1997) *Adv. Mater.* **9**, 392.
Ringsdorf, H., Schlarb, B., Venzmer, J. (1988) *Angew. Chem. Int. Ed. Engl.* **27**, 113.
Rogers, J.A., Bao, Z., Dhar, L. (1998) *Appl. Phys. Lett.* **73**, 294.
Ross, C.B., Sun, L., Crooks, R.M. (1993) *Langmuir* **9**, 632.
Rothman, T. (1988) *Sci. Am.* **August**, 12.
Rouhi, A.M. (2001) *C&E News* **79(32)**, 41.
Rozsnyai, L.F., Wrighton, M.S. (1994) *J. Am. Chem. Soc.* **116**, 5993.
Rozsnyai, L.F., Wrighton, M.S. (1995) *Langmuir* **11**, 3913.
Rubner, M.F., Skotheim, T.A. (1991) In *Conjugated Polymers*, Brédas, J. L., Silbey, R. (eds), Kluwer Academic Publishers, Dordrecht.
Rueckers, T., Kim, K., Joselevich, E., Tseng, G.Y., Cheung, C.-L., Lieber, C.M. (2000) *Science* **289**, 94.
Saito, Y., Hamaguchi, K., Nishino, T., Hata, K., Tohji, K., Kasuya, A., Nishina, Y. (1997) *Jpn. J. Appl. Phys.* **36**, L1340.
Saito, Y., Hamaguchi, K., Uemura, S., Uchida, K., Tasaka, Y., Ikazaki, F., Yumura, M., Kasuya, A., Nishina, Y. (1998) *Appl. Phys. A*, **67**, 95.
Samal, S., Mohanty, B.C., Nayak, B.B. (1994) *Polym. Sci. (India)* **1**, 222.
Satoh, M., Imanishi, K., Yasuda, Y., Tsushima, R., Aoki, S., Yoshino, K. (1985) *Jpn. J. Appl. Phys.* **24**, 1423.
Schäffer, E., Thurn-Albrecht, T., Russell, T.P., Steiner, U. (2000) *Nature* **403**, 874.
Schanze, K.S., Bergstedt, T.S., Hauser, B.T. (1996) *Adv. Mater.* **8**, 531.
Schierbaum, K.D., Weiss, T., Thoden van Velzen, E.U., Engbersen, J.F.J., Reinhoudt, D.N., Gopel, W. (1994) *Science* **265**, 1413.
Schimmel, T., Denninger, G., Riess, W., Voit, J., Schwoerer, M., Schoepe, W., Naarmann, H. (1989) *Synth. Met.* **28**, D11.
Schmid, W., Dankesreiter, R., Gmeiner, J., Vogtmann, T., Schwoerer, M. (1993) *Acta Polym.* **44**, 208.
Shelimov, K.B., Davydov, D.N., Moskovits, M. (2000) *Appl. Phys. Lett.* **77**, 1722.
Shimoda, H., Oh, S.J., Geng, H.Z., Walker, R.J., Zhang, X.B., McNeil, L.E., Zhou, O. (2002) *Adv. Mater.* **14**, 899.
Shirakawa, H., Akagi, K., Katayama, S., Araya, K., Mukoh, A., Narahara, T.J. (1988) *Macromol. Sci., Chem.* **A25**, 643.
Shirakawa, H., Akagi, K., Suezaki, M. (1989) *Synth. Met.* **28**, 1.
Shirakawa, H., Ito, T., Ikeda, S. (1973) *Polym. J.* **4**, 460.
Skotheim, T.A. (ed.) (1986) *Handbook of Conducting Polymers*, Marcel Dekker, New York.
Sohn, J.I., Choi, C., Lee, S., Seong, T. (2001) *Appl. Phys. Lett.* **78**, 3130.
Spatz, J.P., Sheiko, S., Möller, M. (1996) *Adv. Mater.* **8**, 513.
Steiger, D., Smith, P., Weder, C. (1997) *Macromol. Rapid Commun.* **18**, 643.
Stockton, W.B.; Rubner, M.F. (1997) *Macromolecules* **30**, 2717.
Suh, J.S., Lee, J.S., Jin, S., (1999) *Appl. Phys. Lett.* **75**, 2047.
Sukwattanasinitt, M., Wang, X., Li, L., Jiang, X., Kumar, J., Tripathy, S.K., Sandman, D.J. (1998) *Chem. Mater.* **10**, 27.
Tanaka, K., Nishio, S., Matsuura, Y., Yamabe, T. (1993) *J. Appl. Phys.* **73**, 5017.
Tanaka, K., Yamabe, T., Takeuchi, T., Yoshizawa, K., Nishio, S. (1991) *J. Appl. Phys.* **70**, 5653.
Tang, C.W., VanSlyke, S.A. (1987) *Appl. Phys. Lett.* **51**, 913.
Tanemura, M., Iwata, K., Takahashi, K., Fujimoto, Y., Okuyama, F., Sugie, H., Filip. V. (2001) *J. Appl. Phys.* **90**, 3.
Tans, S.J.; Verschueren, A.R.M.; Dekker, C. (1998) *Nature* **393**, 49.

Teo, K.B.K., Chhowalla, M., Amaratunga, G.A. I., Milne, W.I., Hasko, D.G., Pirio, G., Legagneux, P., Wyczisk, F., Pribat, D. (2001) *Appl. Phys. Lett.* **79**,1534.
Terfort, A., Bowden, N., Whitesides, G. M. (1992) *Science* **257**, 1380.
Terfort, A., Bowden, N., Whitesides, G. M. (1997) *Nature* **386**, 162.
Terrones, M., Grobert, N., Olivares, J., Zhang, J.P., Terrones, H., Kordatos, K., Hsu, W.K., Hare, J.P., Townsend, P.D., Prassides, K., Cheetham, A.K., Kroto, H.W., Walton, D.R.M., (1997) *Nature* **388**, 52.
Thakur, M. (1988) *Macromolecules*, **21**, 661.
Theophilou, N., Aznar, R., Munardi, A., Sledz, J., Schue, R., Naarmann, H. (1987) *Synth. Met.* **17**, 223.
Theophilou, N.; Manohar, S. *US Patent No. 4935181*, 1993.
Tokito, S., Smith, P., Heeger, A. J. (1991) *Polymer* **32**, 464.
Townsend, P.D., Pereira, C.M., Bradley, D.D.C., Horton, M.E., Friend, R.H. (1985) *J. Phys. C: Solid State Phys.* **18**, L283.
Tsai, S.H., Shiu, C.T., Lai, S.H., Chan, L.H., Hsieh, W.J., Shih, H.C. (2002) *J. Mater. Sci. Lett.* **21**, 21.
Tsang, S.C., Chen, Y.K., Harris, P.J.F., Green, M.L.H. (1994) *Nature* **372**, 159.
Ulman, A. (1996) *Chem. Rev.* **96**, 1533.
Ungerank, M., Winkler, B., Eder, E., Stelzer, F. (1995) *Macromol. Chem. Phys.* **196**, 3623. (1997) *ibid.* **198**, 1391.
Vargo, T.G., Calvert, J.M., Wynne, K.J., Avlyanov, J.K., MacDiarmid, A.G., Rubner, M.F. (1995) *Supramol. Sci.* **2**, 169.
Vargo, T.G., Thompson, P.M., Gerenser, L.J., Valentini, R.F., Aebischer, P., Hook, D.J., Gardella, J.A. (1992) *Langmuir* **8**, 130.
Väterlein, C., Ziegler, B., Gebauer, W., Neureiter, H., Sokolowski, M., Bäuerle, P., Weaver, M.S., Bradley, D.D.C., Umbach, E. (1996) *Synth. Met.* **76**, 133.
Venugopal, G., Quan, X.; Johnson, G.E., Houlihan, F.M., Chin, E., Nalamasu, O. (1995) *Chem. Mater.* **7**, 271.
Vicentini, F., Barrouillet, J., Laversanne, R., Mauzac, M., Bibonne, F., Parneix, J.P. (1995) *Liq. Cryst.* **19**, 235.
Wallraff, G.M., Hinsberg, W.D. (1999) *Chem. Rev.* **99**, 1801.
Weaver, M.S., Bradley, D.D.C. (1996) *Synth. Met.* **83**, 61.
Weder, C., Sarwa, C., Montali, A., Bastiaansen, C., Smith, P. (1998) *Science* **279**, 835.
Wei, P.K., Hsu, J.H., Hsieh, B.R., Fann, W.S. (1996) *Adv. Mater.* **8**, 573.
Wei, B.Q., Vajtai, R., Jung, Y., Ward, J., Zhang, R., Ramanath, G., Ajayan, P.M. (2002) *Nature* **416**, 495.
Wei, B.Q., Vajtai, R., Zhang, Z.J., Ramanath, G., Ajayan, P.M. (2001) *J. Nanosci. Nanotechnol.* **1**, 35.
Wen, J.G., Huang, Z.P., Wang, D.Z., Chen, J.H., Yang, S.X., Ren, Z.F., Wang, J.H., Wang, J.H., Calvet, L.E., Chen, J., Klemic, J.F., Reed, M.A. (2001) *J. Mater. Res.* **16**, 3246.
Wiederrecht, G.P., Yoon, B.A., Wasielewski, M.R. (1996) *Adv. Mater.* **8**, 535.
Williams, D.J. (ed.) (1983) *Nonlinear Optical Properties of Organic and Polymeric Materials*, ACS, Washington DC.
Winkler, B., Meghdadi, F., Tasch, S., Müllner, R., Resel, R., Saf, R., Leising, G., Stelzer, F. (1998) *Opt. Mater.* **9**, 159.
Winkler, B., Rehab, A., Ungerank, M., Stelzer, F. (1997) *Macromol. Chem. Phys.* **198**, 1417.
Winkler, B., Tasch, S., Zojer, E., Ungerank, M., Leising, G., Stelzer, F. (1996b) *Synth. Met.* **83**, 177.
Winkler, B., Ungerank, M., Stelzer, F. (1996a) *Macromol. Chem. Phys.* **197**, 2343.
Wittmann, J.C.; Smith, P. (1991) *Nature* **352**, 414.
Wrighton, M.S. (1993) *Langmuir* **9**, 1517.
Wu, B., Zhang, J., Wei, Z., Cai, S., Liu, Z. (2001) *J. Phys. Chem. B* **105**, 5075.
Xia, Y., Qin, D., Whitesides, G.M. (1996) *Adv. Mater.* **8**, 1015.

Xia, Y., Whitesides, G.M. (1998) *Angew. Chem., Int. Ed. Engl.* **37**, 550.
Xie, X., Thiele, J.-U., Steiner, R., Oelhafen, R. (1994) *Synth. Met.* **63**, 221.
Xu, S., Liu, G.-Y. (1997) *Langmuir* **13**, 127.
Yang, Y., Huang, S., He, H., Mau, A.W.H., Dai, L., (1999) *J. Am. Chem. Soc.* **121**, 10832-10833.
Yasuda, H. (1978) *Plasma Polymerization*, Academic Press, Orlando, 1978.
Yoshino, K., Kobayashi, K., Myojin, K., Ozaki, M., Akagi, K., Goto, H., Shirakawa, H. (1996) *Jpn. J. Appl. Phys.* **35**, 3964.
Yoshino, K., Nakao, K., Onoda, M. (1989) *Jpn. J. Appl. Phys.* **28**, L1032.
Yoshino, K., Onoda, M., Sugimoto, R. (1988) *Jpn. J. Appl. Phys.* **27**, L2034.
Young, L.J., Lee, B.S. (2002) *Thin Solid Films* **418**, 85.
Zakhidov, A.A., Araki, H., Yoshino, K. (1999) In *Fullerene Polymers and Fullerene Polymer Composites*, Eklund, P.C., Rao, A.M. (eds), Springer-Verlag, Berlin.
Zhang, Z.J.; Wei, B.Q.; Ramanath, G.; Ajayan, P.M. (2000) *Appl. Phys. Lett.* **77**, 23.
Zentel, R., Galli, G., Ober, C.K. (eds) (1997) *Macromol. Symp.* **117**.
Zhao, X.M., Xia, Y., Whitesides, G.M. (1996) *Adv. Mater.* **8**, 837.

Chapter 7

Macromolecular Nanostructures

7.1 Introduction

As the term implies, polymer (or macromolecule) refers to a molecule of extraordinarily large size (typically > 1000 in molecular weight). Although polymer molecules are large, they are made up of individual, repeating units termed monomers. The overall properties of a given polymeric material depend on: i) the constituent monomer units and the way they are arranged in the macromolecule; and ii) the spatial arrangement of the constituent macromolecular chains and the nature of the intermolecular interactions that hold them together (Flory, 1953). These interesting structure-property relationships provide a broad basis to allow for the development of various polymeric materials (*e.g.* polymer fibers, films, powders) with different properties from the same macromolecules.

Polymers have become widely used in all aspects of our daily life. After all, our clothes are made from synthetic fibers, car tires from rubber, and computer chips from plastics. It is now difficult to imagine what our lives would be if there were no polymers. With the recent advances in nanoscience and nanotechnology, various nanostructured polymers have been devised for a wide range of advanced applications. Examples of these advances include the use of polymer nanoparticles as drug delivery devices, polymer nanofibers as conducting wires, polymer thin films and periodically structured polymeric structures for optoelectronic devices. Like most other nanomaterials (Hanna, 1997; Liu T. *et al.,* 2002), nanostructured polymers can also possess mechanical, electronic, optical, and even magnetic properties that are different from those of the bulk materials, depending on their size, shape and composition.

The development of nanostructured polymers has opened up novel fundamental and applied frontiers, which have attracted tremendous interest in recent years. The present chapter provides an overview of the rapidly developing field of polymer nanostructures. I will first discuss the preparation of polymer nanoparticles, with an emphasis on their potential applications for controlled drug delivery. I will then describe the preparation of fiber-like polymers (*i.e.* nanowires, nanotubes and for nanofibers) potential applications in electronic devices. Finally, I will examine the use of polymer thin films (nanofilms or nanosheets) in organic optoelectronic devices, along with the periodically structured polymers as photonic crystals.

7.2 Polymer Nanoparticles

Polymer nanoparticles are of special interest for medical applications. In particular, the recent development of polymer nanoparticles as effective drug delivery devices is revolutionizing the way in which medical treatments are performed. Unlike the current drug therapies that deliver the drug to every cell in the body, encapsulating or incorporating a drug into polymer nanoparticles allows effective delivery of the drug molecules to the target site and their controlled release. This should lead to an increase in the therapeutic benefit and reduction of side effects (Krueger, 1994). Significant advances in polymer science and biomedical engineering have facilitated the development of new synthetic approaches to polymer nanoparticles.

Depending on the method of preparation, polymer nanospheres or nanocapsules can be obtained. These polymer nanoparticles, usually having a diameter ranging from 10 to 300 nm, can be prepared by various techniques, including *in situ* polymerization, dispersion of pre-formed polymers and self-assembling. The *in vitro* and *in vivo* performance of various polymer nanoparticles as drug delivery devices have recently been discussed in several excellent review articles (Allemann *et al.*, 1998; Soppimath *et al.*, 2001). In what follows, I will provide a status summary of the preparation of polymer nanoparticles.

7.2.1 Polymer Nanospheres by Polymerization

While conventional emulsion polymerization has been widely used to produce polymer particles with the size range of 0.1–1 μm, miniemulsion and microemulsion polymerization methods have recently been developed for the preparation of polymer nanoparticles in the ranges of 50–200 nm and 15–50 nm, respectively (Miller and El-Aasser, 1997; Thurmond, *et al.*, 1996). Using polymerizable amphiphilic PEO macromonomers [*i.e.* $CH_3O-(CH_2CH_2O)_n-(CH_2)_{11}-OOCC(CH_3)=CH_2$, n = 10, 15, or 40; designated as: C1-PEO-C11-MA-n] as stabilizers in dispersion/emulsion polymerizations, for example, Gan and co-workers have successfully prepared polymer [*e.g.* polystyrene (PS), poly(methyl methacrylate) (PMMA)] microlatexes with particle sizes ranging from 15 to 200 nm (Liu J. *et al.*, 1997; Xu X.J. *et al.*, 2001a; Xu X.J. *et al.*, 2001b). A large amount of surfactants, however, are normally required for this emulsion polymerization, though high solid-contents up to 40wt% have been reported (Rabelero *et al.*, 1997; Xu X.J. *et al.*, 1999). The relatively high content of surfactant in the dispersions often limits their biomedical applications (Birrenbach and Speise, 1976; Hearn *et al.*, 1981; Couvreur *et al.*, 1979).

Recently, considerable efforts have been devoted to investigate ways to minimize the amount of surfactant used and to maximize the solid content. In this context, both solution and interfacial polymerization methods have been used to encapsulate drug molecules in polymer nanoparticles. For example, Couvreur and co-workers (Couvreur *et al.*, 1979) synthesized poly(alkylcyanoacrylate) nanoparticles by polymerizing methyl or ethyl cyanoacrylate in aqueous acidic medium in the presence of polysorbate-20 as a surfactant. The polymerization follows an anionic mechanism and the schematic representation of the procedure used for the production of poly(alkylcyanoacrylate) nanoparticles is given in Figure 7.1.

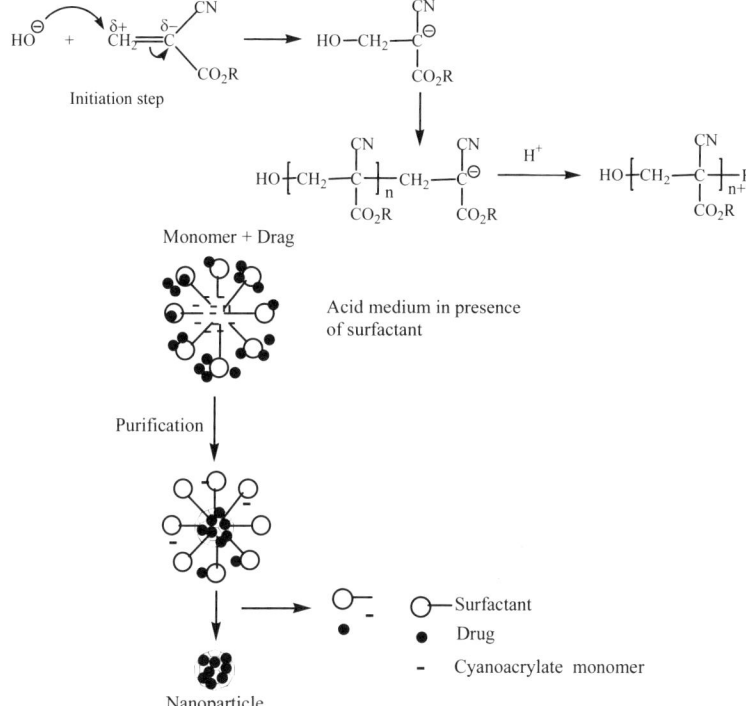

Figure 7.1. Schematic representation of poly(alkylcyanoacrylate) nanoparticle production by anioinc polymerization of alkylcyanoacrylate (After Couvreur *et al.*, 1979, copyright 1979 Kluwer Academic Publishers)

For the purpose of controlled release, drug molecules can be dissolved in the polymerization medium either before or after the polymerization reaction. The resulting suspension of polymer nanoparticles is then purified by ultracentrifugation or by re-suspending the nanoparticles in an isotonic surfactant-free medium. More recently, Lowe and Temple (Lowe *et al.*, 1994) have used the interfacial polymerization technique originally proposed by Al-Khouri Fallouh *et al.* (1986) to prepare poly(isobutyl cyanoacrylate) nanoparticles for encapsulating calcitonin. In this case, isobutyl cyanoacrylate monomers and calcitonin were dissolved in an ethanol solution containing Mygliol@ 812 oil. The oil solution was then added dropwise into an aqueous solution of poloxamer 188 under stirring for the interfacial polymerization to take place at the surface of the Mygliol@ droplets. The resulting poly(isobutyl cyanoacrylate) nanoparticles were separated from other preparation additives by diafiltration (Tishchenko *et al.*, 2001), leading to the incorporation of calcitonin up to an efficiency of 90%.

Along with the above effort in preparing polymer nanoparticles via polymerization, nanoparticles with surface properties (*e.g.* chemical, electrical, mechanical) tailor-made by surface engineering have been reported (Caruso, 2001). For instance, Prakash and co-workers (Dokoutchaev *et al.*, 1999) have developed several methods to deposit metal colloids (*e.g.* Pt, Pd, Ru, Ag, Au) onto the surface of polymer micro- and nanoparticles. Examples of the process include the controlled hydrolysis of surface acetoxy groups to generate hydroxyl functionalities for the specific adsorption of silver and ruthenium nanoparticles (Greci *et al.*, 2001), electrostatic deposition of Au colloids on positively charged polymer micro-/nanoparticles (Greci *et al.*, 2001), and simple adsorption of Pd nanoparticles on poly(vinylpyridine) nanospheres (Pathak *et al.*, 2000). As shown in Figure 7.2, these polymer micro-nanospheres (0.2–3 μm in diameter) functionalized with metal nanoparticles (1–10 nm in diameter) possess a high ratio of surface area to volume and are promising for various potential applications, especially as efficient catalysts, chemical/electronic/optical sensors, active substrates for surface-enhanced Raman scattering and photocatalysts for solar energy conversion.

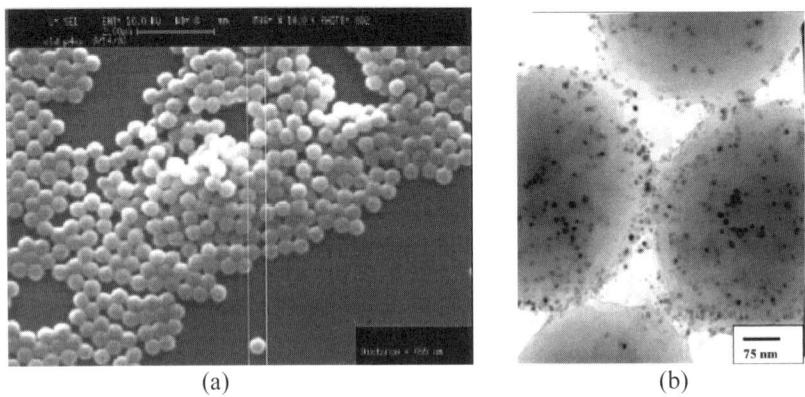

(a)   (b)

Figure 7.2. (a) SEM image of poly(4-vinylpyridine) microspheres and (b) TEM micrographs showing the surface coverage of poly(4-vinylpyridine) microspheres with colloidal palladium nanoparticles (After Pathak *et al.*, 2000, copyright 2000 The American Chemical Society)

7.2.2 Dispersion of Pre-formed Polymers

Several other related methods involving the dispersion of pre-formed polymers, such as those based on the use of emulsifying agents (Birnbaum *et al.*, 2000; Niwa *et al.*, 1993; Wehrle *et al.*, 1995; Zambaux *et al.*, 1998) and supercritical fluids (Turk, 1999; Noyori *et al.*, 2000; Randolph *et al.*, 1993; Tester *et al.*, 2000; Tom and Debenedetti, 1991), have also been used to prepare polymer nanoparticles.

7.2.2.1 Polymer Nanosphere by Emulsifying Dispersion

In this case, a pre-formed polymer is dissolved in an organic solvent, which is then emulsified into an aqueous solution, with or without the presence of drug molecules, to make an oil (O) in water (W) emulsion (O/W emulsion) by using a surfactant/emulsifying agent (Niwa *et al.*, 1993; Zambaux *et al.*, 1998). Upon the formation of a stable emulsion, the organic solvent is evaporated at room temperature by continuous stirring or by heating to produce the polymer nanoparticles.

In order to control the size of the resultant nanoparticles, a modified version of the W/O method has been developed that uses both a water-insoluble organic solvent and a water-soluble solvent (*e.g.* acetone, methanol). In this case, the spontaneous diffusion of the water-soluble solvent could cause an interfacial turbulence between the two phases to create smaller particles. Therefore, the increase in the concentration of water-soluble co-solvent was shown to cause a considerable decrease in the nanoparticle size (Wehrle *et al.*, 1995).

7.2.2.2 Polymer Nanospheres by Supercritical Fluid Method

The above methods inevitably involve the use of organic solvents that are hazardous to the environment and physiological systems (Birnbaum *et al.*, 2000). The need for environmentally safer encapsulation has been a driving force for the development of new methods for the preparation of polymer nanoparticles with the desired physicochemical properties. In this regard, the supercritical fluid technique has been an attractive alternative for generating polymer nanoparticles of high purity without any trace amount of the organic solvent (Tom and Debenedetti, 1991). As detailed discussions on supercritical fluid technology are beyond the scope of this article, interested readers are referred to specialized reviews and monographs (Kropf *et al.*, 2000; Noyori *et al.*, 2000; Tester *et al.*, 2000). Briefly, in a typical supercritical fluid method, the solute of interest is solubilized in a supercritical fluid. The solution is then expanded through a nozzle to reduce the solvent power of the supercritical fluid, resulting in the precipitation of the solute. Although the supercritical fluid method has advantages due to its solvent-free nature, a further research breakthrough is required before any commercial applications will be realized as most polymers exhibit little or no solubility in supercritical fluids.

7.3 Self-assembling of Pre-formed Polymers

Self-assembly involves the aggregation of molecules or macromolecules to thermodynamically stable structures that are held together by weak non-covalent interactions. These weak non-covalent interactions include hydrogen bonding, π-π stacking, electrostatic forces, van der Waals forces, hydrophobic and hydrophilic interactions. Because of the fast dynamic and very specific non-covalent interactions involved, self-assembling processes are usually very fast. The resulting supramolecules could undergo spontaneous and continuous de-assembly and re-assembly processes under certain conditions due to their non-covalent nature. Supramolecular materials can, therefore, select their constituents in response to external stimuli or environmental factors and behave as adaptive materials. Although self-assembling (or complexation) of small molecules has been an active research area for decades, self-assembling of polymeric chains is a recent development. Various supramolecular aggregates with unusual structures (Chen J. *et al.*, 1996; Liu G., 1998; Moffitt *et al.*, 1996; Stupp *et al.*, 1997; Webber, 1998), including shell-core and shell-hollow core, which are difficult to form by conventional chemical reactions, have been successfully prepared.

7.3.1 Shell-core Polymer Nanoparticles

Self-assembling of diblock copolymers in a selective solvent, in which only one block is soluble, can form shell-core nanoparticles (Figure 7.3).

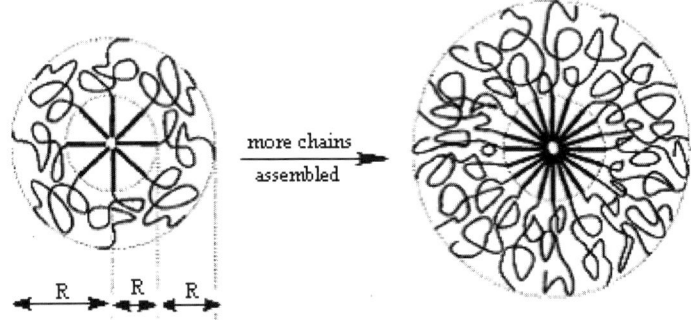

Figure 7.3. Schematic representation of a shell-core nanostructure formed by a self-assembly of coil-rod diblock copolymer chains in a selective solvent (After Zhang G. *et al.*, 2001, copyright 2001 The American Chemical Society)

To demonstrate the above principle, several groups have carried out detailed studies on various amphiphilic copolymers, including poly(ethylene oxide)-*block*-poly(propylene oxide), PEO-PPO, and poly(*N*-isopropylacrylamide)-*block*-poly(ethylene oxide), PNIPAM-*b*-PEO (Harada and Kataoka, 1999; Jenekhe and Chen, 1999; Klok and Lecommandoux, 2001; Thurmond *et al.*, 1996; Zhang *et al.*, 2001; Zhu and Napper, 1999). It was demonstrated that the formation of shell-core architectures in the copolymer systems can be induced by changes in temperature, solvent composition, ionic strength, and/or pH. For example, self-assembled shell-core polymer nanoparticles consisting of PNIPAM core and PEO shell have been prepared in water above ~32 °C. While both PNIPAM and PEO blocks dissolve in water at room temperature, PNIPAM becomes hydrophobic and undergoes a "coil-to-globule" transition at temperatures higher than ~32 °C (Schild, 1992; Zhu and Napper, 2000). By controlling the heating rate, Zhu and Napper (2000) have prepared shell-core nanoparticles with a narrow size distribution from PNIPAM-*b*-PEO block copolymers.

The shell-core formation from appropriate diblock copolymers has also been induced by certain chemical reactions. Of particular interest, Wu C. *et al.* (1999) have used the soluble poly(4-methylstyrene-*block*-phenylvinylsulfoxide), PMS-*b*-PVSO, as the starting material in THF. Thermal elimination of phenylsulfenic acid converted the soluble and flexible PVSO block into an insoluble rod-like polyacetylene, PA, block (Dai, 1999; Kanga *et al.*, 1990), leading to the formation of shell-core nanoparticles with the conducting PA core surrounded by the THF-soluble PMS shell (Figure 7.4). Nanoparticles with conducting polymers either as the core or shell have also been prepared by colloidal chemistry (Liu S. and Armes, 2001).

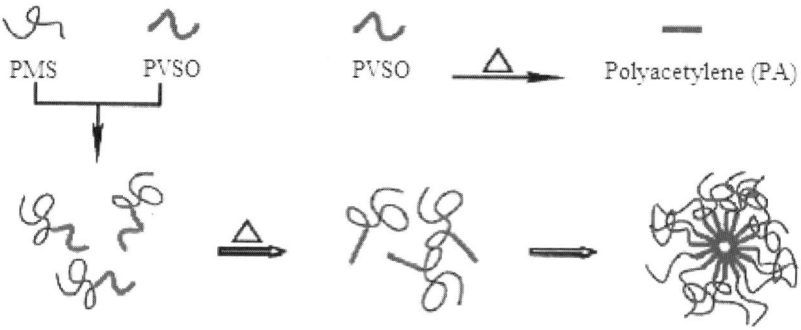

Figure 7.4. Schematic representation of chemical reaction-induced self-assembly of PMS-*b*-PVSO diblock copolymer chains in solution upon heating (After Wu C. *et al.*, 1999, copyright 1999 The American Chemical Society)

Although the block copolymer approach to the shell-core nanoparticles is an effective method and has recently been extended to include the use of triblock copolymers to form polymer nanoparticles with a multilayered shell structure (Liu S. and Armes, 2001), the general application of this method is largely limited by the rather delicate synthesis of diblock copolymers with well-defined chemical and physical structures.

To eliminate the use of block copolymers, Qiu and Wu (1997) investigated the temperature-induced self-assembling of PNIPAM and PEO *graft* copolymers. They have demonstrated that the size of nanoparticles formed by the PNIPAM chains grafted with short PEO chains is inversely proportional to the number of the PEO short chains on individual PNIPAM polymer backbones (Figure 7.5(a)). By suppressing the inter-chain association, these authors have even prepared single-chain shell-core nanoparticles, as schematically shown in Figure 7.5(b) (Qiu and Wu, 1998).

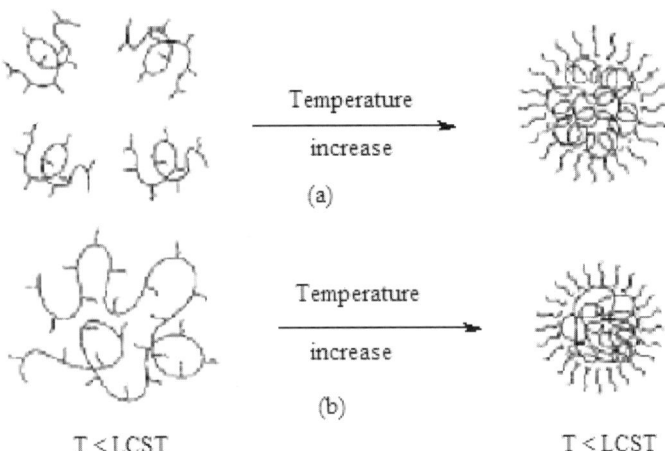

Figure 7.5. Schematic representation of the formation of (a) shell-core polymer nanoparticles from PNIPAM-*g*-PEO graft copolymers (b) a single chain shell-core nanostructure through the coil-to-globule transition of the chain backbone of PNIPAM-*g*-PEO. (After Qiu and Wu, 1998, copyright 1998 The American Physical Society)

Apart from the formation of shell-core nanoparticles from pure grafted polymers, the self-assembling in mixture systems of a homopolymer and a graft copolymer has also been effectively exploited as an alternative "block-copolymer-free" approach to polymer shell-core nanoparticles. Based on the homopolymer and graft polymer approach, Jiang and co-workers (Liu X. *et al.*, 2002) have recently reported a novel approach to hollow nanospheres (*i.e.* shell-hollow core nanoparticles). These authors first prepared shell-core nanoparticles by self-assembling a polymer pair of poly(ε-caprolactone), PCL, and a graft copolymer of methylacrylic acid and methyl methacrylate, MAF, in water. The resulting nanoparticles have a PCL core stabilized by the MAF shell with its short PCL branches anchored onto the core (Figure 7.6(a)). These authors then chemically cross-linked the micellar shell to form the so-called "shell-cross-linked knedel-like nanoparticles", SCKs, (Thurmond *et al.*, 1997), while they biodegraded the core with an enzyme to produce the hollow core (Figure 7.6(a)). The morphologies of the nanoparticles before ((A) and (B)) and after ((C) and (D)) the core degradation are shown by the TEM micrographs in Figure 7.6(b). Careful examination of the nanoparticles shown in Figure 7.6(b) indicates that the particles have expanded significantly to from *ca.* 100 to 300 nm in diameter and a shell thickness from *ca.* 10 to 100 nm, suggesting a swelling effect associated with the core removing process.

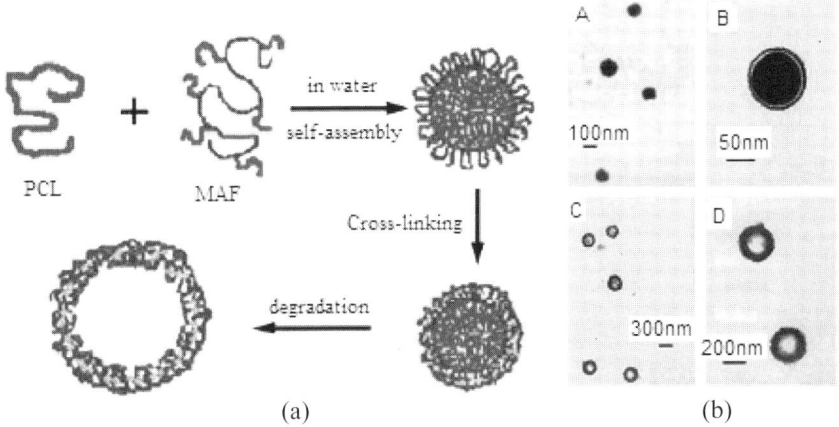

Figure 7.6. (a) A schematic illustration of the processes of self-assembly, cross-linking and degradation; (b) transmission electron microscope, TEM, images of MAF-2/PCL (1:1, w/w) nanoparticles before (A and B) and after (C and D) biodegradation of the PCL core (After Liu X. *et al.*, 2002, copyright 2002 Wiley-VCH Verlag. Reproduced with permission)

Closely related to the above study, Hawker and co-workers (Blomberg *et al.*, 2002) have successfully prepared hollow polymeric nanocapsules through surface-initiated living free-radical polymerization. In particular, these authors first attached trichlorosilyl-substituted alkoxyamine initiating groups onto the surface silanol groups of silica nanoparticles. They then carried out living free-radical polymerization from the surface-initiating groups, resulting in the formation of shell-core morphology (Figure 7.7(a)).

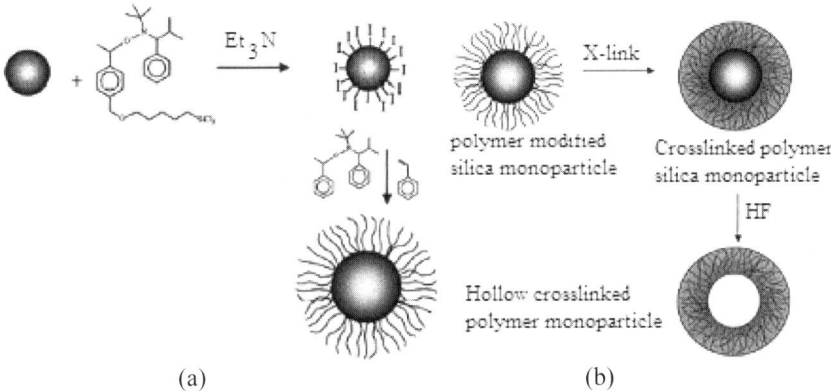

Figure 7.7. Schematic representation of (a) the polymer-modified silica nanoparticles; (b) the formation of hollow, cross-linked polymer nanocapsules from the polymer-modified silica nanoparticles (After Blomberg *et al.*, 2002, copyright 2002 John Wiley & Sons, Inc. Reproduced with permission)

Thermal or chemical cross-linking of the polymeric shell, followed by removal of the silica core with HF, led to stable hollow polymeric nanocapsules (Figure 7.7(b)). Figures 7.8(a) and (b) show field-emission scanning electron microscopic (FESEM) images for cross-linked poly(styrene-*co*-vinylbenzocyclobutene)-modified nanoparticles before and after removal of the silica core, respectively.

(a)

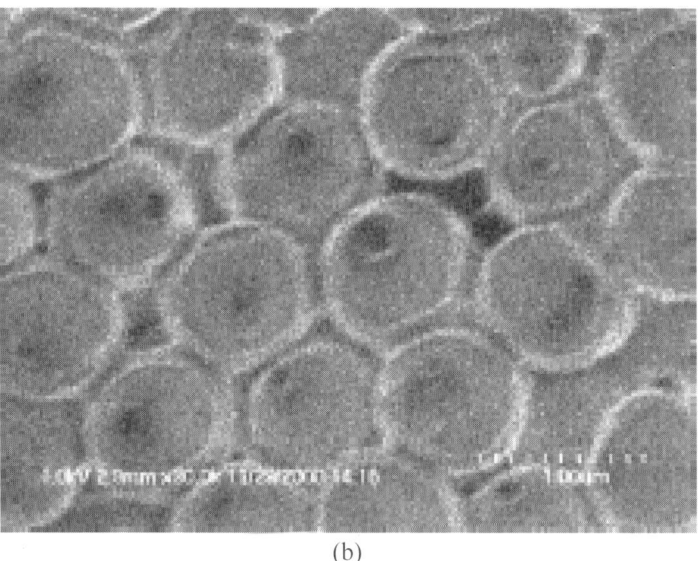

(b)

Figure 7.8. FESEM images for cross-linked poly(styrene-*co*-vinylbenzocyclobutene)-modified nanoparticles (a) before and (b) after removal of the silica core (After Blomberg *et al.,* 2002, copyright 2002 John Wiley & Sons, Inc. Reproduced with permission)

Similar hollow nanospheres have been previously prepared through self-assembling *block* copolymers in selective solvents followed by cross-linking the micellar shell and removing the core via chemical degradation (Jenekhe and Chen, 1999; Stewart and Liu, 1999; Zhang *et al.,* 2000; Zhou *et al.,* 2002), though the "block-copolymer-free" approaches discussed above may be regarded as simpler and more economic. Given that the hollow polymer nanospheres can be used to encapsulate a large variety of guest molecules, including many catalysts and drugs, there has been increasing interest in the research and development of polymer hollow nanoparticles. Clearly, a promising potential for future research and application exists in this area.

7.4 Polymer Nanowires, Nanotubes and Nanofibers

Just as polymer nanoparticles have broadened the potential applications of polymer materials, polymer wires, tubes and fibers of diameters down to a nanometer scale have also been studied for a large variety of potential applications. Polymer nanowires, nanotubes and nanofibers are useful as both "building blocks" for nanodevices and "connecting components" between the nanoscale entities and the macroscale world. In this regard, conducting polymer nanowires, nanotubes and nanofibers are of particular interest as they have been shown to possess the processing advantages of plastics and the optoelectronic properties of inorganic semiconductors or metals. Below, I provide an overview on the syntheses, properties and potential applications of polymer nanowires, nanotubes and nanofibers, with an emphasis on those based on conducting polymers.

7.4.1 Tip-assisted Syntheses of Polymer Nanowires

The use of a scanning tunneling microscope, STM, and scanning electrochemical microscope, SECM, for generation and manipulation of polymer structures as small as a few nanometers has been well documented (Dai, 1999; Yang R. *et al.,* 1995). The polymerization of pyrrole onto specific regions of graphite substrates at a sub-micron resolution was achieved by using the STM tip as an electrode. Polypyrrole strips with a linewidth of 50 µm and length of 1 mm were produced by an SECM (Kranz *et al.,* 1996), as were micron-sized polypyrrole towers (Kranz *et al.,* 1995). By spin-coating a solution of Nafion and anilinium sulfate onto a Pt electrode in a SECM unit, Wuu *et al.* (1989) polymerized aniline into a micron-scale structure. Borgwarth *et al.* (1995) successfully prepared a 20 µm-wide polythiophene line by using the tip of an SECM as an electrode for region-specific oxidation of bromide into bromine, followed by the diffusion of the bromine into a conductive substrate covered with a thiophene derivative to produce localized oxidative polymerization of thiophene monomers.

Aiming for polymer structures at the nanometer scale, Nyffenegger and Penner (1996) produced electrochemically active polyaniline particles with a size ranging from 10 to 60 nm in diameter and 1 to 20 nm in height by using the Pt tip of a scanning tunneling microscope as an electrode for the electropolymerization of aniline. Liu and co-workers (Maynor *et al.*, 2002) used an electrochemical reaction at an atomic force microscope tip for region-specific deposition of conducting polymer *nanowires* with diameters in the range of 50 to 500 nm on semiconducting and insulating substrates.

Bumm *et al.* (1996) demonstrated the use of an STM tip to probe electrical properties of individual conjugated conducting molecules ("*molecular wires*") dispersed into a self-assembled monolayer film of non-conducting alkanethiolate molecules. In a closely related but separate study, Tao and co-workers (He *et al.*, 2001) have electrochemically deposited a conducting polyaniline nanowire bridge between an STM tip and a gold electrode by region-selectively coating the STM tip with an insulation layer so that only a few nm^2 at the tip end were exposed for localized growth of the polyaniline bridge (Figure 7.9).

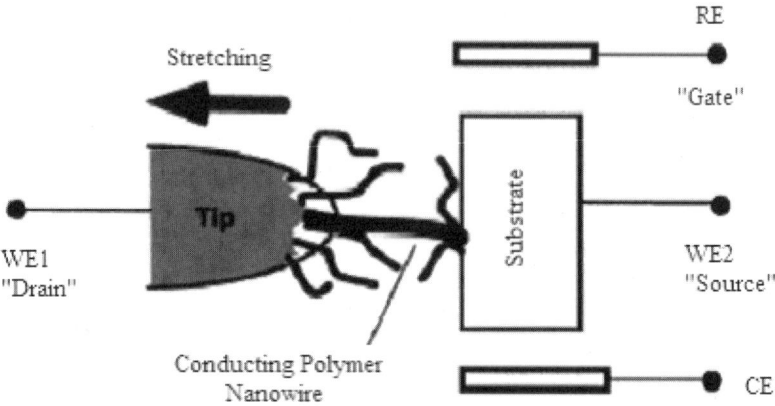

Figure 7.9. Schematic drawing of the experimental set-up. The electrochemical potential of the nanowire was controlled with respect to a reference electrode (RE) in the electrolyte. A counter electrode was used as in a standard electrochemical set-up. In comparison to a field effect transistor, the RE, WE1 and WE2 electrodes are analogous to the gate, source and drain electrodes (After He *et al.*, 2001, copyright 2001 American Institute of Physics)

Upon stretching the polymer nanowire by moving the STM tip away from the Au electrode, these authors observed a stepwise decrease in the conductance (Figure 7.10), in resemblance to what had been reported for metallic nanowires (Greenham and Friend, 1995; Pascual *et al.*, 1993).

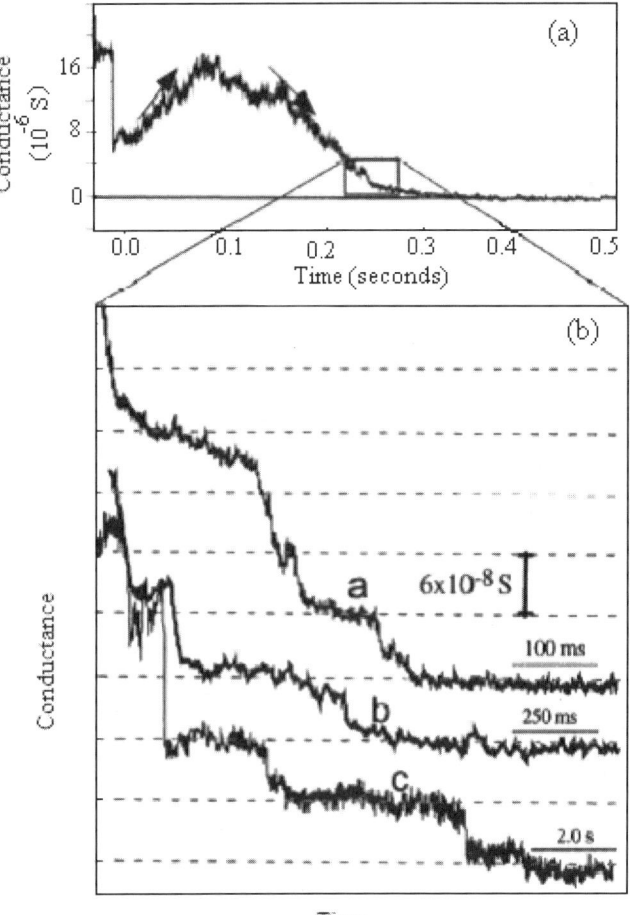

Figure 7.10. (a) Conductance of a polyaniline nanowire during an entire stretching process. The arrows point out that the initial increase in the conductance is followed by a stepwise decrease; (b) a zoom-in of the stepwise decrease. For clarity, curves (a) and (b) are shifted upward by 2 and 3 divisions, respectively. The substrate and tip potentials were held at 0.45 and 0.5 V, respectively (After He et al., 2001, copyright 2001 American Institute of Physics)

The initial increase in the conductance seen in Figure 7.10(a) is attributed to the stretching-induced alignment of the polymer chains in the nanowire, since aligned conducting polymers have been demonstrated to show higher conductivities. The observed smaller conductance step height ($< 2e^2/h$) and wider plateaus than those of metal nanowires may indicate the occurrence of polymer chains sliding in the polyaniline nanowires during stretching. The conductance of the polyaniline nanowires with various diameters were measured as a function of the electrochemical potential by stopping the stretching at different conductance steps.

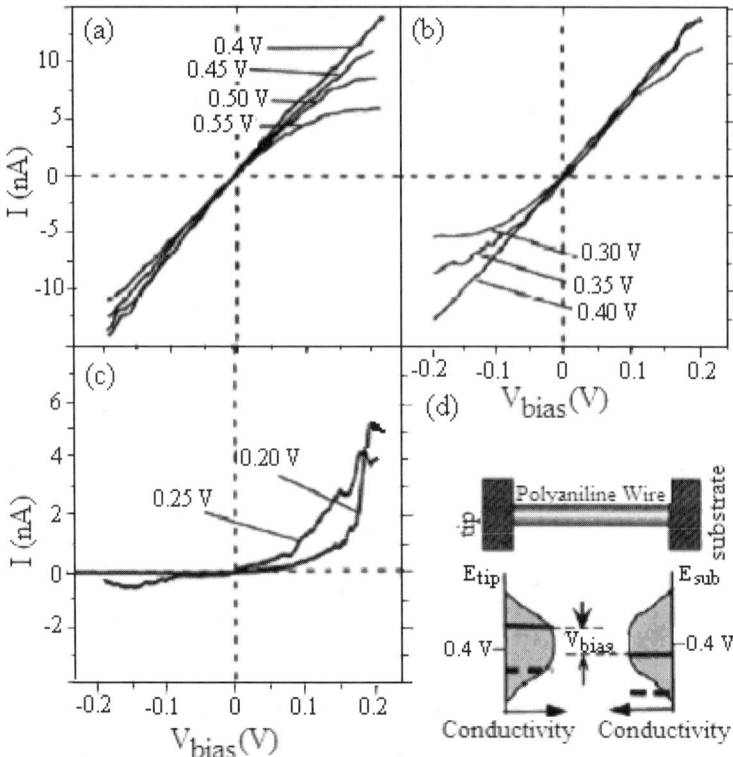

Figure 7.11. (a-c) Current (I) through a polyaniline nanowire as a function of the tip-substrate bias voltage (V_{bias}, drain-source voltage); (d) a model that explains the observed I-V_{bias} characteristics. The shaded bell curves represent conductivity of the nanowire at points near the tip and the substrate electrodes, respectively (After He *et al.*, 2001, copyright 2001 American Institute of Physics)

In analogous to the control of the gate voltage in semiconductor devices (Greenham and Friend, 1995), He *et al.* (2001) measured the current-voltage curves by sweeping the tip-substrate bias voltage (V_{bias}) while maintaining the electrochemical potential of the substrate electrode (or the tip electrode) at different values. The resulting I-V_{bias} curves show a nearly linear relationship when the substrate potential is kept between 0.30 and 0.55 V (Figures 7.11(a) and (b)). As seen in Figure 7.11(c), however, the polyaniline nanowire exhibits rectifying characteristics at 0.25 V; the current is small at negative bias, whereas it increases when the bias is positive. The observed rectifying characteristics become more pronounced when the potential is reduced to 0.2 V. The observed transition from ohmic to rectifying characteristics can be understood by considering the situation described in Figure 7.11(d), which shows that the polyaniline nanowire has one end attached to the tip and the other connected to the substrate. Therefore, the electrochemical potentials of the two ends are fixed at E_{sub} and E_{tip}. As polyaniline is in the oxidized conducting form at the electrochemical potentials close to or slightly higher than 0.3 V (Skotheim et al., 1998), both portions of the polyaniline nanowire near the substrate and tip are conducting at $E_{sub} = 0.40$ V with a small V_{bias} ($E_{tip} = E_{sub} + V_{bias}$) so that the I–V curve is ohmic. At 0.25 V, however, a negative sweep moves the E_{tip} to potentials lower than the redox potential of polyaniline (ca. 0.3 V) so that the polymer is in the non-conducting reduced form while a positive sweep of bias moves E_{tip} to the highly conductive region, leading to the rectifying behavior. The controllable conductance of the conducting polymer nanowires, together with their mechanical flexibility and rectifying characteristics, should have important implications for the use of polymer nanowires in various micro-/nanoelectronic devices and many other applications.

7.4.2 Template Syntheses of Polymer Nanowires, Nanotubes and Nanofibers

Polymer nanowires with nanometer-scale architectures have also been fabricated by using porous membranes or supramolecular nanostructures as templates. For instance, Smith et al. (Smith R. et al., 1997) have recently developed a process for producing nanocomposites with well-defined poly(p-phenylene vinylene), PPV, nanowires within self-organizing liquid-crystal matrixes. This process involves self-assembling of a polymerizable liquid-crystal monomer (e.g. acrylate) into an ordered hexagonal array of hydrophilic channels (ca. 4 nm in diameter) filled with a precursor polymer of PPV, followed by photopolymerzation to lock in the matrix architecture and thermal conversion to form PPV nanofibers in the channels. As a result, significant fluorescence enhancement was observed, most probably due to the much reduced inter-chain-exciton quenching achieved by separating the PPV molecules from the polymer matrix.

More generally, polymer nanowires and nanotubes with improved order and fewer structural defects can be synthesized within a template formed by the pores of a nanoporous membrane (Martin, 1995) or the nanochannels of a mesoporous zeolite (Wu C.G. and Bein, 1994). Template synthesis often allows for the production of polymeric wires or tubules with controllable diameters and lengths (Figure 7.12, Cepak and Martin, 1999). Template synthesis of conjugated polymers, including polyacetylene, polypyrrole, polythiophene, polyaniline and PPV, may be achieved by electrochemical or chemical oxidative polymerization of the corresponding monomers. While the electrochemical template synthesis can be carried out within the pores of a membrane that is pre-coated with metal on one side as an anode (Martin, 1995), the chemical template synthesis is normally performed by immersing the membrane into a solution containing the desired monomer and oxidizing agent, with each of the pores acting as a tiny reaction vessel.

It has been noted with interest that if the pores of polycarbonate membranes are used as the template, highly ordered polymeric tubules are produced by preferential polymerization along the pore walls (Martin, 1991) due to the specific solvophobic and/or electrostatic interactions between the polymer and the pore wall (Martin, 1995). These conducting polymer tubules show a wide range of electrical conductivities, which increase with decreasing pore diameter (Cai et al., 1991; Parthasarathy and Martin, 1994). This is because the alignment of the polymer chains on the pore wall can enhance conductivity, and the smaller tubules contain proportionately more of the ordered material.

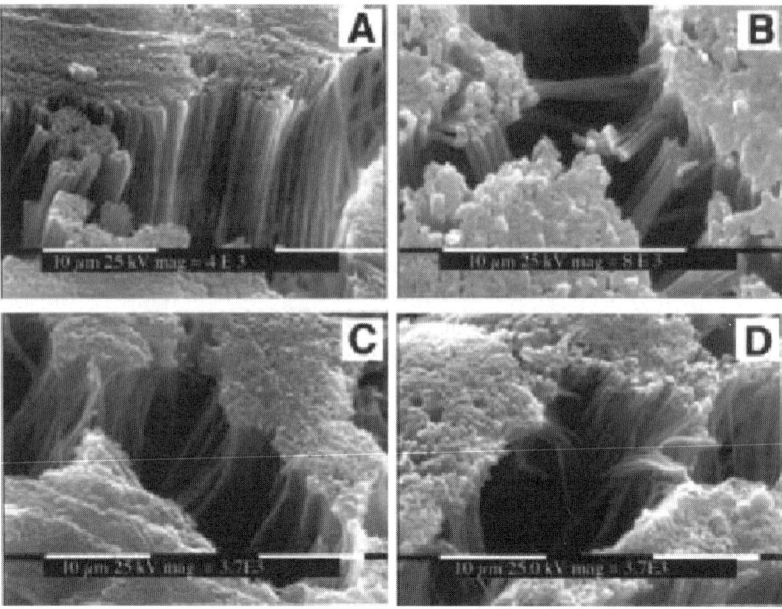

Figure 7.12. Scanning electron micrographs of microfibrils prepared in the alumina template membrane: (A) polystyrene, PS; (B) poly(vinylidene fluoride), PVDF; (C) poly(phenylene oxide); (D) poly(methyl methacrylate), PMMA (After Cepak and Martin, 1999, copyright 1999 The American Chemical Society)

By wetting ordered porous templates with polymer solution or melts, Steinhart et al. (2002) have developed another simple technique for the fabrication of polymer nanotubes with a monodisperse size distribution and uniform orientation. Figure 7.13 shows polymer nanotubes formed by melt-wetting of ordered porous alumina and oxidized macroporous silicon templates with narrow pore size distribution. The method can be used to produce polymer nanotubes of narrow size distribution from almost all melt-processable polymers, their blends or multi-component solutions. More recently, conjugated conducting polymer nanotubes were synthesized using "template-free" polymerization that involves a self-assembled supramolecular template (Qiu H. et al., 2001; Wan and Li, 1999). These studies are of particular interest because high-temperature graphitization of the polymer nanotubes could transform them into carbon nanotubes of superior electronic and mechanical properties (Saito et al., 1998), as demonstrated by polyacrylonitrile (PAN) nanotubules synthesized in the pores of an alumina membrane or in zeolite nanochannels (Kyotani et al., 1997; Parthasarathy et al., 1995).

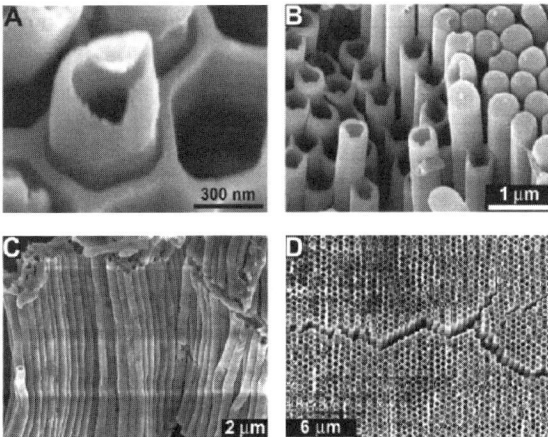

Figure 7.13. Scanning electron micrographs of nanotubes obtained by melt-wetting: (A) Damaged tip of a PS nanotube; (B) ordered array of tubes from the same PS sample after complete removal of the template; (C) array of aligned polytetrafluoroethylene, PTFE, tubes; (D) PMMA tubes with long-range hexagonal order obtained by wetting of a macroporous silicon pore array after complete removal of the template (After Steinhart *et al.*, 2002, copyright 2002 American Association for the Advancement of Science).

On the other hand, Aida and co-workers (Kageyama *et al.*, 1999) prepared crystalline nanofibers of ultrahigh molecular weight polyethylene by extruding the polymer through porous nanoscale reactors as it was synthesized. In particular, these authors coated the hexagonal pores (27 Å) of mesoporous silica fiber reactors with the classic polymerization catalyst titanocene. After activating the catalyst, they then pressurized the vessel containing the reactors with ethylene gas. Having been confined within the narrow pores, the resulting polymer chains were forced to grow out of the framework, much like spaghetti extruding from a pasta maker (Rouhi, 1999) (Figure 7.14).

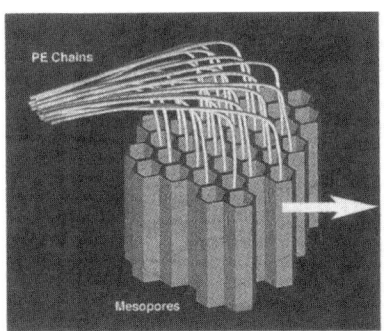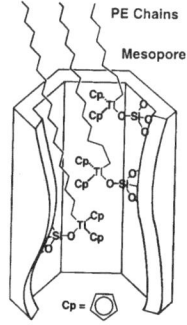

Figure 7.14. Conceptual scheme for the growth of crystalline fibers of polyethylene by mesoporous silica-assisted extrusion polymerization (After Kageyama *et al.*, 1999, copyright 1999 American Association for the Advancement of Science)

7.4.3 Electrospinning of Polymer Nanofibers

While the potential applications of polymer nanowires and nanotubes have yet to be fully exploited, polymer nanofibers have been demonstrated to be attractive for a large variety of potential applications, including high performance filters, scaffolds in tissue engineering, wound dressings, sensors and drug delivery systems (Doshi *et al.,* 1995; Doshi and Reneker, 1995; Reneker *et al.,* 2000; Reneker and Chun, 1996). However, the large-scale fabrication of polymer nanowires/nanofibers at a reasonably low cost has presented a big challenge in the realization of commercial applications of polymer nanofibers. Electrospinning, a technique which was first patented in 1934 (Formhals, 1934), has recently been re-examined, notably by Reneker's group (Doshi *et al.,* 1995; Doshi and Reneker, 1995; Reneker *et al.,* 2000; Reneker and Chun, 1996), as an effective method for the large-scale fabrication of ultrafine polymer fibers.

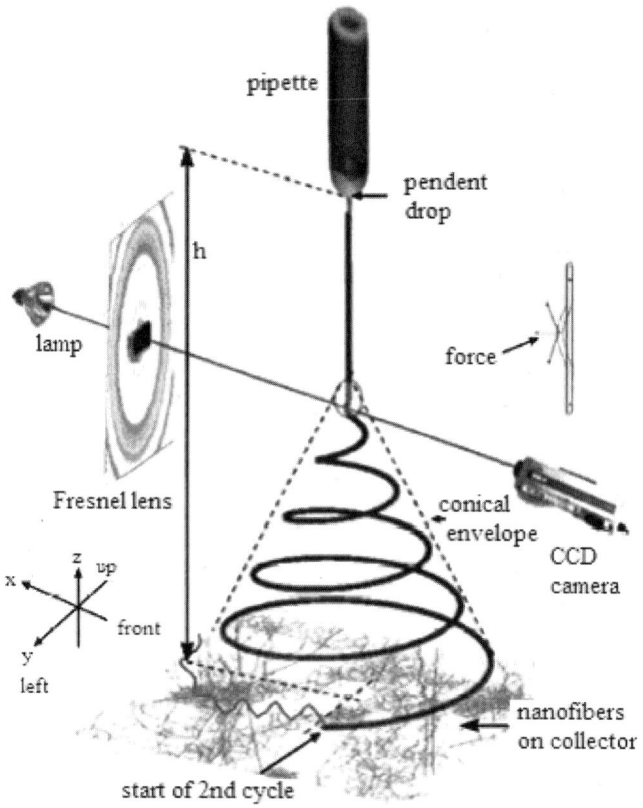

Figure 7.15. Schematic drawing of the electrospinning process (After Reneker *et al.,* 2000, copyright 2000 American Institute of Physics)

As schematically shown in Figure 7.15, the electrospinning process involves the application of a high electric field between a droplet of polymer fluid and a metallic collection screen at a distance of 5–50 cm from the polymer droplet. When the voltage reaches a critical value (typically, ca.5–20 KV) at which the electrical forces overcome the surface tension of the polymer droplet, an electrically charged jet flows from the polymer droplet toward the collection screen. The electrical forces from the charge carried by the jet further cause a series of electrically driven bending instabilities to occur as the fluid jet moves toward the collection screen. The repulsion of the charge on adjacent segments of the fluid jet causes the jet to elongate continuously to form ultrafine polymer fibers. The solvent evaporates and the stretched polymer fiber that remains is then collected on a screen as a non-woven sheet composed of one fiber many kilometers in length. The electrospun polymer fibers thus produced have diameters ranging from several microns down to 50 nm or less. This range of diameters overlaps conventional synthetic textiles and extends through diameters two or three orders of magnitude smaller.

Reneker and co-workers have made significant advances on both the theoretical and experimental fronts. They have carried out detailed theoretical studies on the formation of the electrically charged polymer jet and its bending instability (Spivak et al., 2000; Yarin et al., 2001a, b), and have also prepared a large number of novel polymer fibers with diameters from a few nanometers to microns by electrospinning various synthetic and natural polymers from either solution or melts. Various polymers ranging from conventional synthetic polymers, through conducting and liquid crystalline polymers, to biopolymers have been electrospun into polymer nano-/microfibers (Fong et al., 1999; Reneker and Chun, 1996). A typical microscopic image of the resulting polymer nanofibers is shown in Figure 7.16 (Chun et al., 1996).

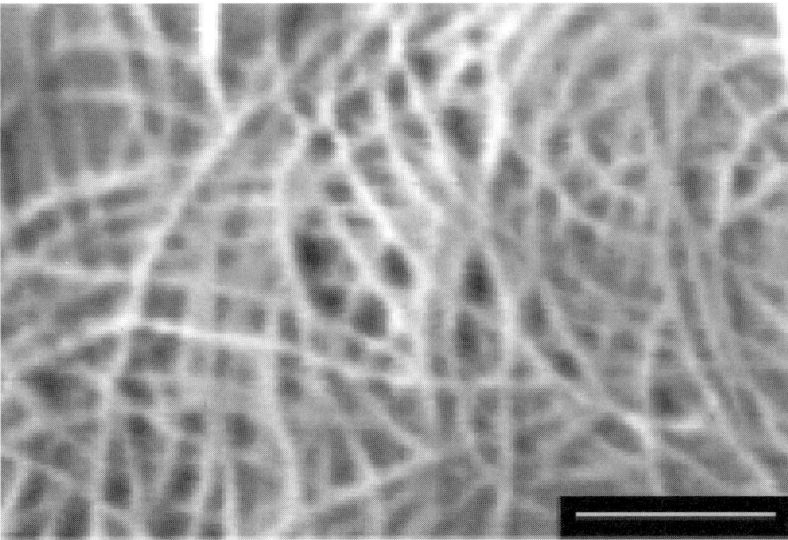

Figure 7.16. Typical SEM image of electrospun polystyrene fibers with an average diameter of 43 nm, Scale bar: 1000 nm (After MacDiarmid et al., 2001, copyright 2001 Elsevier. Reproduced with permission)

The electrospun fiber mats possess a high surface area per unit mass, low bulk density, and high mechanical flexibility. These features make the electrospun fibers very attractive for a wide range of potential applications, as mentioned above. Among them, the use of electrospun polymer fibers for electronic applications is of particular interest. The high surface to volume ratio associated with electrospun *conducting polymer* fibers makes them excellent candidates for electrode materials, since the rate of electrochemical reaction is proportional to the surface area of an electrode and the diffusion rate of the electrolyte. Reneker and co-workers (Norris *et al.*, 2000) have reported the pyrolysis of electrospun polyacrylonitrile nanofibers into conducting carbon nanofibers. They have also prepared conducting polymer nanofibers by electrospinning polyaniline from sulfuric acid into a coagulation bath and characterized them by scanning tunneling microscopy, electron microscopy, and electron diffraction.

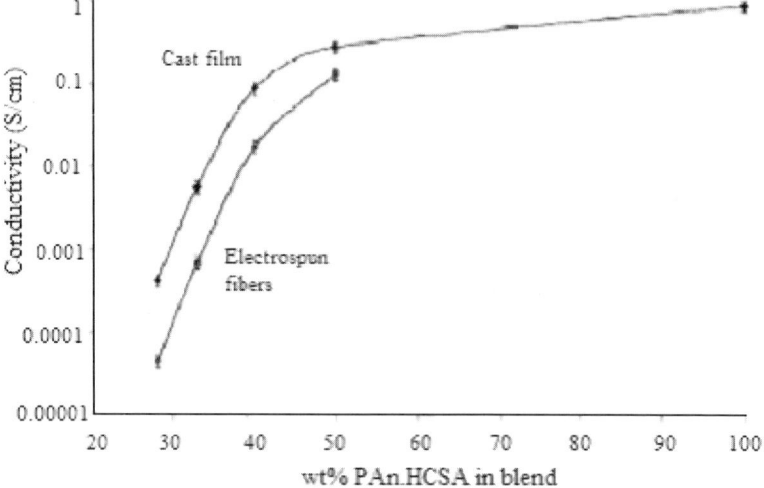

Figure 7.17. The effect of the weight percentage content of HCSA-PANI on the room temperature conductivity obtained from the HCSA-PANI/PEO electrospun fibers and cast films (After Norris *et al.*, 2000, copyright 2000 Elsevier)

MacDiamid and co-workers (Norris *et al.*, 2000) prepared camphorsulfonic acid-doped polyaniline, HCSA-PANI, and polyethylene oxide composite thin fibers (*ca.* 0.95–2 μm in diameter) by the electrospinning technique from a mixed solution. The four-point probe method (van der Pauw, 1958) was used to measure the conductivity of the non-woven fiber *mat* and cross-checked with measurements of the conductivity of cast films produced from the same solution. Figure 7.17 shows the effect of the weight percentage content of HCSA-PANI on the room temperature conductivity obtained from the HCSA-PANI/PEO electrospun fibers and film (Norris *et al.*, 2000). As can be seen in Figure 7.17, the conductivity of the electrospun fiber mat was lower than that for a cast film, though they have very similar UV-visible absorption characteristics. The lower conductivity values for the electrospun fibers compared to those of cast films can be attributed to the porous nature of the non-woven electrospun fiber mat as the four-probe method measures the volume resistivity rather than the conductivity of an individual fiber. Although the measured conductivity for the electrospun mat of conducting nanofibers is relatively low, their porous structures, together with the high surface-to-volume ratio, enables faster de-doping and re-doping in both liquids and vapors.

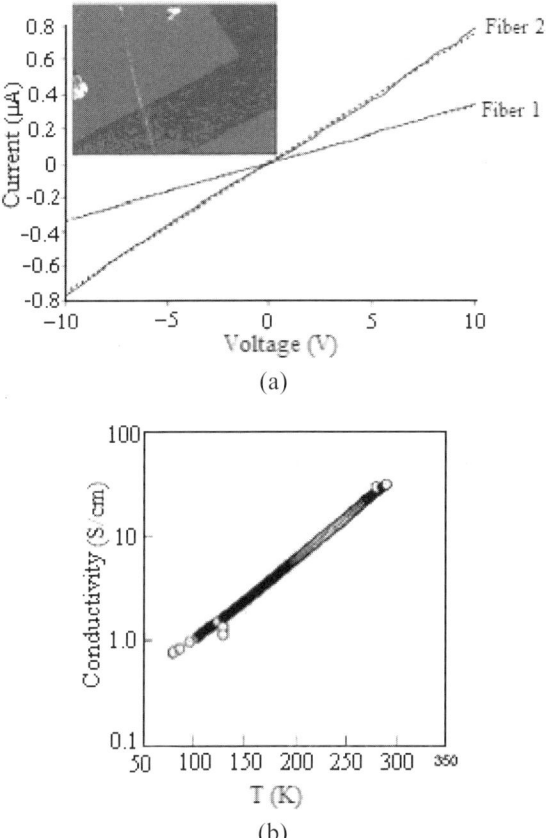

Figure 7.18. (a) Current-voltage curve for a single 50 wt% HCSA-PANI/PEO nanofiber; (b) temperature-dependence of the conductivity for a single 72 wt% HCSA-PANI/PEO nanofiber (After MacDiarmid *et al.*, 2001, copyright 2001 Elsevier. Reproduced with permission)

In order to measure the conductivity of an individual nanofiber, MacDiarmid and co-workers (MacDiarmid *et al.*, 2001) have also collected a *single* electrospun HCSA-PANI/PEO nanofiber on a silicon wafer coated with a thin layer of SiO_2 and deposited two separated gold electrodes at 60.3 µm apart on the nanofiber (two-probe method). The current-voltage (*I–V*) curves thus measured for single 50 wt% HCSA-PANI/PEO fibers with diameters of 600 and 419 nm are shown in Figure 7.18(a) (*i.e.* Fiber 1 and Fiber 2, respectively), which give a more or less straight line with a conductivity of *ca.* 0.1 S/cm. The temperature dependence of the conductivity for a single 72 wt% HCSA-PANI/PEO electrospun fiber with a diameter of 1.32 µm given in Figure 7.18(b) also shows a linear plot with a room-temperature conductivity of 33 S/cm (295 K). This value of conductivity is much higher than the corresponding value of about 0.1 S/cm for a cast film (Figure 7.18), suggesting a highly aligned nature of the PANI chains in the electrospun fiber.

7.5 Polymer Nanofilms

Polymer thin films with a thickness at the nanometer scale (*i.e.* polymer nanofilms or nanosheets) are an important class of materials from the perspective of both fundamental science and practical applications. For instance, the immobilization of a thin layer of biomacromolecules onto the surfaces of polymeric biomaterials could produce bioactivities (Dai and Mau, 2000), while conjugated polymer thin films play vital roles in organic optoelectronic devices, including polymer thin film transistors (Horowitz, 1998; Katz, 1997; Lovinger and Rothberg, 1996), photoconductors (Mylnikov, 1994), photovoltaic cells (Sariciftci and Heeger, 1994), and light-emitting diodes, LEDs, (Dai *et al.*, 2001; Kraft *et al.*, 1998). In this section, we choose polymer thin films formed by solution casting, plasma deposition, Langmuir-Blodgett deposition, and physical adsorption as a few examples to illustrate their importance to various practical applications with a focus on polymer LEDs.

7.5.1 Polymer Nanofilms by Solution Casting

The syntheses of soluble conjugated polymers offer excellent possibilities for advanced device fabrications using various solution-processing methods *(e.g.* spin-coating) (Dai, 1999). The simplest single-layer LED devices consist of a thin solution-cast layer of electroluminescent conjugated polymer (*e.g.* MEH-PPV: poly[2-methoxy-5-(2'-ethyl-hexyloxy)-*p*-phenylene vinylene]) sandwiched between two electrodes (an anode and a cathode). Among the two electrodes, at least one should be transparent or semi-transparent. In practice, a transparent indium-tin-oxide, ITO, coated glass is often used as the anode and a layer of low-work-function metal (*e.g.* aluminum, calcium) acts as the cathode.

Upon the application of an electrical voltage onto the EL polymer layer in a LED device, electrons from the low-work-function cathode (*e.g.* Al) and holes from the high work-function anode (*e.g.* ITO) are injected respectively into the lowest unoccupied molecular orbital, LUMO, and the highest occupied molecular orbital, HOMO, of the polymer, resulting in the formation of a singlet exciton that produces luminescence at a longer wavelength (the Stokes shift) through radiative decay. To maintain a low operation voltage, a very thin polymer layer *(ca.* 100 nm) is usually used. The wavelength (and hence the color) of the photons thus produced depends on the energy gap of the organic light-emitting material, coupled with a Stokes shift. Since conjugated polymers can have π-π^* energy gaps over the range of 1 to 4 eV (1240–310 nm), polymer-based LEDs should, in principle, emit light across the whole spectrum from ultraviolet to near infrared with a good color-tunablity by controlling the energy gap through molecular engineering (Dai *et al,* 2001; Kraft *et al.,* 1998). Indeed, polymer LEDs with a wide range of emission colors and with reasonably high quantum efficiencies (a few percent) are now reported. For detailed discussion on the current state of the art in the development of polymeric LEDs, the interested readers are referred to specialized reviews and conference monographs (Friend *et al.,* 1999; Greenham and Friend, 1995; Pei *et al.,* 1997; Yang and Karasz, 1997).

Although the solution-processing technique provides a simple and attractive approach for polymer thin film formation, it may cause a number of problems related to the quality of polymer films thus formed. These include the difficulty in choosing suitable solvent(s) for the preparation of pinhole-free single-layer or multilayer polymer thin films, and in controlling the film thickness and polymer chain conformation. Consequently, various other methods, including plasma polymerization, Langmuir-Blodgett technique, and layer-by-layer adsorption, for the formation of polymer nanofilms have been developed.

7.5.2 Polymer Nanofilms by Plasma Polymerization

To circumvent those problems intrinsically associated with the solution-processing technique, plasma polymerization has been used as an alternative thin film-formation method (d'Agostino, 1990; Yasuda, 1985). Although plasma polymerization, in most cases, produces an electrically insulating organic film, few approaches to semiconducting plasma polymer films have been reported. Along with others (Kang *et al.*, 1996; Tanaka *et al.,* 1994), Dai and co-workers (Gong *et al.,* 1998) have plasma-polymerized polyaniline thin films (*ca.* 350 nm). The plasma-polyaniline films thus prepared are smooth and free of oxidant and solvents, having improved physicochemical characteristics compared with polyaniline films formed by electrochemical polymerization or chemical synthesis followed by spin-coating, solvent casting or melt extrusion. The presence of conjugated sequences in the plasma-polyaniline films was confirmed by ultraviolet/visible, UV/vis, Fourier transform infrared, FTIR, and electron spin resonance, ESR, measurements, which allow the conductivity of the plasma-polyaniline films to be enhanced by three orders of magnitude through HCl treatment. It was further demonstrated that atomic ratios of C/N both higher and lower than that of aniline (C/N = 6) can be obtained by choosing appropriate discharge conditions, indicating the possibility of tailoring the structure and properties of the plasma films by optimizing the discharge conditions.

Although the advantages of the plasma-generated semiconducting polymer thin films for electronic applications remain to be demonstrated, an oxygen or nitrogen plasma treatment of a PPV layer at the PPV/Al interface in an LED device with the Cr/PPV/Al structure has been demonstrated to cause the disappearance of the rectifying behavior and an increase in the current by many orders of magnitude (Nguyen et al., 1995). Similarly, the efficiency, brightness and lifetime of LEDs were shown to be significantly improved by an air or argon plasma treatment on the ITO surface, presumably due to the removal of contaminants coupled with an increase in the work function of ITO (Furukawa et al., 1997; Wu C.C. et al., 1997). A hydrogen plasma, on the other hand, was found to increase the turn-on voltage and reduce the efficiency largely due to a decrease in the work function of ITO (Furukawa et al., 1997; Hu and Karasz, 1998; Salaneck et al., 1996; Wu C.C. et al., 1997). Thus, the plasma technique shows much promise for interfacial engineering in LEDs (Hu and Karasz, 1998; Salaneck et al., 1996). Furthermore, few studies have reported the use of plasma polymers as emitting layers in LED devices (Xu X. et al., 1995). There is no doubt that the plasma polymerization of semiconducting organic films and plasma surface modification demonstrated above will have potential implications for making novel electronic and photonic devices. The ease with which multilayered polymer nanofilms and micro-sized patterns of organic materials can be made by the plasma technique could add additional benefits to this approach (Dai, 2001).

7.5.3 Polymer Nanofilms by Langmuir-Blodgett Deposition

Although the solution-processing and plasma polymerization techniques discussed above provide simplified approaches for polymer nanofilm formation, they provide no means of controlling the orientation of polymer chains in the resulting films. The Langmuir-Blodgett, LB, method involves the formation of a Langmuir film at the gas/liquid interface, followed by a dipping process to transfer the polymer thin film onto a solid substrate (Roberts, 1990). In addition to the ease with which the number of molecular layers (and hence the thickness of films) can be controlled at the molecular level, the LB technique has another major advantage for controlling the film structure in three dimensions (Schwiegk et al., 1992) as the orientation of polymer chains in each of the layers can be independently controlled by anisotropic compression of the corresponding Langmuir film during the dipping process (Aoki and Miyashita, 1997). The oriented and/or patterned LB films (Suzuki et al., 1993) can be further used as substrates, for example, for controlling liquid crystal alignment (Suzuki et al., 1993). As a consequence, the LB technique has been widely used for making nanometer-thick multilayer films with well-defined structures and ordered molecular organizations from amphiphilic conjugated polymers, non-conjugated precursor polymers, and even certain non-surface-active conjugated polymers.

LB films of highly anisotropic conducting polymers can be prepared through: (a) the manipulation of amphiphilic macromolecules comprising an alkyl chain as the hydrophobic group and with a conjugated polymer being attached to its hydrophilic end (Sagisaka et al., 1993); (b) the manipulation of mixed monolayers consisting of a surface active agent (e.g. dopant) and non-surface active conjugated polymer (Cheung et al., 1992); or (c) the direct synthesis of conducting polymers at the air-water interface or within a preformed multilayer film. To mention but a few examples: Shimidzu et al. (Ando et al., 1989; Iyoda et al., 1987; Sagisaka et al., 1993; Shimidzu et al., 1988) have prepared conducting LB films containing polypyrrole (Shimidzu et al., 1988), polyaniline (Iyoda et al., 1987) or poly(3-alkylthiophene) (Ando et al., 1989). Among them, a multilayer LB film (200 layers) of a substituted polypyrrole showed an anisotropic conductivity as high as 10^{10} (i.e. $\sigma_{||} = 10^{-1}$ S/cm and $\sigma\perp = 10^{-11}$ S/cm) (Iyoda et al., 1987). Well-characterized LB films of various conjugated polymers including poly(3-alkylthiophene) have also been reported by Rubner and co-workers (Cheung et al., 1992) for which the readers are referred to the excellent review by Rubner and Skotheim (Rubner and Skotheim, 1991).

More recently, PPV-type LB films have also been prepared from the sulfonium precursor of PPV (Era et al., 1988; Nishikata et al., 1989) and some of its soluble derivatives (Kamiyama et al., 1990; Remmers et al., 1996; Vahlenkamp and Wegner, 1994). Also, LEDs with polarized electroluminescent, EL, emissions have been fabricated using certain LB films as the EL material (Bolognesi et al., 1997; Cimrová et al., 1996). On this basis, Grüner et al. (1997) have precisely determined the location of the exciton recombination zone (Greenham and Friend) in LEDs by making multilayer LB films of soluble PPP derivatives with two regions of perpendicular macroscopic orientation, followed by quantifying the emission profile through the analysis of the EL anisotropy.

7.5.4 Polymer Brushes by End-adsorption

End-adsorbed polymer brushes comprise another major class of oriented polymer thin films, in which the constituent polymer chains align *perpendicularly* on the substrate surface (Halperin et al., 1992; Luckham and Costello, 1993; Milner, 1991; Patel et al., 1989). Just as the in-plan orientation of polymer chains in the LB films has brought new properties and hence broadened the scope for using polymer thin films in optoelectronic devices, the surface-bound polymer brushes have facilitated polymer thin films for many technological applications (Cantor and Schimmel, 1980; Halperin, et al., 1992; Luckham and Costello, 1993; Milner, 1991; Napper, 1983; Patel et al., 1989; Ratner and Castner, 1996), such as in controlling the stability of colloidal dispersions, regulating the permeability of a cell membrane, metabolism, the transmission of information, the biocompatibility of biomaterials, and molecular separation by liquid chromatography. For instance, capsule membranes with end-grafted temperature-/pH-responsive polymer chains (*e.g.* poly(*N*-isopropylacrylamide) or poly(glutamic acid)) may be used as thermoselective drug release devices and/or pH-sensitive gates via a conformational change of the end-grafted polymer brushes (Ito et al., 1997; Okahata et al., 1986). Furthermore, end-adsorbed polymer chains have also been shown to be useful for regulating the frictional forces between solid surfaces, the spreading properties of liquids and the stability of thin films on solid surfaces (Klein et al., 1994; Martin et al., 1996). Knowledge of macromolecular conformations and conformational transitions of surface-bound polymer chains is essential for a detailed understanding of their performance at the molecular level. Recent developments on both experimental and theoretical fronts (Dai and Toprakcioglu, 1991; Halperin, et al., 1992; Luckham and Costello, 1993; Milner, 1991; Patel et al., 1989; Taunton et al., 1988, 1990; Toprakcioglu et al., 1993) in the investigation of adsorbed polymer have brought the goal of structural testing against theories within reach.

Generally speaking, the conformation of polymer chains adsorbed at the liquid-solid interface differs from that in a bulk solution (Halperin, *et al.*, 1992; Luckham and Costello, 1993; Milner, 1991; Patel *et al.*, 1989). This is particularly true for end-adsorbed polymer chains in a good solvent, in which random coils of the macromolecules in solution may become highly stretched polymer "brushes" at the liquid-solid interface.

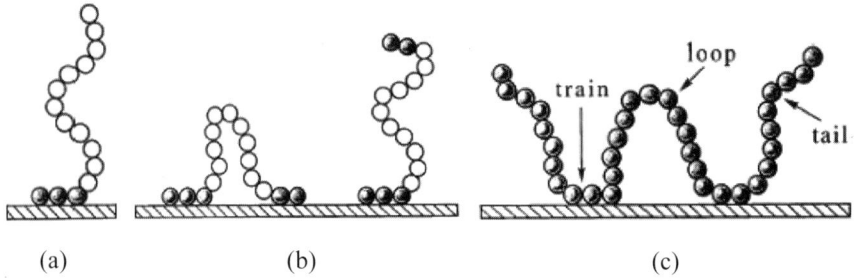

Figure 7.19. Conformations of adsorbed polymer chains: (a) "tail-like" conformation of end-adsorbed homopolymer or diblock copolymer; (b) "loop-like" or "tail-like" conformation of end-adsorbed triblock copolymer; (c) "tail-loop-train" conformation of adsorbed homopolymer

While end-adsorbed homopolymers and diblock copolymers form polymer brushes in which the chains can only exist in a "tail" conformation (Figure 7.19(a)), end-adsorbed triblock copolymers, ABA (the A blocks represent short "sticking" segments, whereas the B block does not adsorb), can form either a "loop" or a "tail" depending on the sticking energy of the A block and the surface coverage at the liquid-solid interface (Dai *et al.*, 1995) (Figure 7.19(b)). Adsorbed homopolymer, on the other hand, forms a more complicated conformational structure (Figure 7.19(c)) due to the random adsorption of its constituent monomers onto the surface.

Along with others, Dai *et al.* (1995) have carried out the structural investigation of adsorbed polymers at the liquid-solid interface using surface force apparatus (Israelachvili, 1992; Klein, 1988; Sarid, 1991), SFA, and neutron reflectometry (Field *et al.*, 1992a, b; Higgins and Benoît, 1994). The surface force apparatus is schematically shown in Figure 7.20, in which two molecularly smooth mica sheets, mounted in a crossed cylinder configuration, are used as substrates for polymer adsorption (Israelachvili, 1992). The surface separation, D, can be varied by mechanical or piezo-electric transducers and is measured with the aid of interferometry to an accuracy of *ca.* 3 Å. The corresponding force, $F(D)$, between the mica substrates is measured directly by observing the deflection of a leaf spring of known spring constant bearing one of the mica sheets.

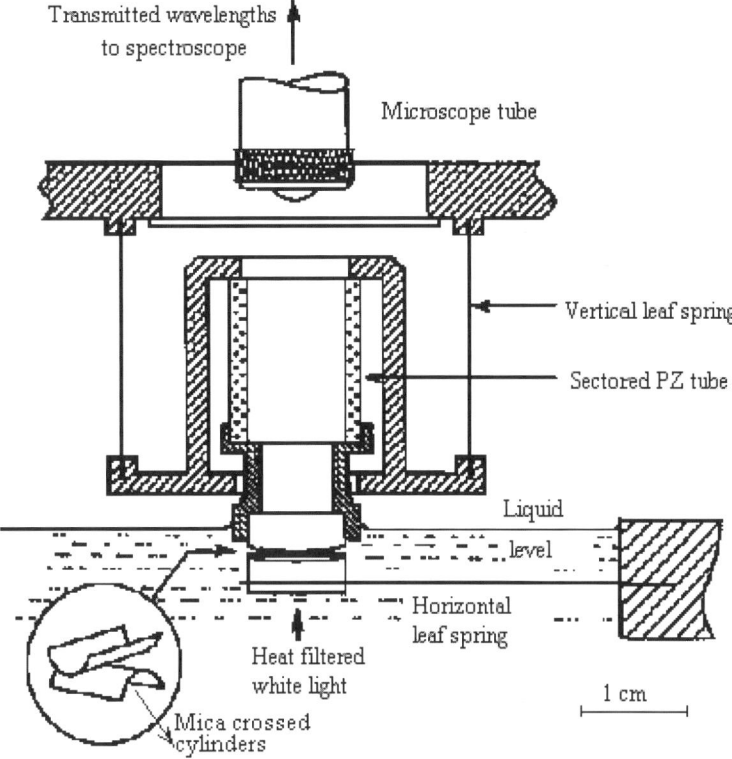

Figure 7.20. Surface force apparatus, SFA, for directly measuring the force laws between surfaces in liquids or vapors with an angstrom resolution (After http://www.weizmann.ac.il/fluids/klein/techniques_sfb.html)

The results are presented in the form $F(D)/R$ vs. D, where R is the geometric mean radius of curvature of the two mica surfaces, and the value of $F(D)/R$ gives the interaction energy per unit area of surface, $E(D)$, via the Derjaguin approximation (Dai *et al.*, 1995; Derjaguin, 1934; Israelachvili, 1992; Sarid, 1991):

$$F(D)/R = 2\pi E(D) \tag{7.1}$$

$E(D)$ is related to the force per unit area, f, as follows:

$$E(D) = \int_{2L_o}^{D} f(D')dD' \tag{7.2}$$

The use of SFA for measuring the interactions between adsorbed homopolymer surfaces at a high surface coverage in a wide variety of solvency conditions has been reviewed elsewhere (Bhushan *et al.*, 1995; Klein, 1988). Without repetition of the numerical discussions, the main results from the SFA studies are summarized in Figure 7.21, which shows only schematic smooth curves instead of the actual $F(D)$ data for reasons of clarity. As can be seen from curve (a) of Figure 7.21, the interaction between surfaces bearing adsorbed polymer chains in a poor solvent (*e.g.* polystyrene in cyclohexane at $T = 24$ °C, $T_\theta = 35$ °C) is characterized by a deep, long-ranged (*ca.* $3R_g$, R_g is the radius of gyration of a polymer coil) (Dai, 1993; Mattoussi and Karasz, 1993; Yamakawa, 1971) attractive force, well-originating from the monomeric attraction in the poor solvent, followed by a repulsion at lower D values due to reduced available configurations for the adsorbed polymer chains in the gap because of excluded volume effects. At the theta temperature, T_θ, the attractive forces disappear (curve (b) of Figure 7.21), reflecting the absence of the segment-segment interactions at the theta point. The onset distance for interactions reduces from $3R_g$ for the poor solvent case to around $2R_g$. In a good solvent, the interaction becomes monotonically repulsive with a longer-ranged onset distance of $6-8R_g$ (region (c) of Figure 7.21), attributable to the repulsive monomer-monomer interactions associated with the positive second virial coefficient in the good solvent.

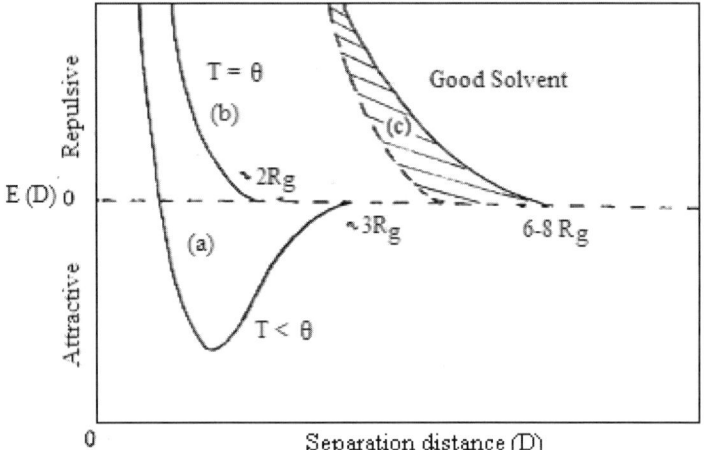

Figure 7.21. Summary of interaction profiles between surfaces with adsorbed homopolymers in various solvency conditions at high surface coverage (After Klein, 1988, copyright 1988 Elsevier)

Using a symmetric polystyrene-polyvinylpyridine (PS-PVP) diblock copolymer, Hadziioannou and coworkers (Hadziioannou *et al.*, 1986) demonstrated that the non-adsorbing PS chains can be held onto the mica surface by the covalently-linked adsorbing PVP blocks, with the PS chains being stretched out to form a polymer brush at the mica/toluene interface. To mimic a layer of end-grafted polymer chains more closely, Taunton *et al.* (1990) used end-functionalized PS-X homopolymers and highly asymmetric PS_m-PEO_n diblock copolymers for the surface force measurements, where PS-X denotes linear polystyrene chains with one end terminated by the zwitterionic group ($-(CH_2)_3N^+(CH_2)_2-(CH_2)_3SO_3^-$) and PS_m-PEO_n denotes polystyrene-poly(ethylene oxide) diblock copolymers (m and n are weight-averaged degrees of polymerization for polystyrene and poly(ethylene oxide) blocks, respectively, and n is small). These authors demonstrated that the strong interaction between the zwitterionic or ethylene oxide groups and the mica surface can also terminally anchor the non-adsorbing PS block, as a dangling chain, at the mica-solvent (*e.g.* toluene) interface. At a higher surface coverage the PS chains are shown to be strongly stretched, to give a monotonic repulsive interaction between two such polymer "brushes", consistent with theoretical predictions (Alexander, 1977; Milner *et al.*, 1988; Taunton *et al.* 1990).

Dai *et al.* (1995) have explored the conformational transitions of the end-adsorbed PS-PEO diblock copolymer chains by directly measuring the interactions of a *single* layer of the end-adsorbed PS-PEO diblock copolymer chains against a *bare* mica surface in pure toluene (good solvent) and pure cyclohexane (poor solvent, T_θ = 34 °C). The results are given in Figure 7.22 on a log-linear scale, in which no attraction is observed and strong repulsive forces are seen.

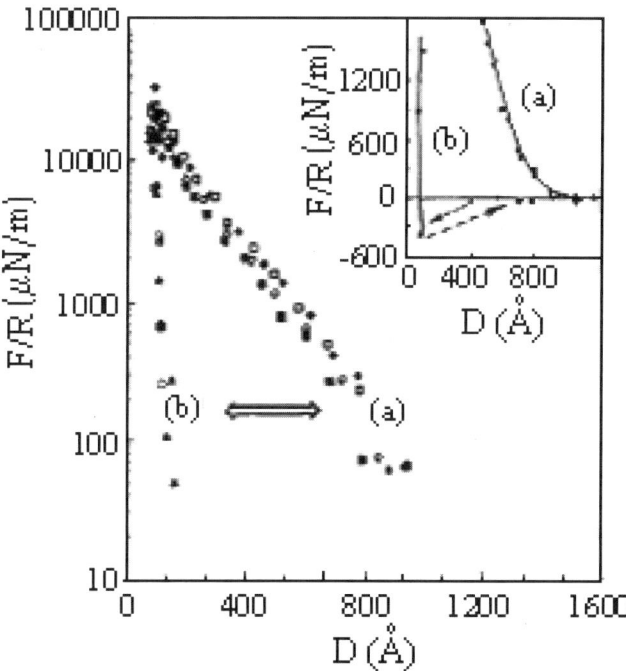

Figure 7.22. Force-distance profiles on a log-linear scale for an end-adsorbed PS_{1420}-PEO_{51} diblock copolymer single layer against a *bare* mica surface during the first few compression-decompression cycles: (a) in toluene before and after having been immersed in cyclohexane; (b) in cyclohexane. The open double-arrow indicates that the transition between (a) and (b) is practically reversible. The inset shows the corresponding force profiles as (a) and (b), respectively, on a linear-linear scale (After Dai *et al.*, 1995, copyright 1995 The American Chemical Society. Reproduced with permission)

The effective layer thickness of *ca.* 1000 Å deduced from Figure 7.22(a) indicates that the PS_{1420}-PEO_{51} chains are terminally attached to the mica surface via the short PEO segments with the PS chains extending into the solution to form a polymer brush. Figure 7.22(b) gives the force-distance curve that was measured on the same end-adsorbed PS layer after the toluene was thoroughly replaced by pure cyclohexane. In comparison with Figure 7.22(a), Figure 7.22(b) shows a steeper repulsive wall commencing at a much shorter separation distance, which demonstrates unambiguously that the end-adsorbed PS brush has collapsed in the poor solvent to form a more compact polymer layer at the cyclohexane/mica interface. Furthermore, the transitions between Figure 7.22(a) and Figure 7.22(b) with changing solvent quality are characterized by fast dynamics, suggesting potential applications, for example, in polymeric valves, sensors and liquid chromatography (Auroy and Auvray, 1992; Lai and Halperin, 1992).

Although the surface force apparatus serves as a good experimental technique for probing conformations and conformational transitions of macromolecular chains within the adsorbed polymer thin film at the liquid-solid interface, the force measurements cannot determine the distribution of segment density normal to the surface. Neutron reflection experiments provide information on the density profile normal to the interface (Higgins and Benoît, 1994; Penfold *et al.*, 1997). This is because the variation in specular reflection with wave vector transfer, (*i.e.* $Q = 4\pi \sin\theta/\lambda$, where λ is the neutron wavelength and θ is the glancing angle of incidence), depends on the neutron refractive index profile in the direction normal to the reflecting interface ((Higgins and Benoît, 1994; Penfold *et al.*, 1997). As shown in Figure 7.23, for radiation approaching the interface of two media with different refractive indices relative to vacuum (n_1 and n_2, respectively), the incident (glancing) and transmission angles, θ_i and θ_t, are related by Snell's Law (O'Shea *et al.*, 1993), in the form,

$$\frac{\cos\theta_i}{\cos\theta_t} = \frac{n_2}{n_1} = n_{12} \qquad (7.3)$$

when $n_1 > n_2$, then $n_{12} < 1$, and Equation (7.3) gives:

$$\cos\theta_t = \frac{\cos\theta_i}{n_{12}} > \cos\theta_i \qquad (7.4)$$

Therefore, a real value for θ_t is only obtained when $(\cos\theta_i)/n_{12} \leq 1$ (*i.e.* $\cos\theta_i \leq n_{12}$). The particular value of θ_c given by $\cos\theta_c = n_{12}$ is called the critical angle. When the glancing angle is less than the critical angle, so-called total external reflection takes place (Penfold *et al.*, 1997). However, it is interesting to note that reflection does not disappear completely when θ_c is exceeded and that the reflectivity *R*, defined as the intensity ratio of the reflected light to the incident light, *vs.* the wave vector transfer, *Q*, should depend on the refractive index profile in the direction normal to the reflecting interface, as mentioned above. On this basis, the segment density profiles of the surface-bound polymers normal to the interface was investigated by reflectometry, choosing neutrons as the radiation because of important contrast effects which can operate at the molecular level.

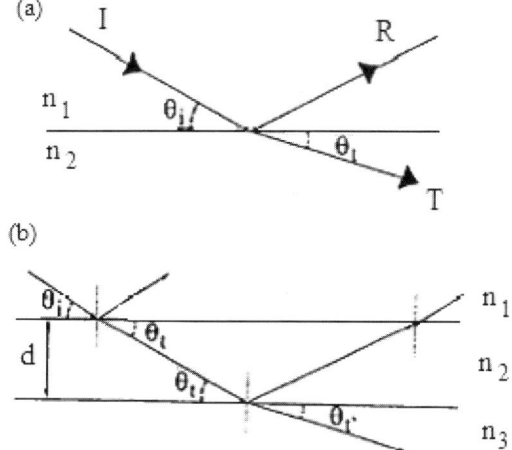

Figure 7.23. (a) A wave passing from a medium of refractive index n_1 to one with refractive index n_2; (b) the wave passes through a thin film of refractive index n_2 between two media of refractive indices n_1 and n_3.

Since neutrons resemble waves in their interaction with matter, they may be reflected or refracted at the boundary between two media with different neutron refractive indices. They may also be totally reflected, when the incident neutron beam is at glancing angles less than the critical angle. At a particular neutron wavelength, λ, the neutron refractive index profile (relative to a vacuum) of a medium of coherent scattering length, b_i (i = 1, 2, ···), is given by (Hughes, 1954):

$$n = 1 - \frac{\lambda^2 \sum_i b_i n_i(z)}{2\pi} = 1 - \frac{\lambda^2 \rho(z)}{2\pi} \tag{7.5}$$

where $n_i(z)$ is the number density of nuclei in the medium and $\rho(z)$ the coherent scattering length density profile. In the kinematic or Born approximation (Crowley et al., 1991), the reflectivity, R, is given by:

$$R(Q) = \frac{16\pi^2}{Q^4} |\rho'(Q)|^2 \tag{7.6}$$

where $\rho'(Q)$ is the Fourier transform of the gradient of the scattering length density profile normal to the interface. Exact calculations of the reflectivity profile for a given density profile can be made using optical matrix methods (Heavens, 1965).

7.5.4.1 Polymer Mushrooms

For terminally anchored polymer chains in a good solvent, two regimes are anticipated. If the grafting density is low, the tethered chains do not interact with each other and as a result exist as separate "mushrooms". If the grafting density is high, with a distance between anchor points less than the Flory radius of the chains (*i.e.* $s < R_F$), the polymers take up a more extended configuration to form a semi-dilute "brush" (Field *et al.*, 1992a, b). In the mushroom regime, each chain may be thought of as occupying roughly a hemisphere, with a radius comparable to the Flory radius for a polymer coil in a good solvent (Field *et al.*, 1992a, b; de Gennes, 1980). The density profile was demonstrated to be given by the Schultz function or a skew Gaussian function (Field *et al.*, 1992a, b):

$$\phi(z) = Az^n \exp(-\alpha z^\beta) \tag{7.7}$$

$$\phi(z) = \phi_o \frac{\exp[-(z-z_0)^2/2\delta_i^2] - \exp[-(1/2\eta_i^2)]}{1-\exp[-(1/2\eta_i^2)]} \tag{7.8}$$

Equation (7.7) becomes a Schultz function when $\beta = 1$. Equation (7.8) was constructed by joining two Gaussians of different widths of δ_1 and δ_2 centered at z_0, where $z_0 = \delta/\eta_1$. The parameter, η_i, allows the density profile to shift parallel to the $\phi(z)$-axis and some of the tail of the Gaussian to be removed so that the profile can look very much like the Schultz function.

Figure 7.24 shows the schematic diagram and model scattering length density profiles used to fit the reflectivity data of PS_{577}-PVP_{283} end-adsorbed chains to Equations (7.7) and (7.8), respectively.

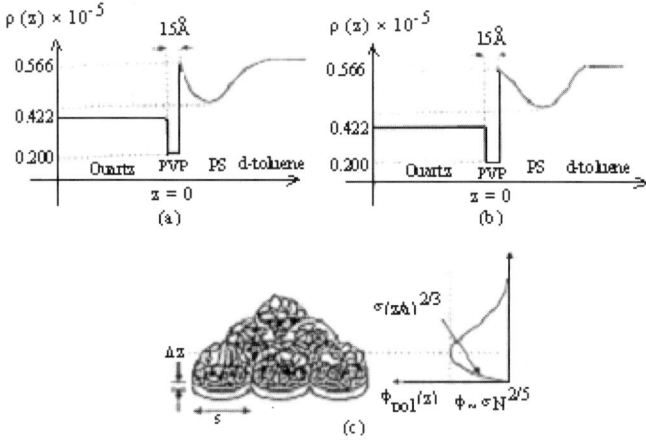

Figure 7.24. Model scattering length density profile used to fit the reflectivity data to (a) a Schultz function defined by Equation (7.7); (b) a skew Gaussian function defined by Equation (7.8); (c) schematic diagram representing PS-PVP block copolymer chains adsorbed at the solid-liquid interface from a selective solvent. Also included is the corresponding polymer density profile (After Field *et al.*, 1992b, copyright 1992 Editions Physique, Les Ulis Cedex)

The typical reflectivity profile, together with the model fitting, is given in Figure 7.25, which shows the reflectivity multiplied by Q^4, as a function of Q, as this gives a better indication of the fit quality and the asymptotic limit of the reflectivity at high Q. Also included in the inset of Figure 7.25 is the reflectivity vs. Q. The Schultz function and Gaussian density profiles for this particular system are almost identical in shape. These data could not be fit to the parabolic profile expected for polymer brushes, which are discussed below.

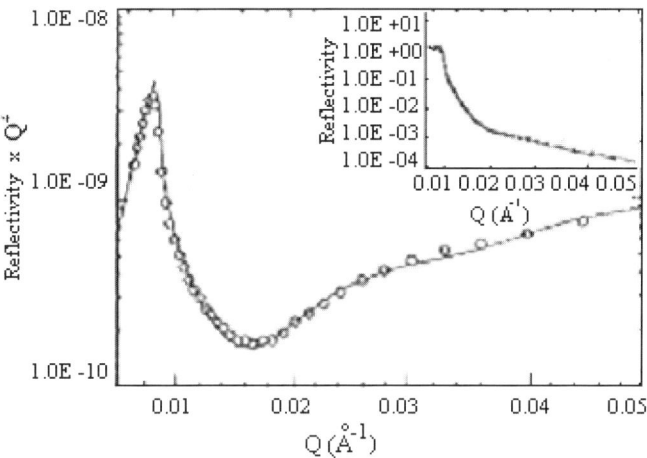

Figure 7.25. Neutron reflectivity data of a PS_{576}-PVP_{283} diblock copolymer and model fit based on the Schultz function defined by Equation (7.7) (After Field *et al.*, 1992b, copyright 1992, copyright 1992 Editions Physique, Les Ulis Cedex)

7.5.4.2 Polymer Brushes

In the brush regime, the end-adsorbed chains take a "tail" conformation to form polymer brushes (Dai *et al.*, 1995; Dai and Toprakcioglu, 1992; Field *et al.*, 1992a). Theories of polymer brushes fall into two main categories: scaling and mean field (de Gennes, 1979; Milner, 1991). In the scaling theory, each of the end-attached polymer chains is assumed to consist of connected semi-dilute "blobs" of diameter s at high surface densities ($s \ll R_F$, Flory radius) (Alexander, 1977; de Gennes, 1979). Chains in a polymer brush are stretched out by the osmotic repulsion between blobs, which is opposed by the increase in the elastic free energy of the chains when overstretched. Based on the scaling concept, Alexander expressed the equilibrium layer thickness of a polymer brush, $L=L_0$, as follows (Alexander, 1977):

$$L_0 = s(R_F/s)^{5/3} = Ns^{-2/3}a^{5/3} \tag{7.9}$$

where $R_F = N^{3/5}a$; a is the monomer size and N is the degree of polymerization.

de Gennes extended the scaling theory further to obtain the force per unit area $f(D)$ between two polymer-brush-coated parallel plates at a distance D apart (de Gennes, 1985):

$$f(D) = k_B T/s^3[(2L_0/D)^{9/4} - (D/2L_0)^{3/4}] \qquad D < 2L_0 \tag{7.10}$$

where k_B and T are Boltzmann's constant and the absolute temperature, respectively. The first term in the square brackets of Equation (7.10) derives from the osmotic repulsion while the second term represents the reduction in free energy due to compression of the overstretched chains.

As mentioned above, the SFA results are normally presented in terms of $F(D)/R$, which corresponds to the interaction energy per unit area, $E(D)$, between two flat parallel surfaces of the same nature and at the same separation, D, via the Derjaguin approximation (Derjaguin, 1934). Therefore, the experimental results can be directly compared with the scaling calculations from Equations (7.1), (7.2) and (7.10). (Dai et al., 1995; Taunton et al., 1990).

Apart from the scaling theory, various mean field calculations of the configuration of adsorbed chains have been reported (Milner, 1991). The main approaches to calculating polymer density profiles of end-attached polymer brushes use essentially a self-consistent mean field (SCF) theory. In particular, Milner et al. (1988) made use of the fact that the chains in the brush are strongly stretched (i.e. in the semi-dilute regime) and may be assumed to behave classically. Relating the mean field, which is the effective chemical potential, to the free energy in the classical limit gives a parabolic density profile, rather than the step function assumed by Alexander and de Gennes in their scaling calculations (Alexander, 1977; de Gennes, 1979). In view of the parabolic density profile predicted by theory, Field et al. (1992a, b) used the following density or volume fraction profiles to fit their neutron reflectivity data:

$$\phi(z) = \phi_0[1 - (z/L_0)^n] \tag{7.11}$$

$$\phi(z) = (\phi_0/2)\{1 - \text{erf}[(z - L_0)/(2\delta_z)]\} \tag{7.12}$$

where erf is the error function and δ_z is the roughness of the layer. The value of n in the polynomial profile defined by Equation (7.11) determines the shape of the profile. n is 1 for a linear profile, 2 for a parabolic profile, and tends to infinity in the case of a step function. For $0 < n < 1$ the profile becomes concave upwards. The parameter ϕ_0 defines the volume fraction at the liquid/solid interface and L_0 defines the thickness of the adsorbed layer where $\phi(L_0) = 0$.

A representative fitting curve for the reflectivity data from PS_{1700}-PEO_{167} is shown in Figure 7.26(a). In the inset of Figure 7.26(a) the reflectivity multiplied by Q^4 is plotted as a function of Q. The density profiles correspond to Equations (7.11) and (7.12) are given in Figure 7.26(b), which shows that the two density profiles are almost identical in the region close to the surface. In the region at the outer extremity of the brush, however, the error function type profile consists of a parabolic shape with an approximately Gaussian tail, indicating that the classical parabolic density profile (Milner, et al. 1988) requires a correction due to the effects of the finite chain length and/or its polydispersity (Field et al., 1992a). Overall, these reflectivity data indicate that the polymer density profile normal to the substrate is well described by either the parabolic or the error function, which is in good agreement with the theoretical prediction for semi-dilute polymer brushes.

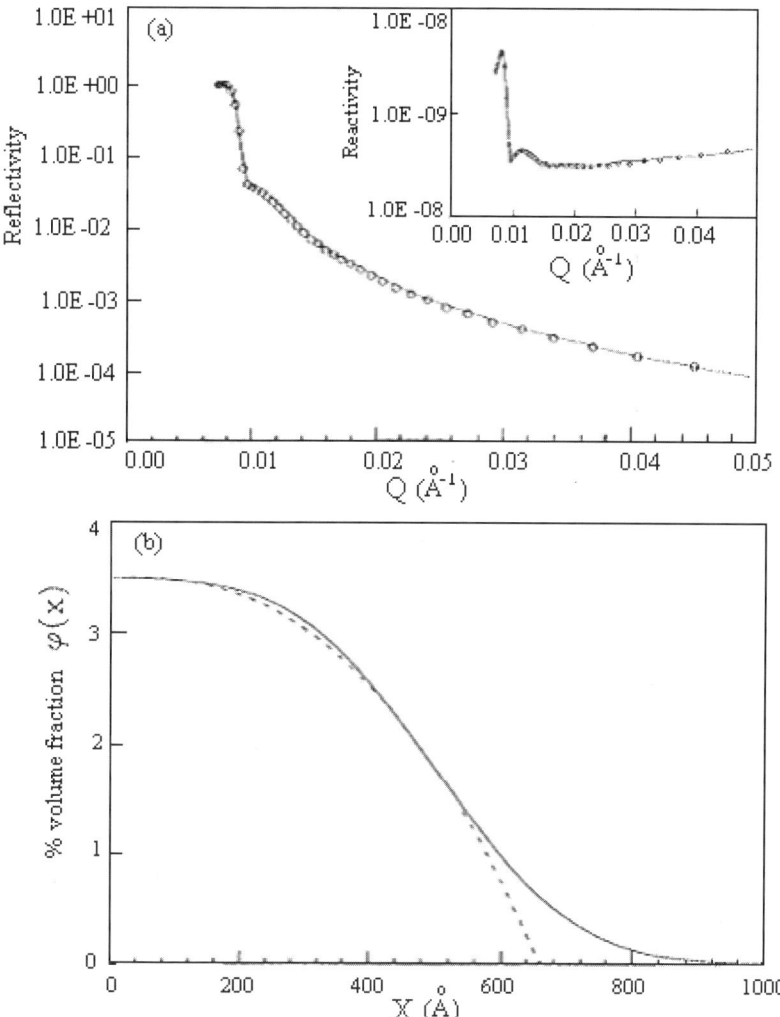

Figure 7.26. (a) Neutron reflectivity profile of an end-adsorbed PS_{1700}-PEO_{167} from toluene-D8 onto quartz fitted with Equation (7.11). The inset shows the same profile as (a) plotted as RQ^4 vs. Q; (b) error function (——) and parabolic (- - -) density with the parameters: $\phi_0 = 0.035$, $t = 500$ Å, and $\delta = 170$ Å (After Field et al., 1992a, copyright 1992 The American Chemical Society. Reproduced with permission)

More recent neutron reflectivity measurements on telechelic *triblock* copolymer chains have demonstrated that (Anastassopoulos et al., 1998), to a very good approximation, ABA type end-tethered chains behave in a manner equivalent to their AB-type counterparts of half their molecular weight and they adsorb predominantly in a loop conformation for copolymers with sufficiently strong sticking A blocks, consistent with the SFA measurements discussed above (Dai *et al.*, 1995; Dai and Toprakcioglu, 1992).

7.6 Nanostructured Polymers with Special Architectures

Just as microchips have revolutionized computers and electronics, nanotechnology has the potential to revolutionize many industrial sectors. As discussed above, various polymer nanostructures (*e.g.* polymer nanoparticles, nanofibers, nanofilms) have been prepared to possess interesting physicochemical properties for device applications. However, considerable efforts are required to design, construct and operate new devices at the nanometer scale, as the technology for making proper connections from nanoscale entities to the macroscale world still remains a big challenge. In particular, electrically conducting nanofibers with diameters at the nanometer scale and lengths in kilometers are both useful "building blocks" for nanodevice construction and ideal connections between the nanoscale entities and the macroscale world. Effective fabrication of *ordered* nanostructured polymers is a pre-requisite for the intergration of nanostructured materials into useful "building blocks" and various functional units/nanodevices of practical significance. As we shall see below, techniques, including colloidal self-assembling, phase separation, and nanopatterning, have been devised to construct polymer materials and functional systems of multi-dimensional *ordered* nanostructures.

7.6.1 Self-assembly of Ordered Nanoporous Polymers

Polymers with periodically modulated mechanical, optical, electronic or magnetic properties have potential applications in many areas, including sensors, displays, microelectronics, data storage devices and non-linear optics (Murray *et al.*, 1995; Tsukruk, 1997). For instance, polymers with periodic dielectric media, like photonic crystals, could be used to manipulate and control light (Markel and George, 2001; Zakhidov *et al.*, 1998). Therefore, the fabrication of thin films with three-dimensional (3D) periodic arrays by incorporation of large quantities of monodisperse domains in an organized fashion within appropriate matrices or using self-organization (*i.e.* self-assembling) of colloid nanoparticles has rapidly become the forefront of development in materials science. Consequently, many novel nanostructured polymers of multi-dimensional periodic structures have been reported. To mention but a few examples, Kumacheva and co-workers (Kumacheva *et al.*, 1999) have used the self-assembling approach to producing nanostructured polymers with a 3D periodic structure (Figure 7.27).

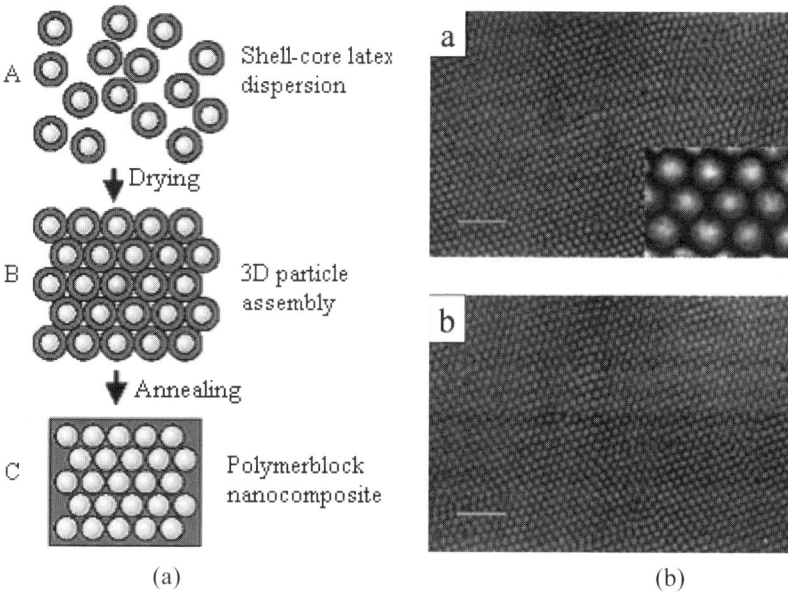

Figure 7.27. (a) Schematic representation of the approach for producing nanostructured polymers with 3D ordered structure: A) synthesis of the shell-core polymer nanoparticles with hard functional cores and soft inert shells; B) self-assembling of the polymer nanoparticles into a 3D compact; C) heat treatment of the compact to induce melting of the soft shells to form polymer nanocomposite with nanostructured spheres. (b) Laser confocal fluorescent microscopy images of the surface and the bulk morphology of the 3D nanostructured polymer films. *Top:* Composite morphology at the surface of the sample, inset shows the same structure with the width of the area being 3.3 μm. *Bottom:* Polymer structure in the layer located 30 mm below the surface. Scale bar is 10 μm (After Kumacheva *et al.*, 1999, copyright 1999 Wiley-VCH Verlag)

As can be seen in Figure 7.27(a), Kumacheva *et al.* (1999) first synthesized a shell-core latex consisting of a hard core-forming polymer and a soft shell-forming polymer. They then packed these shell-core latex particles into thin layers of 3D structures by, for example, precipitation, centrifugation or electrodeposition followed by latex drying. The resulting materials were finally annealed at high temperature to melt the soft shells, leading to the formation of nanostructured polymer thin films consisting of 3D nanoparticle assemblies. The laser confocal fluorescent microscopy images of the surface and the bulk morphology of the 3D nanostructured polymer films thus prepared are given in Figure 7.27(b), which shows a highly ordered structure in the z-direction as well as in the lateral domains. These highly ordered nanostructured polymers with 3D periodic structures have been demonstrated to be promising media for data storage.

On the other hand, Jenekhe and Chen (1999) have prepared multi-dimensional ordered mesoporous polymer materials by self-assembling hollow spherical micelles made by self-organized rod-coil diblock copolymers in a selective solvent. Figure 7.28(a) schematically shows the approach used with the morphology of the resulting materials given in Figure 7.28(b). Although microporous were observed in Figure 7.28(b) in this particular case, the diameter, periodicity and wall thickness of the ordered arrays of spherical holes were demonstrated to depend on the copolymer molecular weight and composition, indicating the possibilities of forming porous polymer films with nanostructured holes for a variety of applications ranging from size-/shape-selective separation to photonic modulation.

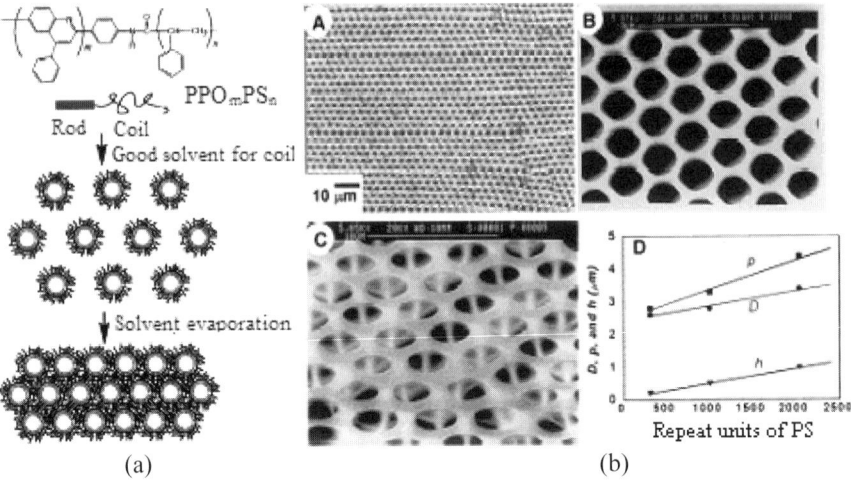

Figure 7.28. (a) Schematic illustration of its hierarchical self-assembly into ordered microporous materials; (b) polarized optical and (B and C) scanning electron micrographs of microporous micellar films obtained from diblock solution by solution casting on a glass slide (A) and an aluminum substrate [B and C]; (D) the molecular weight dependence on geometric parameters of the pores: the diameter, D; minimum wall thickness, h; center-to-center hole periodicity, p ($h = p - D$) (After Jenekhe and Chen, 1999, copyright 1999 American Association for the Advancement of Science)

7.6.2 Coaxial Polymer Nanowires and Nanofibers

The formation of ordered structures from metal nanowires and carbon nanotubes has been an active research area for some years. Various nano- and micro-fabrication techniques have been developed. Examples of these include the use of electric and magnetic fields for aligning suspended metallic nanowires (Duan *et al.*, 2001; Smith *et al.*, 2000; Tanase *et al.*, 2001), the combination of fluidic alignment with patterned surface structures (*e.g.* microchannels) for nanowire patterning (Huang Y. *et al.*, 2001; Messer *et al.*, 2000), and the lithographic patterning of aligned carbon nanotubes (Huang S. and Dai, 2002). The effective fabrication of ordered polymer nanowires or nanofibers is a more recent development. While a few aligned polymer nanowires prepared by synthesizing them at a scanning microscope tip or by using a template were described above, the possibility of forming aligned *electrospun* nanofibers by using a rotating cathode collector has recently been reported (Matthews, *et al.*, 2002; Theron *et al.*, 2001). Non-conducting electrospun polymer fibers have also been used as "templates" for coating with appropriate conducting polymers and for deposition of a metal layer from solution or vapor (Bognitzki et al., 2001). MacDiarmid et al., (2001) reported the uniform coating of an electrospun polyacrylonitrile nanofiber with a layer of conducting polypyrrole (20 to 25 nm thick) by immersing the electrospun fibers in an aqueous solution of polymerizing polypyrrole. These authors also prepared gold coated polyacrylonitrile nanofibers through treatment of the electrospun fibers with a solution of AuS_2O_3 and ascorbic acid.

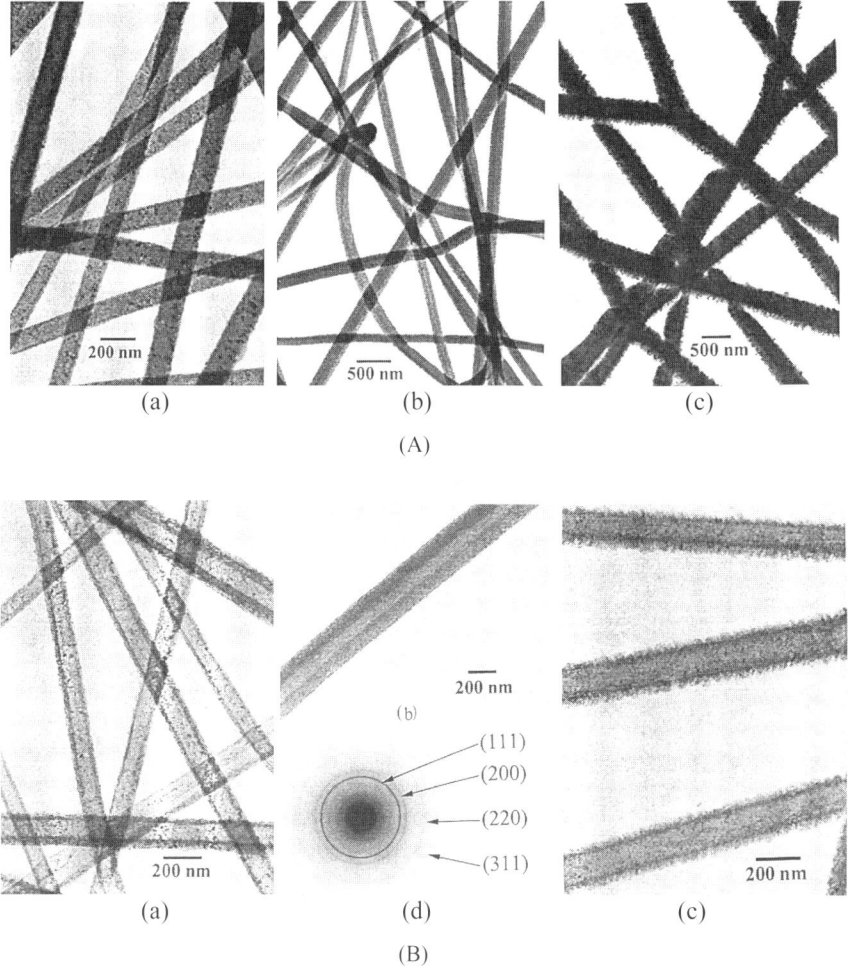

Figure 7.29. (A) (a–c) TEM images of aluminum-coated poly(*meta*-phenylene isophthalamide) electrospun nanofibers with the coating thickness increasing from about 10 nm in (a) to nearly 100 nm in (c); (B) (a–c) TEM images of Al nanotubes prepared by solvent dissolution of the poly(*meta*-phenylene isophthalamide) core. (d) Electron diffraction pattern of nanotubes associated with the aluminum nanotubes shown in Figure 7.29(B)(a) (After Liu, W. *et al.*, 2002, copyright 2002 Materials Research Society)

Reneker and co-workers (Liu W. *et al.*, 2002) coated electrospun poly(*meta*-phenylene isophthalamide) nanofibers with sub-nanometer-thick coatings of various other materials (*e.g.* carbon, Cu and Al) by chemical and physical vapor deposition. They also prepared nanotubes consisting of pure aluminum or mixtures of aluminum and aluminum oxide by coating the electrospun poly(*meta*-phenylene isophthalamide) nanofiber with an aluminum layer, followed by selectively removing the polymer nanofiber core via solvent dissolution or thermal degradation (Liu *et al.*, 2003). During thermal degradation of the poly(*meta*-phenylene isophthalamide) nanofiber cores, the aluminum coated layer was subjected to a limited degree of oxidation, which produced nanotubes of mixed aluminum and aluminum oxide. The aluminum coated layers did not oxidize when the template fiber core was removed by dissolution. Figures 7.29(A)(a-c) show TEM images of aluminum-coated poly(*meta*-phenylene isophthalamide) fibers with the coating thickness increasing from *ca.* 10 to 100 nm. As seen in Figures 7.29(B)(a–c), aluminum nanotubes remained after the dissolution of the fiber cores of poly(*meta*-phenylene isophthalamide) with N,N-dimethylacetamide solvent. The electron diffraction pattern associated with the aluminum nanotubes shown in Figure 7.29(B)(a) is given in Figure 7.29(B)(d), in which all the d-spacings are characteristic of the aluminum crystal unit cell.

Another interesting area closely related to the fabrication of polymer nanowires and nanofibers is the synthesis of coaxial nanowires of polymers and carbon nanotubes. As described in Chapters 5 and 6, Dai and co-workers have recently prepared large-scale aligned carbon nanotubes perpendicular to the substrate surface by pyrolysis of iron (II) phthalocyanine ($FeC_{32}N_8H_{16}$, designated as FePc) under Ar/H_2 at 800-1100 °C and developed a novel approach for chemical modification of aligned carbon nanotubes (Chen Q. *et al.*, 2001). Radio-frequency glow-discharge plasma treatment activated the surface of the nanotubes for subsequent reactions characteristic of the plasma-induced surface groups. These authors successfully grafted polysaccharide chains onto plasma-activated aligned carbon nanotubes through Schiff-base formation, followed by reductive stabilization of the Schiff-base linkage with sodium cyanoborohydride. In addition to the chemical grafting of polymer chains onto the carbon nanotube surface, they have also recently used the aligned carbon nanotubes as nanoelectrodes for making novel conducting coaxial nanowires by electrochemically depositing a concentric layer of an appropriate conducting polymer uniformly onto each of the aligned nanotubes to form the aligned conducting polymer coated carbon nanotube coaxial nanowires (CP-CNT) (Gao M. *et al.*, 2000a, see also Figure 7.25). The presence of the conducting polymer layer was also clearly evident in transmission electron microscopic (TEM) images (Gao M. *et al.*, 2000a) (Figure 7.30). The aligned CP-CNT coaxial nanowires were demonstrated to show much stronger redox responses than the conventional conducting polymer electrode (Gao M. *et al.*, 2000a). The coaxial structure allows the nanotube framework to provide mechanical stability (Gao R. *et al.*, 2000b; Poncharal *et al.*, 1999), and efficient thermal and electrical contact with the conducting polymer layer (Odom *et al.*, 2000; Frank *et al.*, 1998). The large interfacial surface area per unit mass obtained for the nanotube-supported conducting polymer layer is potentially useful in many optoelectronic applications, for example in sensors, organic light-emitting diodes, and photovoltaic cells where the charge injection and separation are strongly limited by the interfacial area available in more conventional devices (Dai, 1999).

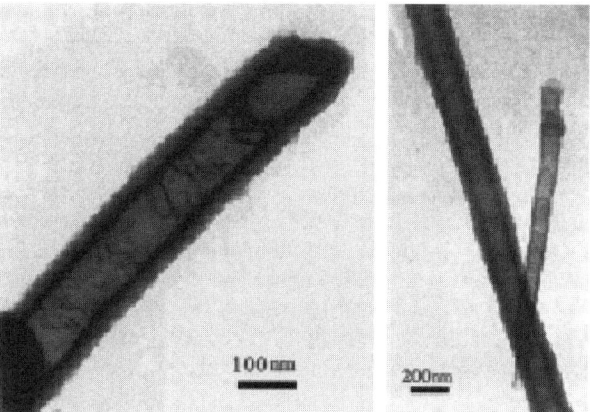

Figure 7.30 TEM images of the CP-CNT coaxial nanowires (After Gao M. *et al.*, 2000a, copyright 2000 Wiley-VCH Verlag. Reproduced with permission)

To demonstrate the potential sensing applications for the CP-CNT coaxial nanowires, Dai and co-workers (Gao M. *et al.*, 2003) have also immobilized glucose oxidase (GO_X) onto the aligned carbon nanotube substrate by electropolymerization of pyrrole in the presence of glucose oxidase. The glucose oxidase-containing polypyrrole-carbon nanotube coaxial nanowires were used to monitor the concentration change of hydrogen peroxide (H_2O_2) generated from the glucose oxidation reaction by measuring the increase in the electro-oxidation current at the oxidative potential of H_2O_2 (*i.e.* the amperometric method) (Gao M. *et al.*, 2003). The amperiometric response was found to be much higher than that of more conventional flat electrodes coated with glucose oxidase-containing polypyrrole films under the same conditions. The CP-CNT nanowire sensors were also demonstrated to be highly selective for glucose with their amperiometric responses being almost unchanged even in the presence of some interference species including ascorbic acid, urea and D-fructose. Therefore, the CP-CNT nanowires could be used for making new glucose sensors with a high sensitivity, selectivity and reliability, which is clearly an area in which future work would be of value.

7.6.3 Multilayered Polymer Nanofilms

Polymer-based multilayer thin films have recently received significant attention due to their potential use as photonic crystals for manipulation and control of light. Indeed, certain multilayer polymer nanostructures with the phase separated lamellar domains of different refractive indices have been used as 1D optical reflectors for reflecting a band of frequencies for both TE and TM polarized light, regardless of angle of incidence (Cumpston et al., 1999; Lansac et al., 1999). Apart from the multilayer film formation by polymer phase separation to be discussed below, the construction of multilayer polymer nanofilms has also been achieved by the LB technique and a layer-by-layer self-assembly process developed by Rubner et al. (Ferreira and Rubner, 1995). The principle of the layer-by-layer process lies in the alternate spontaneous adsorption of oppositely charged polymers from dilute solutions. This technique has been successfully used to generate multilayer thin films comprised of sequentially adsorbed layers of partially doped polyaniline and a polyanion (e.g. sulfonated polystyrene), leading to conductivities comparable to those obtained with spin cast films (0.5–1.0 S/cm) after having been doped with strong acids such as HCl and methanesulfonic acid (Cheung et al., 1997). Based on the layer-by-layer adsorption process, Rubner and co-workers (Ferreira et al., 1996; Fou et al., 1995; Onitsuka et al., 1996) have also constructed multilayer LEDs from PPV and poly(styrene sulfonic acid), SPS, or poly(methacrylic acid), PMA, and found that the type of polyanion used has a significant effect on the performance of the LED devices fabricated with Al and ITO as electrodes. In particular, LEDs fabricated from PMA/PPV multilayers were found to exhibit luminance levels in the range of 20–60 cd/m^2 with a thickness-dependent turn-on voltage and rectification ratios greater than 10^5, whereas the SPS/PPV counterparts showed nearly symmetric I–V curves with a thickness-independent turn-on voltage and much lower luminance levels. The observed difference in the device performance can be attributed to a doping effect associated with the sulfonic acid groups in SPS. Furthermore, these authors have recently extended the layer-by-layer adsorption process to include the hydrogen-bonding interactions between the polyaniline and poly(vinyl pyrrolidone), poly(vinyl alcohol), poly(acrylamide), or poly(ethylene oxide) (Stockton and Rubner, 1997). By using pre-ordered/pre-patterned substrates, the layer-by-layer absorption process should, in principle, lead to the construction of oriented/patterned conjugated polymers.

Another interesting area closely related to the polymer-polymer multilayer structure is the intercalation of polymer chains into the layered structures of clay nanoparticles, leading to the formation of organic-inorganic hybrid multilayer films (Giannelies, 1996). Organic-inorganic hybrid composites constitute a new class of materials, possessing properties characteristic of both constituent components with potential synergetic effects (Fendler and Meldrum, 1995; Mark et al., 1995; Ozin, 1992). Owing to their unusual properties, organic-inorganic hybrid nanocomposites have attracted increased attention in recent years. Consequently, various polymers including poly(ethylene oxides) (Aranda and Ruiz-Hitzky, 1992; Ruiz-Hitzky and Aranda, 1990), poly(olefins) (Johnson et al., 1997), polyimide (Yano et al., 1997), polypyrrole (Ramachandran et al., 1997), and polyaniline (Carrado and Xu, 1998; Dai et al., 2000) have been incorporated into clay nanoparticles through either a solution or a melt intercalation process (Ruiz-Hitzky, 1993). The most widely used clay is the montmorillonite, MMT, which consists of two silica tetrahedral and an alumina octahedral sheet stacked into the layered structure with a gallery gap of ca. 1 nm between the layers. The galleries are normally occupied by cations (e.g. Na^+, Ca^{2+}, Mg^{2+}), which can be easily replaced through an alkylammonium ion-exchange reaction to form the so-called organo-clay. The alkylammonium ions in the organo-clay layered structure facilitates the intercalation of polymer chains. At high intercalation levels, the clay may exfoliate into their nanoscale building blocks to disperse uniformly in the polymer matrices, forming exfoliated polymer-clay nanocomposites. The organic-inorganic hybrid nanocomposites thus prepared have been demonstrated to show improved environmental stability, mechanical strength and lower permeability for gases with respect to corresponding pure polymers (Chang and Whang, 1996; Carlos et al., 1999; De Morais et al., 1999).

On the other hand, certain electroluminescent organic-inorganic hybrid materials have also been prepared by sol-gel chemistry (Brinker and Scherrer, 1989; De Morais et al., 1999). In this context, Dai and co-workers (Winkler et al., 1999) have recently intercalated conjugated conducting and light-emitting polymers into clay nanoparticles at an intercalation level below the critical value required for the clay exfoliation to explore the conformational effects on their optoelectronic properties. As is well known, the band gap energies of conjugated polymers, and hence the related optoelectronic properties, depend strongly on their chain conformations (Skotheim et al., 1998). Therefore, the conformational changes of light-emitting polymer backbones could be exploited as an alternative approach for color tuning, in addition to chemical modification of the polymer structure, in polymer light-emitting diodes. In particular, Dai and co-workers (Winkler et al., 1999) have intercalated light-emitting poly[1,4-(2,5-bis(1,4,7,10-tetraoxaundecyl)phenylene vinylene], EO_3-PPV, into clay nanoparticles for light-emitting measurements.

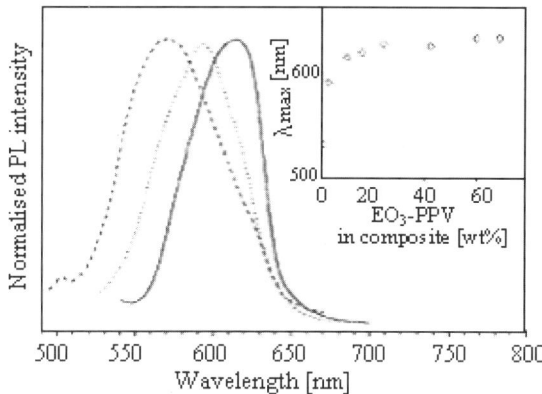

Figure 7.31. PL emissions from the EO_3-PPV intercalated clay nanoparticles at different intercalation levels (After Winkler et al., 1999, copyright 1999 Kluwer Academic Publishers. Reproduced with permission)

The intercalation process is experimentally revealed by a steady increase in the height of the clay galleries as determined by X-ray diffraction (Winkler et al., 1999). The photoluminescent, PL, spectra of the composites at several different intercalation levels are given in Figure 7.31. As can be seen, the PL emission shows a red-shift with increasing intercalation level. Included in the inset of Figure 7.31 is the dependence of the PL peak position (λ_{max}) on the polymer content, which shows that the λ_{max} continuously increases with increasing polymer content to a limiting value of *ca.* 610 nm, characteristic of EO$_3$-PPV, at *ca.* 20% (w/w), then remains unchanged despite further intercalation. The above features could be attributed to an intercalation-induced conformational transition from a "compact coil" to an "expanded coil", which should lead to an increase in the effective conjugation length, and hence a concomitant decrease in the band gap energy. The results from a single-layer LED device based on the EO$_3$-PPV intercalated nanoparticles showed a similar EL spectrum to the PL emission. As expected, the similarity between the EL and PL emissions suggested that the same singlet excitons were generated upon both PL- and EL-excitation (Chapter 9).

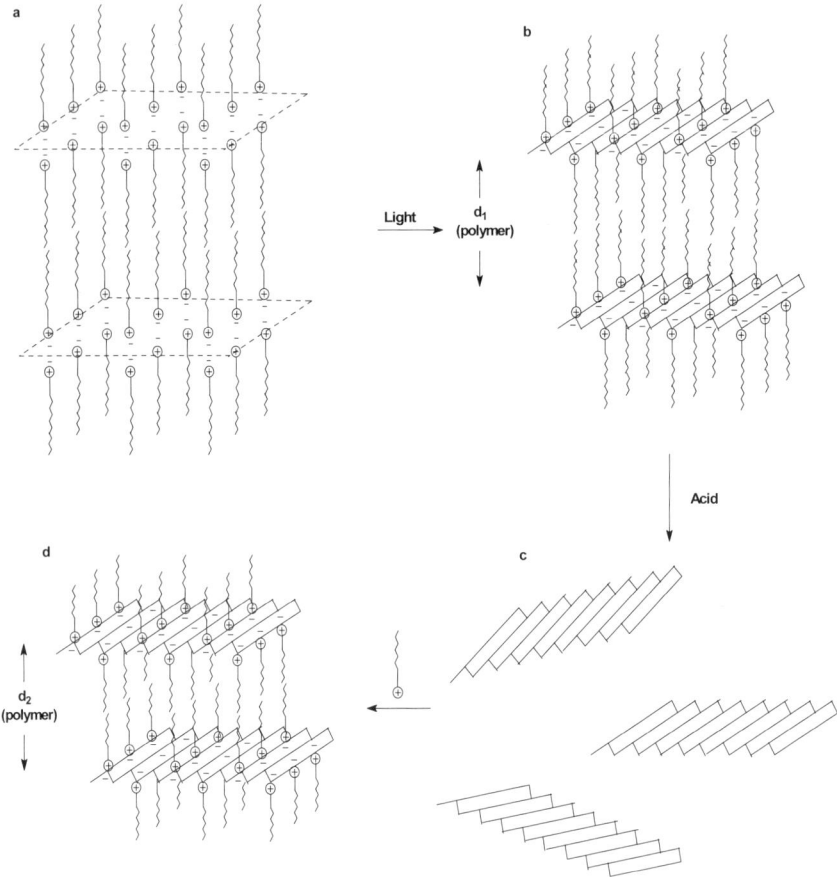

Figure 7.32. Schematic representation showing how to build molecular sandwiches with different fillings (After Matsumoto *et al.,* 2000, copyright 2000 Macmillan Magazines Limited)

Closely related to the preparation of layered nanocomposites consisting of polymer thin films interspersed with clay sheets, Matsumoto and co-workers (Matsumoto *et al.*, 2000) have moved a step further forward by showing novel polymer sandwiches with different fillings. In so doing, these authors first created layered structures consisting of layers of muconate anions (*i.e.* reactive dienes with a carboxylate group at either end) sandwiched between layers of alkylammonium cations (Figure 7.32). They then polymerized the layered muconate anions into molecule-thick polymer sheets by exposing to ultraviolet light. Upon exposure of the "synthetic clay" to acid, the alklammonium cations were removed to free the uncharged polymer sheets. The released polymer sheets can be "glued" together again by one of the many alkylammonium ions for a large variety of potential applications, ranging from molecular recognition to chemical purification.

7.6.4 Nanostructured Polymers by Phase Separation

Apart from the above-discussed *isolated* polymer nanostructures (*e.g.* nanoparticles, nanofibers and nanofilms) and their assemblies, nanostructured polymer domains consisting of an intrinsic part of the polymer morphology, which hence cannot be separated from other features of the polymer material, have also been prepared through, for example, polymer phase separation and nanopatterning. Like the self-assembling technique discussed above, polymer phase separation plays an important role in the fabrication of nanostructured polymers, as it provides the opportunity for nanoscale structuring that otherwise is difficult by lithographic techniques. As discussed in Chapter 6, the conditions necessary for microphase separation in immiscible polymer mixtures depend on their molecular architectures, the nature of monomers, the compositions and their molecular weights. By varying polymer structures and phase separation conditions, various well-defined microdomain structures with sizes typically in the range of a few tens of nanometers have been observed in polymer materials

Among these well-defined microdomain structures observed in phase separated polymer systems, cylindrical microdomains formed by block copolymers are of particular interest as the selective removal of either the minor or major component can transform thin polymer films into an array of nanopores or nanoscopic posts. The resultant polymer nanostructures could be used as membranes for molecular separation and/or as templates for preparing electronic and magnetic nanostructured materials.

The phase separation of block copolymers has been the subject of intensive study over the past several decades (Chapter 6). It has recently been demonstrated that cylindrical microdomains in certain phase separated block copolymer thin films can be oriented normal to the substrate surface by either controlling interfacial interactions between the copolymer and substrate with anchored random copolymers (Huang S. *et al.*, 1998; Mansky *et al.*, 1997) or by applying an external electric field normal to the surface (Morkved *et al.*, 1996; Thurn-Albrecht *et al.*, 2000(a, b). Figures 7.33(A) and (B) show AFM images for a phase separated asymmetric polystyrene and poly(methyl methacrylate) (PS-*b*-PMMA, *ca.* 30% volume fraction of PMMA) block copolymer thin film (40 nm thick) spin-cast onto Si substrates pre-coated with a random copolymer of PS and PMMA having a PS fraction of 0.60 (Thurn-Albrecht *et al.*, 2000c). Cylindrical microdomains oriented in the direction perpendicular to the substrate are clearly evident, especially in the phase image of Figure 7.33(B). The corresponding AFM height and phase images for the same PS-*b*-PMMA film after selectively removing the PMMA component by deep UV radiation and dissolution in a selective solvent (*i.e.* acetic acid for PMMA) are given in Figures 7.33(C) and (D), respectively. Both show an ordered array of circular nanoholes perpendicularly oriented in the polymer film.

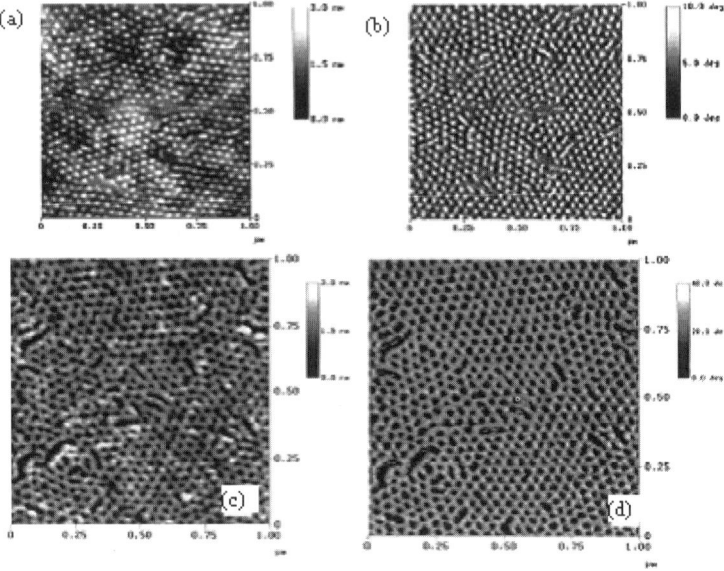

Figure 7.33. AFM image for a PS-*b*-PMMA block copolymer thin film on a neutral substrate after annealing at 170 °C in tapping mode before [(A) shows the height image, (B) the phase image] and after [(C) shows the height image, (D) the phase image] removal of the PMMA cylinders. Holes at the original locations of the cylinders are clearly seen (After Thurn-Albrecht *et al.*, 2000c, copyright 2000 Wiley-VCH Verlag)

By adding PMMA homopolymers with appropriate molecular weights into the PS-*b*-PMMA matrices, Russell *et al.* and Hawker and co-workers (Jeong *et al.*, 2000) were able to tune the size of the cylindrical microdomains (and hence the circular holes) formed in the block copolymer films without perturbing their spatial order and orientation. They found that pore diameters either larger or smaller than that achievable with the pure copolymer could be obtained by selective removal of the homopolymer and corresponding block of the copolymer or the homopolymer only. By so doing, both 6 and 22 nm diameter pores have been prepared from the same block copolymer film.

More interestingly, these authors have recently used the phase separated PS-*b*-PMMA nanoporous films and nanoscopic posts as templates for producing metal nanodots and nanoporous metal films, respectively, by evaporation metal (*e.g.* Cr,

Au) thin films on these polymer nanostructures, followed by selective removal of the polymer templates (Shin *et al.*, 2002).

7.7 References

Alexander, S. (1977) *J. Phys.* (Paris) **38**, 983.
Al-Khouri Fallouh, A.-K. N., Roblot-Treupel, L., Fessi, H., Devissaguet, J.P., Puisieux, F. (1986) *Int. J. Pharm.* **28**, 125.
Allemann, E., Leroux, J.-C., Gurny, R. (1998) *Adv. Drug Delivery Rev.* **34**, 171.
Ando, M., Watanabe, Y., Iyoda, T., Honda, K., Shimidzu, T. (1989) Thin Solid Films **179**, 225.
Anastassopoulos, D.L., Vradis, A.A., Toprakcioglu, C., Smith, G.S., Dai, L. (1998) *Macromolecules* **31**, 9369.
Aranda, P., Ruiz-Hitzky, E., (1992) *Chem. Mater.* **4**, 1395.
Aoki, A., Miyashita, T. (1997) *Adv. Mater.* **9**, 361.
Auroy, P., Auvray, L. (1992) *Macromolecules* **25**, 4134.
Bhushan, B., Israelachvili, J.N., Landman, U (1995) *Nature* **374**, 607.
Birnbaum, D.T., Kosmala, J.D., Henthorn, D.B., Peppas, L.B., (2000) *J. Control. Rel.* **65**, 375.
Birrenbach, G., Speise, P.P. (1976) *J. Pharm. Sci.* **65**, 1763.
Blomberg, S., Ostberg, S., Harth, E., Bosman, A.W., Horn, B.V., Hawker, C.J. (2002) *J. Polym. Sci.*, Part A **40**, 1309.
Bognitzki, M., Czado, W., Frese, T., Schaper, A., Hellwig, M., Steinhart, M., Greiner, A., Wendorff, J.H. (2001) *Adv. Mater.* **13**, 70.
Bolognesi, A., Bajo, G., Paloheimo, J., Östergard, T., Stubb, H. (1997) *Adv. Mater.* **9**, 121.
Borgwarth, K., Ricken, C., Ebling, D.G., Heinze, J. (1995) *Ber. dt. Bunsen. Ges. Phys. Chem.* **99**, 1421.
Brinker, C.J., Scherrer, G. (eds) (1989) *Sol-Gel Science: Physics and Chemistry of Sol-Gel Processing*, Academic Press, San Diego.
Bumm, L.A., Arnold, J.J., Cygan, M.T., Dunbar, T.D., Burgin, T.P., Jones II, L., Allara, D.L., Tour, J.M., Weiss, P.S. (1996) *Science* **271**, 1705.
Cai, Z., Lei, J., Liang, W., Menon, V., Martin, C.R. (1991) *Chem. Mater.* **3**, 960.
Cantor, C. R., Schimmel, P. R. (1980) *Biophysical Chemistry*, W.H. Freeman and Co., New York.
Carlos, L.D., De Zea Bermudez, V., Sa Ferreira, R.A., Marques, L., Assuncaq, M. (1999) *Chem. Mater.* **11**, 581.
Carrado, K.A., Xu, L. (1998) *Chem. Mater.* **10**, 1440.
Caruso, F. (2001) *Adv. Mater.* **13**, 11, and references cited therein.
Cepak, V.M., Martin, C.R. (1999) *Chem. Mater.* **11**, 1363.
Chang, W.P., Whang, W.-T. (1996) *Polymer* **37**, 4229.
Chen, Q., Dai, L., Gao, M., Huang, S., Mau, A.W.H. (2001) *J. Phys. Chem. B* **105**, 618.
Chen, J.T., Thomas, E.L., Ober, C.K., Mao, G.P. (1996) *Science* **273**, 343.
Cheung, J.H., Punkka, E., Rikukawa, M., Rosner, R.B., Royappa, A.T., Rubner, M.F. (1992) *Thin Solid Films* **211**, 246.
Cheung, J.H, Stockton, W.B., Rubner, M.F. (1997) *Macromolecules* **30**, 2712.
Chun, I., Reneker, D.H., Fong, H., Fang, X., Deitzel, J., Tan, N.B., Kearns, K. (1996) *J. Adv. Mater.* **31**, 36.
Cimrová, V., Remmers, M., Neher, D., Wegner, G. (1996) *Adv. Mater.* **8**, 146.
Couvreur, P., Kante, B., Roland, M., Goit, P., Bauduin, P., Speiser, P. (1979) *J. Pharm. Pharmacol.* **31**, 331.

Crowley, T.L., Lee, E.M., Simister, E.A., Thomas, R.K., Penfold, J., Rennie, A.R. (1991) *Colloids Surf.* **52**, 85.
Cumpston, B.H., Ananthavel, S.P., Barlow, S., Dyer, D.L., Ehrlich, J.E., Erskine, L.L., Heikal, A.A., Kuebler, S.M., Lee, I.-Y.S., Sandy, M-M., Qin, D.J., Rockel, H., Rumi, M., Wu, X-L., Marder, S.R, Perry, J.W. (1999) *Nature* **398**, 51.
Dai,, L. (1993) *Eur. Polym. J.* **29**, 645, and references cited therein.
Dai, L. (1999) *J. Macromol. Sci., Rev. Macromol. Chem. Phys.* **39**, 273.
Dai, L. (2001) *Rad. Phys. Chem.* **62**, 55.
Dai, L, Mau, A.W.H. (2000). *J. Phys. Chem. B* **104**, 1891.
Dai, L., Toprakcioglu, C. (1991) *Europhys. Lett.* **16**, 331.
Dai, L., Toprakcioglu, C. (1992) *Macromolecules* **25**, 6000.
Dai, L., Toprakcioglu, C., Hadziioannou, G. (1995) *Macromolecules* **28**, 5512.
Dai, L., Wang, Q., and Wan, M. X. (2000) *J. Mater. Sci. Lett.* **19**, 1645.
Dai, L., Winkler, B., Dong, L., Tong, L., Mau, A.W.H. (2001) *Adv. Mater.* **13**, 915.
D'Agostino, R. (ed.) (1990) *Plasma Deposition, Treatment, and Etching of Polymers*, Academic Press, New York.
de Gennes, P.G. (1979) *"Scaling Concepts in Polymer Physics"*, Cornell University Press, Ithaca.
de Gennes, P.G. (1980) *Macromolecules* **13**, 1069.
de Gennes, P.G. (1985) *C.R. Hebd. Seances Sci.* **300**, 839.
De Morais, T.D., Chaput, F., Lahlil, K., Boilot, J.-P. (1999) *Adv. Mater.* **11**, 107.
Derjaguin, B.V. (1934) *Kolloid-Z.* **69**, 155.
Dokoutchaev, A., James, J.T., Koene, S.C., Pathak, S., Prakash, G.K.S., Thompson, M.E. (1999) *Chem. Mater.* **11**, 2389.
Doshi, J., Reneker, D.H. (1995) *J. Electrost.* **35**, 151.
Doshi, J., Srinivasan, G., Reneker, D.H. (1995) *Polym. News* **20**, 206.
Duan, X., Huang, Y., Wang, J., Lieber, C.M. (2001) *Nature* **409**, 66.
Era, M., Shinozaki, H., Tokito, S., Tsutsui, T., Saito, S. (1988) *Chem. Lett.* 1097.
Fendler, J.H., Meldrum, F.C. (1995) *Adv. Mater.* **7**, 607.
Ferreira, M., Onitsuka, O., Fou, A.F., Hsieh, B.R., Rubner, M.F. (1996) *Mat. Res. Soc. Symp. Proc.* **413**, 49.
Ferreira, M., Rubner, M.F. (1995) *Macromolecules* **28**, 7107.
Field, J.B., Toprakcioglu, C., Ball, R.C., Stanley, H.B., Dai, L., Barford, W., Penfold, J., Smith, G., Hamilton, W. (1992a) *Macromolecules* **25**, 434.
Field, J.B., Toprakcioglu, C., Dai, L., Hadziioannou, G., Smith,G., Hamilton, W.J. (1992b) *J. Phys. II* (France), **2**, 2221.
Flory, P.J. (1953) *Principle of Polymer Chemistry*, Cornell University Press, Ithaca, New York.
Formhals, A. (1934) US patent No. 1, 975, 504.
Fong, H., Chun, I., Reneker, D.H. (1999) *Polymer* **40**, 4585.
Fou, A.C., Onitsuka, O., Ferreira, M., Rubner, M.F., Hsieh, B.R. (1995) *J. Appl. Phys.* **79**, 7501.
Frank, S., Poncharal, P., Wang, Z.L., de Heer, W.A. (1998) *Science* **280**, 1744.
Friend, R.H., Gymer, R.W., Holmes, A.B., Burroughes, J. H.,Marks, R.N., Taliani, C., Bradley, D.D.C, Dos Santos, D.A., Brédas, J.L., Lögdlund, M., Salaneck, W.R. (1999) *Nature* **397**, 121.
Furukawa, K., Terasaka, Y., Ueda, H., Matsumura, M. (1997) *Synth. Met.* **91**, 99.
Gao, M., Dai, L., Wallace, G. G. (2003*) Electroanal.* **15**, 1089.
Gao, M., Huang, S., Dai, L., Wallace, G., Gao, R., Wang, Z. (2000a) *Angew. Chem. Int. Ed.* **39**, 3664.
Gao, R., Wang, Z.L., Bai, Z., de Heer, W.A., Dai, L., Gao, M. (2000b) *Phys. Rev. Lett.* **85**, 622.
Giannelies, E.P. (1996) *Adv. Mater.* **8**, 29.

Gong, X., Dai, L., Mau, A.W.H., Griesser, H.J. (1998) *J. Polym. Sci., Part A: Polym. Chem.* **36**, 633.
Greci, M.T., Pathak, S., Mercado, K., Prakash, G.K.S., Thopson, M.E., Olah, G.A. (2001) *J. Nanosci. Nanotechnol.* **1**, 3.
Greenham, N.C., Friend, R.H. (1995) *Solid State Phys.* **49**, 1.
Grüner, J., Remmers, M., Neher, D. (1997) *Adv. Mater.* **9**, 964.
Halperin, A., Tirrell, M., Lodge, T.P. (1992) *Adv. Polym. Sci.* **100**, 31.
Hanna, S.N. (1997) in *Handbook of Nanophases Materials*, Goldstein, A.N. (ed.), Marcel Dekker, Inc., New York.
Harada, A., Kataoka, K. (1999) *Science* **283**, 65.
Hadziioannou, G., Patel, S., Granick, S., Tirrell, M. (1986) *J. Am. Chem. Soc.* **108**, 2869.
He, H.X., Li C.Z., Tao, N.J. (2001) *Appl. Phys. Lett.* **78**, 811.
Hearn, J., Wilkinson, M.C., Goodall, A.R. (1981) *Adv. Colloid. Interface Sci.* **14**, 173.
Heavens, O.S. (1965) *Optical Properties of Thin Solid films*, Dover Publications, New York.
Higgins, J.S., Benoît, H.C. (1994) *Polymers and Neutron Scattering*, Clarendon Press, Oxford.
Horowitz, G. (1998) *Adv. Mater.* **10**, 365.
http://www.weizmann.ac.il/fluids/klein/techniques_sfb.html
Hu, B., Karasz, F.E. (1998) *Chem. Phys.* **227**, 263.
Huang, S., Dai, L. (2002) *J. Nanoparticle Res.* **4**, 145.
Huang, Y., Duan, X., Wei Q., Lieber, C.M. (2001) *Science* **291**, 630.
Huang, E., Rockford, L., Russell, T.P., Hawker, C.J., Mays, J. (1998) *Nature* **395**, 757.
Hughes, D.J. (1954) *Neutron Optics*, Interscience Publishers, Inc., New York.
Israelachvili, J. (1992) *Intermolecular and Surface Forces*, Academic Press, London.
Ito, Y., Ochiai,Y., Park, Y. S., Imanishi, Y. (1997) *J. Am. Chem. Soc.* **119**, 1619.
Iyoda, T., Ando, M., Kaneko, T., Ohtani, A., Shimidzu, T., Honda, K. (1987) *Langmuir* **3**, 1169.
Jenekhe, S.A., Chen, S.A. (1999) *Science* **283**, 372.
Jeong, U., Kim, H.-C., Rodriguez, R.L., Tsai, I.Y., Christopher, C.M., Stafford, M., Kim, J.K., Hawker, C.J., and Russell, T.P. (2000) *Adv. Mater.* **14**, 274.
Johnson, S.A., Brigham, E.S. , Ollivier, P.J., Mallouk, T.E. (1997) *Chem. Mater.* **9**, 2448.
Kageyama, K., Tamazawa, J.-I., Aida, T. (1999) *Science* **285**, 2113.
Kamiyama, K., Era, M., Tsutsui, T., Saito, S. (1990) *Jpn. J. Appl. Phys.* **29**, L840.
Kang, E.T., Kato, K., Uyama, Y., Ikada, Y. (1996) *J. Mater. Res.* **11**, 1570.
Kanga, R.S., Hogen-Esch, T.E., Randrianalimanana, E., Soum, A., Fontanille, M. (1990) *Macromolecules* **23** 4235 and 4241.
Katz, H.E. (1997) *J. Mater. Chem.* **7**, 369.
Klein, J. (1988) in *Molecular Conformation and Dynamics of Macromolecules in Condensed Systems*, Nagasawa, M. (ed.), Elsevier Science Publishers, Amsterdam.
Klein, J., Kumacheva, E., Mahalu, D., Perahia, D., Fetters, L.J., (1994) *Nature* **370**, 634.
Kranz, C., Gaub, H.E., Schuhmann, W. (1996) *Adv. Mater.* **8**, 634.
Kranz, C., Ludwig, M., Gaub, H.E., Schuhmann, W. (1995) *Adv. Mater.* **7** 38, 568.
Klok, H.A., Lecommandoux, S. (2001) *Adv. Mater.* **13**, 1217 and references cited therein.
Kraft, A., Grimsdale, A.C., Holmes, A.B. (1998) *Angew. Chem. Int. Ed.* **37**, 402.
Krueger, J. (ed.) (1994) *Colloidal Drug Delivery Systems*, Marcel Decker, New York.
Kumacheva, E., Kalinina, O., Lilge, L. (1999) *Adv. Mater.* **11**, 231.
Kyotani, T., Nagai, T., Inoue, S., Tomita, A. (1997) *Chem. Mater.* **9**, 609.
Lai, P-Y., Halperin, A. (1992) *Macromolecules* **25**, 6693.
Lansac, Y., Glaser, M.A., Clark, N.A., Lavrentovich, O.D. (1999) *Nature* **398**, 54.
Liu, G.J. (1998) *Curr. Opin. Colloid. Inter.* **3**, 200.
Liu, S., Armes, S.P. (2001) *J. Am. Chem. Soc.* **123**, 9910.
Liu, J., Gan, L.M., Chew, C.H., Teo, W.K., Gan, L.H. (1997) *Langmuir* **13**, 6421.

Liu, W., Graham, M., Satola, B.V., Evans, E.A., Reneker, D.H. (2002) *J. Mater. Res.* **17**, 3206.
Liu, T., Tang, J., Zhao, H.Q., Deng, Y.P., Jiang, L. (2002) *Langmuir* **18**, 5624.
Liu, X., Jiang, M., Yang, S., Chen, M., Chen, D., Yang, C., Wu, K. (2002) *Angew. Chem. Int. Ed.* **41**, 2950.
Lovinger, A.J., Rothberg, L.J., (1996).*J. Mater. Res.* **11**, 1581.
Lowe, P.J., Temple, C.S. (1994) *J. Pharm. Pharmacol.* **46**, 547.
Luckham, P.F., Costello, B.A. (1993) *Adv. Colloid. Interface Sci.* **44**, 183.
MacDiarmid, A.G., Jones, W.E., Jr., Norris, I.D., Gao, J., Johnson, A.T., Jr., Pinto, N.J., Hone, J., Han, B., Ko, F.K., Okuzaki, H., Llaguno, M. (2001) *Synth. Met.* **119**, 27.
Mansky, P., Liu, Y., Huang, E., Russell, T.P., Hawker, C.J. (1997) *Science* **275**, 1458.
Mark, J.E., Lee, C.Y.-C., Bianconi, P.A. (eds), (1995) *Hybrid Organic-Inorganic Composites*, ACS Symp. Ser. **585**, ACS, Washington DC.
Markel, V., George, T. (eds) (2001) *Optics of Nanostructured Materials*, John Wiley & Sons, Inc., New York.
Martin, C.R., (1991) *Adv. Mater.* **3**, 457.
Martin, C.R. (1995) *Acc. Chem. Res.* **28**, 61, and references cited therein.
Martin, J.I., Wang, Z.-G., Schick, M. (1996) *Langmuir* **12**, 4950.
Matsumoto, A., Odani, T., Sada, K., Miyata, M., Tashiro, K. (2000) *Nature* **405**, 328.
Matthews, J.A., Wnek, G.E., Simpson, D.G., Bowlin, G.L. (2002) *Biomacromolecules* **3**, 232.
Mattoussi, H., Karasz, F. E. (1993) *J. Chem. Phys.* **99**, 9188.
Maynor, B.W., Filocamo, S.F., Grinstaff, M.W., Liu, J. (2002) *J. Am. Chem. Soc.* **124**, 522.
Messer, B., Song, J.H., Yang, P. (2000) *J. Am. Chem. Soc.* **122**, 10232.
Miller, C.M., El-Aasser, M.S. (1997) *Recent Advances in Polymeric Dispersions, NATO ASI Series E, Applied Science*, **335**, Kluwer Academic Publishers, Dordrecht.
Milner, S.T. (1991) *Science* **251**, 905, and references cited therein.
Milner, S.T., Witten, T., Cates, M. (1988) *Macromolecules* **21**, 2610.
Moffitt, M., Khougaz, K., Eisenberg, A. (1996) *Acc. Chem.* Res. **29**, 95.
Morkved, T.L., Lu, M., Urbas, A.M., Ehrichs, E.E., Jaeger, H.M., Mansky, P., Russell, T.P. (1996) *Science* **273**, 932.
Murray, G.A.C.B. Kagan, C.R., Bawendi, M.G. (1995) *Science* **270**, 1335.
Mylnikov, V. S. (1994) *Photoconducting Polymers, Adv. Polym. Sci. Vol.* 115, Springer-Verlag, Berlin.
Napper, D.H. (1983) *Polymeric Stabilization of Colloidal Dispersions*, Academic Press, London.
Nguyen, T.P., Le Rendu, P., Amgaad, K., Cailler, M., Tran, V.H. (1995) *Synth. Met.* **72**, 35.
Nishikata, Y., Kakimoto, M.-A., Imai, Y. (1989) *Thin Solid Films* **179**, 191.
Niwa, T., Takeuchi, H., Hino, T., Kunou, T.N., Kawashima, Y. (1993) *J. Control. Rel.* **25**, 89.
Norris, I.D., Shaker, M.M., Ko, F.K., MacDiarmid, A.G. (2000) *Synth. Met.* **114**, 109.
Noyori, R., Ikariya, T. (2000) in *Stimulating Concepts in Chemistry,* Vgötle, F., Stoddart, J.F., Shibasaki, M. (eds.), Wiley-VCH, Weinheim.
Nyffenegger, R.M., Penner, R.M. (1996) *J. Phys. Chem.* **100**, 17041.
Odom, T.W., Huang, J.-L., Kim, P., Lieber, C.M. (2000) *J. Phys. Chem. B* **104**, 2794, and references cited therein.
Okahata, Y., Noguchi,Y., Seki, T. (1986) *Macromolecules* **19**, 494.
Onitsuka, O., Fou, A.C., Ferreira, M., Hsieh, B.R., Rubner, M.F. (1996) *J. Appl. Phys.* **80**, 4067.
O'Shea, S.J., Welland, M.E., Rayment, T. (1993) *Langmuir* **9**, 1826.
Ozin, G.A. (1992) *Adv. Mater.* **4**, 612.
Parthasarathy, R.V., Martin, C.R. (1994) *Chem. Mater.* **6**, 1627.
Parthasarathy, R.V., Phani, K.L.N., Martin, C.R. (1995) *Adv. Mater.* **7**, 896.

Pascual, J.I., Mendez, J., Gomez-Herrero, J., Baro, A.M., Garcia, N., Binh, V.T. (1993) *Phys. Rev. Lett.* **71**, 1852.
Patel, S., Tirrell, M. (1989) *Annu. Rev. Phys. Chem.* 40.
Pathak, S., Greci, M.T., Kwong, R.C., Mercado, K., Prakash, G.K.S., Olah, G.A., Thompson, M.E. (2000) *Chem. Mater.* **12**, 1985.
Pei, Q., Yang, Y., Yu, G., Cao Y., Heeger, A.J. (1997) *Synth. Met.* **85**, 1229.
Penfold, J., Richardson, R.M., Zarbakhsh, A., Webster, J.R.P., Bucknall, D.G., Rennie, A.R., Jones, R.A.L., Cosgrove, T., Thomas, R.K., Higgins, J.S., Fletcher, P.D.I., Dickinson, E., Roser, S.J., McLure, I.A., Hillman, A.R., Richards, R.W., Staples, E.J., Burgess, A.N., Simister, E.A., White, J.W. (1997) *J. Chem. Soc., Faraday Trans.* **93**, 3899.
Poncharal, P., Wang, Z.L., Ugarte, D., de Heer, W.A. (1999) *Science* **283**, 1513.
Qiu, H., Wan, M., Matthews, B., Dai, L. (2001) *Macromolecules* **34**, 675.
Qiu, X., Wu, C. (1997) *Macromolecules* **30**, 7921.
Qiu, X., Wu, C. (1998) *Phys. Rev. Lett.* **80**, 620.
Rabelero, M., Zacarias, M., Mendizábal, E., Puig, J.E., Dominguez, J.M., Katime, I. (1997) *Polym. Bull.* **38**, 695.
Ramachandran, K., Lerner, M.M. (1997) *J. Electrochem. Soc.* **144**, 3739.
Randolph, T.W., Randolph, A.D., Mebes, M., Yeung, S. (1993) *Biotechnol. Prog.* **9**, 429.
Ratner, B.D., Castner, D.G. (1996) (eds.), *Surface Modification of Polymeric Biomaterials*, Plenum Press, New York.
Remmers, M., Schulze, M., Wegner, G. (1996) *Macromol. Rapid Commun.* **17**, 239.
Reneker, D.H., Chun, I. (1996) *Nanotechnology* **7**, 216.
Reneker, D.H., Yarin, A.L., Fong, H., Koombhongse, S. (2000) *J. Appl. Phys.* **87** 4531.
Roberts, G.G. (ed.), (1990) *Langmuir-Blodgett Films*, Plenum Press, New York.
Rouhi, M. (1999) *C&E News*, **Sept. 27**, 10.
Rubner, M.F., Skotheim, T.A. (1991) In *Conjugated Polymers*, Brédas, J.L., Silbey, R. (eds), Kluwer Academic Publishers, Dordrecht.
Ruiz-Hitzky, E. (1993) *Adv. Mater.* **5**, 334.
Ruiz-Hitzky, E., Aranda, P. (1990) *Adv. Mater.* **2**, 545.
Sagisaka, S., Ando, M., Iyoda, T., Shimidzu, T. (1993) *Thin Solid Films* **230**, 65.
Saito, R., Dresselhaus, G., Dresselhaus, M.S. (1998) *Physical Properties of Carbon Nanotubes*, Imperial College Press, London.
Salaneck, W.R., Stafström, S., Brédas, J.L. (1996) *Conjugated Polymer Surfaces and Interfaces*, Cambridge University Press, Cambridge, UK.
Sariciftci, S,. Heeger, A. J. (1994) *Int. J. Mod. Phys. B* **8**, 237.
Sarid, D. (1991) *Scanning Force Microscopy: with Applications to Electric, Magnetic and Atomic Forces*, Oxford University Press, New York.
Schild, H.G. (1992) *Prog. Polym. Sci.* **17**, 163.
Schwiegk, S., Vahlenkamp, T., Xu, Y., Wegner, G. (1992) *Macromolecules* **25**, 513.
Shimidzu, T., Iyoda, T., Ando, M., Ohtani, A., Kaneko, T., Honda, K. (1988) *Thin Solid Films* **160**, 67.
Shin, K., Leach, K.A., Goldbach, J.T., Kim, D.H., Jho, J.Y., Tuominen, M., Hawker, C.J., Russell, T.P. (2002) *Nano Lett.* **2**, 934.
Skotheim, T.A., Reynolds, J., Elsenbaumer, R. (eds) (1998) *Handbook of Conducting Polymers*, 2nd edn, Marcel Dekker, New York.
Smith, P.A., Christopher, D.N., Thomas, N.J., Mayer, T.S., Martin, B.R., Mbindyo, J., Mallouk, T.E., (2000) *Appl. Phys. Lett.* **77**, 1399.
Smith, R.C., Fischer, W.M., Gin, D.L. (1997) *J. Am. Chem. Soc.* **119**, 4092.
Soppimath, K.S., Aminabhavi, T.M., Kulkarni, A.R., Rudzinski, W.E., (2001) *J. Control. Rel.* **70**, 1, and references cited therein.
Spivak, A.F., Dzenis, Y.A., Reneker, D.H. (2000) *Mech. Res. Commun.* **27**, 37.
Steinhart, M., Wendorff, J.H., Greiner, A., Wehrspohn, R.B., Nielsch, K., Schilling, J., Choi, J., Gösele, U. (2002) *Science* **296**, 1997.

Stewart, S., Liu, G.J. (1999) *Chem. Mater.* **11**, 1048.
Stockton, W.B., Rubner, M.F. (1997) *Macromolecules* **30**, 2717.
Stupp, S.I., LeBonheur, V., Walker, K., Li, L.S., Huggins, K.E., Keser, M., Amstutz, A. (1997) *Science* **276**, 384.
Suzuki, M., Ferencz, A., Iida, S., Enkelmann, V., Wegner, G. (1993) *Adv. Mater.* **5**, 359.
Tanaka, K., Matsuura, Y., Nishio, S., Yamabe, T. (1994) *Synth. Met.* **63**, 221.
Tanase, M., Bauer, L.A., Hultgren, A., Silevitch, D.M., Sun, L., Reich, D.H., Searson, P.C., Meyer, G.J. (2001) *Nano Lett.* **1**, 155.
Taunton, H.J., Toprakcioglu, C., Fetters, L.J., Klein, J. (1988) *Nature* **332**, 712.
Taunton, H.J., Toprakcioglu, C., Fetters, L.J., Klein, J. (1990) *Macromolecules* **23**, 571.
Tester, J.W., Danheiser, R.L., Weinstein, R.D. (2000) in *Green Chemical Syntheses and Processes,* Anastas, P.T., Heine, L.G., Williamson, T.C. (eds), ACS, Washington DC.
Theron, A., Zussman, E., Yarin, A.L. (2001) *Nanotechnology* **12**, 384.
Thurn-Albrecht, T., Derouchey, J., Russell, T.P., Jaeger, H.M. (2000a) *Macromolecules* **33**, 3250.
Thurn-Albrecht, T., Schotter, J., Kästle, A., Emley, N., Shibauchi, T., Krusin-Elbaum, L., Guarini, K., Black, C.T., Tuomine, M.T., Russell, T.P. (2000b) *Science* **290**, 2126.
Thurn-Albrecht, T., Steiner, R. , DeRouchey, J., Stafford, C.M., Huang, E., Bal, M., Tuominen, M. , Hawker, C.J., and Russell, T.P. (2000c) *Adv. Mater.* **12**, 787.
Thurmond, K.B., Koalewski, T., Wooley, K.L. (1996) *J. Am. Chem. Soc.* **118**, 7239.
Thurmond, K.B., Kowalewski, T., Wooley, K.L. (1997) *J. Am. Chem. Soc.* **119**, 6656.
Tishchenko, G., Luetzow, K., Schauer, J., Albrecht, W., Bleha, M. (2001) *Separation Purification Technol.* **22**, 403.
Tom, J.W., Debenedetti, P.G. (1991) *J. Aerosol Sci.* **22**, 555.
Toprakcioglu, C., Dai, L., Ansarifar, M.A., Stamm, M., Motschmann, H. (1993) *Prog. Colloid Polym. Sci.* **91**, 83.
Tsukruk, V.V. (1997) *Prog. Polym. Sci.* **22**, 247.
Turk, M. (1999) *J. Supercrit. Fluid* **15**, 79.
Vahlenkamp, T., Wegner, G. (1994) *Macromol. Chem. Phys.* **195**, 1933.
van der Pauw, L.J. (1958) *Philips Res. Rep.* **13**, 1.
Wan, M. X., Li, J.C. (1999) *J. Polym. Sci., Part A: Polym. Chem.* **37**, 4605.
Webber, S.E. (1998) *J. Phys. Chem. B* **102**, 2618.
Wehrle, P., Magenheim, B., Benita, S. (1995) *J. Pharm. Biopharm.* **41**, 19.
Winkler, B., Dai, L., Mau, A.W.H. (1999) *J. Mater. Lett.* **18**, 1539.
Wu, C.G., Bein, T. (1994) *Science* **266**, 1013.
Wu, C., Niu, A., Leung, L.M. (1999) *J. Am. Chem. Soc.* **121**, 1954.
Wu, C.C., Sturm, J.C., Kahn, A. (1997) *App. Phys. Lett.* **70**, 1348.
Wuu, Y.-M., Fan, F.R.F., Bard, A.J. (1989) *J. Electrochem. Soc.* **136**, 885.
Xu, X.J., Chew, C.H., Siow, K.S., Wong, M.K., Gan, L.M. (1999) *Langmuir* **15**, 8067.
Xu, X., Luo, X., Xie, Y., Zhou, H. (eds), (1995) *Proc. Int. Workshop Electroluminescence (EL'94)*, Science Press, Beijing.
Xu, X.J., Siow, K.S., Wong, M.K., Gan, L.M. (2001a) *Colloid Polym. Sci.* **279**, 879.
Xu, X.J., Siow, K.S., Wong, M.K., Gan, L.M. (2001b) *Langmuir* **17**, 4519.
Yamakawa, H. (1971) *Modern Theory of Polymer Solutions*, Harper & Row Publishers, New York.
Yang, R., Evans, D.F., Hendrickson, W.A. (1995) *Langmuir* **11**, 211.
Yang Z., Karasz, F. (1997) *Macromol. Symp.* **83**, 83.
Yano, K., Usuki, A. , Okada, A. (1997) *J. Polym. Sci. A: Polym. Chem.* **35**, 2289.
Yarin, A.L., Koombhongse, S., Reneker, D.H. (2001a) *J. Appl. Phys.* **89**, 3018.
Yarin, A.L., Koombhongse, S., Reneker, D.H. (2001b) *J. Appl. Phys.* **90**, 4836.
Yasuda, H. (1985) *Plasma Polymerization*, Academic Press, New York.
Zakhidov, A.A., Baughman, R.H., Iqbal, Z., Cui, C., Khayrullin, I, Dantas, Marti, S.O.J., Ralchenko, V.G. (1998) *Science* **282**, 897.

Zambaux, M.F., Bonneaux, F., Gref, R.R., Maincent, P., Dellacherie, E., Alonso, M.J., Labrude, P., Vigneron, C. (1998) *J. Control. Rel.* **50**, 31.

Zhang, G., Niu, A., Peng, S., Jiang, M., Tu, Y., Li, M., Wu, C. (2001) *Acc. Chem. Res.* **34**, 249.

Zhang, Q., Remsen, E.E., Wooley, K.L. (2000) *J. Am. Chem. Soc.* **122**, 3642.

Zhou, J., Li, Z., Liu, G. (2002) *Macromolecules* **35**, 3690.

Zhu, P.W., Napper, D.H. (1999) *Macromol. Chem. Phys.* **200**, 1950.

Zhu, P.W., Napper, D.H. (2000) *Langmuir* **16**, 8543.

Part III
Smart Devices

Chapter 8

Electronic Devices

8.1 Introduction

Until now, organic materials have been widely used as passive components, such as insulators or photoresists for fabricating silicon chip circuitry, in the microelectronics industry. This is because the relatively high level of structural disorder and band gap energy of organic semiconducting materials often leads to much lower charger-carrier mobilities than those of inorganic semiconductors. As a consequence, it is difficult, if not impossible, for organic semiconductors to achieve the switching frequencies required for certain microelectronic devices. However, the recent advancements in organic polymer light-emitting diodes (to be discussed in Chapter 9), DNA conductors and carbon nanotube nanoelectronics have attracted considerable attention for the commercial development of smart devices based on active organic materials. Fabricating organic devices is often less expensive than for those made of inorganic semiconductors, such as silicon, gallium arsenide and gallium phosphide, as films of organic materials can be more easily deposited over a large area. Besides, organic materials are often flexible and light. Two major classes of organic materials based on solution-cast macromolecular films (*e.g.* conjugated polymers, carbon nanotubes, DNA) and small organic molecular coatings (*e.g.* conjugated oligomers, organic crystals) formed by sublimation are under investigation for applications in organic optoelectronic devices. This chapter will summarize the recent developments in the field of macromolecular electronics.

8.2 Conjugated Polymer Devices

Conjugated polymers have been used in electronic devices as both passive and active components. Examples of the use of conjugated conducting polymers as passive components include conducting wires, antistatic coatings and electromagnetic shielding layers. Semiconductor devices made from active components based on conjugated polymers may be divided into two categories depending on whether or not a current or light is produced. The former includes thin film transistors (Horowitz, 1998; Lovinger *et al.*, 1996), photoconductors (Mylnikov, 1994) and photovoltaic cells (Sariciftci and Heeger, 1994), while the latter encompasses light-emitting diodes, LEDs, (Greenham and Friend, 1995) and optically/electrically pumped lasers (Hide *et al.*, 1997). As polymer LEDs and photovoltaic cells will be discussed in detail in the next chapter, we will focus our attention here on applications of organic macromolecules in electronic applications. Although an emphasis is given in this book to active devices, the use of conjugated conducting polymers for electromagnetic shielding is described below due to its importance to electronic devices.

8.2.1 Electromagnetic Shielding

Electronic devices are subjected unavoidably to electromagnetic interference (EMI) from a variety of sources. Housings and enclosures for electronic devices, therefore, must provide shielding from EMI to prevent device malfunction or the loss of information stored on magnetic media. The term "shielding" usually refers to a metallic enclosure that seals an electronic device or a portion of it to prevent either the electromagnetic emissions of the electronic device from radiating outside or externally radiated emissions from entering into the electronic device, or both. Depending on the electromagnetic source-to-shield distance, EMI shielding can be categorized into far-field and near-field shielding. While the former refers to the situation where the electromagnetic source is far from the shield at a distance comparable to the wavelength of radiation, the latter applies to the case in which the electromagnetic field is in close proximity to the shield. Far away from sources, almost all electromagnetic waves behave as uniform plane waves. In the case of the near field, any geometric changes will change the dependence of shielding effectiveness (SE) on frequency, conductivity and source-to-shield distance. For a fixed source-to-shield distance, a transition from near-field to far-field region could occur with an increase in the radiation frequency.

Thermoplastic materials have been widely used for making such housing, but the insulating nature of plastics often makes them transparent to the electromagnetic interference. In order to provide the necessary EMI shielding, plastics must be rendered conductive through coating or mixing with conductive fillers (Chapter 2). Although the mixture system can possess good mechanical properties, it is not easy to ensure a uniform distribution of the conductive fillers throughout the matrix during molding. Consequently, the deficiencies of fillers in the corners, edges and surfaces of the housing could significantly reduce the effectiveness of the EMI shielding.

Carbon black and stainless steel fibers are commonly used conductive fillers for making conventional thermoplastics, such as nylon and PVC, conductive. However, intrinsically conducting conjugated polymers and their blends normally show much higher conductivities than the polymeric carbon black or stainless steel composites. Apart from their lightweight, corrosion resistance and relatively good mechanical properties, conjugated conducting polymers and their thermoplastic blends also show high uniformity of conductivity, leading to superior EMI shielding characteristics. Unlike metals, conjugated conducting polymers can not only reflect but also selectively absorb electromagnetic radiation (Unsworth *et al.*, 1993). If the absorption is strong, the conjugated conducting polymers and their derivatives may also become polymeric stealth materials that can avoid, for example, radar detection (Willey, 1986).

As shown in Figure 8.1, a beam of an electromagnetic wave incident on to the surface of conjugated polymer-based shielding materials can be resolved into: a) direct reflected wave (R); b) an absorbed wave (A); c) an internally reflected wave (B); d) a transmitted wave.

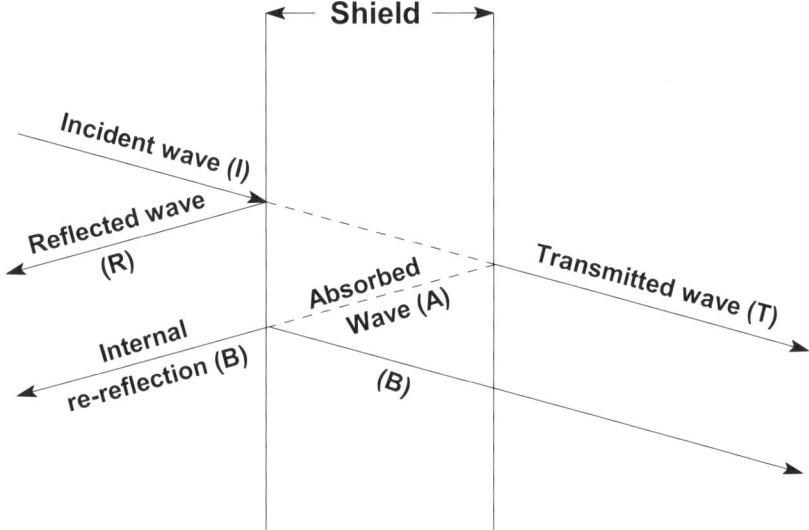

Figure 8.1. Splitting of electromagnetic wave on passing through a shield (After Dhawan *et al.*, 2002, copyright 2002 Elsevier. Reproduced with permission)

The shielding effectiveness (*SE*) is closely related to the reflection loss, absorption loss and multiple reflections, as shown in Equation (8.1) (Paul, 1992):

$$SE = R + A + B \tag{8.1}$$

On the other hand, *SE* is also defined as the ratio of the magnitude of the electric or magnetic field that is incident on the shield to the magnitude of the field strength that is transmitted through the barrier:

$$SE = 10 \log(P_i/P_t) = 20 \log (E_i/E_t) \tag{8.2}$$

The multiple reflections and transmissions may be neglected for the shield thickness much greater than skin depth (*vide infra*) and only the initial reflection and transmission at the left and right interfaces need to be considered.

Adsorption loss for the electric field in a sample of thickness t is given by:

$$A = 20 \log (e^{-t/\delta}) \tag{8.3}$$

where

$$\delta = \text{skin depth} = [2/(\omega\mu\sigma)]^{1/2} \tag{8.4}$$

with σ being the electrical conductivity in S/cm and μ the permeability of the shield material (H/m).

Figure 8.2(a and b) shows the dependence of far-field *SE* on conductivity and field frequency, respectively, for polypyrrole films with different doping levels. As can be seen in Figure 8.2, the SE increases with increasing conductivity (Figure 8.2(a)) and decreases with increasing frequency (Figure 8.2(b)).

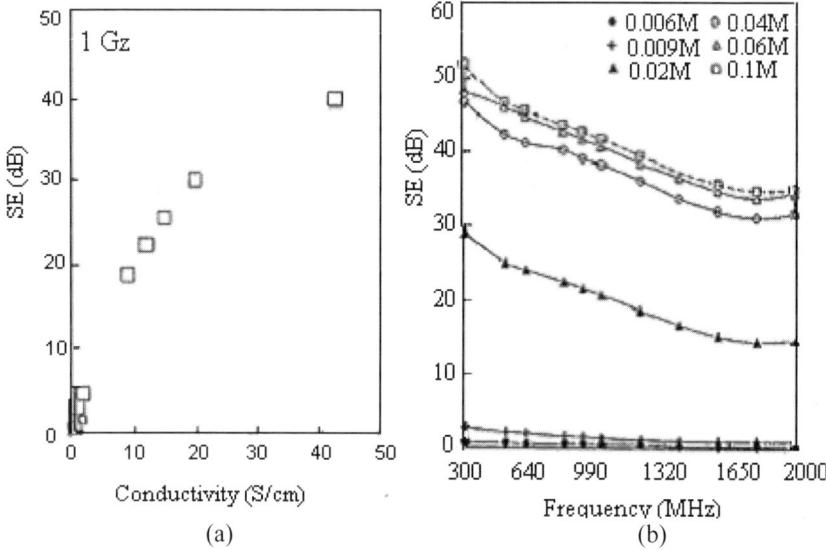

Figure 8.2. Plane wave shielding effectiveness of (a) a polypyrrole film versus dc conductivity; (b) polypyrrole films of different doping levels as a function of frequency (After Kaynak, 1996, copyright 1996 Elsevier. Reproduced with permission)

Similar trends have also been observed for Al-epoxy composites (Musamesh, 1992), a woven metal plated textile and a sintered material (Catrysse, 1990). The corresponding results for near-field EMI are given in Figure 8.3.

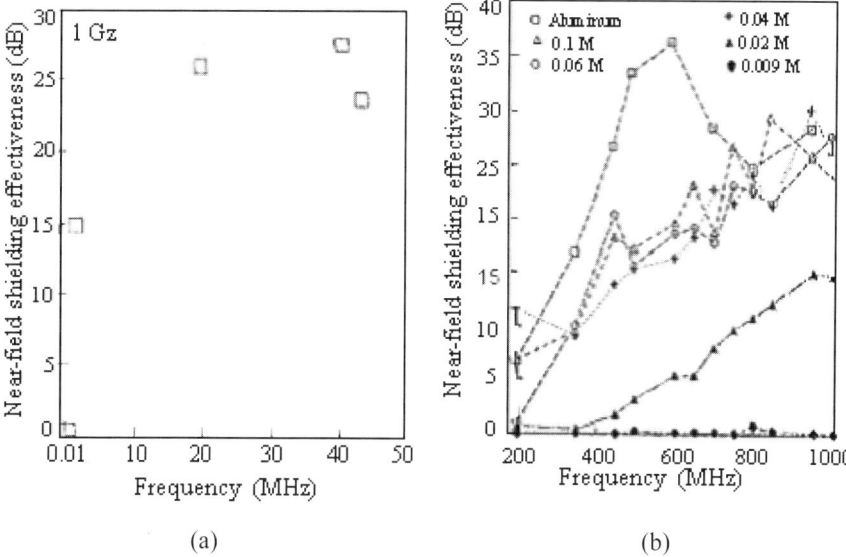

Figure 8.3. Near-field shielding effectiveness of (a) a polypyrrole film versus dc conductivity; (b) polypyrrole films of different doping levels as a function of frequency (After Kaynak, 1996, copyright 1996 Elsevier. Reproduced with permission)

Furthermore, it was found that heavily doped polypyrrole samples were very reflective in the microwave region while lightly doped films were almost transparent in the same region, suggesting that the shielding properties of conducting polymers may be tailor-made simply by controlling doping levels.

Recently, many groups have exploited the potential applications of conjugated conducting polymers for further EMI shielding. For example, Hong et al. (Hong, 2001; Lee et al., 2001) chemically and electrochemically prepared insulating poly(ethylene terephthalate), PET, woven fabric complexes coated with conductive polypyrrole (PPy) and Ag with the Ag/PPy-AQSA (anthraquinone-2-sulfonic acid)/PPy-NSA (naphthalene sulfonic acid)/PET multilayer structure. They found that the SE and the absorbance/reflectance of the shielding material can be controlled by regulating the area and the array structure of the highly conductive Ag and low-conducting PPy layers. On the other hand, Pomposo et al. (1999) have reported the shielding applications of hot melt adhesives based on polypyrrole blends. Similarly, several groups have investigated the EMI shielding properties of conductive polyaniline (Chandrasekhar and Naishadham, 1999), polyaniline blends (Koul et al., 2000), and polyaniline multiplayer films (Lee et al., 1999). Besides, carbon materials, such as carbon composites as well as colloidal and flexible graphite, have also been demonstrated to be attractive for EMI shielding, especially after electroplating with nickel (Chung, 2001). In this regard, highly conductive carbon nanotubes could be used as both the EMI active component and reinforcement filler for polymeric composites.

8.2.2 Schottky Barrier Diodes and Field-effect Transistors

8.2.2.1 Schottky Barrier Diodes

Figure 8.4 shows a typical Schottky-barrier diode structure with polyacetylene sandwiched between two metal contact layers, in which one of the metal layers (*i.e.* Al) forms the Schottky barrier with polyacetylene and the other (gold) provides an ohmic contact.

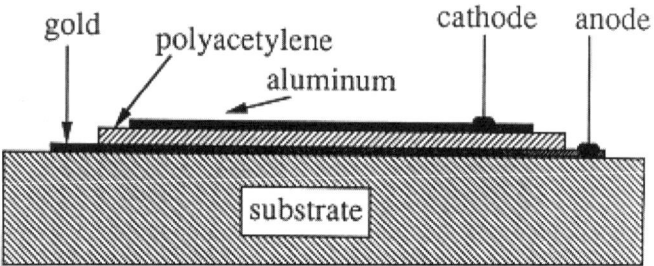

Figure 8.4. A typical Schottky-barrier diode structure with a layer of polyacetylene (500–1000 nm) sandwiched between a thin layer of aluminum, chromium or indium (20 nm) and gold (20 nm) supported by a flat glass substrate (After Burroughes and Friend, 1991, copyright 1991 Kluwer Academic Publishers)

Although the fabrication of this type of Schottky-barrier diode is straightforward, it provides a very useful structure for studying the electrical and electro-optical properties of semiconducting materials, including conducting polymers. For a *p*-type semiconductor, the ideal Schottky barrier is formed with a metal of a work-function (Φ_m) lower than that of the semiconductor (Φ_s). As shown in Figure 8.4, Φ_s is given by:

$$\Phi_s = \chi + \xi \tag{8.5}$$

where χ and ξ are the electron affinity (energy difference between the bottom of the conduction band and the vacuum level) and the energy gap between the Fermi level and bottom of the conduction band, respectively. The formation of the Schottky junction between the semiconductor and the low-work-function metal leads to the equalization of the Fermi energies, at zero bias, by forming a dipole charge layer with positive charge density in the metal and a region of negative charge density (*i.e.* the depletion layer) at the interface. As the charge concentration in a metal is normally much higher than that in a semiconductor, the semiconductor only undergoes the band bending (Figure 8.4). This sets up a barrier (*i.e.* Schottky barrier, Φ_b), which is given by $\chi + E_g - \Phi_m$, for the flow of holes towards the metal contact.

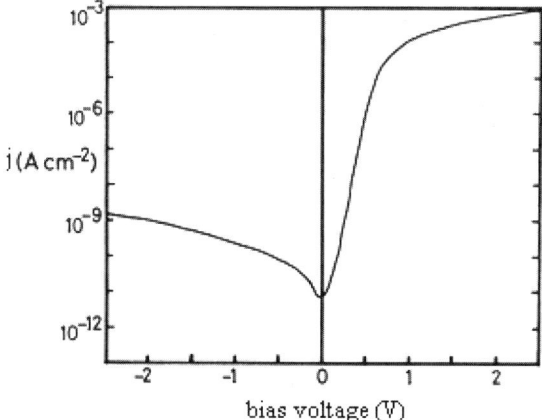

Figure 8.5. Forward and reverse characteristics for a Schottky diode with the Al/polyacetylene/Au structure, which gives a very good diode characteristic with a forward to reverse ratio as high as 5×10^{10} at a bias of ± 1.5 V (After Burroughes and Friend, 1991, copyright 1991 Kluwer Academic Publishers)

In addition to the $J(V)$ characteristics of the junction (Figure 8.5), detailed information on the nature of the junction may be obtained from the complex impedance. The capacitance of the junction in the depletion regime at reverse bias depends strongly on the variation in the width of the depletion regime with bias and is given by:

$$C/A = (q\varepsilon_0\varepsilon_r N_a/2)^{1/2} [1/(V_{d0} + V)]^{1/2} \qquad (8.6)$$

where A is the area of the junction, q is the electronic charge, $V_{d0} = \Phi_b - (E_F - E_V)$ is the diffusion voltage at zero bias, N_a is the acceptor dopant concentration and ε_r and ε_0 are the relative dielectric constant and the dielectric constant in vacuum, respectively. As can be seen from Equation (8.6), the slope of $1/C^2$ versus bias voltage should allow direct measurement of N_a and V_{d0}. From these data, information on the dependence of the depletion width (w) with bias voltage can be obtained by:

$$w = [2\varepsilon_0\varepsilon_r(V_{d0} + V)/qN_a]^{1/2} \qquad (8.7)$$

The increase in width of the depletion regime with increasing reverse bias has resulted from the movement of the extrinsic charge carriers to leave the depletion region with a space charge density (qN_a) due to the acceptors. The increase in absorbance (δ_α) by an increase in the reverse bias voltage is then related with the increase in the depletion width (δ_w) by:

$$\delta_\alpha = -\Delta T/T = \delta_w N_a \sigma \qquad (8.8)$$

where ΔT is the fractional change in the transmission (T) and σ is the optical cross-section of the soliton.

Therefore, the Schottky diodes can also be used for optical measurements since thin metal contacts (*e.g.* the 20 nm thick Al and Au films) allow adequate optical transmission through the device. As shown in Figure 8.6, it is possible to obtain the high sensitivity spectrum of the change in optical transmission with bias voltage.

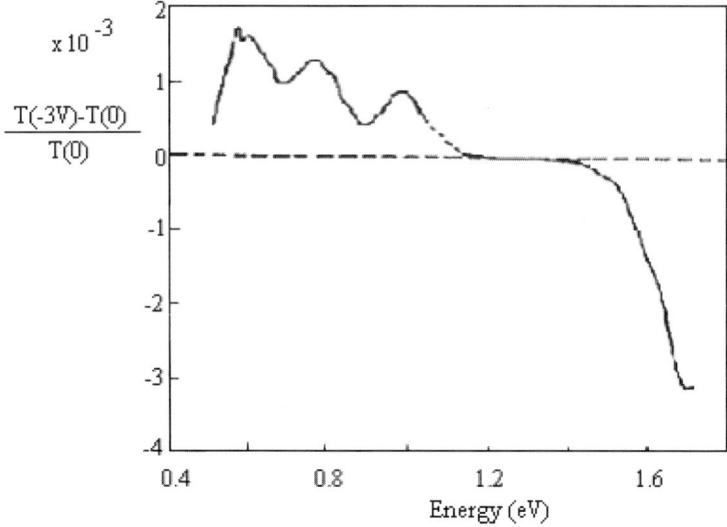

Figure 8.6. Bias voltage modulated optical transmission for an Al/polyacetylene Schottky diode of thickness 900 nm. $T(V)$ is the optical transmission at bias voltage V (After Burroughes and Friend, 1991, copyright 1991 Kluwer Academic Publishers)

The plot of $[T(V) - T(0)]/T(0)$ obtained by taking the bias voltage between 0 and a large negative value (−35 V) gives a positive signal for a bleaching induced by the reverse bias condition. The bleaching has been attributed to the loss of soliton levels associated with the extrinsic charge carriers that were swept out of the semiconductor to form the depletion layer. The oscillations at higher energies were considered to be due to modulation of interference fringes (Burroughes and Friend, 1991).

One of the major advantages in the use of Schottky diode structures for such optical measurements is that it is straightforward to obtain a quantitative measurement of the optical change and of the number of charges that are responsible for it. For example, the capacitance versus bias voltage (not shown) for the device corresponding to Figure 8.7 shows the expected behavior in reverse bias up to −35 V with a value for the acceptor concentration of $N_a = 2.1 \times 10^{16}$ cm^{-3}. Then, the change in width of the depletion layer (w) with the change in bias voltage from 0 to −35 V is calculated by Equation (8.3) to be 470 nm and the optical cross-section per injected charge at the peak of the "mid-gap" absorption band (σ) to be 1.6×10^{16} cm^2 [Equation (8.8)].

8.2.2.2 Field-effect Transistors

The addition of an insulating layer between the semiconductor and one of the metal layers of the Schottky diode structure and the replacement of the top electrode by a source and drain contacts onto the insulator layer shown in Figure 8.7 gives the metal-insulator-semiconductor field effect transistor (MISFET). Figure 8.7 shows a typical MISFET device fabricated on a silicon wafer.

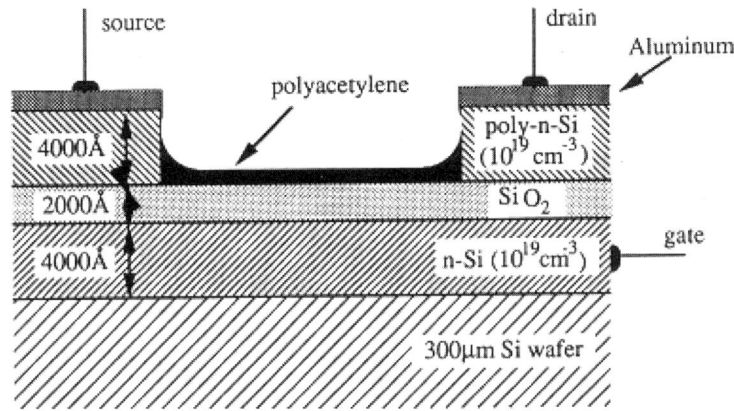

Figure 8.7. A MISFET device fabricated on a silicon wafer substrate, using polyacetylene as the semiconductor, silicon dioxide as the insulator layer, heavily *n*-doped silicon as the bottom metallic layer (the gate), the source and drain contacts. Aluminum is used as a blocking contact (After Burroughes and Friend, 1991, copyright 1991 Kluwer Academic Publishers)

In field-effect transistors, the flow of carriers between the source and drain contacts is controlled by the application of an electric field onto the gate electrode. Current flow along the main conduction path of the semiconductor in a field-effect device is dominated by the majority carriers, leading to their performance being relatively unaffected by external conditions (*e.g.* ambient temperature changes, radiation degradation). The variation of the channel conductance with gate voltage (V_{gs}) for constant drain-source voltage (V_{ds}) defines the transconductance (g_m):

$$g_m = \Delta I_{ds}/\Delta V_{gs} \tag{8.9}$$

For low values of V_{ds}, g_m gives a direct measure of the carrier mobility through Equation (8.10):

$$g_m = w/l(\mu C_i V_{ds}) \tag{8.10}$$

where w is the depletion width, *l* is the mean free path of the carriers in the semiconductor (which is estimated to be 3 nm for polyacetylene), μ is the charge carrier mobility, and C_i is the gate capacitance of the insulator layer (Rhoderick and Williams, 1988).

Furthermore, the saturation current (I_{sat}) for the channel at high values of V_{ds} is given by:

$$I_{sat} = \mu C_i w/2l(V_{gs} - V_{th})^{1/2} \qquad (8.11)$$

where V_{th} is the threshold voltage at which the accumulation channel forms.

While Figure 8.8 shows the plots of I_{ds} against V_{ds} at various negative values of V_{gs} (Figure 8.8(a)) and the variation of I_{ds} with V_{gs} for $V_{ds} = +5$ V (Figure 8.8(b)), the plot of $(I_{ds})^{1/2}$ against $V_{ds} = V_{gs}$ is given in Figure 8.8(c). From the slope of the linear region found at the higher values of V_{ds} in Figure 8.8(c) (the broken line), the values of V_{th} and μ are estimated to be -12 V and 9×10^{-6} cm^2/Vsec, respectively.

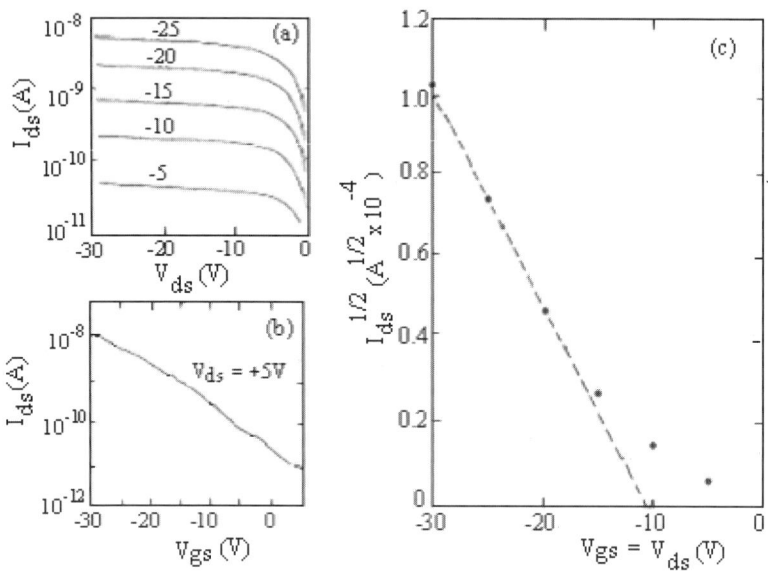

Figure 8.8. Characteristics of MISFET transistor with Au source and drain contacts. (a) I_{ds} versus V_{ds} for various negative values of V_{gs}; (b) I_{ds} versus V_{gs} for $V_{ds} = +5$ V; (c) $(I_{ds})^{1/2}$ versus $V_{ds} = V_{gs}$ (After Burroughes and Friend, 1991, copyright 1991 Kluwer Academic Publishers)

8.3 C_{60} Superconductivity

The Nobel-Prize-winning discovery of high-temperature superconductivity ($T_c = 38$ K) from a layered copper oxide (cuprate) made by Bednorz and Mueller (Bednorz and Mueller, 1986) in 1986 suggested that the large-scale application of superconductive materials might be within reach. Many other groups quickly picked up the significance of this discovery and the T_c of cuprate superconductors had subsequently been dramatically raised up to 133 K by 1993 (Figure 8.9). Further efforts in trying to increase the T_c for cuprate superconductors, however, have not been fruitful.

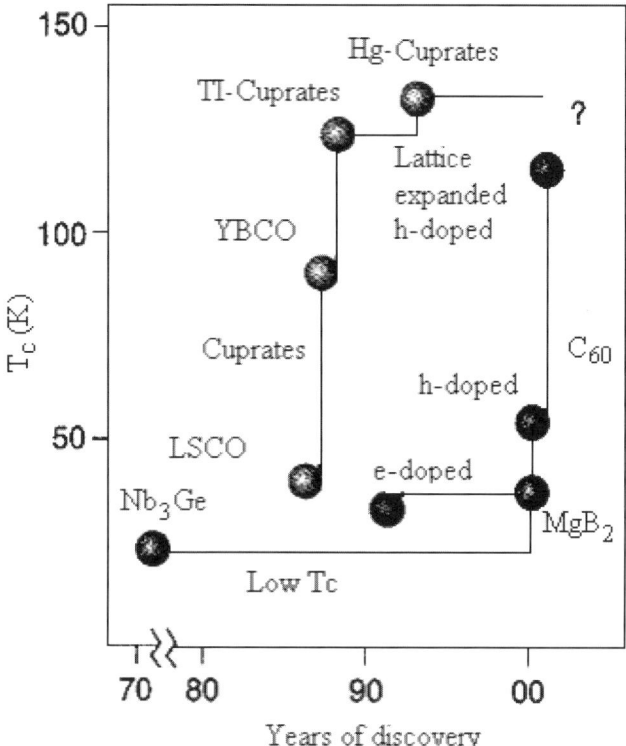

Figure 8.9. T_c versus year of discovery for some superconducting materials (After Dagotto, 2001). The question mark suggests the doubt on the data for the h-doped C_{60} (Dagotto, 2001, copyright 2001 American Association for the Advancement of Science)

The π-system in conjugated polymers provides a continuous conducting pathway for the transport of charge carriers. In molecular organic conducting systems, however, the overlapping of the intermolecular π-orbitals plays an important role in their transport properties. A planar molecular configuration is required for most molecular organic conductors to achieve intermolecular π-orbital overlap. As a consequence, three-dimensional isotropic molecular conductors are difficult to realize with these materials (Hammond and Kuck, 1992).

As discussed in the preceding chapters, fullerene C_{60} has a unique electronic structure and possesses a high electron affinity. This, together with the fact that their π–orbitals orient in all directions, has made fullerenes ideal systems for producing 3D isotropic organic conductors (Haddon et al., 1991). For instance, fullerene thin films have been doped by alkali metals (Hammond and Kuck, 1992), leading to a dramatic increase in conductivity. Surprisingly, the K_xC_{60} (typically, x = 3) combination even showed superconductivity (SC) below 18 K (Haddon, 1992; Hebard et al., 1991).

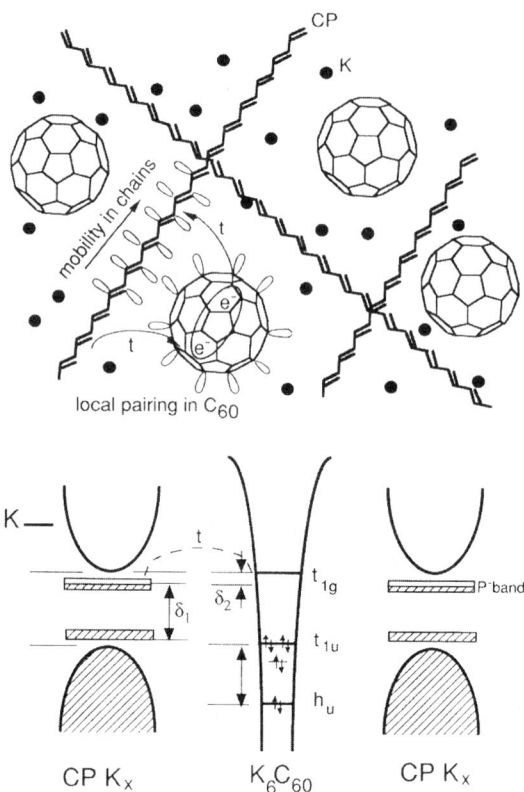

Figure 8.10. Schematic representation of C_{60}-induced superconductivity in an ideal CP-C_{60} system. The intercalated C_{60} provides the SC pairing of electrons in its local level, while CP chains provide their mobility due to good hybridization. Inset below shows hypothetical optimal energy-matching with resonating states and doubly degenerate regime of negative-U level in C_{60} (After Zakhidov *et al.*, 1999, copyright 1999 Springer-Verlag)

Furthermore, doping a poly(3-alkylthiophene)-C_{60}, P(3AT)-C_{60}, composite with an alkali metal, A (*e.g.* K vapor, n-doping), has been suggested to cause extrinsic charge transfer from A to both C_{60} and the conducting polymer (Figure 8.10), leading to superconductivity for the CP-$(C_{60})_yA_x$ ternary composite ($T_c = 17$ K for 5 mol% C_{60}) (Araki *et al.*, 1995, 1996; Yoshino *et al.*, 1996; Zakhidov *et al.*, 1995). Given that superconductivity requires electrons to combine to form the so-called Cooper pairs (Harrison, 1979; Kirtley and Tsuei, 1996), it may be that C_{60} induces the SC pairing of electrons in conducting polymer chains in the CP-$(C_{60})_yA_x$ composite (Araki *et al.*, 1995, 1996; Yoshino *et al.*, 1996; Zakhidov *et al.*, 1995). However, it is more likely that the superconductivity in the CP-$(C_{60})_yA_x$ system results from the domains of crystalline K_3C_{60} (*vide supra*) formed by phase separation. The mechanism of SC in this system deserves further investigation.

Owing to its high electron affinity, fullerene C_{60} has also been used as a *p*-type dopant for certain conducting polymers (Wei *et al.*, 1993; Zakhidov *et al.*, 1996). As discussed in Chapter 2, most dopants are either strong acceptors (*e.g.* I_2, AsF_5) or strong donors (*e.g.* Na, K). For strong acceptors (donors) the LUMO (HOMO) of the dopant is located lower (higher) in energy than the top (bottom) of the conducting polymer's valence (conduction) band. When doping conducting polymers with C_{60}, however, the LUMO of C_{60} is located between the top of the valence band and the bottom of the conduction band (*i.e.* in the forbidden gap) of most conducting polymers (Schlebusch *et al.*, 1996), indicating that fullerenes can hardly have doping effects (Zakhidov, 1991). Consequently, the dark conductivities of the conducting polymers could only be marginally improved by doping with fullerene C_{60} (Wei *et al.*, 1993; Zakhidov *et al.*, 1996). In order to improve electronic properties, however, Dai and co-workers (Dai *et al.*, 1998) have recently used sulfonated fullerene derivatives with multiple $-(O)SO_3H$ groups as protonic acid dopants for polyaniline emeraldine base. The resulting materials have room-temperature conductivities of up to 100 (S/cm), about six orders of magnitude higher than the typical value for fullerene-doped conducting polymers (Wei *et al.*, 1993).

Recent reports by Schön and co-workers on the application of the charge-injection doping technique (Chapter 2) in the FET configuration to the observation of a superconductivity at 52 K from a hole-doped C_{60} crystal film as well as a record high T_c = 117 K for C_{60} films co-crystallized with trichloromethane or tribromomethane (Schön *et al.*, 2000, 2001; Service, 2001) had generated considerable excitement. However, most of Schön's data turned out to be faked (Jacoby, 2002).

8.4 Polymer Batteries and Carbon Nanotube Supercapacitors

8.4.1 Conducting Polymer Batteries

Conducting polymers have been investigated for various applications in polymer batteries. While conjugated conducting polymers have been widely used as the anode and/or cathode materials in Li secondary (rechargeable) batteries, polymeric electrolytes have been used as the electrolytes in solid-state polymer batteries. Compared with conventional Li batteries, polymer batteries possess many advantages, including light weight, good processibility for the formation of unusual shapes (*e.g.* ultrathin, button cells), flexibility and anticorrosive characteristics.

Figure 8.11 shows the working principle for a typical Li battery with Cu cathode and Al anode (Chandrasekhar, 1999). As can be seen in Figure 8.11(b), Li ions migrate from the anode through the Li ion-conducting electrolyte (liquid or solid) to the cathode during discharge (*i.e.* use). During charge, the Li ions are re-transported to the anode to reduce into Li metal (Figure 8.11(a)).

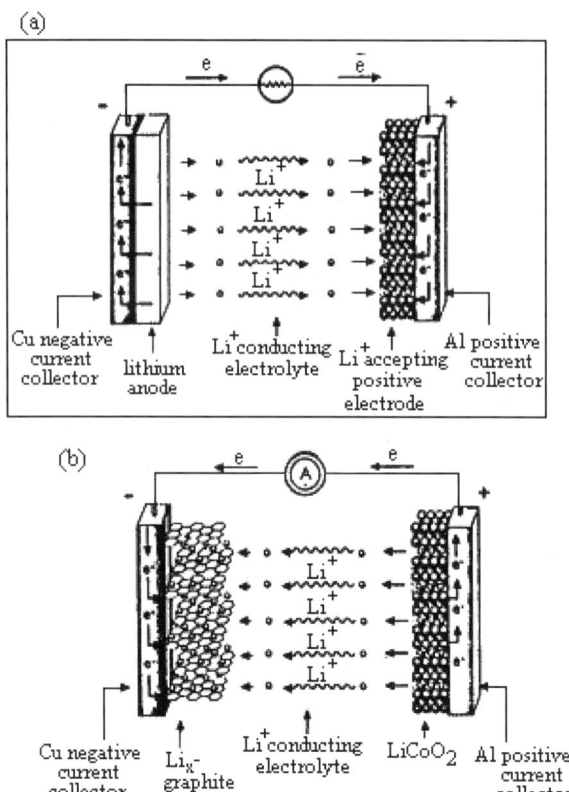

Figure 8.11. A schematic representation of a typical Li secondary battery during (a) charge and (b) discharge processes (After Chandrasekhar, 1999, copyright 1999 Kluwer Academic Publishers)

The major parameters used to measure battery performance include *discharge voltage* (*i.e.* open circuit voltage), *energy density* (milliWatt-hours/gram, mWh/g), *specific capacity* (mAh/g), *cyclability* (number of cycles), *coulombic efficiency* (ratio of charge consumed during charging and discharging, in percent), *discharge current density* and *discharge rate*. A battery with a high discharge rate is defined as a "high-rate" battery, which is normally used for high-power applications (*e.g.* transportation) whilst "low-rate" batteries, including most conducting polymer batteries, are often used for low-power applications, for example, in consumer electronics.

To demonstrate the working principle for conducting polymer Li batteries, MacDiarmid and co-workers (Nigrey *et al.*, 1981) constructed a battery with an Li anode, a *cis*-polyacetylene [$(CH)_x$] cathode and 1 M $LiClO_4$ in propylene carbonate (PC) as the liquid electrolyte. Equation (8.12) shows the doping (charge) and dedoping (discharge) reactions for this battery, which gives an initial discharge voltage of 3.7 V and 6% doped polyacetylene.

Positive electrode:

$$(CH)_x + xyClO_4^- - xye^- \xrightleftharpoons[discharge]{charge} [CH^{+y}(ClO_4)_y]_x$$

Negative electrode:

$$xy\,Li^+ + xye^- \xrightleftharpoons[discharge]{charge} xyLi$$

Total:

$$(CH)_x + xyClO_4^- \xrightleftharpoons[discharge]{charge} [CH^{+y}(ClO_4)_y]_x \tag{8.12}$$

Similar conducting polymer solid-state batteries of the Li/LiClO$_4$-PEO/polypyrrole-ClO$_4$ structure have also been reported (Osaka *et al.*, 1994). These batteries have been claimed to have the coulumbic efficiency of 90% at a discharge current density of 0.1 mA/cm^2, and cyclability up to 1400 cycles at high coulombic efficiency. However, like most solid-electrolyte batteries, these polyelectrolyte batteries must be operated at elevated temperatures (> 60 °C).

Based on the above principle, various polymer batteries of practical significance have been devised. Figure 8.12(a) schematically shows the Li-Al-alloy(anode)/LiBF$_4$-PC-DMF/polyaniline-BF$_4$ "button" battery, which was first commercialized by Bridgestone Corp. in collaboration with Seiko Electronic Components Ltd. mainly for the use as a back-up power source in intelligent telephones, FM receivers, VCR timers, fax machines, solar-powered calculators and watches. Weighing from 0.4 to 2.6 g, these tiny button cells have an operating voltage of 2 to 3 V, nominal specific capacity of 0.5 to 8 mAh, discharge currents in the region of 1 µA to 5 mA, a cyclability of 1000 cycles and a self-discharge rate of 3.3% per month.

Based on LiBF$_4$/ClO$_4$-PC electrolytes, BASF also developed conducting polymer batteries of the structures, such as the one shown in Figure 8.8(b), which show a discharge voltage of 3.3 V, cyclability of 500 cycles, energy density of 300 mWh/g and low discharge rate (Münstedt *et al.*, 1987). Although the flexibility of conducting polymer electrodes was exploited to advantage in BASF's cylindrical battery, the marketing of this battery appears not to be as successful as with Bridgestone's battery.

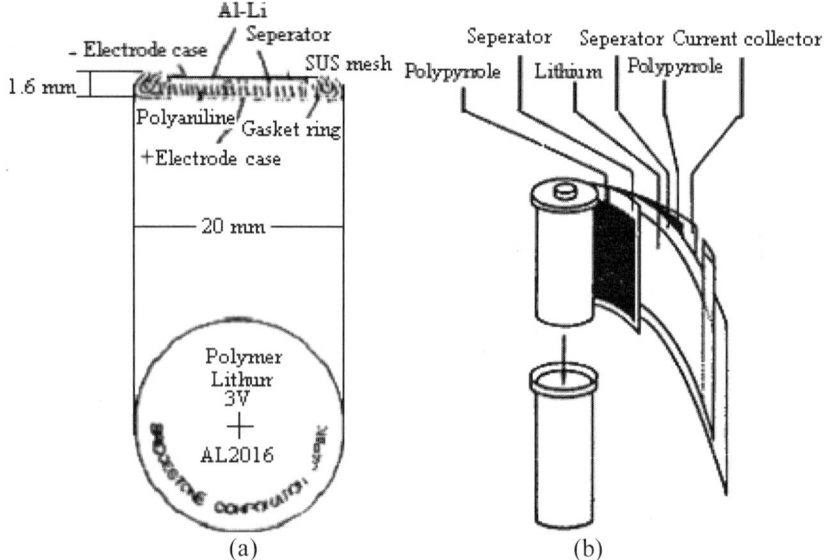

Figure 8.12. Schematic representation of (a) Bridgestone's commercialized Li/Conducting polymer battery (After Nakajima and Kawagoe, 1989, copyright 1989 Elsevier); (b) BASF's cyclindrical polypyrrole battery (After Münstedt *et al.*, 1987, copyright 1987 Elsevier).

8.4.2 Biofuel Cells

More recently, Heller and co-workers (Chen *et al.*, 2001) developed the smallest power source ever built based on enzyme-catalyzed redox reactions. These tiny enzyme-based biofuel cells were made from two 7 μm diameter carbon fiber electrodes coated with electrocatalysts, which were placed in a 1 mm grooves machined into a polycarbonate support as shown in Figure 8.13.

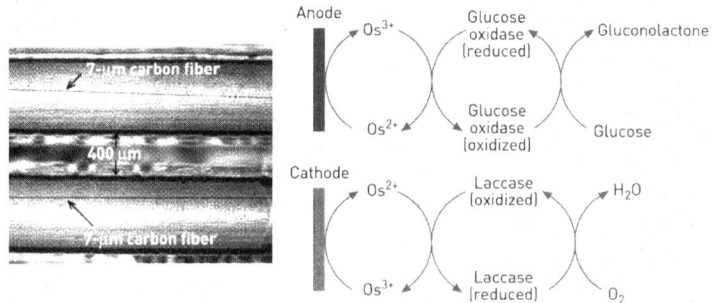

Figure 8.13. A section of the biofuel cell (left) and the associated redox electron-transferring steps of glucose oxidation and O_2 reduction (right) (After Ritter, 2001, copyright 2001 The American Chemical Society).

As can been seen in Figure 8.13, one of the electrodes is a glucose-oxidizing anode, while the other is an oxygen-reducing cathode. These biofuel cells have an output of 600 nW at 37 °C, sufficient to power micro-sized electronics, biosensor-transitters, and/or other implanted medical devices. To reduce the cell size, these authors used electroactive-enzyme-coated microfibers rather than flat electrodes, which eliminated the need for an electrode-separating membrane. The coatings of the electrodes consist of conducting copolymers grafted with enzymes (glucose oxidase for the anode and laccase the for cathode) and osmium complexes that form the redox centers to electrically "connect" the enzymatic reactions to carbon fibers.

There has been significant progress in miniaturizing biological sensors (Chapter 10), as for example, with the development of a subcutaneous glucose sensor to manage diabetes. However, the practical applications of the micro-sized biosensors have been severely hampered by the size of the power sources needed to run such implanted devices. The micro-sized biofuel cell developed by Heller's group sets a benchmark for future efforts in developing miniaturized sensors and microfluidics.

8.4.3 Carbon Nanotube Supercapacitors

Owing to their relatively high power and long cycle life (>10 000), electrochemical supercapacitors (also called ultracapacitors) have attracted a great deal of attention (Huggins, 1996). The resistivity of the electrode materials and the resistivity of the electrolyte within the porous structure of the electrode have been recognized as two key factors in determining the power and response frequency of an electrochemical capacitor. Carbon materials, including activated carbon and carbon fiber, have been widely used as electrodes for electrochemical capacitors. However, these high-surface-area carbons often have a wide pore size distribution ranging from micropores (< 20 Å) to macropores (> 500 Å). While most of the surface areas associated with the micropores are incapable of supporting an electrical double layer, ion migration into the macropores inevitably increases the electrolyte resistance. Consequently, the energy stored in conventional carbon-based electrochemical capacitors can be withdrawn only at low frequency or as DC power.

Having a well-defined pore size, highly accessible surface area, low resistivity and high mechanical/chemical stability, carbon nanotubes could work well as the electrode materials in electrochemical capacitors. In this context, Niu et al. (Niu et al., 1997) have developed high-power supercapacitors based on carbon nanotube electrodes. In particular, these authors prepared electrode sheets from the acid oxidized (Chapter 5), catalytically grown carbon nanotubes of narrow diameter distribution (centered at 80 Å). A typical supercapacitor device was fabricated with two 0.001 inch thick carbon nanotube electrodes having diameters of 0.5 inch and a surface area of 430 m^2/g, which are separated by a 0.001 inch thick polymer separator (Celgard) using 38 wt% H_2SO_4 as the electrolyte. These carbon nanotube supercapacitors possess specific capacitance of 102 and 49 F/g at 1 and 100 Hz, respectively, and most of the stored energy is accessible at frequencies below 100 Hz. An overall power density of more than 8 kW/kg can be easily obtained with these carbon nanotube supercapacitors.

8.5 Carbon Nanotube Nanoelectronics

8.5.1 Carbon Nanotube Nanowires

As discussed in Chapters 5 and 8, a few different approaches have been reported for experimentally measuring the electrical properties of individual nanotubes, which have allowed for the observation of the quantum of conductance. Aligned carbon nanotubes have been demonstrated to possess some novel collective properties with respect to their non-aligned counterparts (Chapter 6). In particular, the aligned structures should facilitate their effective incorporation into devices and their property characterization by conventional physicochemical techniques. Using standard four- and two-probe methods, Wang X. *et al.* (2001) constructed devices (Figure 8.14) for the measurement of electrical resistances of aligned carbon nanotube arrays in a current direction perpendicular (Figure 8.14(a)) and parallel (Figure 8.14(b)) to the nanotube axis.

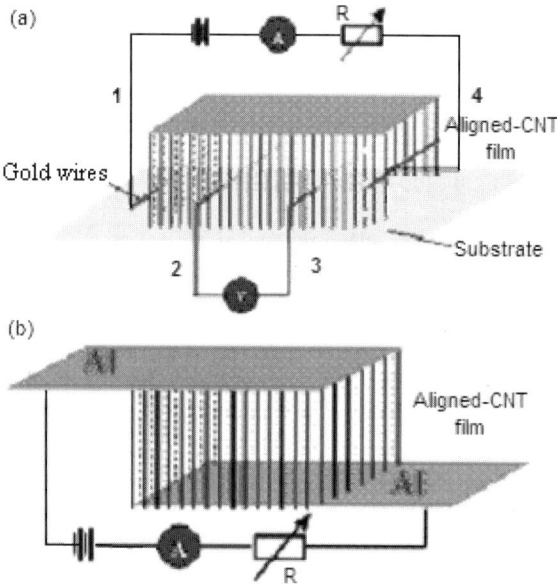

Figure 8.14. Schematic representation of the devices used for measuring electrical resistance of aligned carbon nanotube films in a direction perpendicular (a) and parallel (b) by a four- and two-probe method, respectively (After Wang X. *et al.*, 2001, copyright 2001 The American Chemical Society. Reproduced with permission)

These authors found that the aligned carbon nanotube films prepared by pyrolysis of iron (II) phthalocyanine are conducting in both parallel ($\rho_\parallel$ = 31.2 Ωcm) and perpendicular ($\rho_\perp$ = 4.6 Ωcm) directions, with resistivities that are about two orders of magnitude higher than that of their disordered counterpart (1.3×10^{-2} Ωcm).

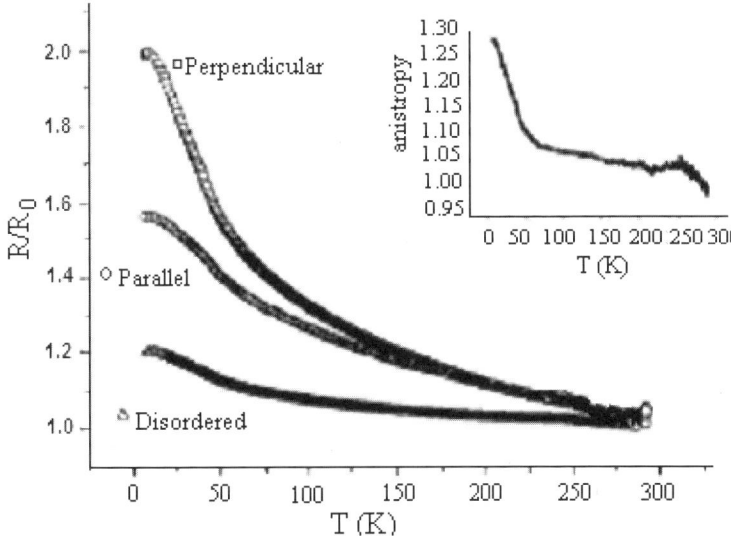

Figure 8.15. Temperature dependence of relative electrical resistances in current directions: perpendicular (□), parallel (O) to the nanotube axis, and the disordered carbon nanotube films (△). The inset shows the temperature dependence of anisotropy ($R_\perp/R_{\parallel}$) of aligned carbon nanotubes (After Wang et al., 2001, copyright 2001 The American Chemical Society. Reproduced with permission)

As shown in Figure 8.15, the anisotropy ($R_\perp/R_{\parallel}$) of electrical resistances increases with decreasing temperature (T), as do the relative resistances (R/R_0, here R_0 is the room temperature resistance) in both the parallel and perpendicular directions. The temperature dependence can be attributed to a three-dimensional variable range conduction (Mott and Daris, 1979). They further found that annealing and Br_2-doping could decrease the resistivities of the aligned carbon nanotube films by about two orders of magnitude due to reduced defect sites and increased carry density.

Although individual carbon nanotubes could behave as semiconductors or metallic conductors, current synthetic methods can only produce mixtures of the types of nanotubes. Semiconducting applications like transistors are therefore complicated by the co-existence of the metallic counterparts. To solve this problem, Avouris and co-worker (Collins et al., 2001) have devised procedures for separating metallic carbon nanotubes from those that are semiconductors by electrically oxidizing just the metallic nanotubes. By so doing, these authors first deposited ropes of mixed carbon nanotubes on a silicon oxide support that sits on a silicon wafer. They then made contact electrodes around the nanotube bundles by lithography. By applying a higher voltage to the circuit, these authors finally oxidized just the metallic nanotubes, leading to pure semiconducting nanotubes. Using a similar procedure, these authors have fabricated an array of field-effect transistors from SWNT ropes and further demonstrated that MWNTs can be peeled controllably shell-by-shell (Figure 8.16).

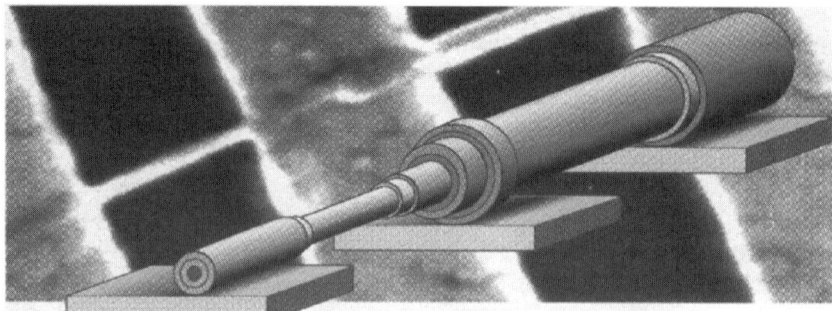

Figure 8.16. Schematic representation of a segment of a multiwall nanotube is stripped of three shells (green) while another portion (purple) is 10 shells thinner than the original thickness (red) (After Jacoby, 2001; Collins *et al.*, 2001. Reproduced with permission from Avouris, IBM Watson Research Center)

8.5.2 Carbon Nanotube Superconductors

Since the discovery of carbon nanotubes, scientists have been trying hard to search for superconducting properties. While there was no superconducting behavior observed for carbon nanotubes with relatively larger diameters (~1.6 nm), several groups reported the superconducting proximity effect in SWNTs connected to superconducting electrodes (Kasumov *et al.*, 1999; Morpurgo *et al.*, 1999). The superconducting proximity effect refers to a phenomenon by which the electronic properties of a normal conductor are modified by the presence of a nearby superconductor. However, it was demonstrated that carbon nanotubes and/or nanotube bundles with exceptionally small diameters exhibit superconductivity at relatively high temperatures (Tang *et al.*, 2001).

In particular, Sheng and co-workers (Tang *et al.*, 2001) demonstrated that 4 Å thick SWNTs synthesized inside the channels of zeolite crystals (Wang N. *et al.*, 2000) became superconducting at about 15 K. The observed superconductivity for small nanotubes could be attributed to the fact that the greater curvature associated with small nanotubes can significantly increase the interaction between electrons and lattice vibrations known as phonons – a property essential for superconductivity. This is consistent with theoretical predictions (Louie, 2001), that the smaller the nanotube diameter, the higher the superconducting temperature.

8.5.3 Carbon Nanotube Rings

As mentioned in Chapter 5, Smalley and co-workers (Liu *et al.*, 1997) were the first to observe the *in situ* formation of carbon nanotube rings (or fullerene crop circles) during the nanotube growth. Later, Avouris and co-workers (Martel *et al.*, 1999) demonstrated that similar nanotube rings could also be produced by ultrasonicating certain preformed carbon nanotubes at a relatively high yield (~50%). Furthermore,

Avouris and co-workers investigated the electronic properties of carbon nanotube rings and found that a nanotube ring's resistance increased with decreasing temperature – the opposite of what's expected for the metallic carbon nanotube. To understand this unusual behavior, these authors placed a nanotube ring across a pair of gold electrodes (Figure 8.17), and then injected an electron into the nanotube on one side and recovered it on the other side.

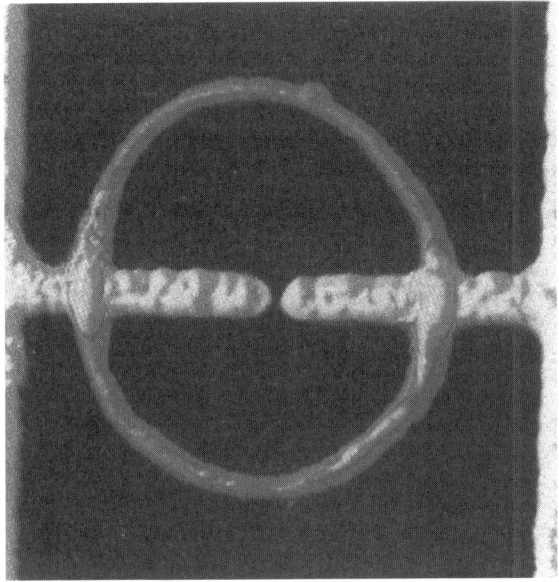

Figure 8.17. AFM image of a carbon nanotube ring (~600 nm in diameter) across a pair of gold electrodes for investigating electrical transport properties (After Jacoby, 1999; Martel *et al.*, 1999. Reproduced with permission from Avouris, IBM Watson Research Center)

Based on arguments from quantum mechanics, there is an equal probability for the electron to traverse the ring in a clockwise or anticlockwise fashion, which can be viewed as the electron wave splitting into two waves that travel in opposite directions. If each wave completes its traversal with its phase unchanged from its initial value, they then would meet at the starting point and interfere constructively to produce a standing wave. As standing waves do not participate in electrical conduction, electrical resistance of the carbon nanotube ring would be greater in their presence compared to that for their absence. If the electrons undergo energy changes during their trip around the ring due to collisions, however, their phases would change. This, in turn, would lead to a decreased interference-induced resistance. Since the probability of these phase-changing collisions diminishes with decreasing temperature, the nanotube ring's resistance should increase with decreasing temperature, consistent with the above experimental observation. These results indicate that carbon nanotubes are ideal waveguides for electrons.

8.5.4 Carbon Nanotube Nanocircuits

Dekker and co-workers (Postma *et al.*, 2001) have recently devised nanotube single-electron transistors (SETs) that can work at room temperature. They started with a metallic SWNT lying across two gold electrodes on an insulating surface, and then bent the nanotube sharply at two sites along its length with an AFM tip (Figure 8.18). This leads to two buckles separated by 20 nm, as shown in the circle in Figure 8.18.

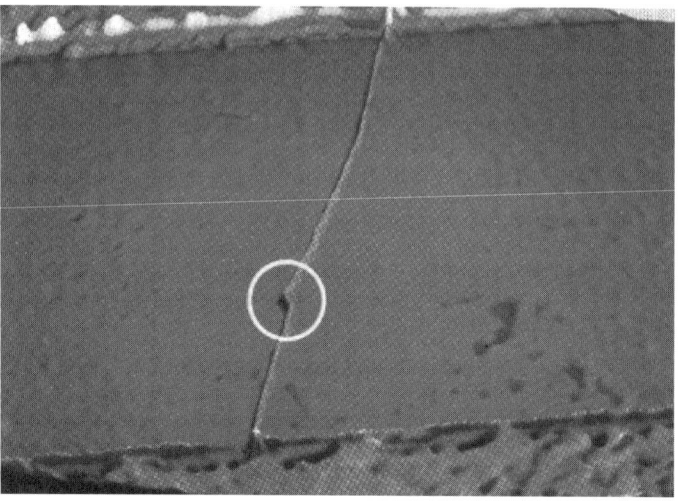

Figure 8.18 AFM image of the carbon nanotube single-electron transistor (After Dagani, 2001; Postma *et al.*, 2001. Reproduced with permission from Dekker and Postma, Delft University Technology)

When an appropriate voltage is applied to the "gate" underneath the island, electrons can only tunnel one at a time through the buckles as they act as electron barriers. As these nanoscale transistors can be switched between "on" and "off" with a single electron and work efficiently at room temperature, they have important implications for the use of carbon nanotubes as active components in molecular electronics.

As can be seen from the above discussion, carbon nanotubes are promising elements in various electronic circuits. In most of the cases, however, the difficulty associated with the circuit integration has prevented carbon nanotubes from being used in practical applications. More recently, Dekker and co-workers (Bachtold *et al.*, 2001) have gone a step further towards the integration of single-nanotube devices onto a chip for digital logic operations. In particular, these authors deposited a SWNT across two Au electrodes on an aluminum wire that serves as one of the FET electrodes (the gate, Figure 8.19), which is covered by a thin, insulating Al_2O_3 layer. As different local Al gates can easily be patterned to address nanotube transistors independently, the layout shown in Figure 8.19 allows for the integration of multiple nanotube transistors on a single chip (Figures 8.18 and 19).

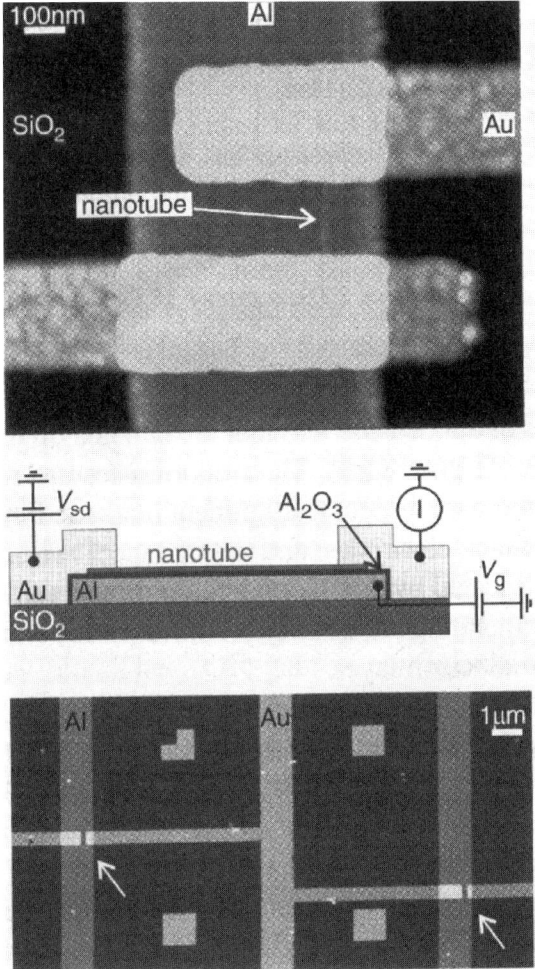

Figure 8.19. Device layout: (A) An AFM height image of a SWNT transistor; (B) schematic side view of the device, in which a semiconducting nanotube is contacted by two Au electrodes while an Al wire coated by a few-nanometer thick oxide layer is used as a gate; (C) height-mode atomic force microscope image of two nanotube transistors connected by an Au interconnect wire. The arrows indicate the position of the transistors. Four alignment markers are also seen (After Bachtold *et al.*, 2001, copyright 2001 American Association for the Advancement of Science)

These authors have demonstrated that all of the logic circuits thus prepared are fully reversible with a bias voltage of −1.5 V across the nanotube to be sufficient for logic applications. A voltage of 0 V can be used to represent a logical 0, while a voltage of −1.5 V represents a logical 1. Figure 8.20 shows schematic representations of one-, two-, three-transistor logic circuits based on the carbon nanotube FETs.

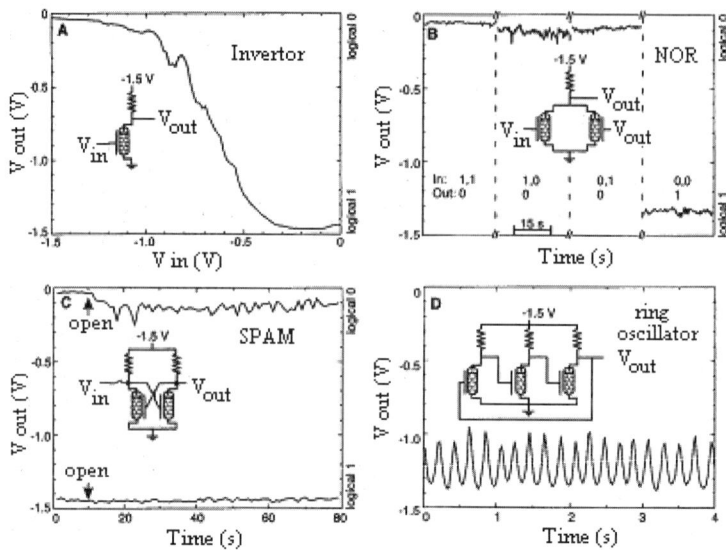

Figure 8.20. One-, two-, three-transistor logic circuits with carbon nanotube FETs. Inset shows schematic representations of the electric circuits: (A) output voltage as a function of the input voltage of a nanotube inverter. The resistance is 100 megaohms; (B) output voltage of a nanotube NOR for the four input states (1,1), (1,0), (0,1) and (0,0). The resistance is 50 megaohms; (C) output voltage of a flip-flop memory cell (SRAM) composed of two nanotube FETs. The output logic stays at 0 or 1 after the switch to the input has been opened. The two resistances are 100 megaohms and 2 gigaohms, respectively; (D) output voltage as a function of time for a nanotube ring oscillator. The three resistances are 100 megaohms, 100 megaohms and 2 gigaohms, respectively (After Bachtold *et al.*, 2001, copyright 2001 American Association for the Advancement of Science)

As shown in Figure 8.20(A), the output voltage is 0 V (logical 0) when the input voltage is set at –1.5 V (logical 1) because the nanotube resistance is much lower than the bias resistance. When V_{in} is 0 V (logical 0), the nanotube becomes non-conducting, setting the output voltage to –1.5 V (logical 1). It has been further found that the output voltage changes three times faster than the input voltage in the transition region with a voltage gain of 3.

By simply replacing the single transistor in the nanotube inverter with two transistors in parallel (Figure 8.20(B)), a NOR gate can be constructed that sets the output to 0 V (logical 0) when either or both of the inputs are a logical 1 ($V_{in} = -1.5$ V). This is because at least one of the nanotubes is conducting in this case. Only when both inputs are 0 V (logical 0) is the output a logical 1 as neither nanotube is conducting then. Besides, a flip-flop memory element (SRAM) that can memorize the output logical 0 or 1 even when the switch to the input has been opened (Figure 8.20(C)) and a three-transistor device that can be used to generate an oscillating ac voltage signal (Figure 8.20(D)) have been fabricated. By varying the device circuitry, various other logical gates including AND, OR, NAND and XOR can, in principle, be constructed.

8.5.5 Carbon Nanotube-based Random Access Memory (RAM) For Molecular Computing

As an alternative way of using carbon nanotubes as active components and molecular wires in molecular-scale electronics, Lieber and co-workers (Rueckers *et al.*, 2000) have recently devised a criss-crossed carbon nanotube relay which consists of a set of parallel SWNTs on a substrate and a set of perpendicular SWNTs suspended over the parallel nanotube array (Figure 8.21).

Figure 8.21. Three-dimensional view of a suspended cross-bar array of carbon nanotubes of four junctions, showing two elements in the ON (contact) state and two elements in the OFF (separated) state. The substrate is made of a highly doped silicon conducting layer (dark gray) coated with a thin dielectric SiO_2 layer (light gray). While the lower nanotubes lie down directly on the dielectric film, the upper nanotubes are suspended on a periodic array of non-conducting inorganic or organic supports (gray blocks) (After Rueckers *et al.*, 2000, copyright 2000 American Association for the Advancement of Science)

As shown in Figure 8.21, each cross point in the criss-crossing nanotube array corresponds to a device element with two switchable ON (a contact upper-to-lower nanotube junction with a low resistance) and OFF (a separated upper-to-lower nanotube junction with a high resistance). A device element could be switched between these well-defined ON and OFF states by applying a certain voltage with alternatively changing bias to the two nanotubes. Theoretically, calculations have demonstrated that these reversible and bistable nanotube device elements could be used to construct non-volatile RAM and logic function tables (Rueckers *et al.*, 2000). Owing to the tiny size and good electrical and mechanical properties intrinsically characteristic of SWNTs, an integration level approaching 10^{12} elements per square centimeter and an element operation frequency in excess of 100 gigahertz are achievable.

8.6 DNA Molecular Wires and DNA Computing

8.6.1 DNA Molecular Wires

Soon after Watson and Crick's discovery of the double-helix structure of DNA in 1953 (Watson and Crick, 1953), scientists made a bold suggestion that deoxyribonucleic acid (DNA), the carrier of genetic information in all living things, might serve as a conductor for electrical charges.

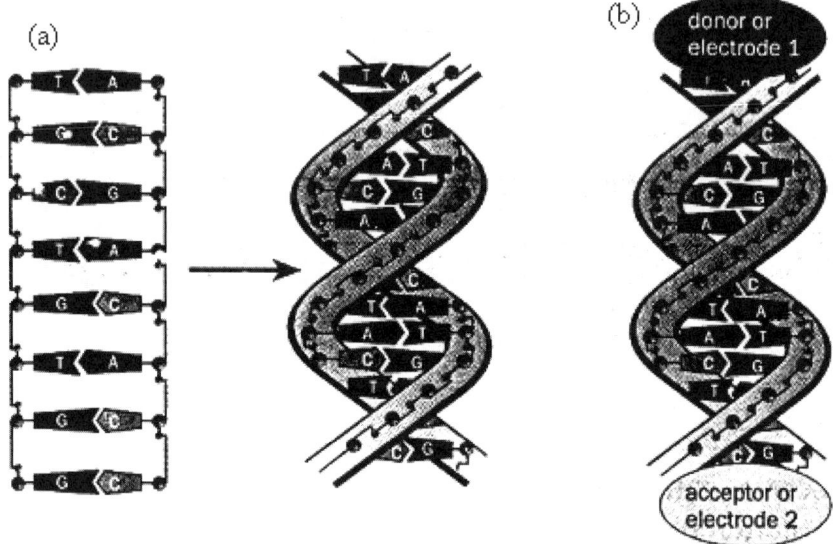

Figure 8.22. A typical molecular structure for the DNA double-helix (After Dekker and Ratner, 2001, copyright 2001 Physics World. Reproduced with permission. See also Chapter 1 this book)

As shown in Figure 8.22, DNA consists of two linear polymer chains of repeating sugar molecules and phosphate groups that form the double-helix structure via hydrogen bonding. Each sugar group is attached to one of four bases – guanine (G), cytosine (C), adenine (A) and thymine (T). In the double-stranded DNA, adenine (A) bases on one strand pair up with the complementary thymine (T) bases on the other strand, while guanine (G) pairs up with cytosine (C). The base pairs look like the rungs of a helical ladder with some of the electron orbitals belonging to the bases overlapping each other along the long axis of the DNA. These so-called stacking interactions resemble those of one-dimensional molecular conductors (*e.g.* TTF-TCNQ) described in Chapter 2. It is these internal tunnels of stacked π-orbitals that make the DNA structure ideal for electron transfer. On the other hand, the DNA chains are usually surrounded by positive "ions", as the phosphate groups on the backbone are negatively charged. As such, DNA could also be a good ionic conductor.

Although provocative, the idea of conducting electricity by DNA had not attracted too much attention in the scientific community until about a decade ago, when Barton and co-workers (Murphy *et al.*, 1993) reported the evidence that DNA chains might indeed be molecular wires. In their ground-breaking experiment, Barton and co-workers (Murphy *et al.*, 1993) measured the fluorescence generated by an excited molecule and found that the fluorescence was effectively quenched when it was attached to a DNA molecule. This observation suggested that the charge on the excited donor molecule had been transferred through the DNA structure to a nearby acceptor molecule, indicating that DNA is a conducting molecular wire. For an extended molecule with a donor unit at one end and an acceptor at the other end, electron transfer can take place by two different mechanisms. The single-step electron-tunneling process involves the transfer of electrons from the donor to the acceptor directly without any energy exchange with the molecule, and the electron is never localized during the transfer. The rate at which such electron-tunneling processes occur decays exponentially with the distance between donor and acceptor, eliminating the possibility for electron transfer over very long distances on any reasonable timescale. The possible mechanism for long-distance electron transfer is thus generally referred to as "thermal hopping", in which electron transfer from donor to acceptor proceeds in a multi-step process with the electron being localized on and exchanging energy with the molecule at each of the steps. Compared with the above coherent tunneling process, the hopping process allows for the transfer of the charge over long distances in a diffusive manner.

As the conduction of charges through DNA has important implications for many issues including DNA damage, genetic detection, and microelectronics, Barton's report provoked considerable interest worldwide. Since then, many other scientists throughout the world have been trying to set up creative experimental systems for directly probing DNA electronic properties and for elucidating the conduction mechanisms. However, the results from early experiments were full of controversy (Dekker and Ratner, 2001; Diederichsen, 1997; Wilson, 1999). While it seems clear that charge carriers can hop along the DNA over at least a few nanometers, the results from electronic transport through DNA molecules over a larger distance are still wildly divided. For example, Fink and Schönenberger (Fink and Schönenberger, 1999) showed efficient conduction with resistances as low as a few MΩ for DNA ropes as long as 600 nm, whilst Dekker and co-workers (Storm *et al.*, 2001) observed no conduction with an infinite resistance (> 10^7 MΩ) for DNA chains even at a length of 40 nm.

The recent significant developments, both at the experimental and theoretical frontiers, have led to substantial clarification of the electrical properties of DNA. Now it is recognized that the sequence of the base pairs along the DNA molecule plays an important role in regulating the charge transfer in DNA, and that it is the extensive variability of DNA sequences that led to those inconsistent results in various early experiments. It has been shown that a hole is more stable on a G–C base pair than on an A–T base pair. Consequently, a hole will localize on a particular G–C base pair, whilst the A–T pairs act as a barrier to a hole transfer (Figure 8.23).

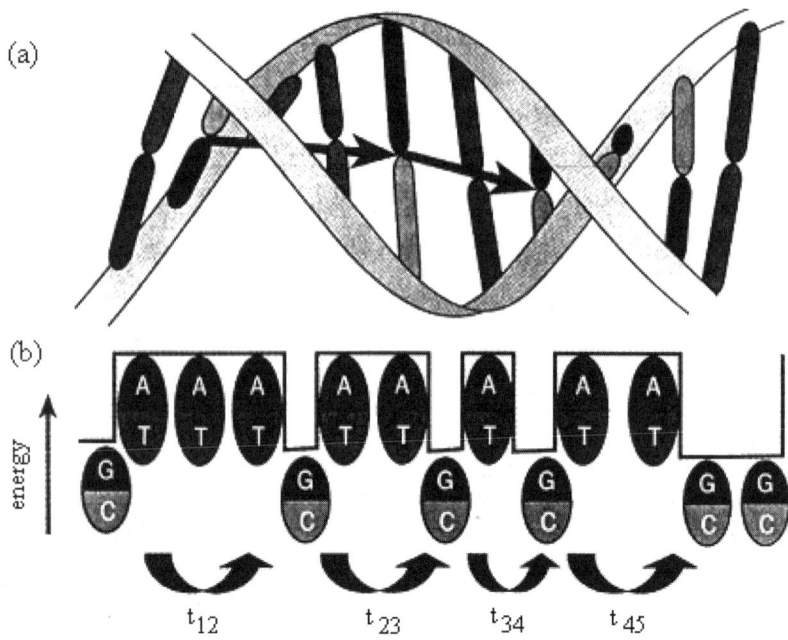

Figure 8.23. (a) Schematic representation of the hopping mechanism of the charge carriers along the length of the DNA molecule from one G–C base pair to the next; (b) the relative energies of G–C and A–T base pairs in this DNA sequence, showing that a positively charged hole has a lower energy on the G–C sites and moves from one G–C pair to the next by coherent tunneling through the A–T sites. The overall motion from the first base pair to the last, however, is an incoherent hopping mechanism, in which the charged carrier is localized along the path. The total time it takes to move along the path is the sum of the times it takes to jump between individual G–C base pairs, $t_{12} + t_{23} + \ldots + t_{45}$ (After Dekker and Ratner, 2001, copyright 2001 Physics World. Reproduced with permission)

As mentioned earlier, the rate of coherent charge transfer decreases exponentially with the distance traveled. When the distance between G–C base pairs becomes too long for coherent charge carriers to jump, thermal hopping becomes the dominant mechanism with which charge carriers transfer along the DNA chain (Giese and Spichty, 2000). In some cases, both charge-tunneling and thermal-hopping mechanisms have been demonstrated to interplay. In particular, Giese and co-workers (Giese *et al.*, 2001) have recently demonstrated that the rate of charge transfer between two G–C base pairs decreases with increasing separation only if there are no more than three A–T base pairs between them. If the two G–C base pairs are separated by more A–T base pairs, only a weak distance dependence is observed (Figure 8.24).

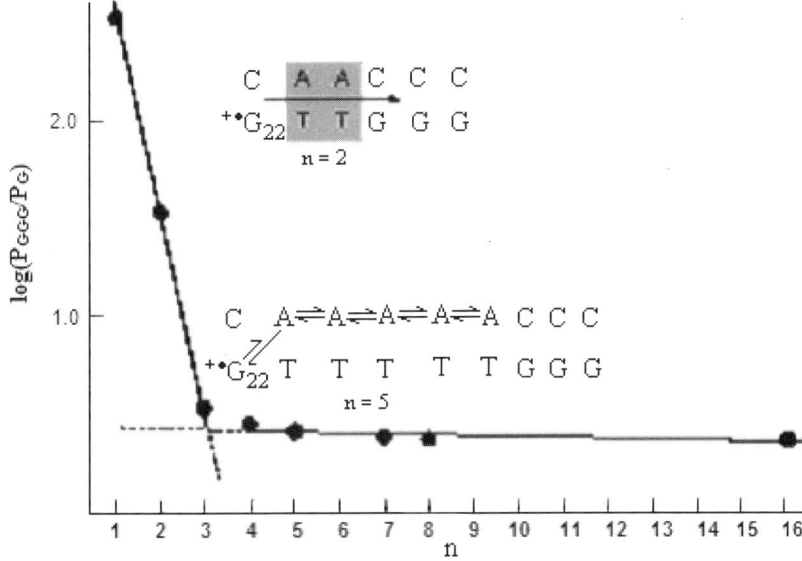

Figure 8.24. Plot of the relative efficiency of the charge transfer [log(P_{GGG}/P_G)] against the number of the A–T base pairs. The steep line corresponds to the coherent tunneling charge transfer while the flat line shows a weak distance dependence associated with the thermal hopping mechanism (After Giese *et al.*, 2001, copyright 2001 Macmillan Magazines Limited. Reproduced with permission)

These authors attributed the above observation to a shift from coherent tunneling over a short distance to a process mediated by thermal hopping of charges among the bridging A–T base pairs over long distances. Given that suitable substitution can rationally change the sequence of DNA molecules, the above results clearly indicate new opportunities for DNA nanoelectronics.

While the electronic transport through long DNA chains is still uncertain, Braun and co-workers have used a DNA polyanion as a template for the assembly of electronic materials. As shown in Figure 8.25, Braun *et al.* (1998) first attached 12-base oligonucleotides end-derivatized with a disulfide group onto two gold electrodes 12–16 μm apart through sulfur gold interactions. They then bridged the gap between the electrodes with a single DNA strand of length 16 μm having 12-base sticky ends with complementary bases to the two 12-base oligonucleotides already attached to the gold electrodes. The Na^+ counterions were then replaced by Ag^+ ions so that the DNA chain could be used as a seed to grow a thin silver-ion wire. The silver-ion-exchanged DNA was finally reduced to form a DNA conducting wire coated with nanometer-sized metallic silver aggregates (Figure 8.25).

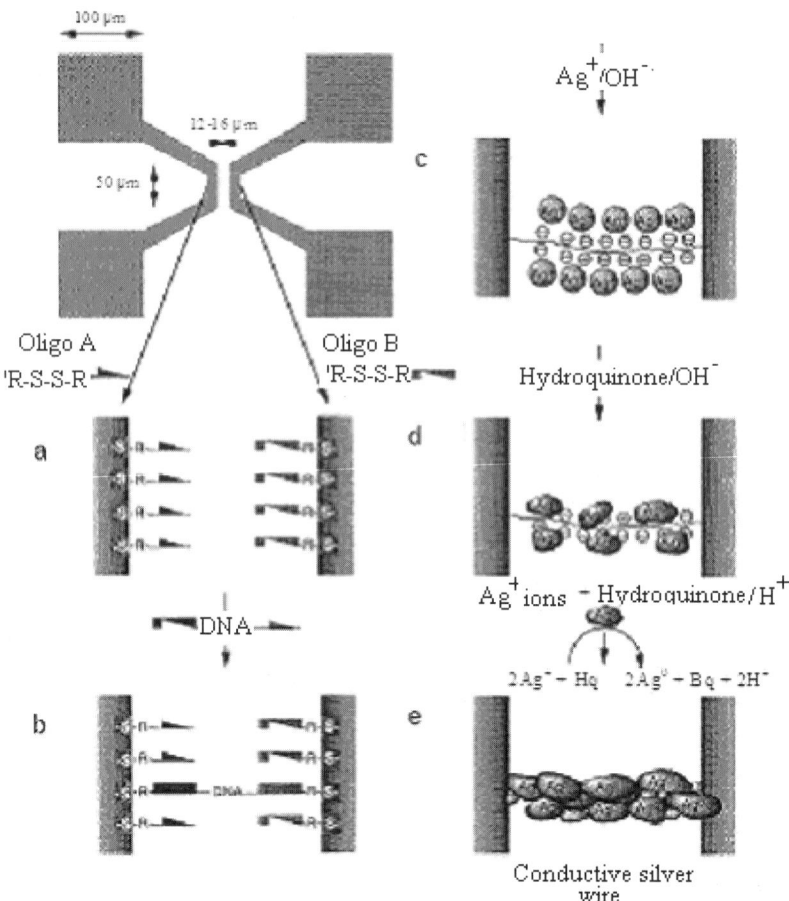

Figure 8.25. Top left image schematically shows the electrode pattern (0.5 × 0.5 mm) used for the formation of a long conducting DNA molecular wire: (a) oligonucleotides with two different sequences attached to the electrodes; (b) DNA bridge connecting the two electrodes; (c) silver-ion-loaded DNA bridge; (d) silver ions were reduced into metallic silver nanoparticles, covering the DNA bridge; (e) fully developed silver wire with the DNA template (After Braun et al., 1998, copyright 1998 Macmillan Magazines Limited. Reproduced with permission)

The resultant I–V curves are shown in Figure 8.26. A zero-current plateau of the order of ca. 0.5 V and a differential resistance of ca. 7 MΩ is observed. The plateau could be permanently eliminated to give ohmic behavior by applying a bias voltage higher than ca. 50 V. As shown by the two I–V curves in the two insets, both the DNA bridge without silver coating and the silver deposition without a DNA bridge are non-conducting with a resistance higher than 10^{13} Ω.

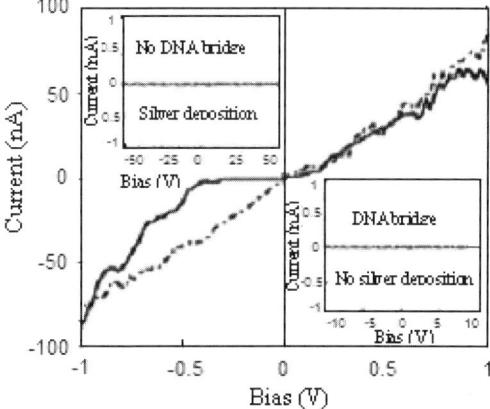

Figure 8.26. Experimentally observed *I–V* curves of a DNA-templated silver wire. The extensively grown silver wire shows a narrow current plateau (solid curve), which can be permanently eliminated to give an ohmic behavior by applying a 50 V bias voltage (dashed-dotted line). Insets give the *I–V* curves of a DNA bridge without silver deposition (bottom) and silver deposition without a DNA bridge (top), showing an insulating behavior in both cases. Note that the current scale in both insets is 100 times smaller than in the main graph (After Braun *et al.*, 1998, copyright 1998 Macmillan Magazines Limited. Reproduced with permission)

Therefore, the silver-deposited DNA bridge is responsible for the electric current transferred through the two electrodes. The unique binding properties that control the DNA bridge formation, coupled with the different metal types, growth conditions and/or post-growth treatments that can be used, could make this particular approach to DNA electronics very useful in the development of future electronic circuits.

8.6.2 DNA Computing on Chips

Although DNA has been demonstrated to be a potential component for molecular electronics, the practical application of DNA molecular wires will be servely limited without assembling them into appropriate integrated circuits. While the majority of research into DNA during the last half-century or so has been focused on its biological properties, the enormous capability of the DNA sequence for information storage has just recently been exploited for molecular computing applications.

Just like the electronic computers using electrons to carry information and transistors to process it, the DNA molecule can store in its sequence of four bases vast quantities of data that can be manipulated in a highly parallel manner by natural enzymes. In 1994, Adleman (Adleman, 1994) reported the first DNA chip that could serve as a DNA computer.

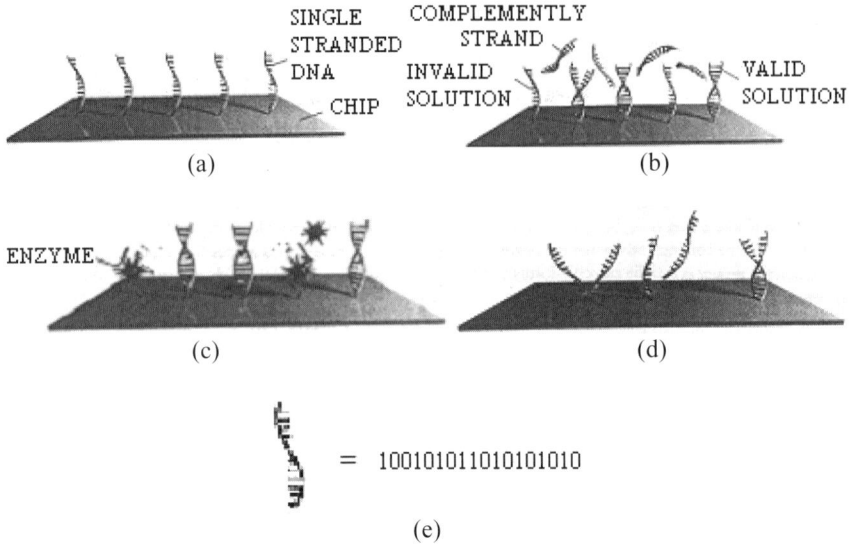

Figure 8.27. The principle for DNA computers: (a) single DNA strands are attached to a silicon chip. They encode all possible values of the variables in an equation that the researchers want to solve; (b) copies of a complementary strand – which encodes the first clause of the equation – are poured onto the chip. These copies attach themselves to any strand that represents a valid solution of the clause. Any invalid solutions remain a single strand; (c) an enzyme removes all the single strands; (d) other processes melt away the added complementary strands. These steps are repeated with all the clauses of the equation; (e) the DNA strand that survives this whole process represents the solution to the whole equation (After Lieber, 2001, copyright 2001 Scientific American)

Adleman's key insight was that all possible values of the variables in and all the clauses of an equation could be encoded by specific sequences in strands of DNA. After having immobilized those single DNA strands (Watson strands), representing all possible values of the variables in an equation onto a chip surface (Figure 8.27(a)), Adleman then started to pour copies of complementary strands (Crick strands) that encodes the first clause of the equation onto the chip. The Crick strands attached themselves to Watson strands that represent a valid solution of the first clause. The remaining single-stranded DNA chains are those that do not satisfy the first clause (Figure 8.27(b)), which are destroyed by enzymes (Figure 8.27(c)). The chip was then heated to melt away the added complementary strands, followed by washing to produce a fresh collection of Crick strands already satisfying the first clause. These steps are repeated with each of the remaining clauses of the equation (Figure 8.27(d)). At the end of the cycles, only the DNA strand that represents the solution to the equation survived.

Although the DNA computer is still in early development, it has been demonstrated that a DNA-based computer could even be used to solve problems (*e.g.* the Hamiltonian path problem, see Adleman, 1998) that are particularly difficult for ordinary computers. Further progress in DNA computing should lead to the development of fully organic computing devices.

8.7 References

Adleman, L.M. (1994) *Nature* **266**, 1021.
Adleman, L.M. (1998) *Sci. Am.* **August**, 54.
Araki, H., Zakhidov, A.A., Tada, K., Yakushi, K., Yoshino, K. (1996) *Synth. Met.,* **77**, 291.
Araki, H., Zakhidov, A.A., Saiki, E., Yamasaki, K., Yakushi,, K., Yoshino, K. (1995) *Jpn. J. Appl. Phys.,* **34**, LI041.
Bachtold, A., Hadley, P., Nakanishi, T., Dekker, C. (2001) *Science* **294**, 1317.
Bednorz, J.G., Mueller, K.A. (1986) *Z. Phys.* **64**, 189.
Bitter, S. (2001) *C&E News* **September 3**, 10.
Braun, E., Eichen, Y., Sivan, U., Ben-Yoseph, G. (1998) *Nature* **391**, 775.
Burroughes, J.H., Friend, R.H. (1991) in *Conjugated Polymers*, Brédas, J.L., Silbey, R (eds), Kluwer Academic Publishers, Dordrecht.
Catrysse, J. (1990) in *Proceedings of 7th International Conference Electromagnetic Compatibility*, York, UK.
Chandrasekhar, P. (1999) *Conducting Polymers, Fundamentals and Applications: A Practical Approach*, Kluwer Academic Publishers, Dordrecht.
Chandrasekhar, P., Naishadham, K. (1999) *Synth. Met.* **115**, 120.
Chen, T., Barton, S.C., Binyamin, G., Gao, Z., Zhang, Y., Kim, H.-H., Heller, A. (2001) *J. Am. Chem. Soc.* **123**, 8630.
Chung, D.D.L. (2001) *Carbon* **39**, 279.
Collins, P.C., Arnold, M.S., Avouris, P. (2001) *Science* **292**, 706.
Dagani, R. (2001) *C&E News* **July 9**, 10.
Dagotto, E. (2001) *Science* **293**, 2410.
Dai, L., Lu, J., Matthews, B., Mau, A. W. H. (1998) *J. Phys. Chem.* **102**, 4049.
Dekker, C., Ratner, M.A. (2001) *Phys. World* **August**, 29.
Dhawan, S.K., Singh, N., Venkatachalam, S. (2002) *Synth. Met.* **125**, 389.
Diederichsen, U. (1997) *Angew. Chem. Int. Ed.* **36**, 2317.
Fink, H.-W., Schönenberger, C. (1999) *Nature* **398**, 407.
Giese, B., Amaudrut, J., Köhler, A.K., Spormann, M., Wessely, S. (2001) *Nature* **412**, 318.
Giese, B., Spichty, M. (2000) *ChemPhysChem* **1**, 195.
Greenham, N.C.; Friend, R.H. (1995) *Solid State Phys.* **49**, 1.
Haddon, R.C. (1992) *Acc. Chem. Res.* **25**, 127.
Haddon, R.C., Hebard, A.F., Rosseinsky, M.J., Murphy, D.W., Duclos, S.J., Lyons, K.B., Miller, B., Rosamilia, J.M., Fleming, K.M., Kortan, A.R., Alarum, S.H., Makhija, A.V., Muller, A.J., Eick, R.H., Zahurak, S.M., Tycko, R., Dabbagh, A., Thiel, F.A. (1991) *Nature* **350**,320.
Hammond, G.S., Kuck, V.J. (eds) (1992), *Fullerenes: Synthesis, Properties, and Chemistry of Large Carbon Clusters*, ACS Symposium Series 481, American Chemical Society, Washington DC.
Harrison, W.A. (1979) *Solid State Theory*, Dover Publications, Inc., New York.
Hebard, A.F., Rosseinsky, M. J., Haddon, R.C., Murphy, D.W., Glarum, S.H., Palstra, S.H., Ramirez, T.T.M., Kortan, A.R. (1991) *Nature* **350**,600.
Hide, F., Díaz-García, M.A., Schwartz, B.J., Heeger, A.J. (1997) *Acc. Chem. Res.* **30**, 430.
Hong, Y.K., Lee, C.Y., Jeong, C.K., Sim, J.H., Kim, K., Joo, J., Kim, M.S., Lee, J.Y., Jeong, S.H., Byun, S.W. (2001) *Curr. Appl. Phys.* **1**, 439.
Huggins, R.A. (1996) *Philos. Trans. R. Soc. Ser. A* (London), **354**, 1555.

Jacoby, M. (1999) *C&E News* Oct. **4**, 31.
Jacoby, M. (2001) *C&E News* April **30**, 13.
Jacoby, M. (2002), *C&E News* **November 4**, 31.
Horowitz, G. (1998) *Adv. Mater.* **10**, 365.
Kasumov, A.Y., Deblock, R., Kociak, M., Reulet, B., Bouchiat, H., Khodos, II., Gorbatov, Y.B., Volkov, V.T., Journet, C., Burghard, M. (1999) *Science* **284**, 1508.
Kaynak, A. (1996) *Mater. Res. Bull.* **31**, 845.
Kirtley, J.R., Tsuei, C.C. (1996) *Sci. Am.* **August**, 50.
Koul, S., Chandra, R., Dhawan, S.K. (2000) *Polymer* **41**, 9305.
Lee, C.Y., Lee, D.E., Joo, J., Kim, M.S., Lee, J.Y., Jeong, S.H., Byun, S.W. (2001) *Synth. Met.* **119**, 429.
Lee, C.Y., Song, H.G., Jang, K.S., Oh, E.J., Epstein, A.J., Joo, J. (1999) *Synth. Met.* **102**, 1346.
Lieber, C.M. (2001) *Sci. Am.* **285**, 51.
Liu, J., Dai, H., Hafner, J.H., Colbert, D.T., Smalley, R.E., Tans, S.J., Dekker, C. (1997) *Nature* **385**, 780.
Louie, S.G. (2001) *Top. Appl. Phys.* **80**, 113, and references cited therein.
Lovinger, A.J., Rothberg, L.J. (1996) *J. Mater. Res.* **11**, 1581.
Martel, R., Shea, H.R., Avouris, P. (1999) *J. Phys. Chem. B* **103**, 7551.
Morpurgo, A.F., Kong, J., Marcus, C.M., Dai, H. (1999) *Science* **286**, 263.
Mott, N.F., Daris, E.A. (1979) *Electronic Processes in Non-Crystalline Materials*, Clarendon Press, Oxford.
Münstedt, H., Köhler, G., Möhwald, H., Naegele, D., Bitthin, R., Ely, G., Meissner, E. (1987) *Synth. Met.* **18**, 259.
Murphy, C.J., Arkin, M.R., Jenkins, Y., Ghatlia, N.D., Turro, N.J., Barton, J.K. (1993) *Science* **262**, 1025.
Musamesh, S.M., Ahmud, M.S., Zihlif, A.M., Abdelazeez, M., Malinconico, M., Martuscelli, E., Ragosta, G. (1992) *Mater. Sci. Eng.* **B14**, 1.
Mylnikov, V.S. (1994) *Photoconducting Polymers, Adv. Polym. Sci. Vol.* **115**, Springer-Verlag: Berlin.
Nakajima, T., Kawagoe, T. (1989) *Synth. Met.* **28**, C629.
Nigrey, P.J., MacInnes, D., Jr., Nairns, D.P., MacDiarmid, A.J. (1981) *J. Electrochem. Soc.* **128**, 1651.
Niu, C., Sichel, E.K., Hoch, R., Moy, D., Tennent, H. (1997) *Appl. Phys. Lett.* **70**, 1480.
Osaka, T., Momma, T., Nishimura, K., Kakuda, S., Ishii, T. (1994) *J. Electrochem. Soc.* **141**, 1994.
Paul, C.R. (ed.) (1992) *Introduction of Electromagnetic Compatibility*, John Wiley & Sons, Inc., New York.
Pomposo, J.A., Rodríguz, J., Grande, H. (1999) *Synth. Met.* **107**, 111.
Postma, H.W.C., Teepen, T., Yao, Z., Grifoni, M., Dekker, C. (2001) *Science* **293**, 76.
Rhoderick, E.H., Williams, R.H. (1988) *Metal-Semiconductor Contacts*, Oxford University Press, Oxford.
Ritter, S. (2001) *C&E News* **Sept. 3**, 10.
Rueckers, T., Kim, K., Joselevich, E., Tseng, G.Y., Cheung, C.-L., Lieber, C.M. (2000) *Science* **289**, 94.
Sariciftci, S., Heeger, A.J. (1994) *Int. J. Mod. Phys. B* **8**, 237.
Schlebusch, C., Kessler, B., Cramm, S., Eberhardt, W. (1996) *Synth. Met.* **77**, 151.
Schön, J.H., Kloc, Ch., Batlogg, B. (2000) *Nature* **408**, 549.
Schön, J.H., Kloc, Ch., Batlogg, B. (2001) *Science* **293**, 2432.
Service, R.F. (2001) *Science* **293**, 1570.
Storm, A.J., van Noort, J., de Vries, S., Dekker, C. (2001) *Appl. Phys. Lett.* **79**, 3881, and references cited therein.

Tang, Z.K., Zhang, L.Y., Wang, N., Zhang, X.X., Wen, G.H., Li, G.D., Wang, J.N., Chan, C.T., Sheng, P. (2001) *Science* **292**, 2462.
Unsworth, J., Kaynak, A., Lunn, B., Beard, G.E. (1993) *J. Mat. Sci.* **28**, 3307.
Wang, X., Liu, Y., Yu, G., Xu, C., Zhang, J., Zhu, D. (2001) *J. Phys. Chem. B* **105**, 9422.
Wang, N., Tang, Z.K., Li, G.D., Chen, J.S. (2000) *Nature* **408**, 50.
Watson, J.D., Crick, F.H.C. (1953) *Nature* **171**, 737.
Wei, Y., Tian, J., MacDiarmid, A.G., Masters, J.G., Smith, A. L., Li, D. (1993) *J. Chem. Soc., Chem. Commun.* 603, and references cited therein.
Willey, R. (1986) *Emittance and Reflectance of Various Materials in the 2 to 20 Micrometers Spectral Region*, **643**, SPIE Press, Washington DC.
Wilson, E.K. (1999) *C&E News* **August 23**, 43.
Yoshino, K., Tada, K., Yoshimoto, K., Yoshida, M., Kawai, T., Araki, H., Hamaguchi, M., Zakhidov, A.A. (1996) *Synth. Met.* **78**,301.
Zakhidov, A.A. (1991) *Synth. Met.* **41**, 3393.
Zakhidov, A.A., Araki, H., Tada, K., Yakushi, K., Yoshino, K. (1995) *Phys. Lett. A* **205**,317.
Zakhidov, A. A., Araki, H., Tada, K., Yoshino, T. K. (1996) *Synth. Met.* **77**, 127, and references cited therein.
Zakhidov, A.A., Araki, H., Yoshino, K. (1999) in *Fullerene Polymers and Fullerene Polymer Composites*, Eklund, P.C., Rao, A.M. (eds), Springer-Verlag, Berlin.

Chapter 9

Photonic Devices

9.1 Introduction

As can be seen in the preceding chapters, semiconducting polymers have shown a number of commercial advantages over their inorganic counterparts, such as silicon and gallium arsenide, for use as active components in various optoelectronic devices. Although liquid crystal displays (LCDs) are dominating the flat panel display market, for example, the emissive displays could change the display marketplace of the new millennium (Tullo, 2000). Unlike LCDs, which are essentially passive light valves modulating the transmitted luminance of a white backlight through color filters (Tremblay, 1999), emissive flat panel displays based on engineered materials, such as light-emitting semiconductors and phosphor pixels, can emit light directly upon electronic excitation (Korczynski, 1999). The development of electroluminescent devices based on inorganic semiconductors and conjugated organic molecules has, for decades, been an active research area (Williams and Hall, 1987). The use of polymeric electroluminescent diodes for flat panel displays offers many advantages, including their versatility for fabrication (especially over a large area), lightness, flexibility, low operating voltage and the ease with which color tuning of light emission can be achieved (Bradley and Tsutsui, 1995; Burroughes et al., 1990; Dai *et al.*, 2001; Greenham and Friend, 1995; Hide *et al.*, 1997; Kraft *et al.*, 1999). Therefore, the report from Cambridge in 1990 of electroluminescent light emission from poly(p-phenylene vinylene), PPV (Burroughes *et al.*, 1990), provoked considerable excitement (Heeger and Long, 1996; Dagani, 1995). Many other groups around the world quickly picked up the significance of this discovery (Conwell and Miller, 1993; Ohnishi *et al.*, 1995; Xie, *et al.*, 1995; Staring, 1995). Various polymer light-emitting devices of novel features (*e.g.* patterned multi-color emissions) to be attractive for practical applications have been made through the use of some clever device design and construction (Dai, 1999a). Along with these developments, Moses observed the laser action of MEH-PPV in solution in 1992 (Moses, 1992). Since then, the laser actions of conjugated polymer films under both optical and electrical pumping also attracted considerable attention (Hide *et al.*, 1996a).

358 Intelligent Macromolecules for Smart Devices

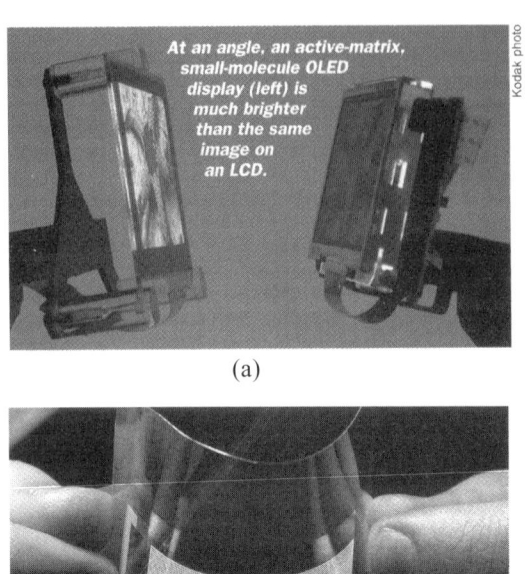

(a)

(b)

Figure 9.1. (a) An organic light-emitting display (left) in comparison with a liquid crystal display (right) (After Tullo, 2000, copyright 2000 Kodak. Reproduced with permission); (b) flexible organic light-emitting display (After Freemantle, 2001, copyright 2001 Universal Display Corporation. Reproduced with permission)

On the other hand, electron gun-based cathode ray tubes or electrons generated by field emission from sharp tips have been widely used for numerous applications (*e.g.* in TV sets or electron microscopies). In these cases, thermionically emitted electrons from hot tungsten wires or other heated materials with low work functions are normally utilized to excite phosphorescent materials under ultrahigh vacuum for emitting light (de Heer *et al.*, 1997). Although carbon fibers a few μm in diameter have been used as electron emitters, their practical applications have so far been hampered by poor reproducibility and rapid deterioration of the tip (de Heer *et al.*, 1997). Carbon nanotubes, consisting of concentric shell(s) of graphitic sheet(s) with nanometric tips, have been demonstrated to be ideal electron emitters, though the emission current from a single tube is still somewhat limited (de Heer *et al.*, 1997). To overcome this limitation, various large-scale well-aligned nanotube arrays have been prepared for flat panel display applications (Dai and Mau, 2000).

On another front, the interaction of buckminsterfullerenes with light has attracted considerable interest among scientists in the exploration of applications related to photophysical, photochemical and photo-induced charge transfer properties of fullerenes. The observation of photovoltaic effects arising from the photo-induced charge transfer at the interface between conjugated polymers as donors and a C_{60} film as an acceptor (Kraabel et al., 1993, 1996; Sariciftci et al., 1992; Sariciftci and Heeger, 1994) suggests interesting opportunities for improving the energy conversion efficiencies of photovoltaic cells based on conjugated polymers.

In this chapter, recent progress in research on the device construction/performance of polymeric light-emitting diodes, polymer lasers, C_{60} and carbon nanotube photonic devices is briefly reviewed. As there are many excellent reviews and specialized monographs on liquid crystal polymers (Collyer, 1992; Isayev et al., 1995; Wu and Yang, 2001) and non-linear optical polymers (Messier et al., 1989; Pellegrino et al., 1996; Prasad and Ulrich, 1988; Prasad and Nigam, 1991; Williams D., 1983) already published, interested readers are referred to other appropriate literature.

9.2 Light-emitting Polymer Displays

9.2.1 Device Construction

Since the discovery of electroluminescent light emission from poly(*p*-phenylene vinylene), PPV, in 1990 (Burroughes et al., 1990), various other electroluminescent conjugated polymers have been synthesised (Chapters 2 and 9) (Bradley, 1992; Burn et al., 1992). Figure 9.2 schematically shows a typical structure representative of the simplest single-layer LED devices using PPV as the light-emitting polymer. Among the two electrodes, at least one should be transparent or semi-transparent. In practice, a transparent indium-tin-oxide, ITO, coated glass is often used as the anode and a layer of low-work-function metal (*e.g.* aluminum, calcium) acts as the cathode.

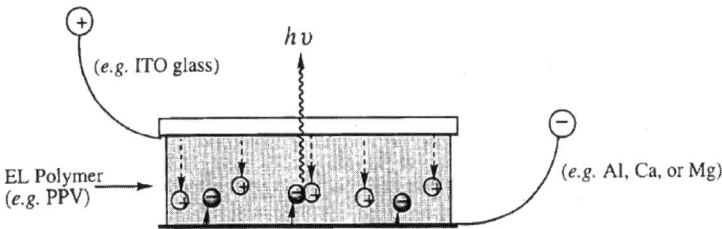

Figure 9.2 Schematic representation of a single-layer polymer LED

PPV prepared by the sulfonium precursor route has an energy gap between the π and π* states of about 2.5 eV. Photoexcitation of an electron from the highest occupied molecular orbital (HOMO) to the lowest unoccupied molecular orbital (LUMO)

creates a singlet exciton (Figure 9.3(a)), which produces luminescence at a longer wavelength (the Stokes shift) through radioactive decay.

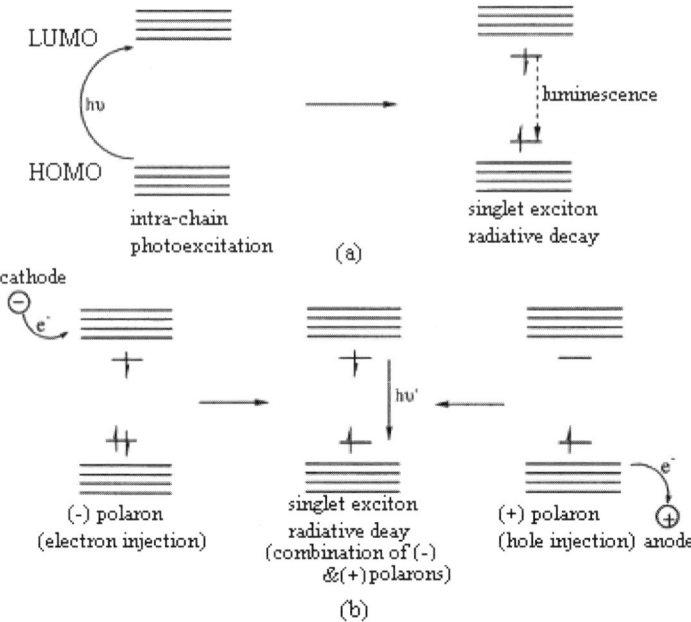

Figure 9.3. The working principle for (a) PL and (b) EL (After Dai, 1999b, copyright 1999 John Wiley & Sons, Inc. Reproduced with permission)

Polymer LEDs work under a similar general principle. Upon the application of an electrical voltage onto a LED device, electrons from the low-work-function cathode (*e.g.* Al in Figure 9.2) are injected into the LUMO of the PPV layer leading to the formation of negatively charged polarons, whereas holes from the high-work-function anode (*i.e.* ITO in Figure 9.2) are injected into the HOMO, producing positively charged polarons. These negatively and positively charged polarons migrate under the influence of the applied electric field and combine in the band gap of the PPV layer (Figure 9.3(b)), resulting in the formation of the same singlet exciton as is produced by the photoexcitation. The singlet exciton thus produced can also decay radiatively with the emission of light. The wavelength (and hence the color) of the photons thus produced depends on the energy gap of the organic light-emitting material, coupled with a Stokes shift (Greenham and Friend, 1985). To maintain a low operation voltage, a very thin polymer layer *(ca.* 100 nm) is usually used.

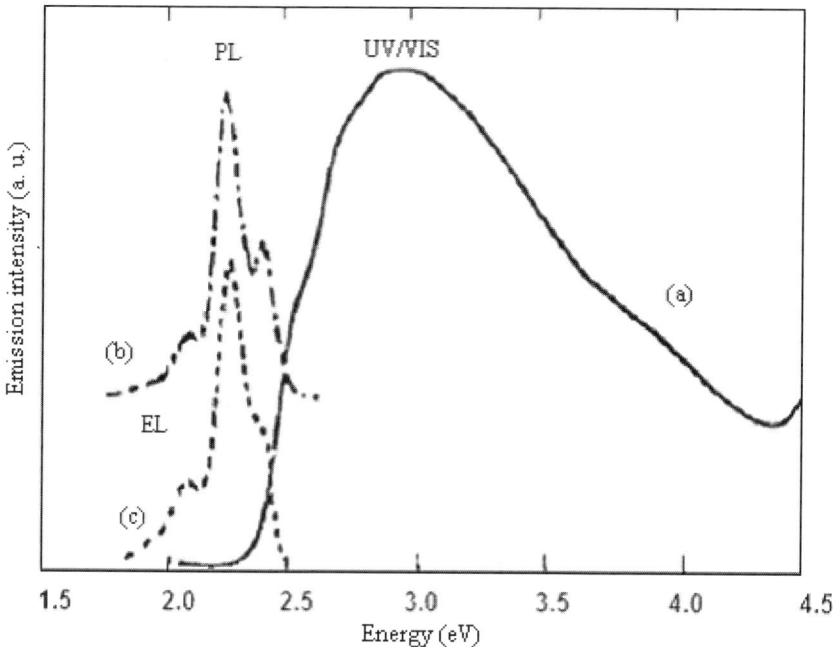

Figure 9.4. The absorption (curve a) and emission (curves b and c) spectra of PPV. The similarity of the emission spectra generated by photoexcitation (Figure 9.3(a)) and by charge injection (Figure 9.3(b)) suggests that the same excited state is responsible for both photoluminescence (PL) and EL (Greenham and Friend, 1995, copyright 1995 Academic Press).

9.2.2 Quantum Efficiency

Although the same working principle applies to both PL and EL, the theoretical maximum EL efficiency may only reach one quarter of the best PL efficiency (27% for PPV) as electron-hole capture is expected to result in the loss of 75% of the electron-hole pairs to triplet excitons, which do not decay radiatively with high efficiency (Greenham and Friend, 1995). Recently, however, there have been arguments that the EL efficiency could be higher than this limit due to the high quantum yield of spatially indirect excitons (*i.e.* bound polaron pairs — see, for example: Yan *et al.*, 1995; Pauck *et al.*, 1995; Graupner *et al.*, 1996). The introduction of species that allow efficient triplet luminescence has also been shown to be an attractive approach for improving EL efficiency (Baldo *et al.*, 1998; Cleave *et al.*, 1999; Hertel *et al.*, 2001).

For comparison purposes, it is worthwhile to mention here that the first LED device with the Al/PPV/ITO structure reported by Burroughes *et al.* (1990) had an efficiency of only 0.05%. A few years later, this group obtained a very high internal efficiency (*ca.* 4%) from both a single active layer device based on the PPVdimethoxy-substituted PPV copolymer (Burn et al., 1992a, b) and a PPV/cyano-substituted PPV bilayer device (Brown et al., 1992 and 1996; Greenham et al., 1993). Although the synthesis of EL copolymers with alternating conjugated and non-conjugated segments has since been demonstrated to be a useful method for improving the EL efficiency through exciton confinement (Hiberer et al., 1997; Yang Z. et al., 1993 and 1998), the work functions of the cathode and anode need to be matched with the corresponding LUMO and HOMO levels of the EL polymer in a particular LED in order to achieve efficient emission.

The internal quantum efficiency, η_{int}, defined as the ratio of the number of photons produced within an EL device to the number of electrons flowing through the external circuit (Greenham and Friend, 1995), is given by:

$$\eta_{int} = gr_{st}q \tag{9.1}$$

where g is the number of excitons formed per electron flowing through the external circuit, r_{st} is the fraction of excitons formed as singlets and q is the efficiency of radiative decay of these singlet excitons. Therefore, in order to improve the EL efficiencies, various issues concerning the formation of excitons and their decay need to be considered. The process of exciton formation, in turn, involves carrier injection, carrier transport and carrier combination. As we shall see later, the polymer-electrode interface has an important impact on the carrier injection and carrier transport.

Balanced charge injection and transport is a prerequisite for LEDs with a high quantum efficiency (Greenham and Friend, 1995). Balanced injection may be achieved using two polymer-electrode interfaces with equal barriers, which must be small for low operating voltages (Greenham and Friend, 1995). The work functions for some of the commonly used electrode materials are given in Table 9.1.

Table 9.1. The work function of electrode materials and barriers to injection of electrons and holes[a]

Electrode	Work function (eV)[b]	Barrier to injection (eV)
Al	4.2	1.6
Ca	2.9	0.3
Mg	3.7	1.1
Au	5.3	2.7
In	4.1	1.5
Cu	4.7	2.1
Cr	4.5	1.9
ITO	4.6	0.5

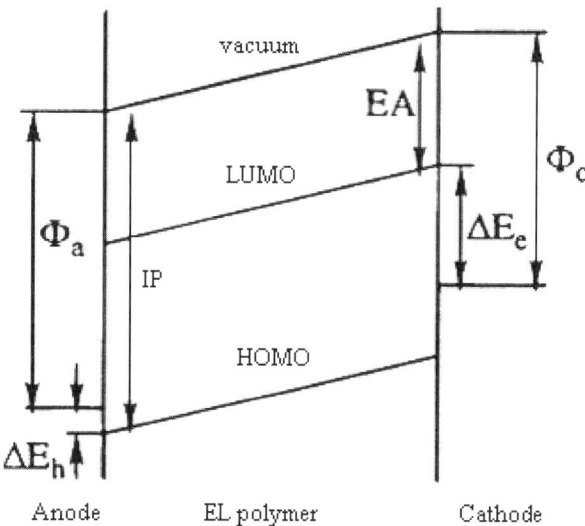

[a] The above schematic energy level diagram shows the ionization potential (*IP*) and electron affinity (*EA*) of the light-emitting polymer, the work function of the anode (ϕ_a) and cathod (ϕ_c), and the barriers to injection of electrons (ΔE_e) and holes (ΔE_h). By taking the *IP* and the band gap (*i.e.* LUMO–HOMO) of PPV as 5.1 and 2.5 eV, respectively, we should have:

$$EA = 5.1 - 2.5 = 2.6 \, eV \tag{9.2}$$

$$\Delta E_e = \phi_c - 2.6 \, (eV) \tag{9.3}$$

and

$$\Delta E_h = IP - \phi_a \tag{9.4}$$

[b] See Lide, 1996

It can be seen that the Ca/PPV/ITO provides a better match in the barriers for electron and hole injection than Al/PPV/ITO. Consequently, the quantum efficiency of a PPV/ITO based LED was significantly improved when calcium was used in place of aluminum (Holmes *et al.*, 1993; Jeon, *et al.*, 1992; Nakayama *et al.*, 1993). However, metals of a low work function are generally oxygen-reactive. The use of highly reactive metals with a low work function (*e.g.* calcium) often requires careful encapsulation, which makes this method difficult, if not impossible, for practical applications. To circumvent this difficulty, modifications of the electrode(s) (Bradley and Tsutsui, 1995; Hu and Karasz, 1998; Leyden and Collins, 1988) and the charge injection characteristics (Salaneck *et al.*, 1996) by surface and interface control of the polymer-metal (electrode) and/or the polymer-polymer interfaces have been exploited as alternative avenues for improving the EL efficiencies.

9.2.3 Interface Engineering

9.2.3.1 Chemical Derivatization of the Metal Electrodes

Many recent experimental/theoretical investigations have shown that metal may interact, either physically or chemically, with EL polymers in polymeric LEDs (Dannetun *et al.,* 1993b; Gao *et al.,* 1993; Konstadinidis *et al.,* 1995; Nguyen *et al.,* 1995a; Nguyen and de Vos, 1996). For instance, it was reported that calcium atoms, like other metal atoms (*e.g.* Al, K, Rb or Na) (Salaneck and Brédas, 1996), diffused from the cathode into the near surface region of a PPV film to form a 20–40 Å thick interface of the *p*-doped PPV (Dannetun *et al.,* 1994), leading to emission quenching (Choong *et al.,* 1996). Aluminum was even found to chemically break the conjugation sequences of PPV by covalently bonding onto the vinylene units (Dannetun *et al.,* 1993a). Furthermore, ITO electrodes were shown to not only chemically interact with conjugated polymers (Scott *et al.,* 1996) but also with the HCl released from a sulfonium precursor polymer of PPV upon thermal conversion [Equation (2.6), Chapter 2]. Dai and co-workers (Winkler *et al.,* 1999a) have followed the thermal conversion shown in Equation (2.8) for oligo(ethylene oxide) grafted PPVs by XPS measurements.

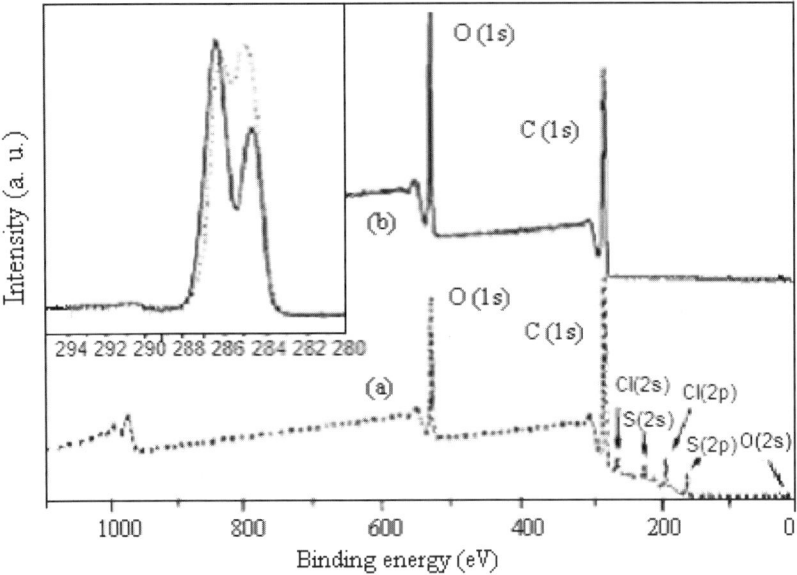

Figure 9.5. XPS survey spectra of (a) EO_1-PPV precursor polymer and (b) EO_1-PPV. The inset shows the high-resolution C(1s) spectra of EO_1-PPV precursor polymer (…) and EO_1-PPV (——) (After Winkler *et al.,* 1999a, copyright 1999 American Chemical Society. Reproduced with permission)

Figure 9.5 shows the XPS survey spectra of the freshly prepared EO_1-PPV precursor polymer [*i.e.* product (4–1) of Equation (2.8)] spin-cast on a *quartz* plate before and after the thermal conversion. Also included in the inset of Figure 9.5 are the corresponding high resolution C(1s) spectra. As expected, Figure 9.5(a) shows the presence of C, O, Cl and S in the EO_1-PPV precursor polymer. However, XPS analyses on the precursor polymer after the thermal transformation gave a survey spectrum with no signals attributable to Cl and S (Figure 9.5(b)). The large decrease in the aliphatic carbon peak region at 285 eV seen for the C(1s) spectrum of the EO_1-PPV precursor polymer upon thermal treatment (inset of Figure 9.5) arises from the concomitant loss of the tetrahydrothiophene units. The calculated carbon atomic ratio of *ca.* 8 to 6 from the signals characteristic of the ether carbons at *ca.* 286.6 eV and carbons of the conjugated backbone at *ca.* 284.7 eV is consistent with the structure of the monomer unit of EO_1-PPV [*i.e.* the final product of Equation (2.8)].

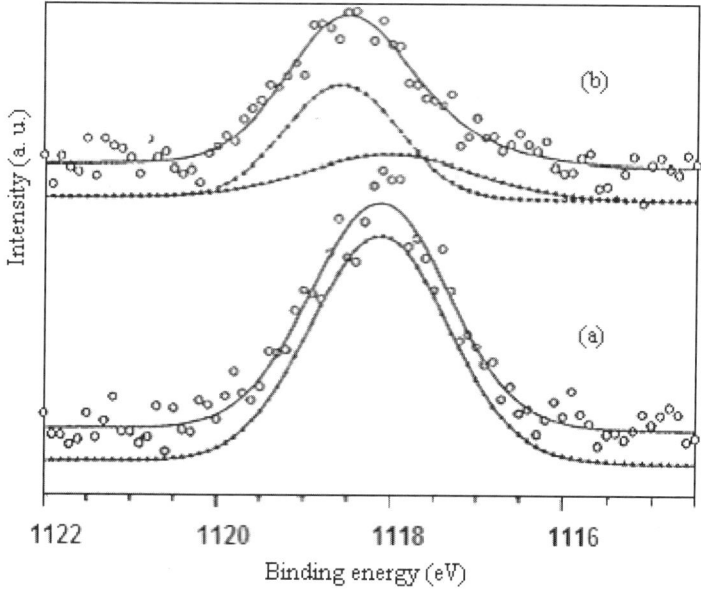

Figure 9.6. XPS Ga(2*p*) spectra of (a) an uncovered, Ga-substituted ITO substrate; (b) a ultrathin film of precursor-PPV spin-coated onto Ga-substituted ITO. The experimental peaks were fitted with two peak components at binding energies of 1118.6 and 1118.0 eV (After Kugler, 1999, copyright 1999 Elsevier. Reproduced with permission)

A separate XPS study on a thin film of the sulfonium precursor polymer of PPV on an ITO substrate, however, indicated the formation of $InCl_3$ during the thermal conversion (Kugler *et al.*, 1999). In this study, the indium ions in the uppermost surface layer of ITO were partially substituted by gallium ions in order to detect the subtle chemical interactions at the ITO/precursor-PPV interface. Given the chemical similarity between gallium and indium, the chemical reactivity of a Ga-substituted ITO surface is taken to be the same as a pristine ITO substrate. The Ga($2p$) spectra, together with the curve fits, for the pristine and sulfonium precursor-coated Ga-substituted ITO substrate, are given in Figure 9.6. It was found that the Ga($2p_{3/2}$) peak from the uncoated, Ga-substituted ITO substrate was located at 1118.1 eV (Figure 9.6(a)). The Ga($2p_{3/2}$) spectrum for the as-cast precursor-PPV on the Ga-substitute ITO, however, showed a main component at 1118.6 eV and a minor component at 1118.0 eV (Figure 9.6(b)). The up-shift of the gallium binding energy upon deposition of precursor-PPV has been taken as an indication for chemical interactions of the ITO surface with HCl to produce $GaCl_3$ (and $InCl_3$), as HCl was shown to be present in the precursor-PPV even at room temperature (Kugler *et al.*, 1999).

It seems, therefore, that some form of physicochemical interaction always occurs at the conjugated polymer-metal interface. The metal-induced emission quenching, however, could be recoverable, for example, by intentional partial oxidation of the metal surface (Park *et al.*, 1997). Also, the insertion of a thin layer of insulating materials [*e.g.* functionalized oligophenylene (Nuesch *et al.*, 1997), carbon (Gyoutoku *et al.*, 1997) or certain polymers (Service, 1998)] between the EL polymer and metallic cathode can significantly improve the device performance (Hung *et al.*, 1997; Jabbour *et al.*, 1997; Kim Y. *et al.*, 1996; Tang H. *et al.*, 1997; Yang and Heeger, 1994a). In particular, an oxygen or nitrogen plasma treatment of a PPV layer at the PPV/Al interface in an LED device with the Cr/PPV/Al structure has been demonstrated to cause a disappearance of the rectifying behavior, along with an increase in the current by many orders of magnitude (Nguyen *et al.*, 1995b).

The plasma treatment is believed to be useful for creating a charge carrier confinement layer to facilitate the charge injection, as is the case with the electron-transporting layer (Berggren *et al.*, 1994a). Furthermore, a few studies have reported on the use of the plasma polymerization technique for the fabrication of LED devices (Sun *et al.*, 1996; Xu X. *et al.*, 1995; Wu C. *et al.*, 1997). In view of the fact that plasma treatment including plasma etching can produce a large variety of surface functionalities and surface topologies (Dai *et al.*, 1997a, b; Liang *et al.*, 1992) for the modification of the metal electrodes (Ohkawa *et al.*, 1991; Yasuda, 1985), there is no doubt that the plasma technique will have potential implications for making fascinating optoelectronic devices. The ease with which micro-patterns of organic materials can be made by the plasma technique (Chapter 6) could add additional benefits to this approach.

9.2.3.2 Polymer-polymer Interface

Unlike the single-layer LED shown schematically in Figure 9.2, multilayer LED devices consist of multiple polymer layers sandwiched between two metal electrodes, in which the polymer-polymer interfaces also play an important role in regulating the performance of the device. Although quantitative structural description of the polymer-polymer interfaces (*e.g.* in terms of interdiffusion, sharpness of the interface) in multilayer LEDs is still limited (Hong *et al.*, 1998; Orita *et al.*, 1997), the general use of neutron reflectometry for polymer-polymer interfacial characterization has been demonstrated to a significant extent during the past decade (Higgins and Benoît, 1994; Penfold *et al.*, 1997; Richards *et al.*, 1991). An example of the neutron reflectivity profiles from a bilayer film of polystyrene(deutero), PS(D), and PS(H) on a glass substrate is given in Figure 9.7 (Stamm, 1992; Stamm *et al.*, 1991).

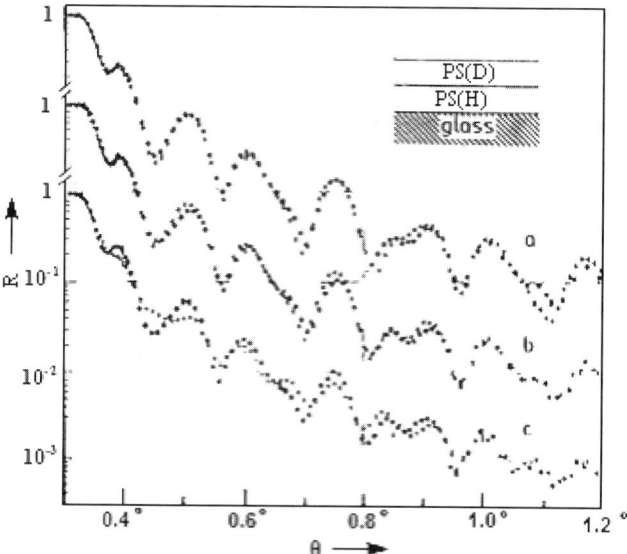

Figure 9.7. Neutron reflectivity profiles of PS(D)/PS(H) on a glass substrate during the initial stages of interdiffusion. Different times of interdiffusion t after annealing are compared: (a) unannealed (+) and 2 min at 120 °C (□); (b) 2 min (+) and 910 min (•); (c) 910 min (•) and 11750 min (+). Annealing times are reduced to 120°C. The neutron wavelength is 0.43 nm (After Stamm, 1992, copyright 1992 Springer-Verlag. Reproduced with permission)

In this study, all interfaces are assumed to be Gaussian, with the neutron reflectivity curves being fitted by simple error functions (see Chapter 7). The variance δ_z of the error functions is then taken as a measure of the interface width. As can be seen from curve (a) in Figure 9.7, the significant change at larger angles upon 2 minutes of annealing at 120 °C indicates a fast diffusion and/or relaxation of the polymer chains over a small distance. Thereafter, no obvious change in the reflectivity was observed over a considerable period of annealing time (*e.g.* 2 to 910 min, curve (b) of Figure 9.7). Further annealing caused a significant change in the reflectivity curves at small angles (curve (c) of Figure 9.7), suggesting segment diffusion over large distances. The interfacial broadening due to the segment interdiffusion is clearly seen in Figure 9.7. Clearly, neutron reflectometry is a promising tool for characterizing polymer-polymer interfaces in multilayer LEDs.

9.2.4 Modification of the Charge Injection Characteristics

Yang Y. and Heeger (1994b) have significantly reduced the operating voltage and increased the quantum efficiency for a polymer LED based on MEH-PPV {*i.e.* poly[2-methoxy-5-(2'-ethyl-hexyloxy)-*p*-phenylene vinylene]} by using a combination of doped polyaniline and ITO as the anode. The improvement was attributed to a higher electronegativity, and hence a smaller hole-injection barrier of polyaniline than ITO. In addition, the polyaniline layer was reported to stabilize the chemistry of the polymer-electrode interface, possibly by acting as a barrier to the passage of oxygen out of the ITO (Karg *et al.*, 1996). LEDs with a five-layer structure, in which a polyaniline emeraldine-base film was sandwiched between the emitter and the two electrodes at each side, have also been reported (Wang Y. *et al.*, 1996a, Wang H. *et al.*, 1996b). Due to changes in the charge injection/transport characteristics, these so-called symmetrically configured AC light-emitting (SCALE) devices are insensitive to the electrode materials used and work equally well in forward- and reverse-dc bias modes, indicating that they can run on ordinary household current (Dagani, 1995; Wang Y. *et al.*, 1996a, Wang H. *et al.*, 1996b). The change of the charge injection characteristics at the polymer-electrode interface in the SCALE devices is clearly evidenced by a decrease in the total resistance with an increasing number of insulating polymer layers (Wang H. *et al.*, 1996c). The charge transport can also be regulated by the use of electron/hole transporting materials (Hu *et al.*, 1994; Kim D. *et al.*, 1995; Zhang *et al.*, 1995). Of particular interest, dendrimers containing triarylamine (TAA) hole-transporting peripheral groups and highly luminescent cores have been used, together with electron-transporting 2-(4-biphenyl)-5-(4-tert-butylphenyl)-1,3,4-oxadiazole (PBD), for light-emitting applications (Freeman, 2000).

9.2.5 Light-emitting Electrochemical Cells (LECs)

Another relevant interesting development is the subsequent report on the light-emitting electrochemical cells, LECs (Hide *et al.*, 1997), in which a light-emitting *p-n* junction is believed to form *in situ* by simultaneous *p*-type and *n*-type electrochemical doping of electrolyte-containing conjugated polymer films on opposite electrodes (Pei *et al.*, 1995; 1996) (this operation mechanism is, however, currently challenged – see Dai *et al.*, 2001). Because the carrier injection occurs through ohmic contacts into the *p*- and *n*-doped regions, stable metals can be used as the electrodes in LECs and there is no need to match the work functions of the anode and cathode to the π and π^* energies of the luminescent polymer. Unlike LEDs, LECs may work both at the forward and reverse bias at a low turn-on voltage close to the energy gap, indicating that they can run on ordinary household current. Besides, since the color of the emitted light from an LEC is determined by the energy gap of the luminescent polymer in which the *p-n* junction is formed, multi-color emissions can be achieved by region-specific control of the *p-n* junction in a multilayer LEC device. For instance, a two-layer LEC with the ITO/PPV + PEO(LiCF$_3$SO$_3$)/MEH-PPV + PEO(LiCF$_3$SO$_3$)/Al structure has been shown to emit green light when the induced *p-n* junction was completely located inside the PPV layer at one bias, whereas the same device gave the orange emission when it was biased at the opposite polarity, causing the *p-n* junction to be shifted into the MEH-PPV layer (Yang and Pei, 1996). Similar polarity-dependent multicolor emissions had also been obtained for certain multilayer polymer LEDs by appropriate control of the polymer-electrode and/or polymer-polymer interfaces (Hamaguchi and Yoshino, 1996; Wang Y. *et al.*, 1997; Yoshida *et al.*, 1996) or a single-layer LED based on polymer emitters of a small band gap symmetrically located within the work functions of the electrodes (Yang Z. *et al.*, 1995).

However, LEC devices based on the mixture of polymeric ion conductors (*e.g.* PEO) and light-emitting conjugated polymers often show phase separation, leading to a slow response and short device lifetime. The covalent linkages between the oligo(ethylene oxide)-grafted PPVs [EO_m-PPV, Equation (2.8)] could minimize the phase separation problem associated with LECs. For instance, LECs based on the oligo(ethylene oxide)-grafted PPVs exhibited good device performance. Figure 9.8 shows the current-voltage (I–V) plot for an LEC device made from the EO_3-PPV (Holzer *et al.*, 1999a). Also included in Figure 9.8(a) is the luminescence-V characteristics of the same device. As can be seen in Figure 9.8, the LEC shows symmetric I–V and luminescence-V curves with a turn-on voltage of about 2 V both for the current and electroluminescence. This value is very close to the polymer energy gap (*ca.* 2.1 eV) and smaller than that of a corresponding LED of comparable thickness with a similar configuration (*i.e.* 120 nm; Al and ITO electrodes), which turns on at *ca.* 3 V (the inset of Figure 9.8(a)). Both the LED and LEC based on EO_3-PPV are characterized by a red-orange emission, as shown in Figure 9.8(b). The change in the emission intensity seen for the LEC under continuous operation is independent of the wavelength, leading to no change in the emission color. This differs from those conventional LECs based on physically mixed PEO and light-emitting polymers, which often show changes in the emission color during operation due to the presence of segregated polar PEO environments (Tasch *et al.*, 1999; Wenzl *et al.*, 1998).

The minimized phase separation and good ionic conductivity associated with EO_m-PPVs ensure an effective conduction of ions to the opposite electrodes in the LEC devices, and hence a short response time. A response time of *ca.* 480 µs and brightness of around 35 cd/m^2 at 3 V were determined for the LEC made from EO_3-PPV operating with a square waveform pulse (Figure 9.8(c)) (Holzer L. *et al.*, 1999a; Tasch *et al.*, 1999; Wenzl *et al.*, 1998). Due to the complexion between metal ions and the covalently-linked glyme-like side chains, EO_3-PPVs showed an ionochromic effect in both the adsorption and electroluminescence spectra, providing a new base toward chemical sensors (Holzer L. *et al.*, 1999b).

Although the PEO-LiCF$_3$SO$_3$ system has been widely investigated as a prototype for other ionically conducting polymers, crown ether-substituted polymers/oligomers coupled with appropriate cations were also demonstrated to show sufficient ionic conductivities for certain applications (Collie *et al.*, 1993). In addition to the gain of ionic conductivities, the crown ether substitution can serve as an additional means for color tuning since crown ether rings could form a complex with various metal ions of different electron affinities. This feature may also be utilized for chemical sensing (Nakatsuji et al., 2000; Prodi *et al.*, 2000). In order to explore these intriguing features, Dai and co-workers (Winkler *et al.*, 2000) have synthesized PPV derivatives substituted with crown ether rings [*i.e.* CE-OPV, Equation (2.9)].

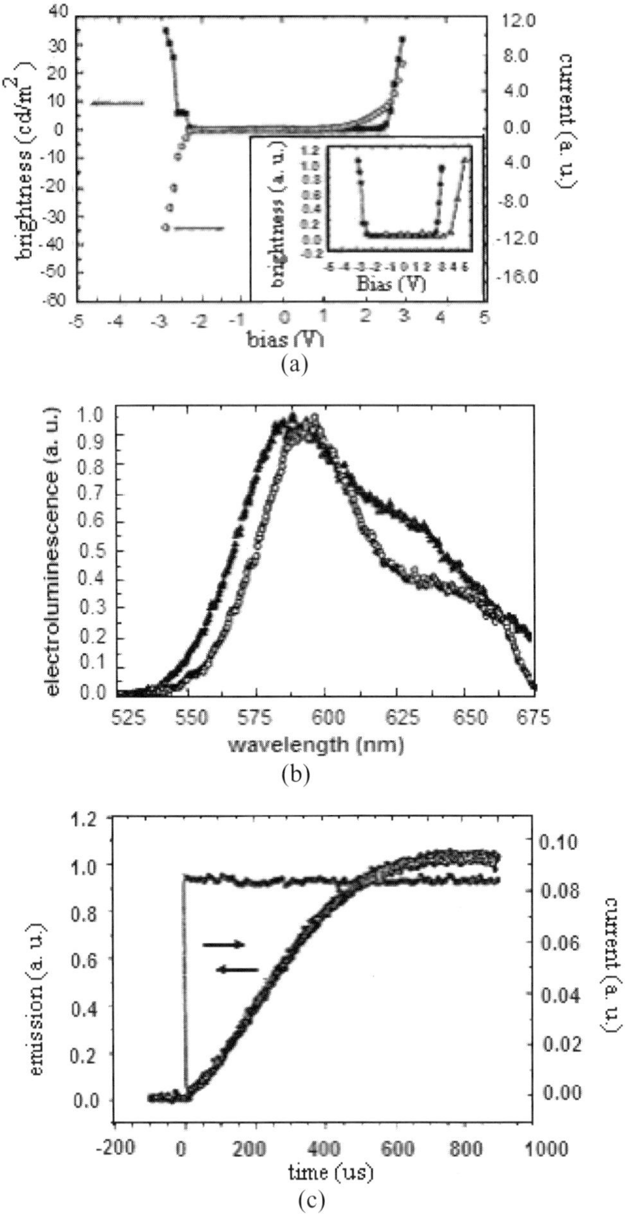

Figure 9.8. (a) $I-V$ (circles) and brightness-V characteristics of an LEC based on EO_3-PPV and Li triflate. Inset shows brightness-V characteristics of the LEC (closed circles) and corresponding LED (open triangles); (b) electroluminescence emission spectra for LED (open circles) and LEC (closed triangles) made from EO_3-PPV; (c) time response of the EL emission from the LEC based on EO_3-PPV and Li triflate under pulsed operation conditions (70 Hz, 50% duty cycle) (After Holzer L. *et al.,* 1999a, copyright 1999 American Institute of Physics. Reproduced with permission)

UV/vis absorption and PL spectra of the CE-OPV in solution and in the solid state are given in Figure 9.9(a), which shows a small red-shift for the solid spectra with respect to their solution counterparts. Briefly, the CE-OPV film showed an absorption peak at *ca.* 365 nm with an optical band gap energy of *ca.* 2.58 eV. On photoexcitation (λ_{ex} = 365 nm), the CE-OPV film gave a bluish-green emission peak at 515 nm with a full width at half maximum (FWHM) of *ca.* 77 nm.

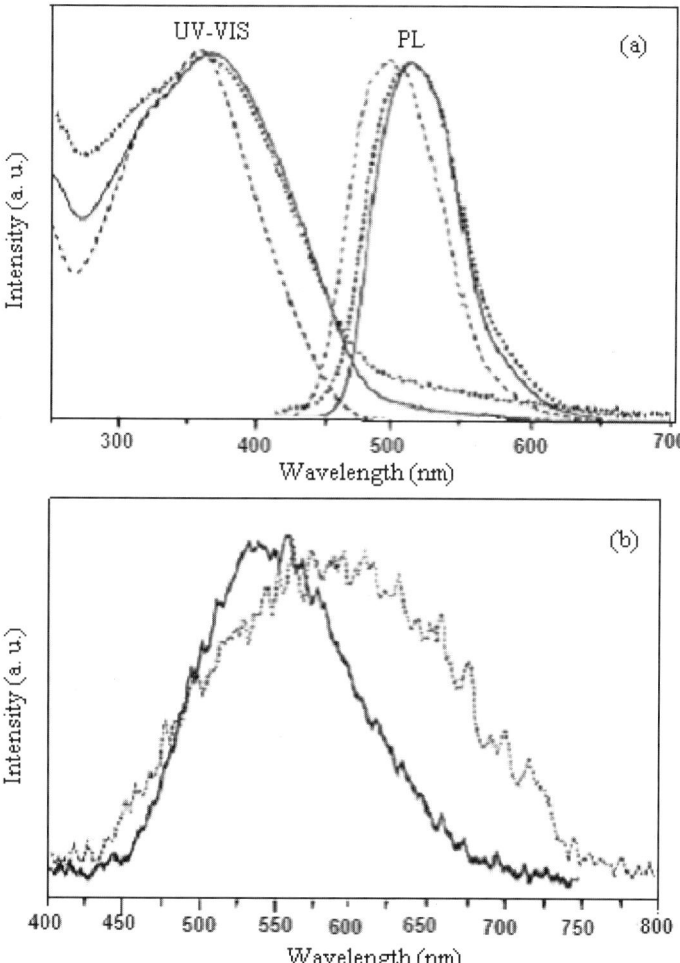

Figure 9.9. (a) UV/vis and PL ($\lambda_{ex} \approx$ 370 nm) spectra of CE-OPV in MeCN (---), a bulk film (—) and a bulk film mixed with LiCF$_3$SO$_3$ (...); (b) EL spectra of an LED with the structure ITO/CE-OPV/Al (—) at 15V and an LEC with the structure ITO/CE-OPV + LiCF$_3$SO$_3$/Al (---) at 10 V. Inset shows the I–V characteristics of the LED (open squares) and LEC (open circles) devices (After Winkler *et al.*, 2000, copyright 2000 The Royal Society of Chemistry. Reproduced with permission)

PL quantum efficiency of the CE-OPV was determined in MeCN solution to be as high as 80%, suggesting some potential applications in electroluminescent devices. Blending with $LiCF_3SO_3$ in the solid state, as for LECs, caused a very subtle blue-shift in optical absorbance and slightly broadened the PL spectrum. This is probably because of the formation of complexes between the crown ether rings and the Li ions, which could increase the band gap energy of the conjugated backbone, and hence the blue-shift, due to the electron-withdrawing action of the positively charged Li ions. On the other hand, possible scattering of the absorbed and/or emitted light by domains associated with the metal complexes could broaden the optical spectra, as seen in Figure 9.9(a).

Single-layer LEDs with the structure ITO/CE-OPV/Al showed a similar emission (Figure 9.9(b)) to the PL emission (Figure 9.9(a)). The observed similarity between the PL and EL emissions indicates the formation of the same singlet excitons in both cases. LECs based on the structure ITO/CE-OPV + $LiCF_3SO_3$/Al showed a similar EL emission to that of LEDs, but with a much broader envelope (Figure 9.9(b)) due, most probably, to the scattering effect mentioned above. As shown in the inset of Figure 9.9(b), the LED exhibited typical diode characteristics with a turn-on voltage of *ca.* 9 V (EL layer thickness is *ca.* 100 nm) at the forward bias. In contrast, its LEC counterpart gave an asymmetrical $I-V$ curve with a rather low turn-on voltage of *ca.* 3 V both at the forward and reverse bias, indicating a good ionic conduction arising from the chemically grafted crown ether rings in the presence of $LiCF_3SO_3$. Therefore, the combination of the ionically conducting crown ether side groups and the EL-active poly(*p*-phenylene vinylene) backbones presents unique advantages for the use of CE-OPV as a class of promising EL materials. Furthermore, the preliminary results obtained by Dai and co-workers (Winkler *et al.*, 2000) indicated that the ease with which the grafted crown ether rings on the CE-OPV chains can form complex compounds with various metal ions of different electron affinities should have potential implications in color-tuning for the photo/electroluminescent emissions. This would also open up possibilities for various practical applications ranging from chemical sensors to multicolor displays.

9.2.6 Color Tuning

Since conjugated polymers can have π-π* energy gaps over the range of 1 to 4 eV (1240–310 nm), polymer-based LEDs should, in principle, emit light across the whole spectrum from ultraviolet to near-infrared (Berggren *et al.*, 1994; Gruner *et al.*, 1994; Horowitz *et al.*, 1994; Huber *et al.*, 1994; Lee J.-K. *et al.*, 1995; Ohmori *et al.*, 1991) with a good color-tunablity by controlling the energy gap through molecular engineering (Burn *et al.*, 1993). Indeed, polymer LEDs with a wide range of emission colors and with reasonably high quantum efficiencies (a few percent) have now been reported.

Using a *stretching-oriented* polythiophene derivative as the electroluminescent material, Inganäs and co-workers (Dyreklev *et at.*, 1995) reported, in 1995, the first polymer LED that can emit polarized light under a voltage as low as 2 V. Since then, LEDs with polarized EL emissions have also been fabricated based on certain rubbing-aligned (Hamaguchi and Yoshino, 1995; Peeters *et al.*, 1997), liquid crystalline (Grell *et al.*, 1997), and Langmuir-Blodgett (LB) deposited (Cimrova *et* al., 1996; Bolognesi et al., 1997) conjugated polymer films. Furthermore, Grüner et at. (Grüner et al., 1997) have precisely determined the location of the exciton recombination zone (Greenham and Friend, 1995) by quantifying the anisotropic EL emission profiles from multilayer LB films of soluble poly(p-phenylene) derivatives with two regions of perpendicular macroscopic orientation.

Blue-light emission requires a wide band gap semiconductor, and efficient blue emission has only recently been achieved by LEDs based on inorganic molecular materials (Jeon et al., 1992; Jin and Kim, 1997; Nakayama et al., 1993; Ohkawa et al., 1991). One strategy to obtain blue emission from organic LEDs made of conjugated polymers is to reduce the average conjugation length. Thus, graft polymers with molecular emitters (Sokolik et al., 1993) attached as pendant side groups (Aguiar et al., 1995, 1996; Baigent et al., 1995), alternating copolymers containing a distyrylbenzene unit as an emissive group and a triphenylamine unit as a charge transporting group (Kim D. et al., 1995), poly(1,20-(10,13-didecyl)distyrylbenzene-co-l,2-(4-(p-ethylphenyl))triazole (Grice et al., 1997), and poly(2,5-diheptyl-l,4-phenylene-alt-2,5-thienylene) (PDHPT) (Fahlman et al., 1995) have been used to produce blue emission, as have some doped polymers (Gautier et al., 1996). In addition, blue light emission has also been produced by LEDs using certain polymer blends (e.g., PPV/PVK (PVK: poly(9-vinylcarbazole) – a hole-transporting polymer with aromatic pendant groups) (Zhang et al., 1994; Hu et al., 1994), PDHPT/PDPP (PDPP: poly(2,5-diheptyl-2'5'-dipentoxybiphenylene)) (Birgerson et al., 1996), and ladder-type poly(para-phenylene)/para-hexaphenyl) (Leising et al., 1996) or homopolymers [e.g., poly(alkylflourenes) (Uchida et al., 1993), poly-p-phenylene (Grem et al., 1993), and poly(p-pyridine) (Gebler et al., 1995)] as the active emitting layer. On the other hand, *white* light emission (nearly as bright as a fluorescent lamp – ca. 2000 candelas/m^2) was achieved with a multilayer polymer LED device, in which three emitter layers with different carrier transport properties, each emitting blue, green or red light, are used to generate the white light (Kido et al., 1995). A similar multilayer device containing poly[3-(4-octylphenyl)-2,2'-bithiophene] (PTOPT) and 2-(4-biphenylyl)-5-(4-*tert*-butylphenyl)-1,3,4-oxadiazole (PBD) also emits white light with an emission spectrum composed of an ultra-violet, a green and a red band (Berggren et al., 1994a). By the modification of the latter device, Inganäs and co-workers (Berggren et al., 1995) were able to enhance the UV peak and suppress the red and green peaks, leading to the emission of UV light. However, near-UV emission has previously been demonstrated for a single-layer LED with the ITO/PVK/Al (or Ca) structure (Hu et al., 1994). More recently, single-layer LEDs based on guest-host systems (Christ et al., 1997), non-conjugated and conjugated polymer blends (Shim et al., 1997), and certain PPV derivatives grafted with pendant anthracene moieties (Chung et al., 1997) have also been reported to emit white light. The availability of white emission, in conjunction with the use of appropriate filters, should, in principle, allow for the generation of any monochromatic light source. Besides, voltage-controlled multiple-color emissions have recently been reported for LEDs based on conjugated polymer blends (Berggren et al., 1994b).

The conformational change of light-emitting polymer backbones could be exploited as an alternative approach for color tuning, in addition to the chemical modification of the polymer structure. This is because the band gap energies of conjugated polymers, and hence the related optoelectronic properties, depend strongly on their chain conformations (Skotheim, 1986). As discussed in Chapter 2, a "compact coil" to "expanded coil" transition for clay-intercalated polyaniline chains has been directly observed, which is accompanied by a concomitant increase in conductivity by up to several orders of magnitude (Dai et al., 2000). By extension, Winkler et al. (1999b) have also intercalated certain conjugated polymers (e.g. EO_3-PPV) into clay nanoparticles (e.g. Bentone 34, B34) for light-emitting applications. The intercalation process is schematically shown in Figure 9.10(a) (Winkler et al., 1999b) and experimentally revealed by a steady increase in the height of the clay galleries as determined by X-ray diffraction. The PL spectra of the composites at several different intercalation levels were already given in Figure7.31 of Chapter 7, which shows a red-shift for the PL emission with increasing intercalation level, indicating an intercalation-induced conformational transition from a "compact coil" to an "expanded coil".

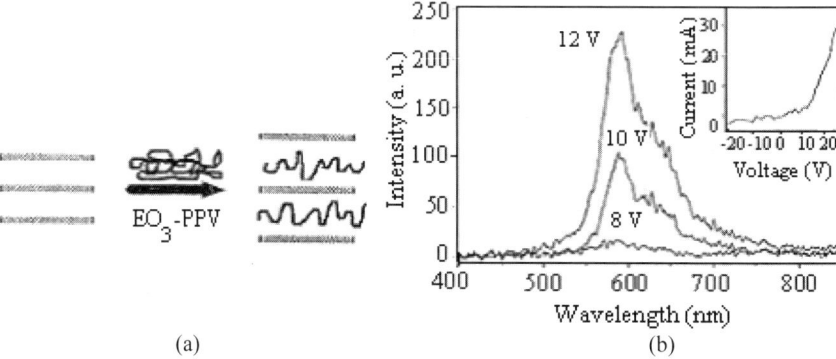

Figure 9.10. (a) A schematic representation of the intercalation process; (b) electroluminescent emissions from a single layer LED device with the structure ITO/EO-PPV-B34(0.95:0.05)/Al. The inset shows the corresponding current-voltage characteristics of the LED device (After Winkler et al., 1999b, copyright 1999 Kluwer Academic Publishers. Reproduced with permission)

The results from a single-layer LED device with the structure ITO/EO_3-PPV-B34 (0.95:0.05)/Al showed a similar EL spectrum (λ_{max} = 590 nm, Figure 9.10(b)) to the PL emission. As expected, the similarity between the EL and PL emissions suggested that the same singlet excitons were generated upon both the PL- and EL-excitation. Besides, the LED showed a typical diode characteristic with a turn-on voltage of ca. 8 V at the forward bias (inset of Figure 9.10(b)). Furthermore, Dai and co-workers (Dai et al., 2001) have constructed multilayer LEDs based on the layer-by-layer adsorption of the positively charged sulfonium PPV precursor polymer and negatively charged acid-oxidized carbon nanotubes (Chapter 5), followed by the thermal transformation of the non-conjugated precursor polymer to the conjugated structure (Chapter 2). It was found that the multilayer structure showed also a significant red-shift both in PL and EL emissions of PPV, presumably caused by conformational changes of the light-emitting polymer chains due to the interaction with carbon nanotubes (Chapter 5). These interesting results indicate that polymer composites may be a class of very promising new light-emitting materials of good color-tunability and environmental stability for both the PL and EL applications. In fact, a single-layer LED based on the poly(*m*-phenylene vinylene-*co*-2,5-dioctoxy-*p*-phenylene vinylene) and carbon nanotube composite has been constructed by Curran *et al.* (1998), which was demonstrated to show lower current densities and better thermal stabiltites than the corresponding pure polymer devices. More recently, carbon nanotubes and diamond-like carbon films (Lmimouni *et al.*, 2001) have been used as electron-injection layers in polymer LEDs to reduce the driving voltages. In a separate, but somewhat related experiment, Lieber and co-workers (Duan *et al.*, 2001) have developed the smallest LEDs by crossing a *p*-type-doped InP *nanowire* over its *n*-type counterpart and applying a voltage to one of them; light is emitted where the two nanowires cross.

Apart from the light-emitting-polymer intercalated clay particles, certain electroluminescent organic-inorganic hybrid materials have been prepared, for example, by sol-gel chemistry (Carlos *et al.*, 1999; Chang and Whang, 1996; de Morais *et al.*, 1999). These organic-inorganic hybrid composites have been demonstrated to show improved environmental stability, mechanical strength and lower permeability for gases (*e.g.* O_2, moisture) with respect to corresponding pure polymeric materials.

9.2.7 Patterned Emission

In addition to conventional thermolysis [Equation (2.6), Chapter 2], the transformation from the unsubstituted PPV precursor polymer to the conjugated state was found also to occur upon UV irradiation (Schmid *et al.*, 1993). This finding led Schmid *et al.* (1993) to produce microstructured PPV, by UV interferometry, suitable for EL applications. Recently, using an ink-jet printer, Yang *et al.* (Yang and Bharatham, 1998; Yang Y. *et al.*, 2000) prepared a polymer-based light-emitting display by pattern-wise printing of a water-soluble polythiophene (Figure 9.11), which provides a patterned electrical connection from the ITO electrode to an overlying layer of poly[2-methoxy-5-(2'-ethyl-hexyloxy)-*p*-phenylene vinylene] (MEH-PPV).

Figure 9.11. Micro-patterned polymer LEDs made by the ink-jet printing technology. The inset shows the active light-emitting pixels (After Yang *et al.*, 2000, copyright 2000 Kluwer Academic Publishers. Reproduced with permission)

On the other hand, Friend and co-workers (Sirringhaus et al., 2000) have demonstrated ink-jet printing of all-polymer transistor circuits with an unprecedentedly high resolution down to the micrometer scale.

Although the relatively short device lifetime still remains as a major problem to be solved before any commercial applications will be realized, several methods, including suitable encapsulation, operating under high vacuum or in an inert atmosphere, chemical derivatization of electrodes and the use of conjugated oligomers, have led to an operating lifetime from 1000 to 10 000 hours (Freemantle, 2001). The simplicity of the polymer LED display devices and their excellent characteristics described above, together with strong commitments of many multinational companies, including Kodak, DuPont, Dow Chemical, Motorola, Pioneer, Sanyo and TDK, for the commercial development, will surely spur their market acceptance, especially for alphanumeric display applications in devices such as mobile phones, car radios and video displays associated with camcorders and digital cameras (Tullo, 2000).

9.3 Laser Action of Conjugated Polymers

To make a laser, the excited atoms or molecules in an active material must emit photons when stimulated, and the number of photons travelling through the laser medium must be amplified. Since the discovery of the laser in 1960, various materials including gases, liquids and solid dyes have been used as working media (Bloembergen, 1993). Solid-state dye lasers have many obvious advantages over gas or liquid-dye lasers, including compactness and a lack of toxicity and flammability. Therefore, interest in the use of polymer matrices containing laser dyes to produce solid-state dye lasers has a long history (Sastre and Costela, 1995). The use of pure organic polymers as lasing materials, however, is a very recent development. Since Moses reported the laser action of MEH-PPV in solution in 1992 (Moses, 1992), laser action of poly(2,2',5,5'-tetraoctyl-p-terphenyl-4,4'-xylene-vinylene-p-phenylene vinylene) (TOP-PPV) (Brouwer et al., 1995), and poly(m-phenylenevinylene-co-2,5-dioctoxy-p-phenylene vinylene) in various solvents have been studied (Holzer W. et al., 1996). In the meantime, the laser action of MEH-PPV "diluted" in a polystyrene film containing titanium dioxide nanoparticles was observed (Hide et al., 1996b).

Very recently, Friend et al., Heeger et al. and Vardeny et al. have independently achieved similar stimulated emissions of light in thin films from a dozen different conjugated polymers when excited by an optical laser (Denton et al., 1997; Dfaz-Garcfa et al., 1997; Frolov et al., 1997; Hide et al., 1990, 1997; Schwartz et al., 1997; Tessler et al., 1996, 1997). As shown in Figure 9.12, the PL of a poly(2-butyl-5(2'-ethylhexyl)-1,4-phenylene vinylene) (BuEH-PPV) film is characterized by a broad emission band, possibly due to an inhomogeneously broadened density-of-states (Kersting et al., 1993; Pauck et al., 1995). A gain-narrowed peak, however, was found to rise out of the broad emission spectrum when the photoexcitation energy increased over a threshold value (Figure 9.12). The observed energy threshold (≤ 1 μJ/pulse, in Figure 9.12) is well below the onset of any optical damage, implying that conjugated polymers are of promise as laser materials.

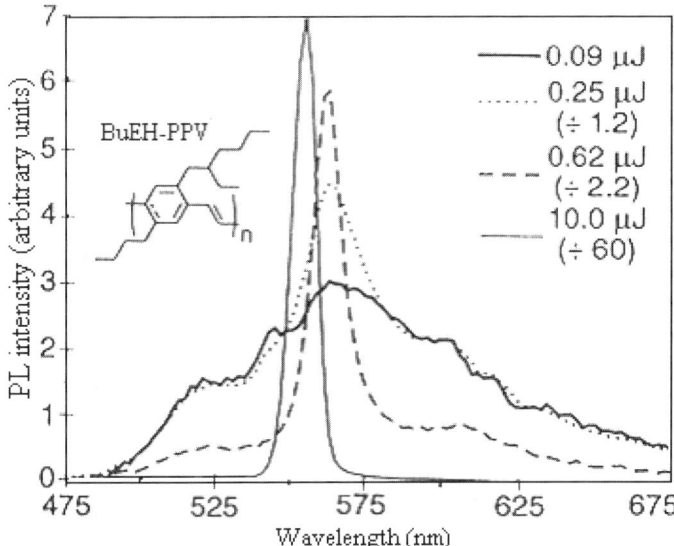

Figure 9.12. The PL spectrum of a spin cast (from *p*-xylene solution) 210 nm thick BuEH-PPV film on glass at various pump pulse energies (After Hide *et al.*, 1996a, copyright 1996 American Association for the Advancement of Science)

For a particular polymer, however, a minimum thickness is required for laser action, below which gain-narrowing does not occur (a typical thickness for the lasing polymers is from a sub-micrometer to 100 μm film). The thickness dependence of gain-narrowing suggests that the observed laser action arises from a wave guiding effect; that is, the emitted photons are confined and amplified in the film because of differences in the refractive indexes of the polymer and its surrounding media (Hide *et al.*, 1996b). Friend *et al.* (Tessler *et al.*, 1996), however, were able to achieve similar stimulated emissions through the microcavity effect by sandwiching a PPV film between two mirrors, against which the light was bounced back and forth. In all these cases, however, the conclusion of laser action should be treated with some caution, as superradiance and/or superfluorescence can also increase emission intensity and produce emission-line-narrowing above a threshold energy (Kumar *et al.*, 1997). Nevertheless, the observed high solid-state PL quantum yields (≥ 70%) indicate that lasing is highly possible. Furthermore, the first solid-state tunable cavity lasing from conjugated polymers has recently been demonstrated for poly[1,8-octanedioxy-2,6-dimethoxy-1,4-phenylene-1,2 ethenylene-1,4-phenylene-1,2-ethenylene-3,5-dimethoxy-1,4-phenylene] dispersed in PVK (Kumar *et al.*, 1997).

The polymer-based lasers have many advantages over conventional laser materials such as gallium arsenide, organic dyes or gas mixtures. For instance, soluble conjugated polymers can be easily processed into desired shapes onto various substrates, or cast into free-standing films. More importantly, the semiconducting properties of conjugated polymers provide the possibility for producing electrically pumped polymer laser diodes, and chemical modification of the polymer structure should permit the color of a polymer diode laser to be readily tuned, as is the case for LEDs. However, current densities of more than thousands of amperes per square centimetre are required to surpass the gain-narrowing threshold for producing an electrically pumped plastic laser with the same photon densities as in the optically pumped films (Hide *et al.*, 1996a). Although the thicker films of conjugated polymers required for lasing as opposed to those for light-emitting offer advantages for device construction, they require much higher turn-on voltages for electrically pumped polymer laser diodes. Recent results from tunneling-induced EL experiments indicate possibilities for the construction of electrically pumped polymer lasers (Lidzey *et al.*, 1997) and conjugated polymers have been demonstrated to possess the two basic requirements (*i.e.* supporting optically pumped lasing and current densities of kA/cm^2) for such a laser device (Tessler *et al.*, 2000). Although the above prerequisites are sufficient for traditional semiconducting lasers, it remains to be seen whether the conjugated polymers can work well for electrically pumped lasers due to materials-related issues such as charge-induced absorption, charge mobility, exciton generation and device-related properties. Therefore, it is too early to uncork the champagne bottle as the ultimate goal is still not in sight.

9.4 Carbon Nanotube Displays

As seen in Chapters 5 and 8, the interesting physicochemical properties, together with the unique molecular symmetries, have led to diverse applications for carbon nanotubes, including as new electron field emitters in panel displays. The carbon nanotube electron emitters work on a similar principle to a conventional cathode ray tube, but their small size could lead to thinner, more flexible and energy-efficient display screens with a higher resolution. Figure 9.13a shows a schematic drawing of a typical experimental set up used by de Heer *et al.* (1997) to study the nanotube field emissions. As can be seen, the aligned carbon nanotube film is separated from a fine copper grid cover by a $d = 30$ μm thick perforated mica sheet with a 1 mm diameter hole. Upon application of a potential V between the grid and the nanotube film, an average field $E_0 = V/d$ is induced which, in turn, causes the extraction of electrons from the nanotubes. The emitted electrons then pass through the grid, focusing on a fluorescent screen for picture displays or on an anode plate for measuring the current as a function of the applied field.

Figure 9.13(b) reproduces the $I-V$ curve recorded from the nanotube field emitter shown in Figure 9.13(a). By plotting the log of I/E^2 against $1/E$, a straight line was obtained (inset of Figure 9.13(b)), indicating that the $I-V$ data thus measured were found to fit the Fowler-Nordheim (FN) equation (Fowler and Nordheim, 1928):

$$I = a E^2_{\text{eff}} \exp(-b/E_{\text{eff}}) \tag{9.5}$$

where a and b are constants and E_{eff} is the electric field strength at the emitting point. The field enhancement factor γ (i.e. $\gamma = E_{eff}/E_0$, the ratio of the electric field at the emitting point to the applied field) was deduced from the slope to be 1270 in this particular case.

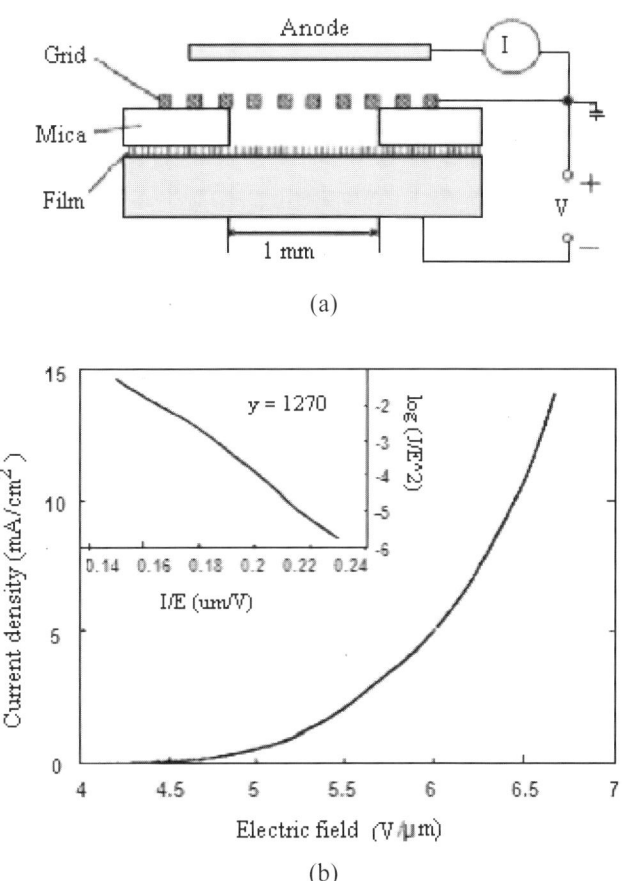

Figure 9.13. (a) Schematic representation of a typical experimental set up for studying the field emission from aligned carbon nanotube films; (b) $I-V$ curve for the aligned nanotube field emitter. Inset shows the Fowler-Nordheim plot of the emission current (After de Heer et al., 1997, copyright 1997 Wiley-VCH Verlag)

Apart from γ, two other important parameters used to characterize field emitters are the turn-on field (E_{t0}), defined as the electric field required to emit a current $I = 10$ $\mu A/cm^2$, and the threshold field (E_{thr}) corresponding to $I = 10$ mA/cm^2.

380 Intelligent Macromolecules for Smart Devices

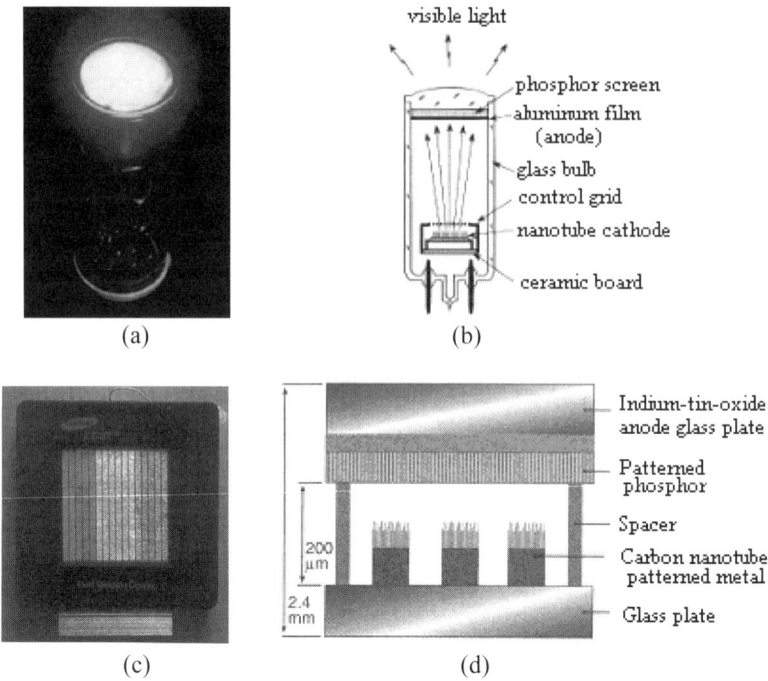

Figure 9.14. (a) The cathode-ray lighting element equipped with a carbon-nanotube field-emitting cathode; (b) a schematic representation of the carbon nanotube field-emission lamps, in which electrons are extracted from the cathode by applying a voltage to the control grid. The electrons are accelerated towards the phosphor screen, which consequently emits light. The color of the light depends on the choice of the phosphor; (c) the flat panel display developed by Samsung, in which the electrons emitted from carbon nanotubes are used to light up the pixels to create images on the screen. This prototype 9 inch flat panel display produces full-color video images; (d) a cross-section view of the flat panel display showing carbon nanotubes that generated the full-color image (After Saito *et al.*, 1998; de Heer and Martel, 2000, (a, b) copyright 1998 Springer-Verlag; (c, d) copyright 2000 Physics World. Reproduced with permission)

The field emission study on single-wall carbon nanotubes carried out by de Heer and co-workers (Bonard *et al.*, 1998) showed some promising results with E_{t0} and E_{thr} in the range of 1.5–4.5 V/μm and 3.9–7.8 V/μm, respectively. These values, like those for most MWNTs, are far lower than corresponding values for other film emitters. Field emissions from an epoxy-carbon nanotube composite have also been reported (Collins *et al.*, 1996). More interestingly, Saito *et al.* (1998) have constructed an electron tube lighting element equipped with MWNT field emitters as a cathode (Figures 9.14(a) and (b)). In this study, stable electron emission, bright luminance and long life suitable for various practical applications have been demonstrated. More recently, some prototypes of carbon-nanotube field-emission flat panel displays were reported, notably by Samsung Advanced Institute of Technology in Korea and Ise Electronics Corp. in Japan (Figures 9.14(c) and (d)) (de Heer, 2000).

9.5 Bucky Light Bulbs and Optical Limiters

9.5.1 C_{60} Light Bulbs

Although the fluorescence of C_{60} and its monoadducts is very weak (Chapter 4), Wudl and co-workers (Hutchison *et al.*, 1999) have recently reported that a C_{60} adduct, T_h-hexapyrrolidine [THP, Equation (9.6)] has a strong yellow-green fluorescence due to a higher excitation energy and larger single-triplet gap caused by a reduction of conjugation within the fullerene core.

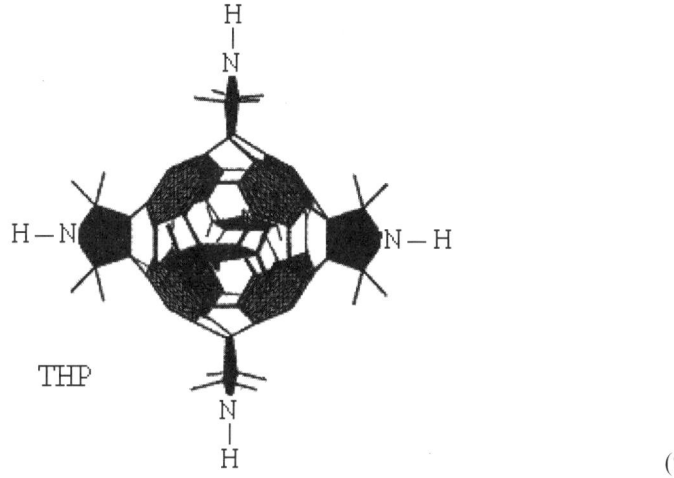

(9.6)

Largely contradicting the conventional belief that C_{60} is an efficient luminescence quencher (Sariciftci *et al.*, 1992) and exhibits electroluminescence with apparent decomposition, as it is slightly *n*-doped C_{60} (Palstra *et al.*, 1997), Wudl and co-workers (Hutchison *et al.*, 1999) have used THP as an emissive material, blended with a hole-transporting poly(9-vinylcarbazole), PVK, and an electron-transporting 2,5-bis-(4-naphthyl)-1,3,4-oxadiazole, BND, to fabricate single-layer LEDs. They found that the Ca/PVK-THF-BND/ITO device emitted strong white light, leading to the first bucky light bulb.

9.5.2 C_{60} Optical Limiters

Photonic technologies, in which photons instead of electrons are used to acquire, store and process information, have received increasing attention. The development of photonic devices requires materials with various non-linear optical (NLO) properties including frequency conversion, optical switching and optical limiting (Prasad and Nigam, 1991). Materials with optical-limiting properties are transparent to low-intensity light and become nearly opaque above a critical intensity (*i.e.*

threshold). The optical limiting effect can be caused by many processes (Bowden *et al.*, 1981) including reverse saturable absorption (RSA) (Guiliano and Hess, 1967), two-photon absorption (TPA) (Sheik-Bahae *et al.*, 1990) and self-focusing and defocusing (SFD) (Soileau *et al.*, 1983). While the RSA at a particular wavelength occurs when a material absorbs weakly at the given wavelength and has a higher energy transition, with a large excited-state absorption cross-section from the state that is populated by the residual ground-state absorption, the TPA takes place if the material has an electronic excited level at twice the frequency of the input beam (Prasad and Williams, 1991). The SFD effect arises from an intensity-induced change in the refractive index of a non-linear medium, which makes the medium act as a positive (negative) lens for focusing (defocusing) the beam. In the case of RSA, the critical intensity is determined by the difference between the excited state and ground state absorption cross-sections (Prasad and Nigam, 1991), and optical limiting will not work if the ground state has no absorption at the frequency range of interest. In contrast, even weakly absorbing samples may provide large refractive index changes due to thermal effects, leading to a strong optical limiting effect through the SFD process (Brèdas *et al.*, 1994). The optical-limiting materials are very useful in many applications, for example as a protective layer for optical sensors and the human eye to prevent damage by bright light sources. For some applications, the optical-limiting materials need to respond rapidly with a low optical-limiting threshold.

Due to their delocalized π-electron orbitals, certain conjugated polymers and fullerenes have been demonstrated to possess large, ultrafast non-linear optical properties (Arbogast *et al.*, 1991; Hann and Bloor, 1989; Messier *et al.*, 1989; Mukamel *et al.*, 1994; Pellegrino *et al.*, 1996; Williams, 1983). The relatively large size of the C_{60} buckyballs can ensure a strong optical transition, and the very long lifetime of the triplet excited state of C_{60} could lead to the low threshold for optical limiting (Arbogast and Foote, 1991; Foote, 1994; Kamat and Asmus, 1996). These features suggest fullerenes to be ideal materials for potential applications as optical limiters. Indeed, fullerene C_{60} has been shown to exhibit third-order optical responses comparable to those of certain conjugated polymers (Flom *et al.*, 1992; Gilmour *et al.*, 1994; Henari *et al.*, 1993; Kajzar *et al.*, 1994; Lindle *et al.*, 1993), and pronounced optical-limiting behavior with a low saturation threshold in toluene and methylene chloride solutions (Hess *et al.*, 1996; Kost *et al.*, 1994; Misra *et al.*, 1994; Tutt and Kost, 1992). Although the optical limiting of C_{60} in toluene for an input power less than 50 J/cm^2 was demonstrated to arise from a combination of RSA and self-defocusing, possible contributions from non-linear refraction and scattering have also been proposed (Kost *et al.*, 1994; Misra *et al.*, 1994). While solutions of C_{60} in toluene and chloronaphthalene have been reported to possess an optical-limiting capability comparable to a carbon-black suspension at 694 nm, the C_{60}/toluene solution was demonstrated to be even a better optical limiter than the carbon-black suspension for microsecond pulses at 514 nm (Hood *et al.*, 1994).

Since many practical applications require solid film-type optical limiters, the NLO effects of pure C_{60} films (Hoshi *et al.*, 1991), C_{60} blended in conjugated polymer (*e.g.* MEH-PPV and poly(3-octylthiophene)) thin films (Flom *et al.*, 1994; Cha *et al.*, 1995), C_{60} mixed with poly(methylmethacrylate) (PMMA) films or sol-gel glasses (Bentivegna *et al.*, 1993; Kost *et al.*, 1993; Smilowitz *et al.*, 1997), and C_{60} covalently attached polystyrene (PS) chains (Chapter 4) in the solid state

(Kojima *et al.*, 1995) have thus been investigated and compared with corresponding solution properties. A comparison study on the optical-limiting performance at 532 nm between C_{60} in PMMA and N-methylthioacridine (CAP), King's complex ($[(C_5H_5)Fe(CO)]_4$) or ruthenium Kin''s complex ($[(C_5H_5)Ru(CO)]_4$) in PMMA indicated that C_{60} is the best reverse saturable absorber (Bentivegna *et al.*, 1993; Kost *et al.*, 1993; Smilowitz *et al.*, 1997). Furthermore, a higher threshold for optical limiting was observed for C_{60} in the PMMA film than in toluene solution (Bentivegna *et al.*, 1993; Kost *et al.*, 1993; Smilowitz *et al.*, 1997), while the optical-limiting effect of the polystyrene-bound C_{60} film was found to be higher than a C_{60} solution by a factor of 5 (Kojima *et al.*, 1995). Enhanced optical-limiting behavior had also been observed for solutions of phenyl-C_{60}-butyric acid cholesteryl ester, which was attributed to the improved solubility and increased ground state population, due to an extension of the ground state absorption into the red, upon grafting the solubilizing chain onto C_{60} (Smilowitz *et al.*, 1996). Continued research efforts in this embryonic field could give birth to a flourishing area of photonic technologies.

9.6 Photovoltaic Cells

9.6.1 Polymer Photovoltaic Cells Containing Fullerenes

The interaction of buckminsterfullerenes with light has attracted considerable interest among scientists in the exploration of applications related to photophysical, photochemical, and photo-induced charge transfer properties of fullerenes. The photo-induced charge transfer properties of fullerenes are largely determined by their electronic structures. Although a HOMO–LUMO energy gap as high as 4.9 e V has been determined for isolated C_{60} molecules (Skotheim, 1986; Lichtenberger *et al.*, 1991), it was found to be diminished in solution or in the condensed solid state due to intermolecular electronic screening (Hammond and Kuck, 1992; Hebard, 1993). In fact, the minimum energy required to create a separated electron and hole in a C_{60} film was found to be of the order of 2.3–2.6 eV (Lof *et al.*, 1992), which is well within the wavelength of UV/vis light.

The photo-induced charge transfer of fullerenes is of importance for the development of polymeric photovoltaic cells, which can be used to store light energy as electron relays for producing electricity. The photovoltaic effect involves the generation of electrons and holes in a semiconducting device under illumination, and subsequent charge collection at opposite electrodes. Inorganic semiconductors, such as amorphous silicon, gallium arsenide and sulfide salts, have been widely used in conventional photovoltaic cells, in which free electrons and holes were produced directly upon photon absorption (Bube, 1992). A major problem associated with inorganic semiconductor-based photovoltaic cells is the high cost for the production of the semiconductors. Therefore, organic dyes (Loutfy and Sharp, 1979; Moriizumi and Kudo, 1981; Stanbery *et al.*, 1985; Yamashita *et al.*, 1987; Hiramoto *et al.*, 1991) and conjugated polymeric semiconductors have received considerable attention in the search for novel photovoltaic cells (Sariciftci, 1997; Skotheim,

1986). Since the fabrication of devices from inorganic semiconductors and/or organic dyes often involves relatively expensive techniques of vacuum evaporation or vapor deposition, photovoltaic cells based on soluble conjugated polymers, which can form thin films even over large areas by solution processing methods, become most attractive. However, the energy conversion efficiency which can be achieved with conjugated polymers is relatively low. Unlike their inorganic counterparts, photon absorption by conjugated polymers at room temperature often creates bound electron hole pairs *(i.e.* excitons). Charge collection, therefore, requires dissociation of the excitons, a process which is known to be favorable at the interface between semiconducting materials with different ionization potentials or electron affinities (Tang C., 1986). Accordingly, devices with a single photoresponsive polymer layer generally have small quantum yields and the typical energy conversion efficiencies of 10^{-3} to 10^{-2}% for pure conjugated polymer-based photovoltaic cells are too low for them to be used in practical applications.

The observation of photovoltaic effects arising from the photo-induced charge transfer at the interface between conjugated polymers as donors and C_{60} film as acceptors (Kraabel *et al.*, 1996; Sariciftci *et al.,* 1992, 1993; Sariciftci and Heeger, 1994) suggests interesting opportunities for improving energy conversion efficiencies of photovoltaic cells based on conjugated polymers (Zakhidov *et al.*, 1995; Zakhidov and Yoshino, 1995). Indeed, increased quantum yields have been obtained by the addition of C_{60} to form heterojunctions with conjugated polymers, such as PPV, MEH-PPV (Kraabel *et al.*, 1996; Lee C., *et al.*, 1993; Sariciftci *et al.,* 1993; Yu *et al.,* 1994), P(3TA)s (Morita *et al.*, 1993) and platinum-polyyne (Kohler *et al.*, 1996). In these conjugated polymer-C_{60} systems, excitons generated in either layer diffuse towards the interface between the layers. Although the photo-induced charge transfer between the excited C_{60} acceptor and a conducting polymer donor can occur very rapidly on a sub-picosecond timescale (Kamat *et al.*, 1996), with a quantum efficiency of close to unity for charge separation from donor to acceptor (Yu *et al.*, 1995), the conversion efficiency of a bilayer heterojunction device is still limited (Kraabel *et al.,* 1996; Sariciftci *et al.,* 1993) by several other factors. Firstly, since the efficient charge separation occurs only at the heterojunction interface, the overall conversion efficiency is diminished by the limited effective interfacial area available in the layer structure. Secondly, because the exciton diffusion range is typically at least a factor of 10 smaller than the optical absorption depth (Tang C., 1986), the photoexcitations produced far from the interface recombine before diffusing to the heterojunction. Finally, the conversion efficiency is also limited by the carrier collection efficiency. In order to overcome these deficiencies, interpenetrating networks consisting of two semiconducting polymers have been developed as ideal photovoltaic materials for a high-efficiency photovoltaic conversion (Figure 9.15). The interpenetrating network structure provides the spatially distributed interfaces necessary for both an efficient photogeneration and a facile collection of the electrons and holes (Halls *et al.*, 1995; Yu *et al.*, 1995; Yu and Heeger, 1995). As expected, significantly improved conversion efficiencies have been reported for photovoltaic cells based on interpenetrating network composites consisting of MEH-PPV and C_{60} derivatives (Yu *et al.*, 1995).

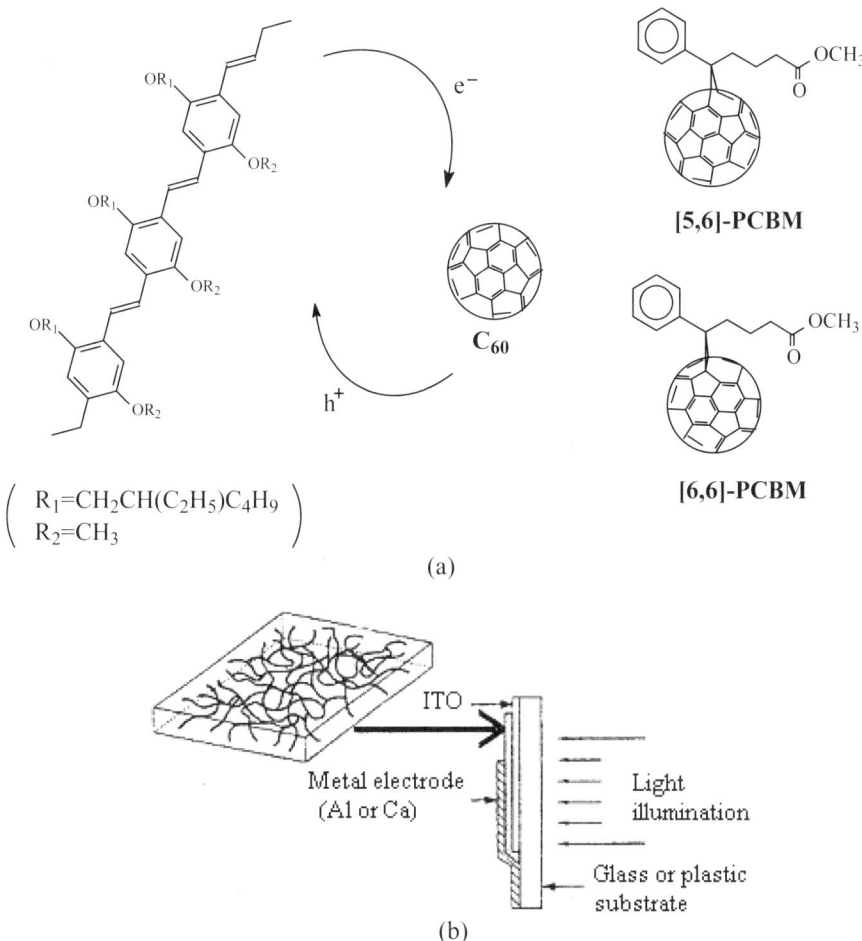

Figure 9.15. Schematic illustrations (a) charge transfer between C_{60} derivatives and MEH-PPV; (b) the interpenetrating conjugated polymer-C_{60} (donor-acceptor) network and the photovoltaic cell (After Yu et al., 1995, copyright 1995 American Association for the Advancement of Science)

9.6.2 Polymer Photovoltaic Cells Containing Carbon Nanotubes

Just as fullerenes have boosted the efficiency of polymer photovoltaic cells, carbon nanotubes have also been used to improved the device performance of polymer photovoltaic cells (Ago et al., 1999; Koprinarov et al., 1996; Kymakis and Amaratunga, 2002). In particular, Kymakis and Amaratunga (2002) have spin-cast composite films of poly(3-octylthiophene), P3OT, and SWNTs on ITO-coated quartz substrates. They found that diodes with the Al/P3OT-SWNT/ITO structure (Figure 9.16(A)) showed photovoltaic behavior with an open circuit voltage of

0.7–0.9 V even at a low nanotube concentration (< 1%). As shown in Figure 9.16(B), the short circuit current is increased by two orders of magnitude compared with the pristine polymer diodes, with an increased fill factor from 0.3 to 0.4 for the nanotube-polymer cell. The improved device performance has been attributed to the good electronic properties of carbon nanotubes and their large surface areas.

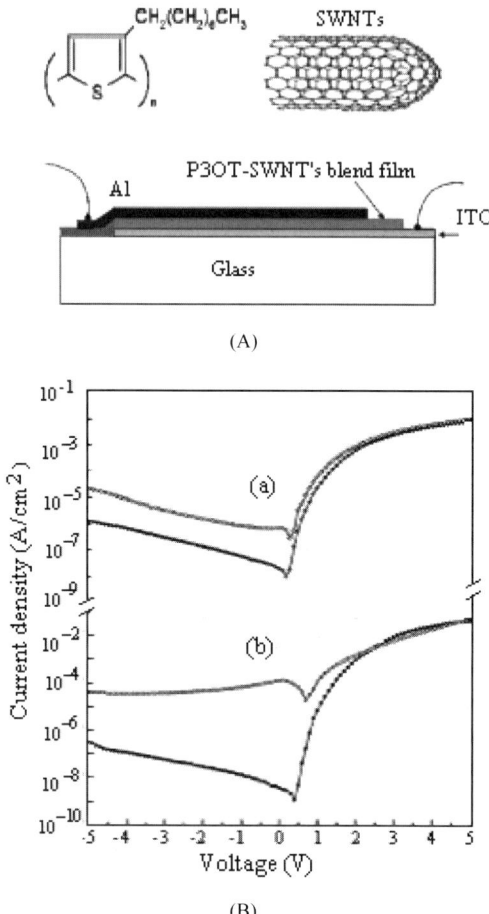

Figure 9.16. (A) A typical device architecture of the P3OT/SWNT photovoltaic cell. (B) (a) I–V characteristics of an ITO/P3OT/Al device in dark (filled circles) and under illumination (open circles); (b) the same data for an ITO/P3OT-SWNT/Al device (After Kymakis and Amaratunga, 2002, copyright 2002 American Institute of Physics. Reproduced with permission)

Similarly, self-assembling techniques have also been used to form polymer (Baur *et al.*, 2001) and small molecular photovoltaic cells (Schmidt-Mende *et al.*, 2001; Zurer, 2001; Hamers, 2001) with increased donor-acceptor interfacial areas, and hence improved device efficiencies. Using a novel processable hybrid polymer of poly(*p*-phenylene vinylene) and poly(*p*-phenylene ethynylene) with covalently linked methanofullerenes as the photoactive material, Ramos *et al.* (Ramos *et al.*, 2001) have also reported the first polymer photovoltaic cell based on a covalently linked donor-acceptor bulk-heterojunction. Clearly, a promising potential for future research exists in this area.

9.7 Light-harvesting Dendrimers

It is well known that natural photosynthetic systems collect sunlight using a vast array of light-harvesting chromophores that channel the absorbed energy to a single reaction center. Adronov and Fréchet (Adronov and Fréchet, 2000) have written an excellent review on light-harvesting dendrimers, in which they discussed the structural features and properties of light-harvesting dendrimers in a close context of natural photosynthetic systems. Some of the main points are summarized below.

The high-resolution X-ray crystal structural study of purple bacteria reveals that the natural photosynthetic unit (PSU) consists of a central reaction center (RC) surrounded by light-harvesting (LH, antenna) complexes (Figure 9.17) (McDermott *et al.*, 1995).

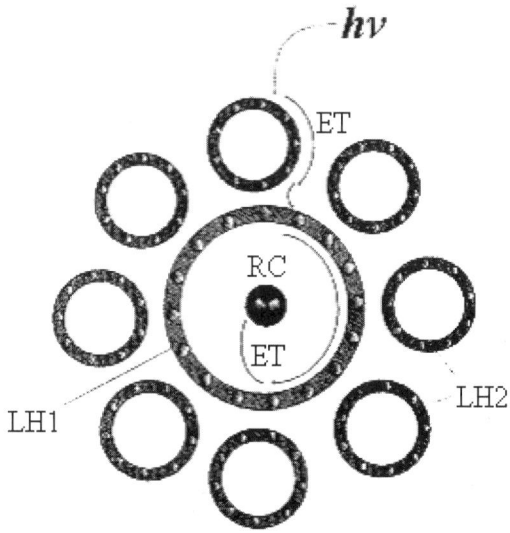

Figure 9.17. Schematic representation of bacterial light-harvesting complexes (LH1 and LH2), in which the circles around the reaction center (RC) represent various protein-embedded light-absorbing porphyrins and the arrows indicate the path of energy transfer (ET) (After Adronov and Fréchet, 2000; McDermott *et al.*, 1995, copyright 2000 The Royal Society of Chemistry. Reproduced with permission)

As can be seen in Figure 9.17, the RC is immediately surrounded by an LH1 complex comprising a ring-shaped assembly of chlorophyll and carotenoid moieties embedded in a protein matrix. Further away from the RC, it is surrounded by similar ring-shaped assemblies made up with the LH2 and LH3 complexes. These chlorophyll-containing assemblies with a relatively large surface area absorb photons, whose energy is then transferred to the RC with almost unit efficiency.

There are two different principles governing the energy between an energy donor (D) and energy donor (A). In the mechanism proposed by Dexter (1953), energy transfer occurs through a chemical bond from a chromophore in its excited state to the other in its ground state. As shown in Figure 9.18, the excited donor returns to its ground state, while the acceptor turns into its excited state upon the energy transfer. This process is accompanied by an electron exchange from the S_1 (S_0) state of the donor (acceptor) to the S_1 (S_0) state of the acceptor (donor).

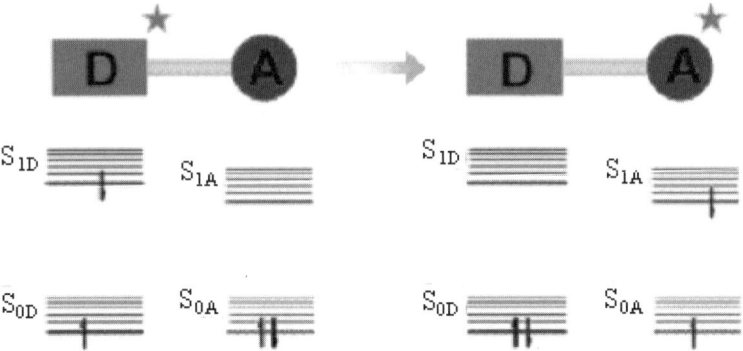

Figure 9.18. The energy transfer process involving the migration of excitation energy from an excited-state donor (D) to a nearby ground-state acceptor (A) (After Adronov and Fréchet, 2000, copyright 2000 The Royal Society of Chemistry. Reproduced with permission)

Such an electron transfer requires a short-range (< 10Å) interaction through strong D–A orbital overlap. The rate, k_{ET}, constant for this process is given in Equation (9.7)

$$k_{ET} = KJ_{exp}(-2R_{DA}/L) \tag{9.7}$$

where K is a constant related to the specific orbital interactions, J is the spectral overlap integral normalized for the extinction coefficient of the acceptor, and R_{DA} is the donor-acceptor separation relative to their van der Waals radii, L (Turro, 1991).

On the other hand, the Förster mechanism does not require electron exchange and the energy transfer occurs via a dipole dipole interaction through space (Förster, 1948). In this case, energy transfer is feasible even with chromophores separated by a relatively large distance (10–100 Å) as the D-A orbital overlap is not necessary. The Förster energy transfer rate constant is described by Equation (9.8):

$$k_{ET} = [9000(\ln 10)\kappa^2 \phi_D J]/(128\pi^5 n^4 N \tau_D R_{DA}^6) \tag{9.8}$$

where κ^2 is the orientation factor (*i.e.* the relative orientation of the donor and acceptor transition dipole moments), ϕ_D is the donor quantum yield in the absence of the acceptor, n is the index of refraction of the solvent, N is Avogadro's number, τ_D

is the donor lifetime in the absence of the acceptor, R_{DA} is the inter-chromophoric distance in cm, and J is the overlap integral (cm^6 mol^{-1}) that is given by Equation (9.9):

$$J = \int f_D(\upsilon)\varepsilon_A(\upsilon)\upsilon^{-4}d\upsilon \tag{12.9}$$

where $f_D(\upsilon)$ is the fluorescence intensity of the donor, $\varepsilon_A(\upsilon)$ is the molar extinction coefficient of the acceptor, and the integral represents the overlap between the donor emission spectrum and the acceptor absorption spectrum, which is closely related to the probability of energy transfer from the donor to the acceptor (Turro, 1991).

Much like the structure shown in Figure 9.19, dendrimers are perfectly branched synthetic macromolecules having numerous chain ends all emanating from a single core (Chapter 4). By means of various advanced synthetic methods (Chapter 4), the locations of chromophores can be precisely controlled. As schematically shown in Figure 9.19, the chromophores can be located at the core, focal point, periphery or even at each branching point of the dendritic structure.

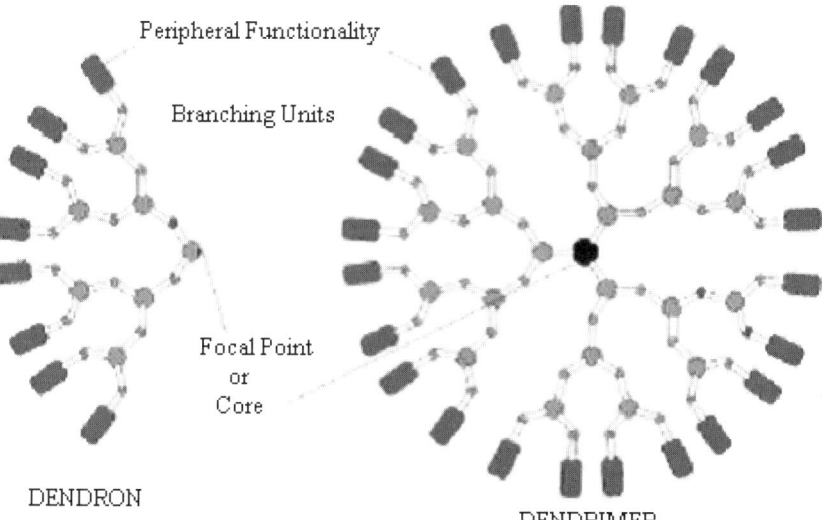

Figure 9.19. Schematic representation of the structure of a dendron and a dendrimer, highlighting the focal point or core (black) surrounded by rings of branching units (gray circles) and end groups (rectangles). Any of them could be chromophoric moieties (After Adronov and Fréchet, 2000, copyright 2000 The Royal Society of Chemistry)

The structure shown in Figure 9.19 resembles the architecture of natural light-harvesting complexes given in Figure 9.17. Adronov and Fréchet (2000) claimed that almost all of the parameters in Equations ((9.17)–(9.19)) can be controlled and varied by regulating the dendritic structures and judiciously choosing monomers. Therefore, the cleave design and synthesis of dendritic structures and assemblies containing various chromphores should allow us to develop many novel photonic devices, including signal amplifiers, fluorescent sensors, LEDs and organic photovoltaic cells, for a better use of the solar energy and reduced dependency on fossil fuels.

Indeed, Xu and Moore (Xu and Moore, 1994) have clearly demonstrated the efficient directional energy transfer from the periphery to the core by making dendrimers which have a directional energy gradient (Figure 9.20).

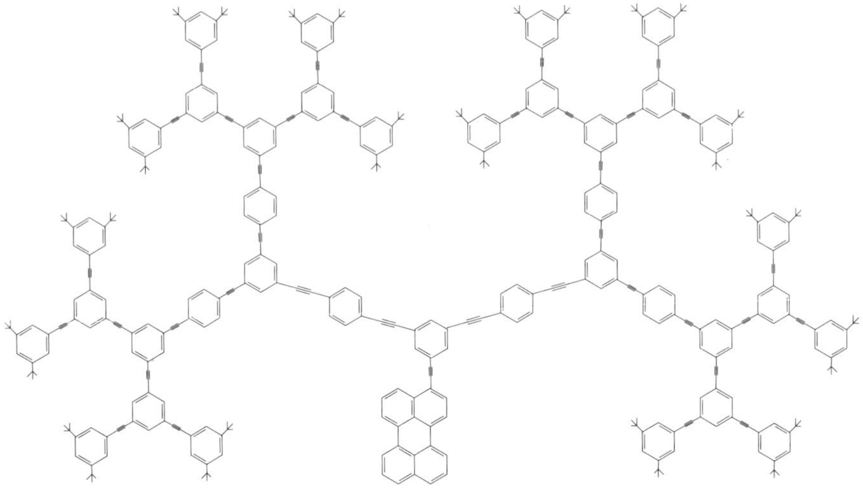

Figure 9.20. Chemical structure of perylene-functionalized phenylacetylene dendrimers with an energy gradient (After Xu and Moore, 1994, copyright 1994 Wiley-VCH Verlag)

As can be seen in Figure 9.20, the band-gap energy of each branch towards the core decreases due to the increase in conjugation length. Consequently, the directional energy transfer from periphery to core can be greatly facilitated by the built-in energy gradient, as indicated by both experimental and theoretical studies (Devadoss et al., 1996; Bar-Haim and Klafter, 1998). However, further study is required in order to understand the mechanism of energy transfer in this system as, apart from the Förster mechanism, orbital overlap contributions from the cross-conjugated dendrimer backbone cannot be ruled out.

More recently, several other groups have studied the encapsulation of luminescent cores by dendritic structures for optical signal amplifications (Jiang and Aida, 1997, 1998; Kawa and Fréchet, 1998; Kimura et al., 1999; Sato et al., 1999). Among them, Kawa and Fréchet (1998) have successfully encapsulated individual Ln^{3+}, Er^{3+}, and Tb^{3+} ions within a dendritic shell, as schematically shown in Figure 9.21.

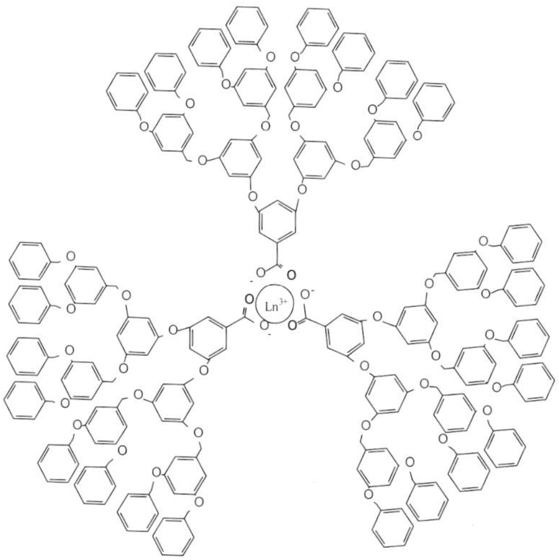

Figure 9.21. Structure of a poly(benzyl ether) dendrimer with a lanthanide ion core (After Kawa and Fréchet, 1998, copyright 1998 The American Chemical Society)

It was found that the side-isolation of metal ions within the dendritic structure led to a much decreased self-quenching effect, and that irradiation at wavelengths where the dendrimer backbone absorbed (280–290 nm) caused strong luminescence from the lanthanide core through the Förster energy transfer from the peripheral dendrimer shell to the luminescent Ln^{3+} at the focal point. This "antenna effect" could have an important implication for the use of these metal-ion-cored dendrimers as signal amplifiers for optical fiber communications, since the poor solubility of metal ions (*e.g.* Ln^{3+}) in silica substrate often leads to the formation of ion clusters and hence self-quenching of fluorescence.

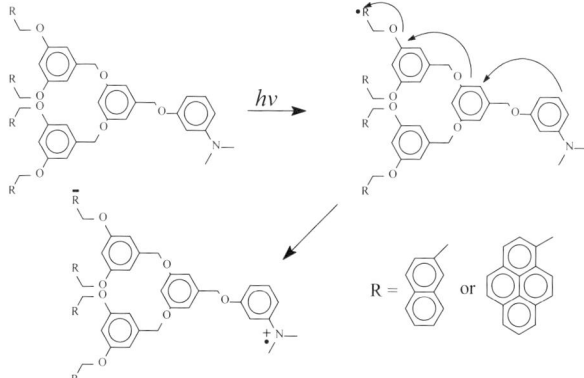

Figure 9.22. Illustration of the electron transfer process via a dendrimer backbone as a result of photoexcitation of the peripheral acceptor chromophores (After Stewart and Fox, 1996, copyright 1996 The American Chemical Society. Reproduced with permission)

Similarly, Stewart and Fox (Stewart and Fox, 1996) have demonstrated the fluorescence quenching of the peripheral naphthyl or pyrenyl chromophores by photo-induced intramolecular electron transfer from the amine core (Fox, 1988) (Figure 9.22).

More interestingly, Fréchet and co-workers (Freeman *et al.*, 1999) have shown the possibility of developing single-layer organic LEDs with multicolor emission features by encapsulating different chromophores in individual constituent dendritic macromolecules in demdrimer mixtures. Separate emission from different dyes otherwise is difficult to achieve, as the light emitted from mixtures of the free dyes is often dominated by that from the dye of the smallest band-gap, due to intermolecular energy transfer.

9.8 Electronic Windows, Electrochromic Displays and Electronic Papers

9.8.1 Electrochromic Windows

An electrochromic window (also termed a "smart window") may be broadly defined as a multilayer thin film electrochromical cell that darkens, with no extraneous curtains, shutters or screens, upon the application of an electrical voltage. A typical multilayer arrangement for smart windows consists of a working electrode coated with an electrochromic layer and a counter electrode that can be either optically passive (*i.e.* colorless whether in the oxidized or reduced state) or electrochromic but in a complementary sense. The two electrodes are separated by transparent ionic conductors, which support the electrochromic reaction. Figure 9.23 shows a schematic diagram of an electrochromic window in which the two electrodes are electrochromic but in a complementary sense.

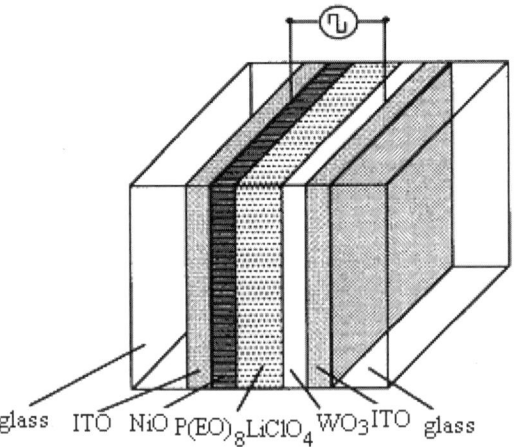

Figure 9.23. Schematic representation of an electrochromic window with electrochromic electrodes in a complementary sense (After Gray, 1991, copyright 1991 VCH Publishers)

In the electrochromic window of the Li$_x$WO$_3$ | P(EO)$_8$LiClO$_4$ | NiO structure, WO$_3$ can be prelithiated and is colored in this form whilst NiO can also undergo reversible lithium intercalation and bleaches with ion insertion. An example of a prototype smart window is shown in Figure 9.24, which clearly reveals a remarkable optical contrast between the "on" and "off" states (Rosseinsky and Mortimer, 2001).

(a) (b)

Figure 9.24. Gentex window 1×2 m^2 (a) clear and (b) darkened (After Rosseinsky and Mortimer, 2001, copyright 2001 Wiley-VCH Verlag. Reproduced with permission)

Although prototype smart windows have been successfully developed, there remains much work to do. For instance, a liquid cell with a gap of the order of 100 μm and tens of square meters made of glass sheets of a few millimeters thick will readily deform. As an alternative, solid electrolytes will offer advantages in eliminating both mechanical deformation and cell leakage. In this regard, the use of polymer electrolyte thin films can offer many advantages over liquid or rigid solid ionic conductors (Chapter 2), including their versatility for large-area deposition, flexibility and mechanically robustness. Apart from the PEO-LiClO$_4$ system, therefore, many other polymer electrolytes have been investigated for applications in smart windows or electrochromic displays (Gray, 1991; Rosseinsky and Mortimer, 2001).

9.8.2 Electrochromic Displays

As shown in Figure 9.23, both electrodes in the smart window may be transparent and the entire window is in the optical path. When one of the electrode is opaque, the same device can be used as a display when it operates in a diffuse reflectance mode (Figure 9.25).

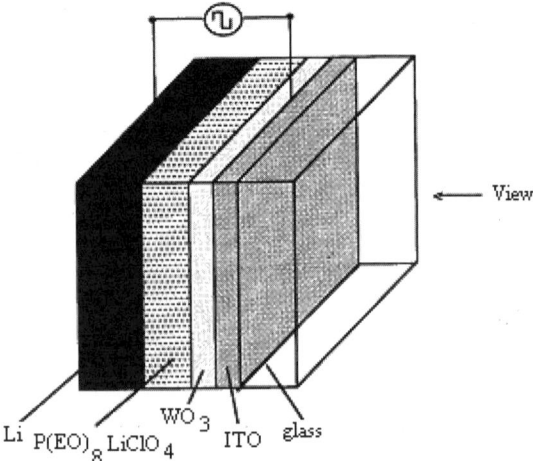

Figure 9.25. Schematic representation of a solid-state electrochromic display using a polymer electrolyte (After Gray, 1991, copyright 1991 VCH Publishers, Inc.)

As discussed in Chapter 3, certain electro responsive polymers exhibit an electrochromism effect to show a reversible color change in response to an electrical voltage. Based on this finding, various conducting polymer displays have been developed (Leventis, 1998; Smela, 1999).

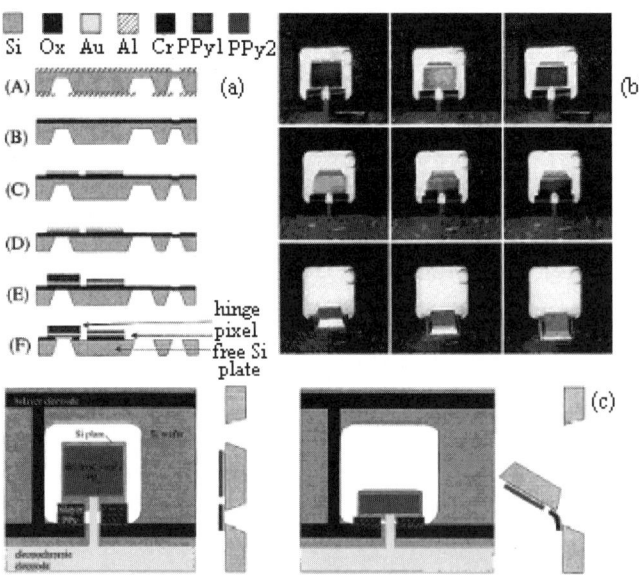

Figure 9.26. Schematica representation of the PPy actuation display: (a) Schematic illustration of the fabrication process; (b) snapshots taken during operation; (c) operating principle (After Smela, 1999, copyright 1999 Wiley-VCH Verlag. Reproduced with permission)

As shown in Figure 9.26, Smela reported (Smela, 1999) a remarkable (color) display system made from pixels of polypyrrole (PPy) on silicon. In this system, each of the PPy/Si platelet pixels is attached to a silicon wafer through a PPy/gold bilayer, which forms a hinge (actuator) between the Si surround and Si platelet. Oxidation and/or reduction in the presence of electrolyte(s) can cause changes in volume of the PPy film on Au (Chapter 11), leading to the rotation of the Si platelet out of the plane of the wafer up to 170°. The angle of rotation, in turn, alters the depth of PPy^+ viewed, and hence changes its perceived color for display applications.

9.8.3 Electronic Papers

Electronic paper represents an emerging plastic technology that combines the best features of ordinary paper (*e.g.* flexible, lightweight, and possessing wide viewing angles) and computer displays (*e.g.* information displayed can be wirelessly updated). In view of the possibility that electronic papers may one day change the way we read books, periodicals and display information, a large number of groups are working on this emerging technology. Notably, the E Ink Corp., in collaboration with Bell Labs and Xerox, have developed prototype products.

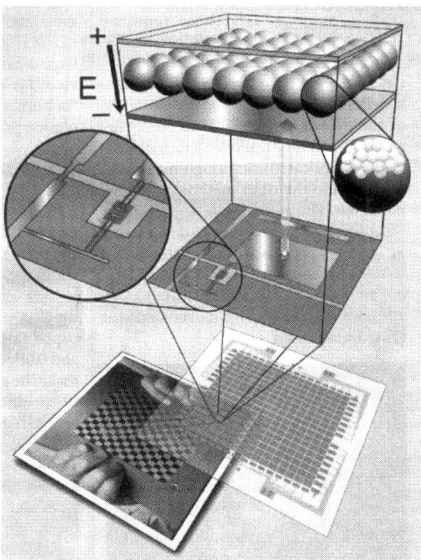

Figure 9.27. Schematic representation of the flexible plastic electronic paper developed by the E Ink Corp. and Lucent Technologies, showing how the key components relate to each other for the final display. The back electrode array is patterned into square pixels by the microcontact printing technology (Chapter 6). Each contains an organic transistor that switches the pixel on or off. The transistor-controlled electric field induces the white particles to migrate either to the top or bottom of the microcapsules, leading to color change for the ink in the region covered by that pixel (After Dagani, 2001, copyright 2001 Lucent Technologies and E Ink Corp.)

As schematically shown in Figure 9.27, the E Ink Corp.'s electronic paper prototype consists of a transparent layer of indium tin oxide-coated thin plastic film (the front electrode), a plastic sheet coated with electronic ink and a plastic sheet (the back electrode) containing an array of field-effect transistors (Chapter 8) with associated wiring connecting from the drain to the pixel electrode. The ink is made from a water-based fluid dispersed with a high concentration of clear plastic microcapsules (50–100 μm in diameter) containing charged particles of white titatnium dioxide suspended in a hydrocarbon oil that is dyed black, for example, for the black and white displays.

When an electric voltage is applied to the gate electrode, current flows from the source to the drain and the pixel electrode becomes charged. Then, the white particles in the plastic microcapsules dispersed in the ink coated between the front and back electrodes migrate to one side of the microcapsules and become visible, for instance, at the side closest to the viewer. By reversing the electric field, the white particles move to the opposite end of the microcapsule and are masked by the black fluid, leading to a dark spot (Comiskey *et al.,* 1998). Like ordinary paper, this electronic paper display is viewed in reflected light without the requirement for backlight. Only a tiny amount of power is required for changing the image, which, once formed, could stay even without a continuous supply of power. One hundred thousand microcapsules could easily be coated on a square inch of surface, providing a resolution of at least 200 dots per inch. Since a sheet of the electronic paper would be packed with many pixels, the size of the transistor becomes a controlling factor towards a high-resolution display. Although the prototype electronic paper discussed above is very promising, a commercial product is still some time away. Since the large-scale production of pixels with dimensions of *ca.* 100×100 μm – the size necessary for high-information-content storage and display is still difficult – the long-term reliability of the electronic paper and its full color display will need to be addressed.

By replacing the titatnium-dioxide-based ink with charged polyethylene microparticles (typically 50–100 μm in diameter) colored with one hemisphere black and the other hemisphere white, Sheridon and coworkers (see Dagani, 2001) have also developed electronic papers by impregnating these black and white polymer balls into a sheet of transparent silicone rubber. When an electric field is applied, each of the doubly colored balls rotates to show either its black or white hemisphere, as each of the hemispheres carries a different charge, which produces a dipole with a specific response to the external field. In this case, the display resolution is determined by the size of the balls, 50 μm particles provide a resolution of 200 dots per inch. Higher resolution display can, in principle, be obtained by using smaller balls. Furthermore, Sheridon and co-workers (see Dagani, 2001) have also demonstrated various two-color displays using balls having different pairs of colors, though full color display is still out of sight. However, further work in the field should bring electronic papers into the market in the foreseeable future.

9.9 References

Adronov, A., Fréchet, J.M.J. (2000) *Chem. Commun.* 1701.
Ago, H., Petritsch, K., Shaffer, M.S.P., Windle, A.H., Friend, R.H. (1999) *Adv. Mater.* **11**, 1281.
Aguiar, M., Hu, B., Karasz, F.E., Akcelrud, L. (1996) *Macromolecules* **29**, 3161.
Aguiar, M., Karasz, F.E., Akcelrud, L. (1995) *Macromolecules* **28**, 4598.
Arbogast, J.W., Darmanyan, A.P., Foote, C.S., Rubin, Y., Diederich, P.N., Alvarez, M.M., Anz, S.J., Whetten, R.L. (1991) *J. Phys. Chem.* **95**, 11.
Arbogast, J.W., Foote, C.S. (1991) *J. Am. Chem. Soc.* **113**, 8886.
Baigent, D.R., Friend, R.H., Lee, J.K., Schrock, R.R. (1995) *Synth. Met.* **71**, 2171.
Baldo, M.A., O'Brien, D.F., You, Y., Shoustikov, A., Sibley, S., Thompson, M.E., Forrest, S.R. (1998) *Nature* **395**, 151.
Bar-Haim, A., Klafter, J. (1998) *J. Phys. Chem. B* **102**, 1662.
Baur, J.W., Durstock, M.F., Taylor, B.E., Spry, R.J., Reulbach, S., Chiang, L.Y. (2001) *Synth. Met.* **121**, 1547
Bentivegna, F., Canva, M., Georges, P., Brun, A. (1993) *Appl. Phys. Lett.* **62**, 1721.
Berggren, M., Gustafsson, G., Inganäs, O., Andersson, M. (1995) *Adv. Mater.* **7**, 900.
Berggren, M., Gustafsson, G., Inganäs, O., Andersson, M., Wennerström, R.O., Hjertberg, T. (1994a) *J. Appl. Phys.* **76**, 7530.
Berggren, M., Inganäs, O., Gustafsson, G., Rasmusson, J., Andersson, M.R., Hjertberg, T., Wennerstrom, O. (1994b) *Nature* **372**, 444.
Birgerson, J., Kaeriyama, K., Barta, P., Bröms, P., Fahlman, M., Granlund, T., Salaneck, W.R. (1996) *Adv. Mater.* **8**, 982.
Bloembergen, N. (1993) *Phys. Today* **October**, 28.
Bolognesi, A., Bajo, G., Paloheimo, L., Östergard, T., Stubb, H. (1997) *Adv. Mater.* **9**, 121.
Bonard, J.-M., Salvetat, J.-P., Stöckli, T., de Heer, W.A., Forró, L., Chatelain, A. (1998) *Appl. Phys. Lett.*, **73**, 918.
Bradley, D.C. (1992) *Adv. Mater.* **4**, 756, and references therein.
Bradley, D.D.C., Tsutsui, T. (Eds.) (1995) *Organic Electroluminescence*, Cambridge University Press, Cambridge.
Brèdas, J.L., Adant, C., Tackx, P., Persoons, A. (1994) *Chem. Rev.* **94**, 243.
Bowden, C.M., Giftan, M., Robl, H.R., (1981) *Optical Bistability,* Plenum, New York.
Brouwer, H.J., Kransnikov, V.V., Hilberer, A., Wildeman, J., Hadziioannou, G. (1995) *Appl. Phys. Lett.* **66**, 3404.
Brown, A.R., Bradley, D.D.C., Burroughes, J.H., Friend, R.H., Greenham, N.C., Burn, P.L., Kraft, A., Holmes, A.B. (1992) *Appl. Phys. Lett.* **61**, 2793.
Brown, A.R., Burroughes, J.H., Greenham, N., Friend, R.H., Bradley, D.D.C., Baigent, D.R., Holmes, A.B., Moratti, S.C., Friend, R.H. (1996) *Synth. Met.* **80**, 119.
Bube, R.H. (1992) *Photoelectronic Properties of Semiconductors,* Cambridge University Press, Cambridge, UK.
Burn, P.L., Bradley, D.D.C., Friend, R.H., Halliday, D.A., Holmes, A.B., Jackson, R.W., Kraft, A. (1992a) *J. Chem. Soc., Perkin Trans.* **1**, 3225.
Burn, P.L., Holmes, A.B., Kraft, A., Bradley, D.D.C., Brown, A.R., Friend, R.H., Gymer, R.W. (1992b) *Nature* **356**, 47.
Burn, P. L., Kraft, A., Baigent, D. R., Bradley, D. D. C., Brown, A. R., Friend, R. H., Gymer, R. W., Holmes, A. B., Jackson, R. W. (1993) *J. Am. Chem. Soc.* **115**, 10117.
Burroughes, J.H., Bradley, D.C.C., Brown, A.R., Mackay, M.K., Friend, R.H., Burn, P.L. (1990) *Nature* **347**, 539.
Carlos, L.D., de Zea Bermudez, V., Sáferreira, R.A., Marques, L., Assunção, M. (1999) *Chem. Mater.* **11**, 581.

Cha, M., Sariciftci, N.S., Heeger, A.J., Hurnrnelen, J.C., Wudl, F. (1995) *Appl. Phys. Lett.* **67**, 3850.
Chang, W.P., Whang, W.-T. (1996) *Polymer* **37**, 4229.
Christ, T., Greiner, A., Sander, R., Stümpflen, V., Wendorff, J.H. (1997) *Adv. Mater.* **9**, 219.
Choong, V.E., Park, Y., Gao, Y., Wehrmeister, T., Mullen, K., Hsieh, B.R., Tang, C.W. (1996) *Appl. Phys. Lett.* **69**, 1492.
Chung, S.-J., Jin, J.-I., Kim, K.-K. (1997) *Adv. Mater.* **9**, 551.
Cimrova, V., Remmers, M., Neher, D., Wegner, G. (1996) *Adv. Mater.* **8**, 146.
Cleave, V., Yahioglu, G., Le Barny, P., Friend, R.H., Tessler, N. (1999) *Adv. Mater.* **11**, 285.
Collie, L., Parker, D., Tachon, C., Hubbard, H.V. St.A., Davies, G.R., Ward, I.M., Wellings, S.C. (1993) *Polymer* **34**, 1541, and references cited therein.
Collins, P.G., Zettl, A. (1996) *Appl. Phys. Lett.* **69**, 1969.
Collyer, A.A. (ed.) (1992) *Liquid crystal polymers: from structures to applications*, Elsevier Applied Science, New York.
Comiskey, B., Albert, J.D., Yoshizawa, H., Jacobson, J. (1998) *Nature* **394**, 253
Conwell, E.M., Miller, M.R. (eds) (1993) *Electroluminescent materials, devices, and large screen displays, SPIE Proc.* **1910**.
Curran, S.A., Ajayan, P.M., Blau, W.J., Carroll, D.L., Coleman, J.N., Dalton, A.B., Davey, A.P., Drury, A., McCarthy, B., Maier, S., Strevens, A. (1998) *Adv. Mater.* **10**, 1091.
Dagani, R. (1995) *C&E News* **April 24**, 42.
Dagani, R. (2001) *C&E News* **Jan. 15**, 40.
Dai, L. (1999a) *J. Macromolecul. Sci, Macromol. Chem. Phys.* **39**, 273, and references cited therein.
Dai, L. (1999b) *Polym. Adv. Technol.* **10**, 357.
Dai, L., Dong, L., Tong , L., Mau, A.W.H. (2001) *Adv. Mater.* **13**, 915, and references cited therein.
Dai, L., Griesser, H., Mau, A.W.H. (1997a) *J. Phys. Chem. B* **101**, 9548.
Dai, L., Mau, A.W.H. (2000) *J. Phys. Chem.* **104**, 1891.
Dai, L., Mau, A.W.H., Gong, X., Griesser, H.J. (1997b) *Synth. Met.* **85**, 1379.
Dai, L., Wang, Q., Wan, M., (2000) *J. Mater. Sci. Lett.*, **19**, 1645.
Dannetun, P., Fahlman, M., Fauquet, C., Kaerijama, K., Sonoda, Y., Lazzaroni, R., Brédas, J.L., Salaneck, W.R. (1994) In *Organic Materials for Electronics: Conjugated Polymer Interfaces with Metals and Semiconductors*, Brédas, J.l., Salaneck, W.R., Wegner, G. (eds), North-Holland, Amsterdam.
Dannetun, P., Löglund, M., Fahlman, M., Boman, M., Stafström, S., Salaneck, W.R., Lazzaroni, R., Fredriksson, C., Brédas, J.L., Graham, S., Friend, R.H., Holmes, A.B., Zamboni, R., Taliani, C. (1993a) *Synth. Met.* **55**, 212.
Dannetun, P., Lögdlund, M., Salaneck, W.R., Fredriksson, C., Stafström, S., Holmes, A.B., Brown, A., Graham, S., Friend, R.H., Lhost, O. (1993b) *Mol. Cryst. Liq. Cryst.* **228**, 43.
de Heer, W.A., Bonard, J.-M., Fauth, K., Chătelain, A., Forró, L., Ugarte, D. (1997) *Adv. Mater.* **9**, 87.
de Heer, W.A., Martel, R. (2000) *Phys. World* **June**, 49.
de Morais, T.D., Chaput, F., Lahlil, K., Boilot, J.-P. (1999) *Adv. Mater.* **11**, 107.
Denton, G.J., Tessler, N., Stevens, M.A., Friend, R.H. (1997) *Adv. Mater.* **9**, 547.
Devadoss, P., Bharathi, P., Moore, J.S. (1996) *J. Am. Chem. Soc.* **118**, 9635.
Dexter, D.L. (1953) *J. Chem. Phys.* **21**, 836.
Dfaz-Garcfa, M.A., Hide, F., Schwartz, B.J., Andersson, M.R., Pei, Q., Heeger, A.J. (1997) *Synth. Met.* **84**, 455.
Duan X.F., Huang, Y., Cui, Y., Wang, J.F., Lieber, C.M. (2001) *Nature* **409**, 66.
Dyreklev, P., Berggren, M., Inganäs, O., Andersson, M.R., Wennerstrom, O., Hjertberg, T. (1995) *Adv. Mater.* **7**, 43.
Fahlman, M., Birgersson, J., Kaeriyama, K., Salancck, W.R. (1995) *Synth. Met.* **75**, 223.

Flom, S.R., Pong, R.G.S., Bartoli, F.J., Kafafi, Z.H. (1992) *Phys. Rev. B: Condens. Matter Mater. Phys.* **46**, 15598.
Flom, S.R., Sarkas, H.W., Pong, R.G.S., Bartoli, F.J., Kafafi, Z.H. (1994) *Polym. Prep.* **35(2)**, 110.
Foote, C. S. (1994) *Top. Curr. Chem.* **169**, 347.
Förster, T. (1948) *Ann. Phys.* **2**, 55.
Fowler, R.H. and Nordheim, L.W. (1928) *Proc. Royal. Soc. A* (London) **119**, 173.
Fox, M.A. (1988) *Photoinduced Electron Transfer*, Elsevier, Amsterdam.
Freemantle, M. (2001) *C&E News* **79(17)**, 49.
Freeman, A.W., Fréchet, J.M.J., Koene, S.C., Thompson, M.E. (1999) *Polym. Prep.* **40**, 1246.
Freeman, A.W., Koene, S.C., Malenfant, P.R.L., Thompson, M.E., Frechet, J.M.J. (2000) *J. Am. Chem. Soc.* **122**, 12385.
Frolov, S.V., Gellermann, W., Vardeny, Z.V., Ozaki, M., Yoshino, K. (1997) *Synth. Met.* **84**, 471.
Gao, Y., Park, K.T., Hsieh, B.R. (1993) *J. Appl. Phys.* **73**, 7894.
Gautier, E., Nunzi, J.-M., Sentein, C., Lorin, A., Raimond, P. (1996) *Synth. Met.* **81**, 197.
Gebler, D.D., Wang, Y.Z., Blatchford, J.W., lessen, S.W., Lin, L.-B., Gustafson, T.L., Wang, H.L., Swager, T.M., MacDiarmid, A.G., Epstein, A.J. (1995) *J. Appl. Phys.* **78**, 4264.
Gibbs, H.M. (1985) *Optical Bistability: Controlling Light with Light*, Academic Press, New York.
Gilmour, S., Marder, S.R., Perry, J.W., Cheng, L.T. (1994) *Adv. Mater.* **6**, 494.
Graupner, W., Leising, G., Lanzani, G., Nisoli, M., De Silvestri, S., Scherf, U. (1996) *Phys. Rev. Lett.* **76**, 847.
Gray, F.M. (1991) *Solid Polymer Electrolytes*, VCH Publishers, Inc., New York.
Greenham, N.C., Friend, R.H. (1995) *Solid State Phys.* **49**, 1.
Greenham, N.C., Moratti, S.C., Bradley, D.D.C., Friend, R.H., Holmes, A.B. (1993) *Nature* **365**, 628.
Grell, M., Bradley, D. C., Inbasekaran, M., Woo, E. P. (1997) *Adv. Mater.* **9**, 798.
Grem, G., Leditzky, G., Ulrich, B., Leising, G. (1992) *Synth. Met.* **51**, 383.
Grem, G., Leising, G. (1993) *Synth. Met.* **55-57**, 4105.
Grice, A. W., Tajbakhsh, A., Burn, P. L., Bradley, D.D.C. (1997) *Adv. Mater.* **9**, 1174.
Gruner, J., Hammer, P.J., Friend, R.H., Huber, H.-J., Scherf, U., Holmes, A.B. (1994) *Adv. Mater.* **6**, 748.
Grüner, J., Remmers, M., Neher, D. (1997) *Adv. Mater.* **9**, 964.
Guiliano, C.R., Hess, L.D. (1967) *IEEE J. Quant. Electron.* **3**, 338.
Gyoutoku, A., Hara, S., Komatsu, T., Shirinashihama, M., Iwanaga, H., Sakanoue, K. (1997) *Synth. Met.* **91**, 73.
Halls, J.J.M., Walsh, C.A., Greenham, N.C., Marseglia. E.A., Friend, R.H., Moratti, S.C., Holmes, A.B. (1995) *Nature* **376**, 498.
Hamaguchi, M., Yoshino, K. (1995) *Appl. Phys. Lett.* **67**, 3381.
Hamaguchi, M., Yoshino, K. (1996) *Appl. Phys. Lett.* **69**, 143.
Hamers, R.J. (2001) *Nature* **412**, 489.
Hammond, G.S., Kuck, V.J. (eds) (1992) *Fullerenes: Synthesis, Properties, and Chemistry of Large Carbon Clusters*, ACS Symposium Series 481, American Chemical Society, Washington DC.
Hann, R.A., Bloor, D. (eds) (1989) *Organic Materials for Nonlinear Optics*, Royal Society of Chemistry, London.
Hebard, A.F. (1993) *Annu. Rev. Mater. Sci.* **23**, 159.
Heeger, A.J., Long, J. Jr. (1996) *Opt. Photon. News* **August**, 24.
Henari, F.Z., MacNamara, S., Stevenson, O, Callaghan, J., Weldon, D., Blau, W.J. (1993) *Adv. Mater.* **5**, 930.
Hertel, D., Setayesh, S., Nothofer, H.G., Scherf, U., Müllen, K., Bässler, H. (2001) Adv. Mater. **13**, 65.

Hess, B.C., Bowersox, D.V., Mardirosian, S.H., Unterberger, L.D. (1996) *Chem. Phys. Lett.* **248**, 141.
Hiberer, A., van Hutten, P.F., Wildeman, J., Hadziioannou, G. (1997) *Macromol. Chem. Phys.* **198**, 2211.
Hide, F., Díaz-García, M.A., Schwartz, B.J., Heeger, A.J. (1997) *Acc. Chem. Res.* **30**, 430.
Hide, F., Dfaz-Garcfa, M.A., Schwartz, J. , Andersson, M.R., Pei, Q., Heeger, A.J. (1996a) *Science* **273**, 1833.
Hide, F., Schwartz, J., Diaz-Garcia, M.A., Heeger, A.J. (1996b) *Chem. Phys. Lett.* **256**, 424.
Higgins, J.S., Benoît, H.C. (1994) *Polymers and Neutron Scattering*; Clarendon Press, Oxford.
Hiramoto, M., Fujiwara, H., Yokoyama, M. (1991) *Appl. Phys. Lett.* **58**, 1062.
Holmes, A.B., Bradley, D.D.C., Brown, A.R., Burn, P.L., Burroughes, J.H., Friend, R.H., Greenham, N.C., Gymer, R.W., Halliday, D.A., Jackson, R.W., Kraft, A., Martens, J.H.F., Pichler, K., Samuel, I.D.W. (1993) *Synth. Met.* **55**, 4031.
Holzer, W., Penzkofer, A., Gong, S.-H., Bleyer, A., Bradley, D. D. C. (1996) *Adv. Mater.* **8**, 974.
Holzer, L., Wenzl, F.P., Tasch, S., Leising, G., Winkler, B., Dai, L., Mau, A.W.H. (1999a) *Appl. Phys. Lett.* **2014**, 75.
Holzer, L., Winkler, B., Wenzl, F.P., Tasch, S., Dai, L., Mau, A.W.H., Leising, G. (1999b) *Synth. Met.*, **100**, 71.
Hong, H., Steitz, R., Kirstein, S., Davidov, D. (1998) *Adv. Mater.* **10**, 1104.
Hood, P.J., Edmonds, B.P., McLean, D.G., *Brandelik,* D.M. (1994) *SPIE Proc.* **2229**, 91.
Horowitz, G., Delannoy, P., Bouchriha, H., Deloffre, F., Pave, J.-L., Gamier, F., Hajlaoui, R., Heyman, M., Kouki, F., Valat, P., Wintgens, V., Yassar, A. (1994) *Adv. Mater.* **6**, 752.
Hoshi, H., Nakamura, N., Maruyama, Y., Nakagawa, T., Suzuki, S., Shiromaru, H., Achiba, Y. (1991) *Jap. J. Appl. Phys.* **30**, L1397.
Hsieh, B.R., Ettedgui, E., Gao, Y. (1996) *Synth. Met.* **78**, 269.
Hu, B., Karasz, F.E. (1998) *Chem. Phys.* **227**, 263.
Hu, B., Yang, Z., Karasz, F.E. (1994) *J. Appl. Phys.* **76**, 2419.
Huber, J., Mtillen, K., Sahlbeck, J., Scherf, U., Stehlin, T., Stem, R. (1994) *Acta Polym.* **45**, 244.
Hung, L.S., Tang, C.W., Mason, M.G. (1997) *Appl. Phys. Lett.* **70**, 152.
Hutchison, K., Gao, J., Schick, G., Rubin, Y., Wudl, F. (1999) *J. Am. Chem. Soc.* **121**, 5611.
Isayev, A.I., Kyu, T., Cheng, S.Z.D. (eds) (1995) *Liquid-Crystalline Polymer Systems: Technological Advances*, ACS Symposium Series 632, ACS, Washington DC.
Jabbour, G.E., Kawabe, Y., Shaheen, S.E., Wang, J.F., Morrell, M.M., Kippelen, B., Peyghambarian, N. (1997) *Appl. Phys. Lett.* **71**, 1762.
Jeon, H., Ding, J., Nurmikko, A.V., Xie, W., Kobayashi, M., Gunshor, R.L. (1992) *Appl. Phys. Lett.*, **60**, 892.
Jiang, D.-L., Aida, T. (1997) *Nature* **388**, 454
Jiang, D.-L., Aida, T. (1998) *J. Am. Chem. Soc.* **120**, 10895.
Jin, M.-S., Kim, W.-T. (1997) *Appl. Phys. Lett.* **70**, 484.
Kajzar, F., Taliani, C., Danieli, R., Rossini, S., Zarnboni, R. (1994) *Chem. Phys. Lett.* **4**, 418.
Kamat, P.V., Asmus, K.-D. (1996) *Electrochem. Soc. Interface* **Spring**, 22.
Karg, S., Scott, J.C., Salem, J.R., Angelopoulos, M. (1996) *Synth. Met.* **80**, 111.
Kawa, M., Fréchet, J.M.J. (1998) *Chem. Mater.* **10**, 286.
Kersting, R., Lemmer, U., Mahrt, R.F., Leo, K., Kurz, H., Bassler, H., Gobel, E.O. (1993) *Phys. Rev. Lett.* **70**, 3820.
Kido, J., Masato, K., Nagai, K. (1995) *Science* **267**, 1332.
Kim, Y.-E., Park, H., Kim, J.-J. (1996) *Appl. Phys. Lett.* **69**, 599.
Kim, D.U., Tsutsui, T., Saito, S. (1995) *Chem. Lett.* 587.
Kimura, M., Shiba, T., Muto, T., Shirai, K., Hanabusa, H. (1999) *Macromolecules* **32**, 8237.
Kohler, A., Wittmann, H.F., Friend, R.H., Khan, M.S., Lewis, J. (1996) *Synth. Met.* **77**, 147.

Kojima, Y., Matsuoka, T., Takahashi, H., Kurauchi, T. (1995) *Matromolecules* **28**, 8858.
Konstadinidis, K., Papadimitrakopoulos, F., Galvin, M., Opila, R.L. (1995) *J. Appl. Phys.* **77**, 5642.
Koprinarov, N., Stefanov, R., Pchelarov, G., Konstantinova, M., Stambolova, I. (1996) *Synth. Met.* **77**, 47.
Korczynski, E. (1999) *Solid State Technol.* **42(1)**, 51.
Kost, A., Jensen, J.E., Klein, M.B., McMahon, S.W., Haeri, M.B., Ehritz, M.E. (1994) *SPIE Proc.* **2229**, 78.
Kost, A., Tutt, L., Klein, M., Dougherty, T.K., Elias, W.E. (1993) *Opt. Lett.* **18**,334.
Kraabel, B., Hummelen, J.C., Vacar, D., Moses, D., Sariciftci, N.S., Heeger, A.J., Wudl, F. (1996) *J. Chem. Phys.* **104**, 4267.
Kraabel, B., Lee, C.H., McBranch, D., Moses, D., Sariciftci, N.S., Heeger, A.J. (1993) *Chem. Phys. Lett.* **213**, 389.
Kraft, A., Grimsdale, A.C., Holmes, A.B., (1998) *Angew. Chem. Int. Ed.* **37**, 402.
Kugler, Th., Andersson, A., M. Lögdlund, Holmes, A.B., Li, X., Salaneck, W.R. (1999) *Synth. Met.* **100**, 163.
Kumar, N.D., Bhawalkar, J.D., Prasad, P.N., Karasz, F.E., Hu, B. (1997) *Appl. Phys. Lett.* **71**, 25.
Kymakis, E., Amaratunga, G.A.J. (2002) *Appl. Phys. Lett.* **80**, 112.
Lee, J.-K., Schrock, R.R., Baigent, D.R., Friend, R.H. (1995) *Macromolecules* **28**, 1966.
Lee, C.H., Yu, G., Moses, D., Pakbaz, K., Zhang, C., Sariciftci, N.S., Heeger, A.J., Wudl, F. (1993) *Phys. Rev. B: Condens. Matter Mater. Phys.* **48**, 15425.
Leising, G., Tasch, S., Meghdadi, F., Athouel, L., Froyer, G., Scherf, U. (1996) *Synth. Met.* **81**, 138.
Leventis, N., Chen, M., Liapis, A.I., Johnson, J.W., Jain, A. (1998) *J. Electrochim. Acta* **145**, L55.
Leyden, D.E., Collins, W.T. (eds) (1988) *Chemically Modified Surfaces in Science and Industry*, Gordon and Breach Science Publishers, New York,.
Liang, W.B., Masse, M.A., Karasz, F.E. (1992) *Polymer* **35**, 3101.
Lide, D.R. (ed.) (1996) *CRC Handbook of Chemistry and Physics*, 77th edn, CRC Press, Boca Raton.
Lidzey, D.G., Bradley, D.D.C., Alvarado, S.F., Seidler, P.F. (1997) *Nature* **386**, 135.
Lichtenberger, D.L., Nebesney, K.W., Ray, C.D., Huffman, D.R., Lamb, L.D. (1991) *Chem. Phys. Lett.,* **176**, 203.
Lindle, J.R., Pong, R.G.S., Bartoli, F.J., Kafafi, Z.H. (1993) *Phys. Rev. B: Condens. Matter Mater. Phys.* **48**, 9447.
Lmimouni, K., Legrand, C., Dufour, C., Chapoton, A., Belouet, C. (2001) *Appl. Phys. Lett.* **78**, 2437.
Lof, R.W., van Veenendaal, M.A., Koopmans, B., Jonkrnan, H.T., Sawatzky, G.A. (1992) *Phys. Rev. Lett.* **68**, 3924.
Loutfy, R.O., Sharp, J.H. (1979) *J. Chem. Phys.* **71**, 1211.
McDermott, G., Prince, S.M., Freer, A.A., Hawthornthwaite-Lawless, A.M., Papiz, M.Z., Cogdell, R.J., Isaacs, N.W. (1995) *Nature* **374**, 517.
Messier, J., Kajzar, F., Prasad, P., Ulrich, D. (eds) (1989) *Nonlinear Optical Effects in Organic Polymers,* Kluwer Academic, Boston.
Misra, S.R., Rawatt, H.S., Joshi, M.P., Mehendale, S.C., Rustagi, K.C. (1994) *SPIE Proc.* **2284**, 220.
Moriizumi, T., Kudo, K. (1981) *Appl. Phys. Lett,* **38**, 85.
Morita, S., Zakhidov, A.A., Yoshino, K. (1993) *Jpn. J. Appl. Phys.* **32**, 873.
Moses, D. (1992) *Appl. Phys. Lett.* **60**, 3215.
Mukamel, S., Rakahashi, A., Wang, H.X., Chen, V. (1994) *Science* **266**, 250.
Nakatsuji, Y., Kita, K., Inoue, H., Zhang, W., Kida, T., Ikeda, I. (2000) *J. Am. Chem. Soc.* **122**, 6307.

Nakayama, T., Itoh, Y., Kakuta, A. (1993) *Appl. Phys. Lett.* **63**, 594.
Nguyen, T.P., de Vos, S. (1996) *Vacuum* **47**, 1153.
Nguyen, T.P., Jonnard, P., Vergand, F., Staub, P.F., Thirion, J., Lapkowski, M., Tran, V.H. (1995a) *Synth. Met.* **75**, 175.
Nguyen, T.P., Le Rendu, P., Amgaad, K., Cailler, M., Tran, V.H. (1995b) *Synth. Met.* **72**, 35.
Normile, D. (1999) *Science* **286**, 2056.
Nuesch, F., SiAhmed, L., Francois, B., Zuppiroli, L. (1997) *Adv. Mater.* **9**, 222.
Ohkawa, K., Uneno, A., Mitsuyu, T. (1991) *Jpn. J. Appl. Phys.* **30**, 3873.
Ohmori, Y., Uchida, M., Muro, K., Yoshino, K. (1991) *Jpn. J. Appl. Phys., Part 2*, **30**, L1938.
Ohnishi, T., Noguchi, T., Doi, S. (1995) *Sumitomo Kagaku (Osaka)*, **1**, 30.
Orita, K., Morimura, T., Horiuchi, T., Matsushige, K. (1997) *Synth. Met.* **91**, 155.
Palstra, T.T.M., Haddon, R.C., Lyons, K.B. (1997) *Carbon* **35**, 1825.
Park, Y., Choong, V.E., Hsieh, B.R., Tang, C.W., Gao, Y. (1997) *Phys. Rev. Lett.* **78**, 3955.
Pauck, T., Hennig, R., Perner, M., Lemmer, U., Siegner, U., Mahrt, R.F., Scherf, U., Müllen, K., Bässler, H., Göbel, E.O. (1995) *Chem. Phys. Lett.* **244**, 171.
Pei, Q., Yang, Y., Yu, G., Zhang, C., Heeger, A.J. (1996) *J. Am. Chem. Soc.* **118**, 3922.
Pei, Q., Yu, G., Zhang, C., Yang, Y., Heeger, A.J. (1995) *Science* **269**, 1086.
Penfold, J., Richardson, R.M., Zarbakhsh, A., Webster, J.R.P., Bucknall, D.G., Rennie, A.R., Jones, R.A.L., Cosgrove, T., Thomas, R.K., Higgins, J.S., Fletcher, P.D.I., Dickinson, E., Roser, S.J., McLure, I.A., Hillman, A.R., Richards, R.W., Staples, E.J., Burgess, A.N., Simister, E.A., White, J.W. (1997) *J. Chem. Soc., Faraday Trans.* **93**, 3899.
Peeters, E., Christiaans, M.P.T., Janssen, R.A.J., Schoo, H.F.M., Dekkers, H.P.J.M., Meijer, E.W. (1997) *J. Am. Chem. Soc.* **119**, 9909.
Pellegrino, V, Radebaugh, R., Mattes, B.R. (1996) *Macromolecules* **29**, 4985.
Prasad, P.N., Nigam, J.K. (eds) (1991) *Frontiers of Polymer Research*, Plenum, New York.
Prasad, P.N., Ulrich, D.R. (eds) (1988) *Nonlinear Optical and Electroactive Polymers*", Plenum, New York.
Prasad, P.N.; Williams, D.J. (1991) *Introduction to Nonlinear Optical Effects in Molecules and Polymers*, John Wiley & Sons, Inc., New York.
Prodi, L., Bargossi, C., Montalti, M., Zaccheroni, N., Su, N., Bradshaw, J.S., Izatt, R.M., Savage, P.B. (2000) *J. Am. Chem. Soc.* **122**, 6769.
Ramos, A.M., Rispens, M.T., van Duren, J.K.J., Hummelen, J.C., Janssen, R.A.J. (2001) *J. Am. Chem. Soc.* **123**, 6714.
Richards, R.W., Penfold, J. (1994) *TRIP* **2**, 5.
Rosseinsky, D.R., Mortimer, R.J. (2001) *Adv. Mater.* **13**, 783.
Saito, Y., Hamaguchi, K., Uemura, S., Uchida, K., Tasaka, Y., Ikazaki, F., Yumura, M., Kasuya, A., Nishina, Y. (1998) *Appl. Phys. A* **67**, 95.
Salaneck, W.R., Brédas, J.L. (1996) *Adv. Mater.* **8**, 48.
Salaneck, W.R., Stafstroem, S., Bredas, J.L. (1996) *Conjugated Polymer Surfaces and Interfaces*, Cambridge University Press, New York.
Sariciftci, N.S. (ed.) (1997) *Primary Photoexcitations in Conjugated Polymers: Molecular Exciton Versus Semiconductor Band Modei,* World Scientific, Singapore.
Sariciftci, N.S., Braun, D., Zhang, C., Srdranov, V., Heeger, A.J., Wudl, F. (1992) *Science* **258**, 1474.
Sariciftci, N.S., Braun, D., Zhang, C., Srdranov, V., Heeger, A.J., Wudl, F. (1993) *Appl. Phys. Lett.* **62**, 585.
Sariciftci, N.S., Heeger, A.J. (1994) *Int. J. Mod. Phys. B* **8**, 237.
Sastre, R., Costela, A. (1995) *Adv. Mater.* **7**, 198, and references cited therein.
Sato, T., Jiang, D.-L., Aida, T. (1999) *J. Am. Chem. Soc.* **121**, 10658.
Schmid, W., Dankesreiter, R., Gmeiner, J., Vogtmann, Th., Schwoerer, M., (1993) *Acta Polym.* **44**, 208.

Schmidt-Mende, L., Fechtenkotter, A., Mullen, K., Moons, E., Friend, R.H., MacKenzie, J.D. (2001) *Science* **293**, 1119.

Schwartz, B.J., Hide, F., Dfaz-Garcfa, M.A., Andersson, M. R., Heeger, A.J. (1997) *Phil. Trans. R. Soc. Lond. A* **355**, 775.

Scott, J.C., Kaufmann, J.H., Brock, P.J., DiPietro, R., Salem, J., Goitia, J.A. J. (1996) *Appl. Phys.* **79**, 2745.

Sheik-Bahae, M., Hagan, D.J., Van Stryland, E.W. (1990) *Phys. Rev. Lett.* **65**, 96.

Shim, H.K., Kang, I.N., Jang, M.S., Zyung, T., Jung, S. D. (1997) *Macromolecules* **30**, 7749.

Service, R.F. (1998) *Science* **279**, 1135.

Sirringhaus, H., Kawase, T., Friend, R.H., Shimoda, T., Inbasekaran, M., Wu, W., Woo, E.P. (2000) *Science* **290**, 2123.

Skotheim, T.A. (ed.) (1986) *Handbook of Conducting Polymers*, Marcel Dekker, New York.

Smela, E. (1999) *Adv. Mater.* **11**, 1343.

Smilowitz, L., McBranch, D., Klimov, V., Grigorova, M. , Robinson, J.M., Weyer, B.J., Koskelo, A., Mattes, B.R., Wang, H., F. Wudl (1997) *Synth. Met.* **84**, 931.

Smilowitz, L., McBranch, D., Klimov, V., Robinson, J.M., Koskelo, A., Gri- gorova, M., Mattes, B. R. (1996) *Opt. Lett.* **21**, 922.

Soileau, M.J., Williams, W.E., Van Stryland, E.W. (1983) *IEEEJ. Quant. Electron.* **19**, 731.

Sokolik, I., Yang, Z., Karasz, F.E., Morton, D.C. (1993) *J. Appl. Phys.* **74**, 3584.

Stamm, M. (1992) *Adv. Polym. Sci.*, **100**, 392, and references cited therein.

Stamm, M., Hüttenbach, S., Reiter, G., Springer, T. (1991) *Europhys. Lett.* **14**, 1441.

Stanbery, B.J., Gouterman, M., Burges, R.M. (1985) *J. Phys. Chem.* **89**, 4950.

Staring, E.G.J. (1995) *Chemisch Magazine (Rijswijk)* **1**, 17.

Stewart, G.M., Fox, M.A. (1996) *J. Am. Chem. Soc.* **118**, 4354.

Sun, R., Peng, J., Kobayashi, T., Ma, H. Zhang, Y., Liu, S. Jpn. J. (1996) *Appl. Phys.* **35**, L1506.

Tang, C.W. (1986) *Appl. Phys. Lett.* **48**, 183.

Tang, H., Li, F., Shinar, J. (1997) *Appl. Phys. Lett.* **71**, 2560.

Tasch, S., Holzer, L., Wenzl, F.P., Gao, J., Winkler, B., Dai, L., Mau, A.W.H., Sotgiu, R., Sampietro, M., Scherf, U., Müllen, K., Leising, G., Heeger, A.J. (1999) *Synth. Met.* **102**, 1046.

Tessler, N., Denton, G.J., Friend, R.H. (1996) *Nature* **382**, 695.

Tessler, N., Denton, G.J., Friend, R.H. (1997) *Synth. Met.* **84**, 475.

Tessler, N., Ho, P.H., Cleave,V., Pinner, D.J., Friend, R.H., Yahioglu, G., Le Barny, P., Gray, J., de Souza, M., Rumbles, G. (2000) *Thin Solid Films* **363**, 64.

Tremblay, J.-F. (1999) *C&E News* **77(29)**, 19.

Tullo, A.H. (2000) *C&E News* **78(26)**, 20.

Turro, N.J. (1991) *Modern Molecular Photochemistry*, University Science Books, Sausalito.

Tutt L.W., Kost, A. (1992) *Nature* **356**, 225.

Uchida, M., Ohmori, Y., Morishima, C., Yoshino, K. (1993) *Synth. Met.* **55–57**, 4168.

Wang, Y.Z., Gebler, D.D., Lin, L.B., Blatchford, J.W., Jessen, S.W., Wang, H.L., Epstein, A.J. (1996a) *Appl. Phys. Lett.* **68**, 894.

Wang, Y.Z., Gebler, D.D., Fu, D.K., Swager, T.M., Epstein, A.J. (1997) *Appl. Phys. Lett.* **70**, 3215.

Wang, H.L., Huang, F., MacDiarmid, A.G., Wang, Y.Z., Gebler, D.D., Epstein, A.J. (1996b) *Synth. Met.* **80**, 97.

Wang, H.L., MacDiarmid, A.G., Wang, Y.Z., Gebler, D.D., Epstein, A.J. (1996c) *Synth. Met.* **78**, 33.

Wenzl, F.P., Tasch, S., Gao, J., Holzer, L., Scherf, U., Heeger, A.J., Leising, G., (1998) *Mater. Res. Soc. Symp. Proc.*, **488**, 57.

Williams, D.J. (ed.) (1983) *Nonlinear Optical Properties of Organic and Polymeric Materials,* American Chemical Society, Washington DC.

Williams, E.W., Hall, R. (1987) *Luminescence and the Light Emitting Diode*, Pergamon Press, Oxford.
Winkler, B., Dai, L., Mau, A.W.H. (1999a) *Chem. Mater.* **11**, 704.
Winkler, B., Dai, L., Mau, A.W.H. (1999b) *J. Mater. Lett.* **18**, 1539.
Winkler, B., Dai, L., Mau, A.W.H. (2000) *Phys. Chem. Chem. Phys.* **2**, 291.
Wu, C.C., Wu, C.I.,Sturm, J.C., Kahn, A. (1997) *Appl. Phys. Lett.* **70**, 1348.
Wu, S.-T., Yang, D.-K. (2001) *Reflective liquid crystal displays*, John Wiley & Sons, Inc., New York.
Xie, D., Xie, Z., Wang, R. (1995) *Polym. Bull. (Chin.)* **3**, 181.
Xu, X., Luo, X., Xie, Y., Zhou, H. (eds) (1995) *Proc. Int. Workshop Electroluminescence (EL'94)*, Science Press, Beijing.
Xu, Z., Moore, J.S. (1994) *Acta Polym.* **45**, 83.
Yamashita, K., Harima, Y., Iwashima, H. (1987) *J. Phys. Chem.* **91**, 3055.
Yan, M., Rothberg, L.J., Kwock, E.W., Miller, T.M. (1995) *Phys. Rev. Lett.* **75**, 1992.
Yang, Y., Bharatham, J. (1998) *ACS Polym. Prep.* **39**, 98.
Yang, Y., Chang, S.C., Bharathan, J., Liu, J. (2000) *J. Mater. Sci.- Mater. Electron.* **11**, 89.
Yang, Y., Heeger, A.J. (1994a) *Nature* **372**, 344.
Yang, Y., Heeger, A.J. (1994b) *Mol. Cryst. Liq. Cryst.* **256**, 537.
Yang, Z., Hu, B., Karasz, F.E. (1995) *Macromolecules* **28**, 6151.
Yang, Z., Hu, B., Karasz, F.E. (1998) *J.M.S. − Pure Appl. Chem.* **A35**, 233.
Yang, Z., Karasz, F.E., Geise, H.J. (1993) *Macromolecules* **26**, 6570.
Yang, Y., Pei, Q. (1996) *Appl. Phys. Lett.* **68**, 2708.
Yasuda, H. (1985) *Plasma Polymerization*, Academic Press, New York.
Yoshida, M., Fujii, A., Ohmori, Y., Yoshino, K. (1996) *Appl. Phys. Lett.* **69**, 734.
Yu, G., Gao, J., Hummelen, J.C., Wudl, F., Heeger, A.J. (1995) *Science* **270**, 1789.
Yu, G., Heeger, A.J. (1995) *J. Appl. Phys.* **78**, 4510.
Yu, G., Pakbaz, V., Heeger, A. J. (1994) *Appl. Phys. Lett.*, **64**, 3422.
Zakhidov, A., Tada, K., Yoshino, K. (1995) *Synth. Met.* **71**, 2113.
Zakhidov, A.A., Yoshino, K. (1995) *Synth. Met.* **71**, 1875.
Zhang, C., von Seggem, H., Pakbaz, K., Kraabel, B., Schmidt, H.-W., Heeger, A.J.. (1994) *Synth. Met.* **62**, 35.
Zhang, C., von Seggern, H., Pakbaz, K., Schmidt, H.-W., Heeger, A.J. (1995) *Synth. Met.* **72**, 185.
Zurer, P. (2001) *C&E News* **79(33)**, 9.

Chapter 10

Sensors and Sensor Arrays

10.1 Introduction

The introduction of standardized systems of measurement represents one of the oldest methods used by human beings to better understand and control the world. Many measurement systems are primarily physical sensors, which measure time, temperature, weight, distance and various other physical parameters. The need for cheaper, faster and more accurate measurements has been a driving force for the development of new systems and technologies for measurements of materials, both chemical and biological.

In fact, chemical and biological sensors (or biosensors) are the evolved products of physical measurement technologies. *Chemical sensors* are measurement devices that convert a chemical or physical change of a specific analyte into a measurable signal, whose magnitude is normally proportional to the concentration of the analyte. On the other hand, *biosensors* are a subset of chemical sensors that employ a biological sensing element connected to a transducer to recognize a physicochemical change and to produce a measurable signal from particular analytes, which may be but are not necessarily biological materials themselves.

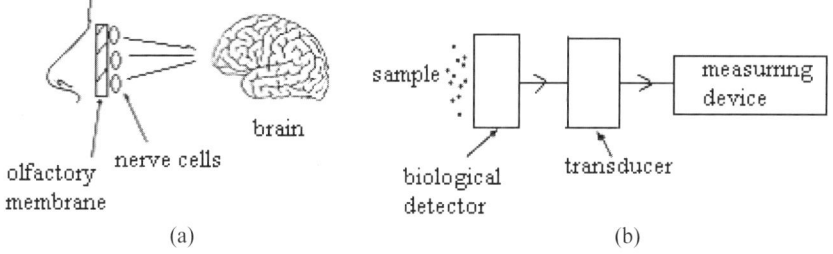

Figure 10.1. Schematic illustration of the nose as a biosensor (After Eggins, 1996, copyright 1996 John Wiley & Sons and B.G. Teubner Publishers)

Our noses and our tongues represent our oldest and most important biosensors, which allow us to distinguish between many different chemical substances qualitatively by smell or taste. Figure 10.1 schematically shows the working principle for the oldest biosensor – the nasal olfactory system – in comparison with the generalized chemical sensor.

As can be seen, the chemical to be detected is sensed by the olfactory membrane, which is the equivalent of the chemical detection element. The responses are then converted by the olfactory nerve cell, which is equivalent to the transducer, into electrical signals to be transmitted to the brain for interpretation. Thus the brain acts as the signal processor that turns the signal into the sensation we know as smell. The tongue operates in a similar way.

Depending on the basis of the transduction principle, chemical and biological sensors can be classified into three major classes with different transducers: sensors with electrical transducers, sensors with optical transducers, and sensors with other transducers (*e.g.* mass change).

The unusual optoelectronic properties of conducting polymers, dendrimers, C_{60}, carbon nanotubes, DNA and proteins discussed in the preceding chapters (Chapters 1, 2, 4, 5 and 8) make them attractive as transducer-active materials in various sensing devices. With so many significant advances in both materials science and device fabrication technology within the last decade or so, the turn of this century witnessed the evolution of early chemical and biological sensors developed in the 1980s into more complex measurement devices with capabilities for various sophisticated analyte detection. In particular, the recent development of sensor arrays (conducting polymer electronic noses, DNA chips, proteome chips, *etc.*) is revolutionizing the way in which many chemical and biomedical tests are performed in both research and clinical diagnostic laboratories. Sensor arrays consist of many different sensors on a single chip and enable us to identify complex mixture systems even without separation.

This chapter provides a status report on the research and development of sensors and sensor arrays based on conducting polymers, dendrimers, C_{60}, carbon nanotubes, DNA or proteins. For convenience, we define sensor devices according to the active materials used for transduction, rather than the underlying transduction principle.

10.2 Conjugated Polymer Sensors

10.2.1 Conjugated Polymer Sensors with Electrical Transducers

Owing to their interesting optoelectronic properties, conjugated polymers have been studied as a major class of sensing elements for the development of sensors with electrical or optical transducers. The electrical transducer system can be further divided into, at least, four basic electrical transduction modes: i) *Conductometry*: monitoring the conductivity changes; ii) *Potentiometric*: monitoring the open circuit potential at zero current; iii) *Amperometry*: monitoring the change in current while the potential is kept constant; and iv) *Voltammetry*: monitoring the change in current while varying the applied potential. The fundamental principle, as well as the pros and cons, for each of the electrical transduction modes are discussed below with an emphasis on the conductometric sensors.

10.2.1.1 Conjugated Polymer Conductometric Sensors

Conductometry is the most commonly used sensing method, in which the change in conductivity of the device is measured (conductivity is the reciprocal of resistivity, which is given by $\rho = RS/L$ with R, S, and L being resistance, sample area and thickness, respectively. For the reason of convenience, however, the resistivity and resistance are used as the same in this book, as are conductivity and conductance). Many processes that lead to changes in carrier density or mobility will cause changes in conductivity. For instance, gas or vapor sorption may lead to polymer swelling, carrier-counterion solvation, and/or disruption of conjugation length. As a result, the inter-chain transport and/or carrier mobility is affected, and hence the change in overall conductivity. The interaction of conjugated polymers with electron acceptors or donors described in Chapter 2, however, causes changes in both carrier density and mobility, leading to more significant changes in conductivity. The early work on the conductivity measurements of polyacetylene films upon doping with vapors of iodine, bromine or AsF_5, and subsequent compensation with NH_3 (Chapter 2), thus constitutes the simplest conducting polymer gas sensors. Indeed, metal (*e.g.* copper or palladium)-doped electrodeposited polypyrrole and poly-3-methylthiophene films have recently been used for the detection of reducing gases such as NH_3, H_2 and CO (Torsi *et al.*, 1998).

More recently, polypyrrole and polyaniline nanotubes prepared by either chemical or electrochemical methods have also been reported as responsive materials in gas sensors and biosensors (Sophie *et al.*, 2000). Other nanostructured conducting polymers (Rouhi, 2001, Qiu *et al.*, 2001, Akagi *et al.*, 1998) could also be very promising as transducer-active materials for sensing applications.

Conjugated Polymer Hydrocarbon Vapor Sensors
As discussed in Chapter 2, the interaction of a doped conjugated conducting polymer (*e.g.* HCSA-doped PANI-EB) with certain organic solvents (*e.g. m*-cresol) could cause a conformational transition of the polymer chain from a "compact coil" to an "expanded coil" through the so-called "*secondary doping*" process, which was found to be accompanied by a concomitant change in conductivity. The conformation-induced conductivity changes, either an increase or a decrease in conductivity, have also been observed when conjugated conducting polymers were exposed to some common organic vapors (methanol, hexane, chloroform, THF, benzene, toluene, acetone, *etc.*), providing the basis for developing conjugated conducting polymer-based sensors for the detection of hydrocarbon vapors (MacDiarmid, 1997). Figure 10.2 shows the response of a dodecylbenzenesulfonate acid-doped polypyrrole, spin-cast on a gold interdigitated electrode, to toluene vapor. As can be seen, about a 5% change in resistance was observed in each cycle with a short response time and good reversibility.

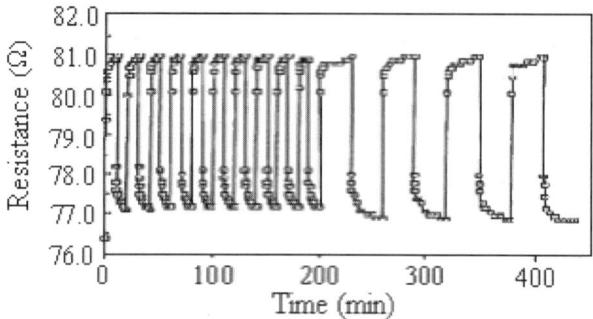

Figure 10.2. Response of a conducting polypyrrole-based sensor to a stream of N_2 (1500 ml/min) saturated with toluene vapor at 0 °C (partial pressure of toluene, 6.8 Torr) passing over the sensor for 10-minute intervals, followed by a 1500 ml/min stream of pure N_2 for 10-minute intervals. The results are given for the first complete ten 20-minute cycles and subsequent cycles with a cycle time of 60 minutes (After MacDiarmid, 1997, copyright 1997 Elsevier. Reproduced with permission)

Figure 10.3 shows the percentage changes in resistance for the polypyrrole sensor after exposure to static air saturated with various analytes at room temperature for 5 seconds in consecutive numerical order. Also included in Figure 10.3 are the corresponding results for a similar sensor based on polypyrrole (molybate anion) synthesized in the presence of polystyrene.

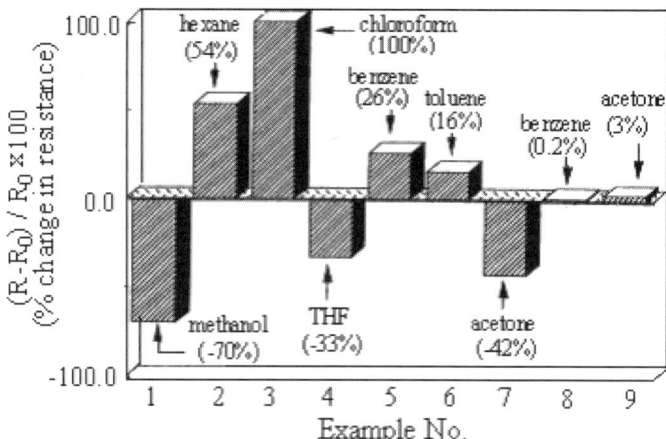

Figure 10.3. Percentage change in resistance of doped polypyrrole sensors on exposure to air saturated with various analyte vapors at room temperature, where R_0 and R are resistance of the sensor before and after 5-second exposure to the analyte/air mixture, respectively (see text) (After MacDiarmid, 1997, copyright 1997 Elsevier)

It can be seen that the sensory response to a given analyte vapor depends strongly on the nature of the dopant anions associated with the conducting polymer, whereas the same sensor can give a large degree of change (either an increase or decrease) in resistance with different analyte vapors. These results clearly indicate conjugated conducting polymers are promising for sensing airborne volatile organic compounds, such as alcohols, ethers, esters, halocarbons, NH_3 (electron donor), NO_2 (electron acceptor), and even certain warfare agent stimulants.

Conjugated Polymer pH Sensors

pH monitoring is very important, as many chemical and biological processes are pH-dependent. Apart from the widely used glass pH electrodes, conjugated conducting polymers have been investigated as potential sensing materials for the monitoring of pH in various systems. As discussed in Chapter 2, protonation of polyaniline emeraldine base (PANI-EB) with certain acids [d,l-camphorsulfonic acid, p-$CH_3(C_6H_4)SO_3H$, $(C_6H_5)SO_3H$, *etc.*] causes an increased conductivity. The conductivity of some other conjugated polymers, such as polypyrrole in the oxidized form, is also pH-dependent (Munstedt, 1986).

To demonstrate the use of the pH-dependent conductivity (or resistance) of certain conjugated conducting polymer films for pH monitoring, Talaie (1997) has potentiostatically deposited (E = 800 mV and Q = 20 mC) a thin polyaniline film onto a platinum plate, followed by applying an potential of 400 mV to achieve the highest initial conductivity for the pH detection test. The polyaniline-coated electrode was then immersed in HCl/NaOH aqueous solutions of different pH values.

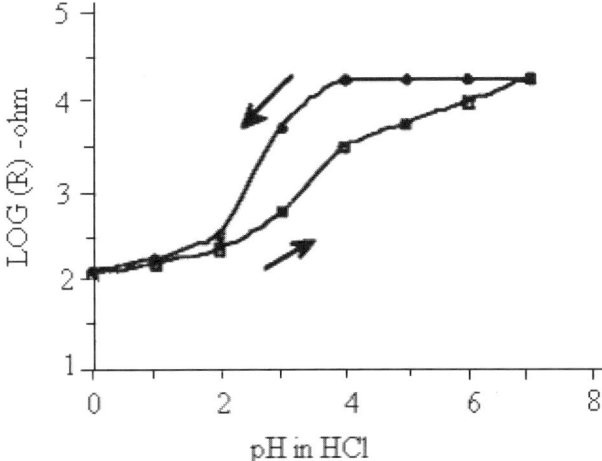

Figure 10.4. The pH-dependent resistance for a polyaniline film (After Talaie, 1997, copyright 1997 Elsevier. Reproduced with permission)

The observed pH dependence of resistance is shown in Figure 10.4. As can be seen, the protonation at low pH results in an increase in polymer conductivity while deprotonation at high pH causes decreasing conductivity. The overall trends of

Figure 10.4 suggest that the polyaniline pH sensor has a good reversibility and sensitivity in an intermediate pH range.

Conjugated Polymer pH-responsive Biosensors

The above understanding of pH-dependent conductivity is also important for constructing new sensor devices from conjugated conducting polymers for probing molecules of biological significance. This is because redox-active enzymes often produce products that can change the local pH so that conjugated conducting polymers can act as both the transducer and immobilization template in these cases.

Uchida and co-workers (Nishizawa *et al.*, 1992) have utilized the pH-dependent conductivity of polypyrrole and polyaniline for sensing glucose and penicillin. The sensor device used in their studies was an interdigitated-type circuit with two sets of Pt arrays on a glass substrate (Figure 10.5(a)). Conjugated conducting polymers were potentiostatically deposited on the array electrode and the whole system was constructed into an electrochemical configuration, as shown in Figure 10.5(b), to allow *in situ* conductivity measurements. Similar devices have previously been reported by Wrighton and co-workers (Jones *et al.*, 1987) for microelectronic applications.

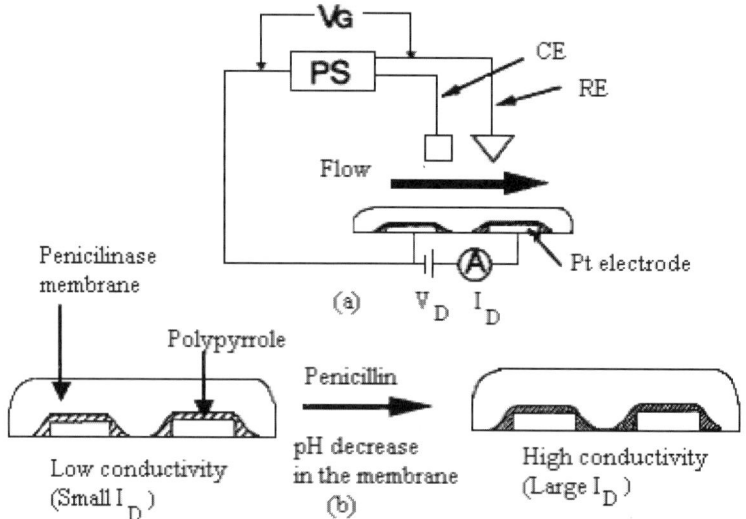

Figure 10.5. Principle of the glucose-penicillin sensors and the configuration of the electrochemical circuit for in-situ conductivity measurement: (a) a circuit diagram: PS, potentiostat; CE, Pt counter electrode; RE, reference electrode (SCE); V_G, gate voltage; V_D and I_D, drain voltage and current; (b) a schematic representation of the sensor electrode (After Nishizawa *et al.*, 1992, copyright 1992 The American Chemical Society. Reproduced with permission).

In the case of glucose sensing, glucose oxidase was immobilized onto the conducting polymer layer. Upon reaction with glucose, the glucose oxidase produced gluconolactone and hydrogen peroxide [Equation (10.1)] to cause the pH change.

Glucose + Glucose Oxidase (FAD)

$\downarrow$ (10.1)

Gluconolactone + Glucose Oxidase (FADH) $\xrightarrow{O_2}$ Glucose Oxidase (FDAH) + H_2O_2

The use of sensor devices of the type shown in Figure 10.5 for sensing penicillin was also demonstrated by depositing a polypyrrole and a penicillinase membrane coating (Nishizawa et al, 1992). In this case, the penicillinase-induced hydrolysis of penicillin to penicilloic acid caused acidification of polypyrrole, and hence an increased conductivity for the conducting polymer. The changes in conductivity were then detected by measuring the current (I_D) between the two array electrodes at a constant applied voltage (V_D) (Figure 10.5(b)).

Figure 10.6 shows the I_D change upon injection of various concentrations of penicillin solutions. Also included in the inset of Figure 10.6 is the variation of the peak height with the penicillin concentration, which indicates a high sensitivity and reliability. Given that most enzymatic reactions are associated with pH changes, the methodology developed in these particular studies should be applicable to the sensing of many other biologically important substances.

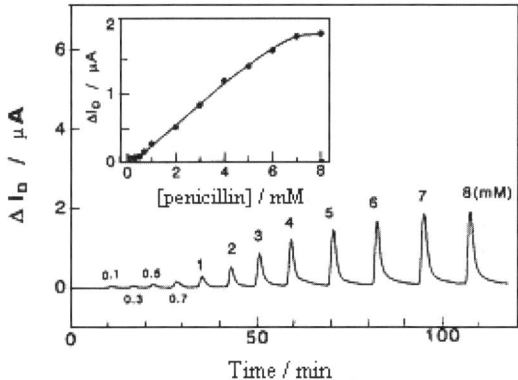

Figure 10.6. The I_D response of the penicillin sensor after injection of penicillin solutions with various concentrations. Inset shows the plot for the peak height against the penicillin concentration (After Nishizawa et al, 1992, copyright 1992 The American Chemical Society. Reproduced with permission)

Conjugated Polymer Enzyme Redox Switches and Remote Sensors
The same device as shown in Figure 10.5, but with a pyrrole-N-methylpyrrole copolymer containing diaphorase (Dp, a flavin enzyme) coating, has previously been used for an "on-off" response to the presence of NADH (nicotinamide adenine dinucleotide – reduced form) (Matsue et al., 1991). The working principle for this particular application lies in the fact that Dp can catalyze the reduction reaction of oxidized dyes, including certain conjugated polymers, by NADH (Matsue et al., 1990). If an "on" state was designated for the device with the conjugated polymer

coating in an oxidized (conductive) form, the addition of Dp should then turn the device to a reduced (non-conducting) "off" state.

Apart from the above applications, Baughman and Shacklette (Baughman and Shacklette, 1991) have demonstrated the use of conductometric sensors based on a radio-frequency antenna coated with conjugated conducting polymer films for remote sensing. Just like the conductivity of conjugated conducting polymers changes in response to external stimuli, such as humidity and oxygen content, the radio-frequency response of the conducting polymer-coated antenna would change with various processes including shorting the antenna, RF shielding and capacitance coupling. Consequently, the conducting polymer-coated antenna can respond to an external radio-frequency interrogation from a remote area. One of the most potentially useful applications for these antennae is to place them inside sensitive or perishable goods to remotely sense container damage or attempted theft without opening the container.

10.2.1.2 Conjugated Polymer Potentiometric Sensors

The potentiometric transduction mode is a very simple form of sensing, in which the conducting polymer serves as a sensing membrane and changes in the open-circuit potential (*i.e.* an electrochemical cell potential at zero current) are measured as the sensor signal. The concerned changes in the open-circuit potential, in turn, mainly result from shifts in the "dopant" anion equilibrium within the polymer film caused either by changes in the solution concentration of dopant anions or ion exchange of the "dopant" anion within the polymer film with other ions in the test solution, or both. Since the conductivity of conjugated conducting polymers may span the range from insulator through semiconductor to metal (Chapter 2), the potentiometric response to small changes in anion or cation concentration depends strongly on the starting redox composition of the polymer electrode. This is a major drawback for this simple transduction mode.

Using electropolymerized poly(thienylpyrrole), PTP, Oubda *et al.* (1998) have constructed miniaturized potentiometric sensors for both *in vitro* and *in vivo* sensing of pH and glucose. In particular, these authors demonstrated that the conjugated conducting polymer-coated microelectrodes with entrapped glucose oxidase exhibited linear behavior up to 20 mM glucose with short response times. Similarly, a linear relationship between the open-circuit potential and pH with a slope of 60mV/pH unit at 37 °C was observed for a PTP-coated microelectrode used for pH sensing. The response time of the pH sensor was found to be within seconds.

10.2.1.3 Conjugated Polymer Amperometric Sensors

In amperometric sensors, the signal is derived from changes in an electrochemical cell current with the change of analyte concentration under an impressed constant voltage on the polymer-modified electrode. In this case, the conjugated conducting polymer may play either an active role by participating in mediation of the redox process or a passive role when the polymer merely provides a site for anchoring an enzyme or other redox-active probe molecules. The use of pulsing techniques such as dc-pulsed amperometry and ac impedance techniques allows kinetic measurements, while the use of current integration to yield a coulometric response can enhance weak signals, but generally does not improve the signal-to-noise ratio.

Through sequential electrochemical polymerization of pyrrole or pyrrole derivatives (*e.g.* ω-carboxyalkylpyrrole) entrapping with tyrosinase (a copper-containing tetrameric enzyme that catalyzes the oxidation of phenolic compounds), Kranz *et al.* (1998) have constructed multilayer amperometric biosensors for the determination of phenolic compounds. The multilayer structure allows, in principle, us to regulate the sensor properties, for example, by sequentially controlling the diffusion of interfering compounds, the immobilizing of enzymes and the electron-transfer pathways between adjacent layers.

10.2.1.4 Conjugated Polymer Voltammetric Sensors

Voltammetric sensors generate the sensing response from the redox peak current characteristic of the analyte under a sweep of the electrode voltage over a range of redox potentials associated with the target redox reaction. As is the case in amperometric sensors, the conjugated conducting polymer in the voltammetric sensors may act either as a redox-active material to reduce the redox potential for the analyte of interest, and hence reduce the influence of background and interfering currents, or merely as the substrate for immobilizing a redox mediator molecule. Unlike the amperometric sensors, however, the voltammetric sensors offer an additional advantage with which redox signals for reference molecules added to the sample can be simultaneously measured to improve accuracy.

Based on the measurements of the NO oxidation current, Fabre *et al.* (1997) have demonstrated voltammetric detection of NO in the rat brain with a carbon fiber electrode coated by a conjugated conducting polymer and Nafion bilayer films. A linear relationship between the oxidation current and the NO concentration in the solution was observed over a wide range from 10^{-7} to 10^{-3} M with a very high sensitivity. This type of sensor has been implanted in the rat brain to detect the NO release in real time and cross-checked by injecting the rat with an NO-synthase inhibitor. Because of their large surface area, excellent electronic properties, and good mechanical and environmental stability (Chapter 6), aligned *carbon nanotubes* may well work better than the carbon fibers for this approach.

10.2.2 Conjugated Polymer Sensors with Optical Transducers

10.2.2.1 Conjugated Polymer Fluorescent Ion Chemosensors

As the recognition of the possible effects of metal (*e.g.* certain transition and post-transition metal ions) is of paramount importance to human health (Foulkes, 1990), there has been considerable effort directed towards the design and development of chemosensors for metal ions. The development of metal ion-selective chemosensors based on low-molecular compounds (*e.g.* crown ether derivatives) that selectively respond to the presence of specific metal ions through an electrochemical (*e.g.* redox potential) or optical (*e.g.* absorption or fluorescence) change induced, for example, by the host-guest inclusion, has been an active research area for many years

(Czarnik, 1992, Inoue and Gokel, 1990, Nakatsuji et al., 2000, Lam, et al., 2000, Prodi et al., 2000). Research on polymeric metal-selective chemosensors, however, is a relatively recent development (Li and Lu, 2000, Swager, 1998). Compared with small molecular monoreceptor systems, polymeric chemosensors offer important advantages in signal amplification via the multi-site receptor-analyte interactions.

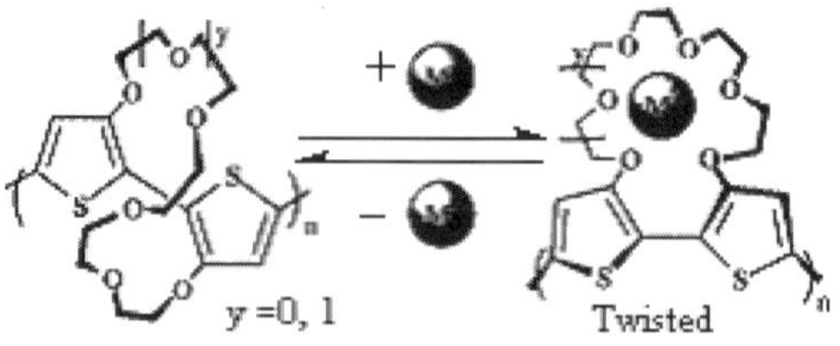

$$\text{(10.2)}$$

By substituting crown-ether macrocycles into the regio-specific polythiophene backbone, Marsella and Swager (1993) have developed chemosensors for sensing alkali metal ions. Since crown-ethers prefer to distribute the constituent oxygen atoms uniformly around a guest metal ion, the ion coordination with the crown ether-substituted polythiophene could cause conformational changes of the polymer backbone [Equation (10.2)], providing both the electrical and optical transduction modes (Marsella and Swager, 1993).

Furthermore, the ring size of the macrocycle substitutes was found to play an important role in regulating the selectivity for metal ions. For instance, the relatively small Na^+ ions caused larger optical effects to polymer **1** (91 nm blue shift) and **2** (63 nm) substituted by 15-crown-5 rings [Equation (10.3)] than K^+ ions, with which the blue shift was considerably smaller. In contrast, polymer **3** with 18-crown-6 analogues [Equation (10.3)] showed a larger blue shift for K^+ (45 nm) than for Na^+ (30 nm).

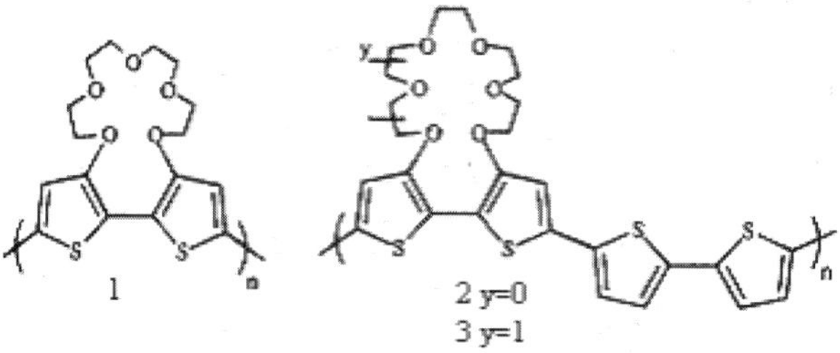

$$\text{(10.3)}$$

Based on these findings, Swager and co-workers (Crawford et al., 1998) used calix[4]arene with a highly organized arrangement of oxygen atoms as a more specific scaffold for the design of ionophores. In this study, the absorption and emission characteristics of a conjugated poly(phenylene bithiophene), **4**, and a monomeric model compound, **5** [Equation (10.4)], were investigated as a function of $[Li^+]$, $[Na^+]$, $[K^+]$ and $[Ca^{2+}]$.

Sensors and Sensor Arrays 415

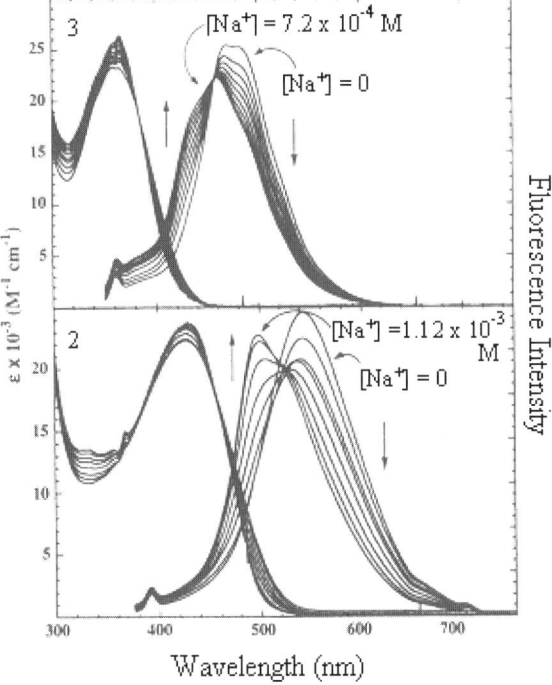

(10.4)

In particular, these authors found that the calix[4]arene bithiphene receptor in both **4** and **5** provides selectivity for Na^+ as the addition of Na^+ reduces the Stokes shift (Chapter 9), whereas Li^+, K^+ or Ca^{2+} does not affect their absorption and emission characteristics.

Figure 10.7. Adsorption and emission spectra of the model compound **5** (top) and polymer **4** (bottom) as a function of increasing $[Na^+]$. For **4**, $[Na^+]$ increased from 0 M to 1.12×10^{-3} M in increments of 1.6×10^{-4} M and $\lambda_{ex} = 350$ nm. For **5**, $[Na^+]$ increased from 0 to 7.2×10^{-4} M in increments of 8×10^{-5} M and $\lambda_{ex} = 320$ nm (After Crawford *et al.*, 1998, copyright 1998 The American Chemical Society. Reproduced with permission)

As seen in Figure 10.7, polymer **4** showed a larger shift in the emission than that for the model compound **5** in response to Na$^+$ due to the multiple binding sites that are present in the former but absence in the latter.

Apart from the aforementioned fluorescent ion chemosensors that involve metal ligation with individual conjugated polymer chains, Swager and co-workers (Kim J. *et al.*, 2000) have also developed chemosensors based on the inter-chain interactions of conjugated polymers. In particular, these authors synthesized the poly(*p*-phenylene ethynylene)s, **6–9** [Equation (10.5)], and exploited a new transduction mechanism related to the aggregation of crown ether-substituted conjugated polymers induced by K$^+$ ions [Equation (10.6)].

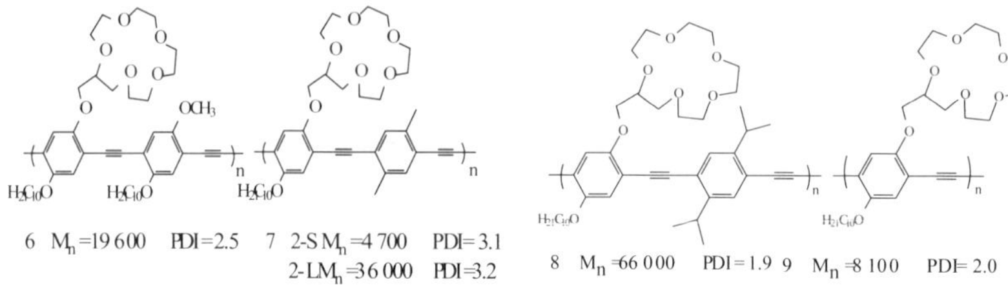

$$\tag{10.5}$$

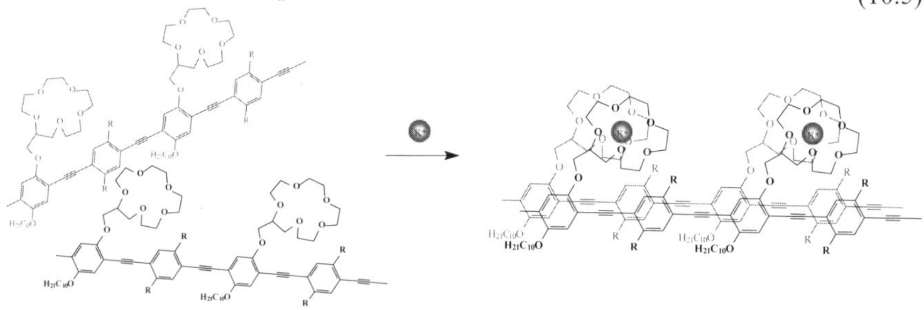

$$\tag{10.6}$$

It was found that this new sensory system displayed a high selectivity for K$^+$ over Na$^+$ with an enhanced sensitivity due to the inter-chain energy transfer. The observed high selectivity for K$^+$ arises from the fact that while K$^+$ ions induce the polymer aggregation via the well-known 2:1 sandwich complex between K$^+$ ions and 15-crown-5, as shown in Equation (10.6), Na$^+$ ions (also Li$^+$) cannot do so due to the formation of a 1:1 complex with 15-crown-5 (Gokel, 1991).

As shown in Figure 10.8, therefore, the addition of K$^+$ ions into a polymer **7** solution in THF produced a new red-shifted peak at 459 nm in the absorption spectra and caused a significant decrease in fluorescence emissions due to the fluorescent quenching associated with the ion-specific aggregation. The sensory responses, in terms of the relative percentage of fluorescence quenching, to K$^+$, Na$^+$ and Li$^+$ clearly show a high sensitivity and selectivity (Kim J. *et al.*, 2000).

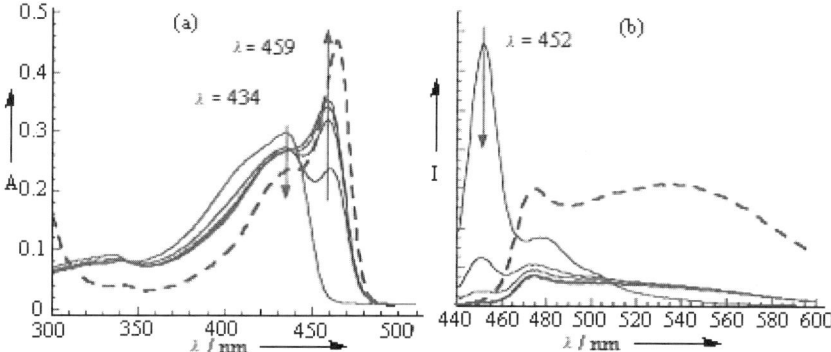

Figure 10.8. The evaluation of (a) absorbance and (b) fluorescence spectra of polymer **7**(2-L, λ_{ex} = 343 nm) with various molar ratios of K^+ ions. A solution KPF_6 (18 mM) in acetonitrile was added gradually into a solution of the polymer **7** (5 µm, 3.6 mL) in THF. The arrows indicate the changes for progressively increasing the concentration ratio of K^+ to 15-crown-5 = 0:1, 0.5:1, 1:1, 2.5:1, 5:1, and the dashed lines are the corresponding spectroscopic data (arbitrarily scaled for clarity) for a monolayer LB film (see Figure 10.9) (After Kim et al., 2000, copyright 2000 Wiley-VCH Verlag. Reproduced with permission)

To prove that the sensory response indeed resulted from the interpolymer π-stacking aggregations, Swager and co-workers (Kim J. et al., 2000) have further prepared a monolayer LB film of polymer **7** on a hydrophobic substrate (Figure 10.9). They found that the changes in the absorption and fluorescence spectra in response to K^+ for the LB film were essentially the same as for solutions (Figures 10.8(a) and (b)).

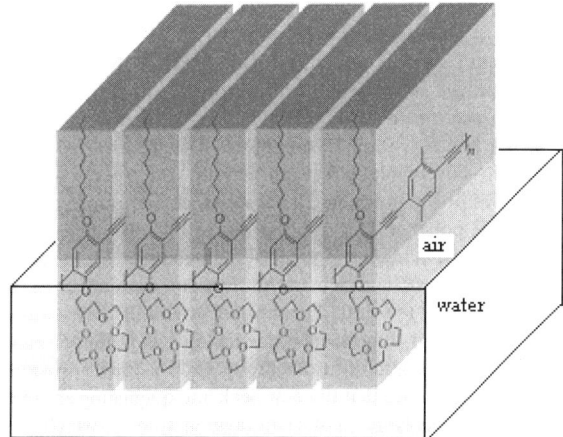

Figure 10.9. Schematic representation of the edge-on structure of the polymer **7**(2-L) LB film at the air-water interface. The dashed lines represent the hydrophobic substrate (After Kim J. et al., 2000, copyright 2000 Wiley-VCH Verlag. Reproduced with permission)

The effectiveness of the interpolymer π-stacking aggregation depends strongly on the bulkiness of the side groups grafted onto the polymer backbone. Generally speaking, bulky side groups will retard the interpolymer aggregation and weaken the response to K^+ ions. In this context, polymer **6** shows a much weaker sensory response than polymer **7**, as the methoxy oxygen atom bound to the phenyl unit in polymer **6** could form a lariat ether-type complexation with 15-crown-5 moieties to compete with the formation of the 2:1 sandwich complex. The lack of sensory response to Na^+, Li^+ or K^+ ions for polymer **8** can be attributed to the presence of isopropyl groups, which are sufficiently bulky to prevent π-stacking aggregation with any ion. However, it was no small surprise to find that polymer **9** with 15-crown-5 groups on every repeating unit did not show any response to Na^+, Li^+ or K^+ ions either. The insensitivity of regio-irregular polymer **9** to metal ions is presumably due to its ability to form intra-macromolecular 15-crown-5-K^+-15-crown-5 bridges, as indicated by molecular modeling studies (Kim J. *et al.*, 2000).

10.2.2.2 Conjugated Polymer Fluorescent TNT Sensors

Although significantly enhanced sensory signals have been observed for some of the above-mentioned fluorescent chemosensors based on conjugated polymers with respect to their small molecular counterparts, it is still a big challenge to develop new chemical sensors for the detection of ultra-trace analytes, including the highly explosive 2,4,6-trinitrotoluene (TNT) and 2,4-dinitrotoluene (DNT). Both TNT and DNT are the principle constituents of *ca.*120 million unexploded land mines worldwide (Maureen, 1997). Along with land mine detection using metal detectors, various methods ranging from neutron activation analysis to electron-capture measurements have been studied for direct TNT detection (Maureen, 1997, Kolla, 1997). However, it is highly desirable to develop *real-time* TNT chemosensory devices that are cheap to construct and easy to use. In this context, Yang and Swager (1998a, 1998b) have recently developed a solid film-type fluorescent chemosensor based on certain tailor-made conjugated polymers by monitoring fluorescence attenuation arising from electron transfer from the excited conjugated polymer to the TNT electron acceptor.

As discussed above, the facile energy migration processes intrinsically associated with conjugated polymers can be used to amplify signals from the conjugated polymer-based fluorescent chemosensors. The enhancement of the sensory signal due to the greater tendency for energy migration in the solid film-type fluorescent TNT chemosensor, however, is often counterbalanced by the decreased fluorescence quantum efficiency due to intermolecular quenching (Chapter 9). Furthermore, the performance of the solid film-type fluorescent TNT chemosensor is expected to depend strongly on the morphology and intermolecular packing of the conjugated polymer chains used as the sensing component. A densely packed polymer film, for instance, will prevent rapid diffusion of analytes into the sensing material. By contrast, a loosely packed polymer film may exhibit unstable (non-reproducible) spectroscopic features due to the continued morphology change. To avoid these possible drawbacks, Yang and Swager (1998a) devised an approach to stabilize fluorescent conjugated polymer films by incorporating rigid three-dimensional pentiptycene moieties into a poly(phenylene ethynylene) backbone, **10** [Equation (10.7a)].

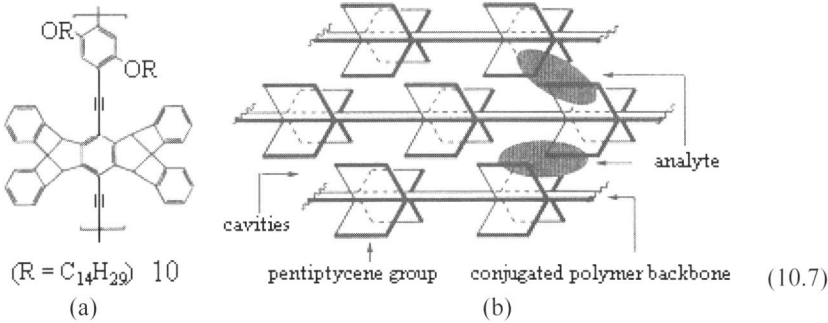

(10.7)

In so doing, these authors introduced, at least two major advantages into the solid-film-type fluorescent chemosensor: 1) the bulkiness of the pentiptycene side groups reduces interpolymer interactions and prevents π-stacking or excimer formation, leading to a high fluorescence quantum yield and spectroscopic stability, and 2) the three-dimensional nature of the pentiptycene side groups allows the formation of sufficiently large cavities between adjacent polymers [Equation (10.7(b))] for analyte binding.

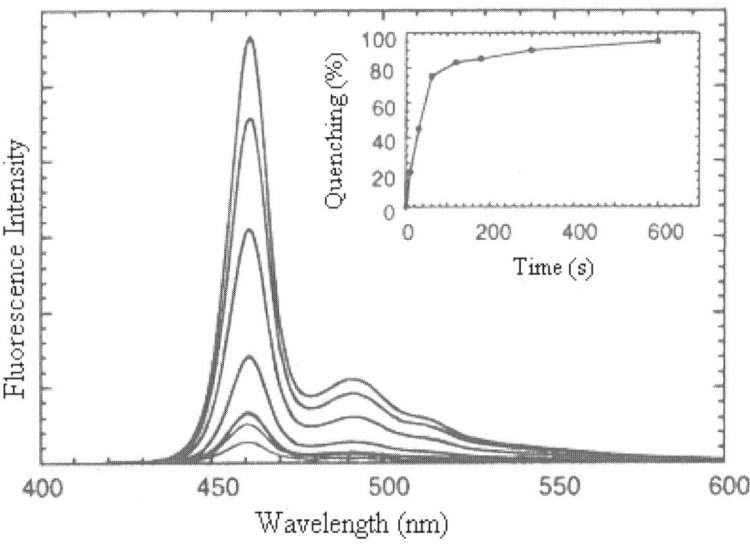

Figure 10.10. The time-dependence of the fluorescence peak intensity of a polymer **10** film (25 ± 5 Å thick) upon exposure to TNT vapor for 0, 10, 30, 60, 120, 180, 300 and 600 s (top to bottom) at room temperature. Inset shows the percentage fluorescence quenching *vs.* time (After Yang and Swager, 1998a, copyright 1998 The American Chemical Society. Reproduced with permission)

Consequently, the time-dependent fluorescence intensity of the pentiptycene grafted poly(phenylene ethynylene) film given in Figure 10.10 shows a fast response to TNT vapor. As seen in the inset of Figure 10.10, the fluorescence quenching increases to 50 ± 5% within 30 s and to 75 ± 5% at 60 s upon exposure to TNT. Furthermore, the lose of the fluorescence intensities shown in Figure 10.10 can be largely recovered by rinsing the TNT-exposed polymer film with methanol followed by gentle air drying, indicating that the pentiptycene-grafted poly(phenyleneethynylene)s are a class of promising materials for developing cheap, reliable and reusable fluorescent TNT chemosensors.

10.2.2.3 Conjugated Polymer Light-harvesting "Turn-on" Sensors

We have so far been concerned only with "turn-off"-type chemosensors with which the analyte binding leads to a dramatic decrease in the initial fluorescence intensity characteristic of conjugated polymers. By contrast, a sensor that can enhance (turn-on) the fluorescence intensity from an initially low level would be much more sensitive. Using a water-soluble cationic poly(p-phenylene ethynylene), **11**, for the multilayer formation with an anionic-polyacrylate-supported pH-sensitive fluoresceinamine (FA) dye, **12**, [Equation (10.8)], Swager and co-workers (McQuade, 2000) have constructed a light-harvesting "turn-on" sensor via the layer-by-layer deposition technique (Chapter 7) through alternate immersion of a glass slide into aqueous solutions of **11** and **12** (1 mg/ml).

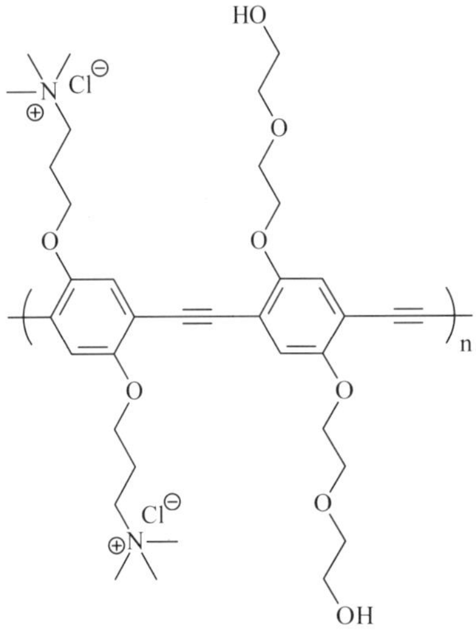

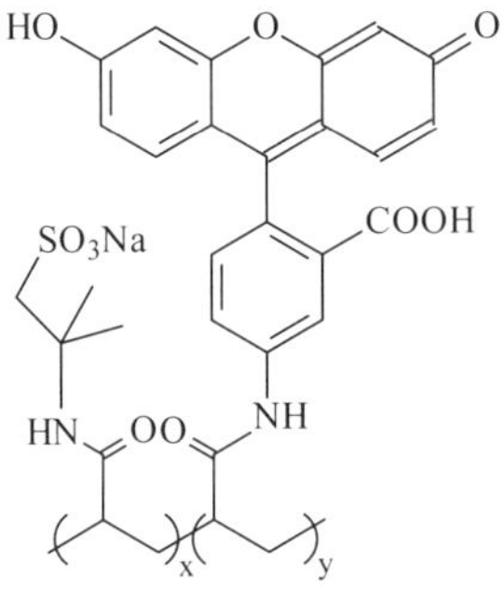

$$(10.8)$$

Since the materials are designed such that the polymer **11** emission overlaps the absorption peak of the dye **12** (Figure 10.11), the fluorescence resonance energy transfer between the polymer and the dye is maximized.

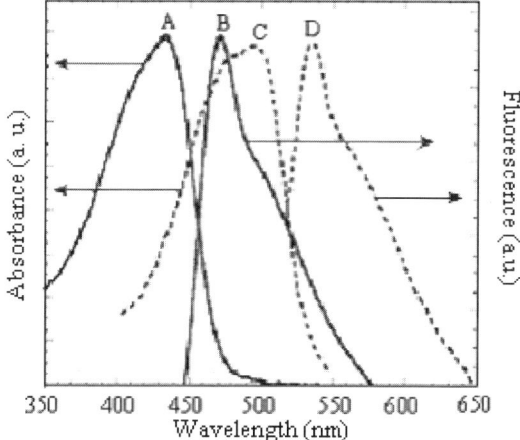

Figure 10.11. Normalized absorbance and emission spectra of polymer **11** and dye **12**. Curve A is for the absorbance of a thin film of polymer **11** on glass (λ_{em} = 439 nm) and curve B is the corresponding emission spectrum (λ_{em} = 471 nm, λ_{ex} = 420 nm). Curve C is for the absorbance of a thin film of polymer **12** (λ_{em} = 490 nm) and curve D is the corresponding emission spectrum (λ_{em} = 535 nm, λ_{ex} = 500 nm) (After McQuade *et al.*, 2000, copyright 2000 The American Chemical Society. Reproduced with permission)

The fluorescence responses for a bilayer sensor composed of one layer of **12** deposited onto a layer of **11** immersed in solutions of varying pH were measured in air on the dried film by selectively exciting conjugated polymer **11** and FA dye **12** at 420 and 500 nm, respectively.

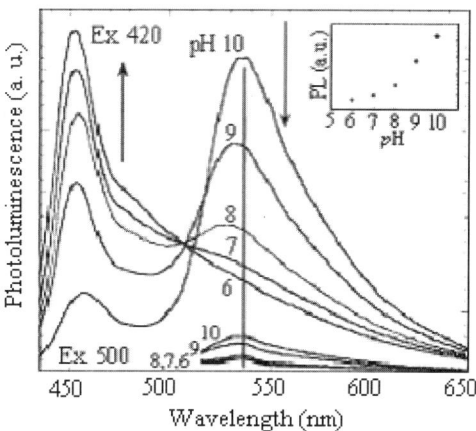

Figure 10.12. The fluorescence emission spectra recorded on a **12/11**/glass multilayer sensor after being dipped into a 0.001 M KHPO$_4$ solution at the denoted pH. The spectra spanning from 435 to 650 nm and those beginning at 515 nm were recorded with the excitation wavelength of 420 and 500 nm, respectively. Inset shows the pH dependence of the emission maximum of the FA band at the excitation wavelength of 420 nm (After McQuade *et al.*, 2000, copyright 2000 The American Chemical Society. Reproduced with permission)

As shown in Figure 10.12, about 90% of the emission associated with the conjugated polymer is transferred to the dye at pH ≥ 10 while the FA's fluorescence is completely lost at pH = 6. The almost 10-fold increase in the FA's emission at 535 nm when excited at 420 nm with respect to that measured by direct excitation at 500 nm seen in Figure 10.12 confirms the fluorescence signal amplification through the fluorescence resonance energy transfer process. The reversible fluorescence changes between pH 6 and 11 seen in Figure 10.12 clearly indicates an "on-off" fluorescent pH sensor. Furthermore, it was demonstrated that placing an additional layer of **11** on top of a bilayer of **11** and **12** could result in the formation of a film that responds differently to pH, implying that there was considerable room for tailoring the device performance by controlling the multilayer structure.

10.3 Charge Transfer Polymer Sensors

Tanaka *et al.* (1999) studied the electrical and optical properties of poly(3-octyloxythiophene), PAOT-8, for potential gas-sensing applications. Due to the relatively high energy associated with the top of its valence band caused by the electron-donating alkoxy group, PAOT-8 shows strong interaction with electron-accepting molecules such as fullerene C_{60}. The conductivity of the charge transfer complexes was found to change reversibly upon exposure to various organic flavor gases with a linear response against the gas concentration, indicating potential gas-sensing applications.

On the other hand, a multilayer biosensor based on the electron transfer between quinone and the electrode surface via an immobilized redox dye has also been reported (Kranz *et al.*, 1998). The multilayer structure was prepared by sequential electropolymerization of pyrrole or pyrrole derivatives. As mentioned earlier (Section 10.2.1.3), the multilayer approach provides additional freedom for the sensor properties to be tailor-made. By covalently binding a redox dye (*i.e.* thionine) on a functionalized-polypyrrole-coated electrode, followed by depositing a second layer of polypyrrole with entrapped tyrosinase on top of the first layer, Kranz *et al.* (1998) were able to optimize the properties of the sensor by controlling the thickness of the polymer layers, the stability of the electrode and the electron transfer between quinone and the electrode surface via the immobilized redox dye.

10.4 Ionically Conducting Polymer Sensors

As mentioned in Chapter 2, ionically conducting polymers, or polymer electrolytes, are normally prepared by dissolving salt species in a solid polymer host and their conductivity results from the ion migration between coordination sites repeatedly generated by the local motion of polymer chain segments. Therefore, the conductivity of ionically conducting polymers may change with any phenomenon that changes the mobility of the polymer chain segments or the solvation state of the ionic charge carriers. These properties have already been utilized for making chemical sensors through some clever designs, as exemplified by the examples spotlighted below.

Based on the finding that the impedance of polyethylene oxide (PEO) doped with various alkali salts (*e.g.* $LiClO_4$) changed over a large range with the humidification and desiccation process without any hysteresis, Sadaoka and Sakai (1986) have constructed humidity sensors driven by the dissolution of counter ions with the adsorbed water. Similarly, Xin *et al.*, (1987) developed a humidity sensor from mixtures of poly(styrene-*co*-quaternized-vinylpyridine) with a perchlorate ($HClO_4$, $LiClO_4$, $KClO_4$, *etc.*) while Huang and Dasgupta (1991) prepared a humidity sensor using a perfluorosulfonic ionomer/H_3PO_4 composite film. The latter was found to be able to measure very low humidity, even down to 2 ppm.

In a separate study, Egashira and co-workers (Egashira *et al.*, 1988, Shimizu *et al.*, 1989, Wu X. *et al.*, 1989) have investigated a CO_2 gas sensor based on polyethylene glycol/K_2CO_3 (Rb_2CO_3 and Cs_2CO_3 were also tried with a lower sensitivity) solution supported on porous alumina ceramics and found that its resistance increased upon exposure to CO_2 under an applied voltage. Furthermore, a linear relationship was observed between the sensitivity (defined as the ratio of resistance in CO_2 to that in air) and the CO_2 concentration from 1 to 9%. By solidifying the sensing layer using a solid polyethylene glycol of high molecular weight doped with a solution mixture of liquid polyethylene glycol and K_2CO_3, Sakai *et al.* (1995) have successfully improved the performance of these types of sensor. In these cases, the change in resistance was demonstrated to arise from the change in the concentration of the charge-carrier K^+ ion, which was governed by Equations ((10.9)–(10.11)) by assuming Equations (10.9) and (10.10) to be at equilibrium:

$$CO_2 + H_2O \underset{}{\overset{k_1}{\rightleftharpoons}} H^+HCO_3^- \underset{}{\overset{k_2}{\rightleftharpoons}} 2H^+ + CO_3^{2-} \quad (10.9)$$

$$K_2CO_3 \underset{}{\overset{k_3}{\rightleftharpoons}} K^+KCO_3^- \underset{}{\overset{k_4}{\rightleftharpoons}} 2K^+ + CO_3^{2-} \quad (10.10)$$

$$[K^+] = \left[\frac{K_3K_4[K_2CO_3^-][H^+]^2}{K_1K_2[H_2O][P_{CO2}]^n} \right]^{1/2} \quad (10.11)$$

Other gas sensors based on ionically conducting polymers have also been developed for detecting O_2 (Yan and Lu, 1989), CO (Otagawa *et al.*, 1990), NO (Buttner *et al.*, 1990), NO_2 (Christensen *et al.*, 1993) and H_2 (Albert S. *et al.*, 1997).

10.5 Conductively Filled Polymer Sensors

As seen in Chapter 2, dispersion of conductive solids in an insulating polymer matrix can impart the conductivity when the volume fraction of the conducting fillers is greater than a value referred to as the percolation threshold. The conductivity of conductively-filled polymers and the precise location of the percolation threshold are affected by many factors, including the nature of the conductive particles and their aggregated state in the polymer matrix. While this could be a hindrance to their uses in certain practical applications, it provides a large scope of possibilities for the design of various sensing devices based on conductively filled polymers.

10.5.1 Conductively Filled Polymer Humidity Sensors

By solution-casting a composite film consisting of poly(*o*-phenylene diamine) [P(*o*-PD)] and polyvinyl alcohol (PVA) onto a comb-shaped Pt electrode, Ogura *et al.* (1995, 1996) have successfully developed an interesting conductometric sensor for sensing humidity. As changes in humidity could affect pH of the composite film by reacting with H_2O molecules in the moisture to produce –OH groups [Equation (10.12)], a humidity-dependent conductivity was observed.

(10.12)

Figure 10.13 shows a linear response of the conductivity to relative humidity (RH) both for the moistening and desiccating processes over a wide range of RH, indicating a high selectivity and reproducibility. The same principle has been applied to the construction of a glucose sensor, in which gluconic acid was produced through certain enzymatic reactions (Hoa *et al.*, 1992).

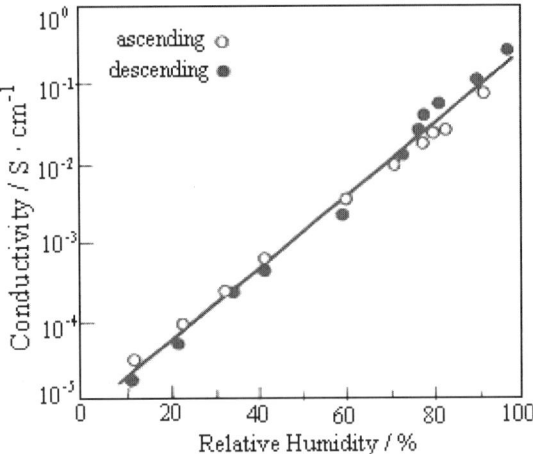

Figure 10.13. The dependence of conductivity on relative humidity (After Ogura *et al.*, 1995, copyright 1995 The Electrochemical Society, Inc. Reproduced with permission)

10.5.2 Conductively Filled Polymer Gas Sensors

A novel gas sensor has also been developed based on polypyrrole/poly(ethylene oxide) composite thin films (Hwang et al., 2001). In this case, the sensor signal was derived from simultaneous measurements of resistance changes (ΔR) on the polypyrrole-based composite film and frequency changes (Δf) on a piezo-electric crystal. Molecular species, such as ethanol and methanol, can be recognized by the ($\Delta R/\Delta f$) values – the higher the $\Delta R/\Delta f$ value, the better is the recognition ability, whereas the quantitative analysis of the vapor molecules can be obtained by measuring ΔR or Δf values.

On the other hand, Albert S. et al. (1997) developed a sensor based on the conducting polymer film coated with Pd on either side for measuring the diffusible H_2 content in steel weldments. The current flowing through the device was found to be directly proportional to the hydrogen content in the gas mixture when one side of the sensor was exposed to an Ar-H_2 gas mixture, while the other side to air.

More recently, palladium "mesowire" arrays supported by a cyanoacrylate film have been reported to also show a rapid (< 75 ms) reversible decrease in the resistance with the change of hydrogen concentration over a range of 2 to 10% (Favier et al., 2001). Similar resistance changes caused by adsorbing analyte molecules (e.g. thiols, amines) onto other metal-nanostructures (e.g. gold nanowires) (Li C. et al., 2000) have been previously observed. The analogous effect on C_{60} and carbon nanotubes will be discussed in Sections 10.7 and 8.

10.5.3 Conducting Polymer-coated Fabric Sensors: Smart Textiles

Using multifunctional materials, including conjugated polymers, as modified cladding materials, MacDiarmid and co-workers (El-sherif et al., 2000) have developed new fiber optic sensors for real-time sensing of various battlefield hazards, such as chemical and biological warfare threats. In this case, the sensing function arises from the ability of conjugated polymers to change the light propagation characteristics of optical fibers. In particular, polyaniline has been used as a photochemical-responsive polymer in the fiber optic sensors, while a segmented polyurethane-diacetylene copolymer was selected, due to its interesting thermochromic properties, for temperature-sensing.

Other conductive polymer-coated fabrics have also been used for chemical-sensing. For example, Collins and Buckley (1996) used conductive polymer-coated poly(ethylene terephathalate) or nylon threads woven into a fabric mesh for sensing several toxic gases. The fabric structure offers a large surface area for high sensitivity, which allows detection limits in parts per million (ppm) for various pollutants and chemical warfare stimulants [dimethyl methylphosphonate (DMMP), NH_3, NO_2, etc.]. A major application for this type of smart fabric is for real-time sensing of chemical warfare stimulants in a battlefield by monitoring resistivity changes of soldiers' clothes.

Because of the fact that mechanical stretching of conjugated conducting polymer thin films could change the polymer conductivity due to the stretching-induced conformation change (Dai, 1999), conductive polymer-coated fabrics could also be useful as mechanical sensors to analyze an athlete's movement and to give indications of the duration and intensity of the exercise.

10.6 Dendrimer Sensors

Dendritic macromolecules are molecular species with a large number of hyperbranched chains of precise length and constitution surrounding a central core (Chapter 4). The ease with which dendritic molecules with well-defined periphery groups (or probes) can be made, together with their huge free volume of precise shapes and sizes for forming host-guest complex structures with analyte molecules (Chapter 4), makes dendrimers and their derivatives attractive materials for sensing applications.

10.6.1 Dendrimer Gas Sensors

10.6.1.1 Dendrimer Iodine (Vapor) Sensor

By modifying the periphery of polyamide dendrimers with oligothiophene groups followed by oxidizing them into cation radicals for π-dimerization, Miller *et al.* (1998) found that the conductivity of the resulting dendrimers increased significantly upon sorption of certain volatile organic compounds. In particular, oxidation with iodine vapor over one day was demonstrated to lead to a conductivity of 10^{-3} S/cm for the dendrimer film, which is stable for several weeks even under ambient conditions.

10.6.1.2 Dendrimer SO_2 Gas Sensors

Certain metallodendrimers, such as dendrimers with the periphery functionalized with platinum(II) metal centers, were found to reversibly adsorb SO_2 to produce macromolecules with significantly enhanced solubility and drastic color changes (Albrecht *et al.*, 1998). Based on this finding, Albrecht and Van (1999) developed highly active sensors from the peripherally functionalized metallodendrimers for detection of milligram quantities of toxic SO_2 gas. More recently, these authors (Albrecht *et al.*, 2001) have further immobilized organoplatinum recognition sites onto a quartz crystal microbalance (QCM) electrode by electrostatic spray (Ward and Buttry, 1990) for detection of the selective recognition of SO_2 by monitoring the mass change.

The principle of QCM for the mass measurements is based on the dependence of the resonance frequency, f, of the quartz resonator on the additional mass according to [Equation (10.13)]:

$$\Delta f/\Delta f_p = -\Delta m/M \qquad (10.13)$$

where Δf is the change in the resonance frequency f caused by the deposition of additional mass Δm on the quartz resonator and M is the weight of the quartz piezo-element.

It can be seen from Equation (10.13) that high mass sensitivity should be achievable by using high-frequency quartz resonators (*e.g.* resonators with thickness-shear vibrations). In the case of AT-cut quartz plates, Equation (10.13) may be rewritten as:

$$\Delta f = -2.26 \times 10^6 / \Delta f^2 \Delta m / S \qquad (10.14)$$

where Δf (in Hz) is the change of the resonance frequency, f (in MHz), caused by the deposition/adsorption of additional mass Δm (in g) on the resonator surface of S (in cm^2). Therefore, the QCM electrode can be used for investigating adsorption characteristics of various gas molecules with a high sensitivity by monitoring the mass increase.

10.6.1.3 Dendrimer CO Gas Sensors

Dendritic carbosilanes with periphery ferrocenyl groups have also been used for sensing CO by monitoring the conductivity change of the dendritic film (Kim C. *et al.*, 2001). Figure 10.14 shows a typical transient response for a spin-cast film of the fourth-generation dendritic carbosilane containing 48 periphery ferrocenyl groups upon exposure to CO carried by a N_2 stream (10–40 vol.% CO). The observed increase in current with increasing CO concentration at the constant voltage of 15 V was attributed to the coordination ability of ferrocene with CO, leading to an increased electron-hole mobility associated with the ferrocene groups.

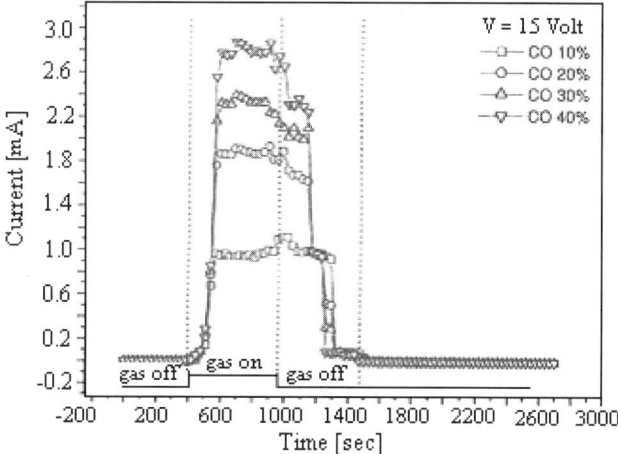

Figure 10.14. A typical transient response of ferrocene end-capped dendritic carbosilane CO sensor at room temperature (After Kim C., *et al.* 2001, copyright 2001 Elsevier. Reproduced with permission)

As seen in Figure 10.14, the current intensity is directly proportional to the CO concentration up to 40 vol.% CO before saturation. This, together with the relatively fast response times of 50 and 420 s for the turn-on and turn-off, respectively, indicates that the ferrocene end-capped dendritic carbosilane can act as a sensitive element for sensing CO. Other dendrimer membranes have also been used for selective recognition and separation of CO from CH_4 or N_2 (Kovvali et al., 2000).

10.6.2 Dendrimer Sensors for Carbonyl Compounds

Highly selective sensors for discriminating carbonyl compounds in the gas phase have also been developed based on host-guest interactions with dendrimer-type materials (Heil et al., 1999). In this case, the dentritic receptors were coated onto the surface of a QCM electrode. Upon contact with carbonyl compounds in the gas phase, reproducible signals were obtained within a short response time. These sensors can be regenerated, as the inclusion of the guests into the host is reversible. Owing to the highly specific host-guest interactions characteristic of dendrimers, these sensors are highly selective, allowing for the discrimination of even individual aldehydes and ketones. Furthermore, a "molecular filter" approach for enhancing the selectivity of dendritic sensors has been reported (Dermody et al., 1999) by chemically immobilizing a thin layer of β-cyclodextrin-functionalized hyperbranched poly(acrylic acid), PAA, film onto an Au electrode. The PAA film, in turn, was capped with a chemically grafted ultrathin polyamine layer, which acts as a pH-sensitive "molecular filter" to allow only suitably charged analytes to pass to the β-cyclodextrin molecular receptors.

10.7 Fullerene C_{60} Sensors

The discovery of carbon cage molecules, such as fullerene C_{60} in 1985, launched an intensive interdisciplinary inquiry into the fundamental properties of the C_{60} molecule (Chapter 4). Since then, its many interesting chemical, physical and photophysical properties have been discovered. These unusual physicochemical properties provide a solid base for the highly symmetric fullerene molecules to be used in a wide range of applications, including as sensing elements described below.

10.7.1 Fullerene Humidity Sensors

By directly sublimating C_{60} onto a QCM quartz resonator, Radeva et al. (1997) prepared a C_{60} sensor for sensing the relative air humidity. This is because the adsorption of water molecules from moisture onto the C_{60} film can be related directly to changes in the resonance frequency of the quartz resonator [Equation (10.13)]. Figure 10.15 shows the dependence of the resonance frequency on the relative air humidity for the C_{60}-coated QCM resonator, recorded at between 10% and 98% relative humidity (RH) at a constant temperature of 30 °C. As can be seen, the sensitivity increases with the increase of the relative air humidity. A humidity sensitivity of ca.11 Hz per %RH over the range of 30–70% RH was observed. This value of sensitivity corresponds to a sensitivity threshold of 0.01% for the humidity measurements, indicating fullerene C_{60} is one of the best water vapor-sensitive materials.

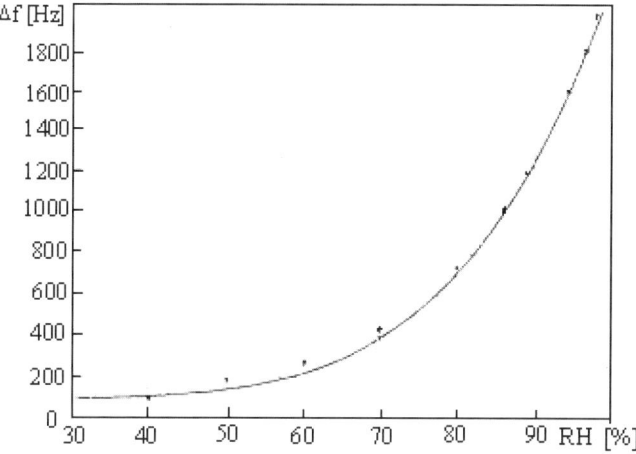

Figure 10.15. The dependence of the resonance frequency on the relative air humidity for a C_{60} coated QCM resonator (After Radeva *et al.*, 1997, copyright 1997 Elsevier. Reproduced with permission)

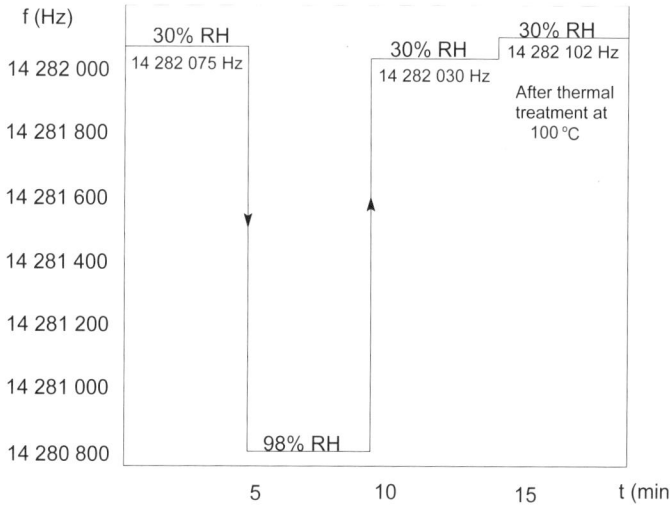

Figure 10.16. The frequency response to a cyclic change of the air humidity in the range of 30–90% for the C_{60} coated QCM resonator (After Radeva *et al.*, 1997, copyright 1997 Elsevier. Reproduced with permission)

The response of the C_{60}-coated QCM resonator to a cyclic change of the air humidity in the range of 30–90% is shown in Figure 10.16, which indicates that the interaction between the fullerene C_{60} layer and water molecules is highly reversible. Furthermore, the response time was found to be less than a second. This value of the response time is considerably shorter than the corresponding values for similar quartz resonator humidity sensors based on other humidity adsorptive materials, including SiO_2 and polymer layers (Radeva et al., 1997). These results indicate that fullerene C_{60} can be used to make advanced sensors for measuring relative air humidity.

10.7.2 Fullerene Gas Sensors

Sakurai and co-workers (Koda et al., 2000) have studied the gas-sensing characteristics of fullerene by measuring the electrical conductivity changes of a fullerene C_{60} film formed on an Al substrate. In particular, these authors found that the conductivity of C_{60}/C_{70}-containing carbon soot increases upon exposure to NH_3, $(C_2H_5)N$ and NO, whereas it decreases when exposed to CH_3CHO (Nosaka et al., 1999). The conductivity of a C_{60} film also changes when exposed to H_2, CH_4, NO_2 and CF_4. These findings allowed them to develop highly selective gas sensors with reversible and fast responses. Their sensitivity to NO_2 can be improved by adding Ni to the carbon soot.

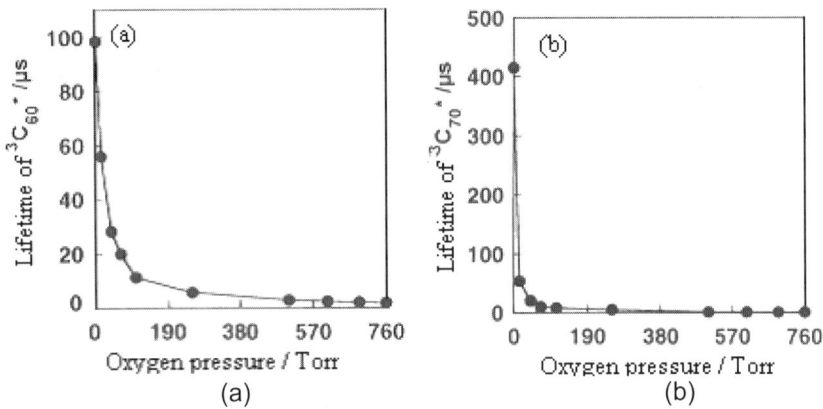

Figure 10.17. Dependence of the photoexcited triplet lifetime of (a) C_{60} and (b) C_{70} dispersed in a polystyrene (PS) film on the oxygen pressure (After Amao et al., 1999, copyright 1999 Chemical Society of Japan)

In addition to the conventional amperometric oxygen electrodes, various optical oxygen-sensing methods based on oxygen-induced luminescence quenching of indicator molecules have been developed (Amao et al., 1999). Among them, fullerene C_{60} and C_{70} were found to be highly desirable materials for sensing oxygen based on their optical property changes. The extremely long lifetime of the photoexcited triplet state of fullerenes (*ca.* 500 µs) indicates a very effective oxygen-quenching effect. Using the laser flash photolysis technique, Amao *et al.* (1999) have measured the photoexcited triplet lifetime for fullerene C_{60} (Figure 10.17(a)) and C_{70} (Figure 10.17(b)) dispersed in a polystyrene (PS) film under various oxygen pressures. As can be seen in Figure 10.17, C_{60}-PS is highly sensitive to oxygen throughout the whole range of the oxygen pressure while C_{70}-PS is particularly sensitive to oxygen in the low-oxygen-pressure region. Furthermore, it was demonstrated that fullerene-PS films are very stable to the irradiations used for the excitation (532 nm), and that the lifetime measurements (740 nm for C_{60} and 860 nm for C_{70}) associated with oxygen-sensing are highly reproducible.

10.8 Carbon Nanotube Sensors

Just as the discovery of C_{60} has created an entirely new branch of carbon chemistry, the subsequent discovery of carbon nanotubes by Iijima in 1991 opened up a new era in material science and nanotechnology (Chapter 5). Like fullerene C_{60}, carbon nanotubes also possess interesting optoelectronic properties. In particular, the electronic properties of carbon nanotubes are such that they may be metallic or semiconducting, depending on their diameter and the arrangement of graphitic rings in the walls. Besides, they show exceptionally good thermal and mechanical properties. The unique properties of carbon nanotubes have led to their use in areas as diverse as sensors, actuators, field-emitting flat panel displays, as well as energy and gas storage (Dai and Mau, 2001). The ability of carbon nanotubes and their derivatives to operate as gas sensors has been demonstrated recently (Collins *et al.*, 2000; Kong *et al.*, 2000, 2001). As we shall see below, the principles for carbon nanotube sensors to detect the nature of gases and to determine their concentrations are based on changes in electrical properties induced by charge transfer with the gas molecules (*e.g.* O_2, H_2, CO_2) or in mass due to physical adsorption.

10.8.1 Carbon Nanotube Gas Sensors

10.8.1.1 Carbon Nanotube Ammonia and Nitrogen Dioxide Sensors

Dai and co-workers (Kong *et al.*, 2000) were the first to demonstrate that semiconducting single-wall carbon nanotubes (SWNTs) act as rapid and sensitive chemical sensors at ambient temperature. By so doing, these authors devised a simple sensing system, in which a single semiconducting SWNT was kept in contact with titanium and/or gold metal pads at the two ends (Figure 10.18). Using the metal pads as the electrodes for electrical measurements, they found that the conductivity of the semiconducting SWNT changed rapidly over several orders of magnitude upon exposure to nitrogen dioxide and ammonia.

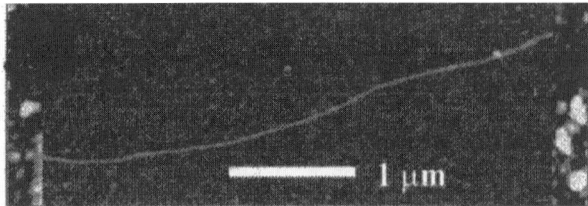

Figure 10.18. Atomic force microscopy image of a metal/SWNT/metal sample used for the gas sensing application (After Kong *et al.*, 2000, copyright 2000 American Association for the Advancement of Science)

In particular, an increase in the conductivity by up to three orders of magnitude was observed within 10 seconds after exposing the semiconducting SWNT to 200 ppm NO_2 (Figure 10.19(a)) while the conductivity decreased by two orders of magnitude within two minutes upon exposure to 1% NH_3 vapor (Figure 10.19(b)).

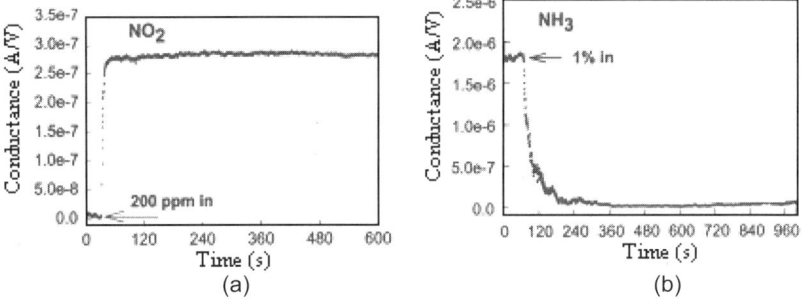

Figure 10.19. Electrical conductance response of a semiconducting SWNT to (a) 200 ppm NO_2 and (b) 1% NH_3 vapor (After Kong *et al.*, 2000, copyright 2000 American Association for the Advancement of Science)

These responses were attributed to the charge transfer between the *p*-type (*i.e.* hole-doped) semiconducting SWNT and electron-donating NH_3 or electron-withdrawing NO_2 gas. While the latter interaction increased the hole carriers in the SWNT, and hence enhanced conductance, the former interaction had an opposite effect. In principle, the semiconducting SWNT sensors could be applied to many other analytes of electron-donating or electron-accepting capabilities (I_2, AsF_5, CO, *etc.*). With appropriate surface modification for gaining specific interactions with analytes of biological significance, they could even be used to detect biological systems in solution.

Commercial sensors based on semiconducting metal oxides have been developed for sensing NO_2 and NH_3, but these sensors need to operate at temperatures of up to 600 °C for high sensitivity. The SWNT sensor, however, shows a strong response, much stronger than is typically recorded for a commercial counterpart, even at room temperature. Besides, the nanotube sensors are relatively simple to make and their production can be easily scaled up. The nanotube sensors, however, need to take several hours to release the analyte at room temperature before they can be reused. Although the slow recovery of the sensor from the nanotube-analyte interaction can be expedited by heating, it remains as a major drawback.

10.8.1.2 Carbon Nanotube Hydrogen Sensors

More recently, Dai and co-workers (Kong *et al.*, 2001) have further demonstrated that certain chemical/physical modifications could enhance the molecular specificity of carbon nanotube sensors. In particular, these authors sputter-coated an individual SWNT and its bundler with Pd nanoparticles. They observed a up to 50% and 33% decrease in conductivity for the Pd-coated individual SWNT and nanotube bundler, respectively, upon exposure to a flow of air mixed with 400 ppm of hydrogen. This was followed by a rapid recovery of the conductivity after the hydrogen flow was turned off (Figure 10.20). The response time, defined as the time required for half of the conductivity change, was found to be about 5 to 10 s, and the time for full recovery was *ca.* 400 s. The interaction between H_2 and the Pd-SWNT is responsible for the sensing mechanism, whereas the corresponding interaction with O_2 is the driving force for the recovery process.

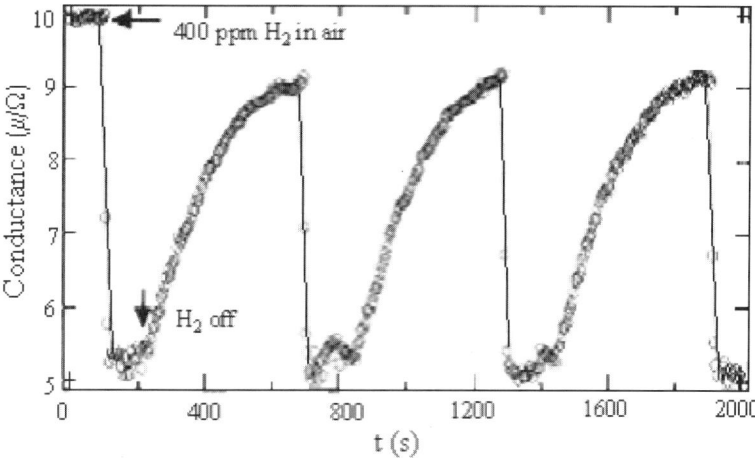

Figure 10.20. Electrical conductance of the SWNT coated with Pd nanoparticles *vs.* time in an air flow with 400 ppm H_2 on and off (After Kong *et al.*, 2001, copyright 2001 Wiley-VCH Verlag. Reproduced with permission)

As is well known, the interaction of the H_2 and Pd surface at room temperature could cause atomic hydrogen to dissolve into Pd with a high solubility, leading to a decrease in the work function for Pd (Mandelis and Christofides, 1993). The decrease in Pd work function, in its turn, caused electron transfer from Pd to the SWNTs to reduce the hole-carriers in the *p*-type nanotube, and hence a decreased conductivity. The recovery of the nanotube's electrical characteristics was made by combining the dissolved atomic hydrogen in Pd with oxygen in air to form H_2O

when the hydrogen was turned off. The high sensitivity, fast response and reversibility observed for the Pd-SWNT sensors at room temperature indicate that carbon nanotubes are also excellent new sensing materials for detecting low concentration of molecular hydrogen.

10.8.1.3 Carbon Nanotube Oxygen Sensors

In addition to the above studies, Zettl and co-workers (Collins *et al.*, 2000) reported that SWNTs are extremely susceptible to oxygen adsorption. As shown in Figure 10.21(a), they found stepwise changes in the electrical resistance of up to 10 to 15% by alternately exposing nanotube samples to air or a vacuum at 290 K. Identical changes were observed if pure dry oxygen were used to replace air, indicating that oxygen was responsible for the observed resistance change.

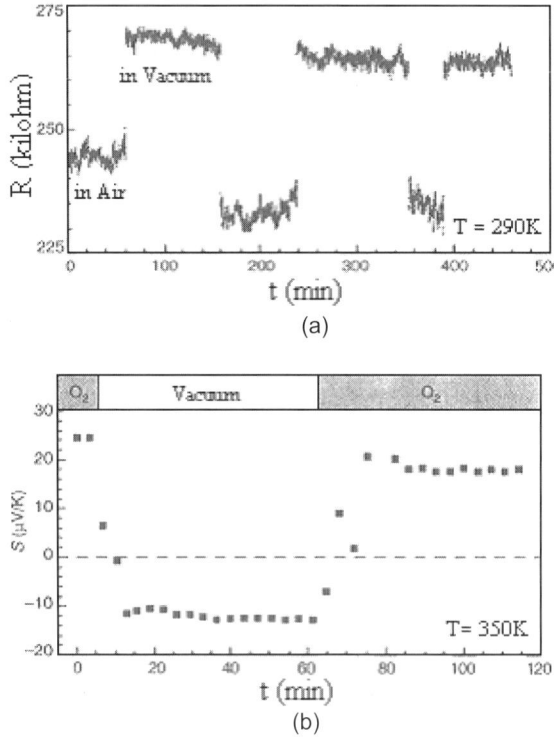

Figure 10.21. (a) Electrical resistance (R) and (b) thermoelectric power responses of SWNT films to gas exposure (After Collins *et al.*, 2000, copyright 2000 American Association for the Advancement of Science)

Similar changes were also observed for the thermoelectric power, S (Barnard, 1972). As seen in Figure 10.21(b), S was positive with a value of *ca.* +20 μV/K for the SWNT in a pure atmospheric pressure of oxygen environment at 350 K. The removal of oxygen from the system caused a continuous change in S from positive to negative, reaching an equilibrium value of *ca.* -10 μV/K. When the system was alternately cycled between pure oxygen and pure vacuum, stepwise changes in S were obtained. While the negative S indicates that the SWNT is an *n*-type semiconductor in a vacuum, the O_2-exposed nanotube with a positive S is a *p*-type semiconductor (see the following subsection for more detailed discussions on S).

Apart from the thermoelectric power measurements discussed above, Collins *et al.* (2000) further used scanning tunneling spectroscopy and transport measurements to confirm that oxygen-exposed SWNTs behave as *p*-type semiconductors and the oxygen's electron affinity could cause these observed changes through charge transfer between the adsorbed oxygen and SWNTs. These results clearly indicate that SWNTs can be used as sensing elements for making new oxygen sensors.

10.8.1.4 Carbon Nanotube Thermoelectric Nanonose

While the aforementioned Dai's gas sensor was made from individual SWNTs, Zettl's oxygen sensor was based on SWNT bundles. As is well known, the as-synthesized SWNTs are often in the form of well-ordered and tightly packed bundles, which contain tens to hundreds of individual SWNTs bound together by a weak van der Waals force (Thess, *et al.* 1996). The sites with which gas molecules can adsorb on the nanotube bundle may be grouped into four categories: the external bundle *surface*; a *groove* formed at the contact between adjacent tubes on the outside of the bundle; inside an interstitial *channel* formed at the contact of three adjacent tubes in the bundle interior; and within an interior *pore* of an individual tube (Figure 10.22) (Adu *et al.*, 2001). However, some of these sites may be inaccessible by certain gases due to size exclusion. Theoretical calculations have predicted that the binding energy (E_B) for hydrogen in these various sites increases in the following order: E_B (surface) < E_B (pore) < E_B (groove) < E_B (channel) (Williams *et al.*, 2000).

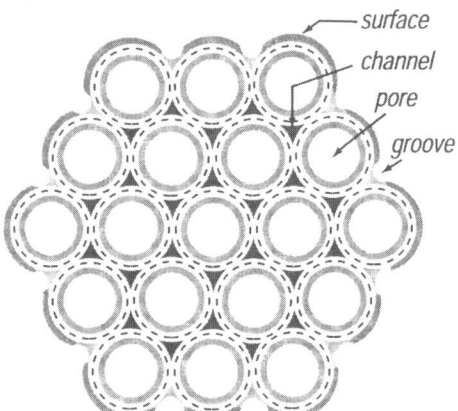

Figure 10.22. Schematic representation of an SWNT bundle showing the pore, groove, channel and surface sites available for gas adsorption (After Adu *et al.,* 2001, copyright 2001 Elsevier. Reproduced with permission)

Based on the above model, Adu *et al.* (2001) have developed a thermoelectric *nanonose* from tangled bundles of SWNTs, which can be used to detect and distinguish gases, such as He, N_2 and H_2, and to tell whether the gas molecule is physisorbed or chemisorbed.

In a typical experiment, these authors prepared mats of tangled SWNT bundles (*ca.* 1 mm × 2 mm × 0.1 mm) and then placed a small, ceramic-coated resistor near one end of a SWNT mat as a heater to induce a temperature difference $\Delta T < 0.5$ K along the length of the mat. The corresponding thermoelectric voltage ΔV was measured to give the thermoelectric power as follows (Barnard, 1972):

$$S = \Delta V / \Delta T \tag{10.14}$$

After degassing in a vacuum (10^{-8} Torr) at 500 K for *ca.* 10 h to remove the adsorbed O_2, the thermoelectric power of the SWNT mat was measured under the vacuum at room temperature and designated as S_0 (typically, -45 µV/K $< S_0 < -40$ µV/K) (Adu *et al.*, 2001). As mentioned above, the negative value of S indicates that the SWNT mat behaves thermoelectrically as an *n*-type metal, most probably arising from the 1/3 metallic nanotubes (Chapter 5) that form conducting pathways through the sample by the percolation process (Sumanasekera *et al.*, 2000, also Chapter 2).

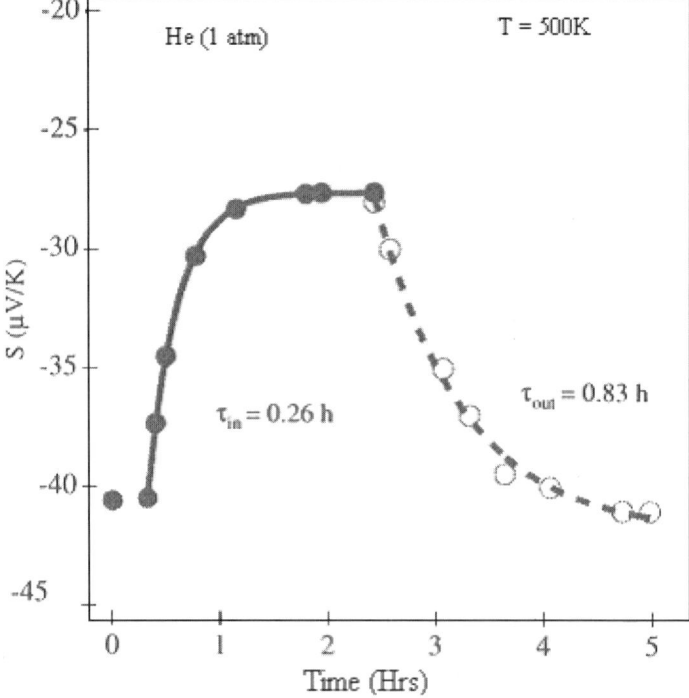

Figure 10.23. Time-dependence of the thermoelectric response of the SWNT mat to 1 atm overpressure of He gas (filled circles) and to the subsequent application of a vacuum over the sample (open circles) (After Adu *et al.*, 2001, copyright 2001 Elsevier. Reproduced with permission)

Figure 10.23 shows the thermoelectric response of the SWNT mat as a function of time to a 1 atm overpressure of He gas and to the subsequent evacuation. As can be seen, the thermopower S increases exponentially with the exposing time to He gas, followed by an exponential decay upon removing the He overpressure. It was noted with interest that the desorption time constant is about three times longer than the adsorption time constant, and that the resistivity ρ exhibited a similar exponential rise and fall. Both the observed changes in S and ρ are completely reversible. The thermoelectric response, in terms of S and ρ, of the SWMT mat to gases, such as He, N_2, H_2, O_2 and NH_3, can be understood by considering both the possible charge transfer-induced change of Fermi energy [*i.e.* the adsorbed molecule donates (removes) an electron to (from) the conduction (valence) band] and the creation of additional defects by adsorption/insertion of gas molecules on the nanotube walls. In view of the fact that the metallic behavior of the SWNT mat results from the percolating pathways formed by the metallic tube components in the sample, Adu *et al.* (2001) rewrote the Mott relation between the thermoelectric power and the diffusion of free carriers in a metal as a logarithmic energy derivative of the electrical resistivity (Barnard, 1972):

$$S = CTd/dE\{\ln\rho(E)\}_{EF} \qquad (10.16)$$

where $C = (\pi^2 k_B^2/3|e|)$, k_B is Boltzmann's constant, T is the temperature, $\rho(E)$ is the energy-dependent resistivity, $|e|$ is the charge on an electron, and the derivative is evaluated at the Fermi energy E_F.

Then, Adu *et al.* (2001) further separated the contributions to the total bundle resistivity (ρ) into the scattering from pre-existing defects in the bundle before gas adsorption (ρ_0) and the extra impurity scattering associated with the adsorbed gas (ρ_a):

$$\rho = \rho_0 + \rho_a \qquad (10.17)$$

When $\rho_0 \gg \rho_a$ as is the case here, substituting Equation (10.17) into Equation (10.16) and approximating $1/\rho \sim \rho_0^{-1}(1 - \rho_a/\rho_0)$ leads to Equation (10.18):

$$S = S_0 + (\rho_a/\rho_0)(S_a - S_0) \qquad (10.18)$$

where $S_j = CTd/dE(\ln\rho_j)_{EF}$ for $j = (0, a)$, S_0 and S_a are the contributions to the thermopower from $\rho_0(E)$ and $\rho_a(E)$, respectively. The Equation (10.18) is the same as the well-known Nordheim-Gorter relation developed for Au and Ag alloys (Barnard, 1972).

From Equation (10.18), it is not difficult to understand that a linear relation of S vs. ρ_a corresponds to constant S_0 and S_a, indicating the band structure and E_F of the host is unaffected by the (gas) doping. Therefore, if the particular gas molecules are *physisorbed* onto the nanotube walls via van der Waals force, they induce little perturbation on the SWNT band structure, leading to a *linear* Nordheim-Gorter plot. On the other hand, a *non-linear* Nordheim-Gorter plot should be observed if the gas molecules are *chemisorbed* onto the nanotube walls. Consequently, Nordheim-

Gorter plots are very useful for identifying the nature of the gas adsorption onto carbon nanotubes.

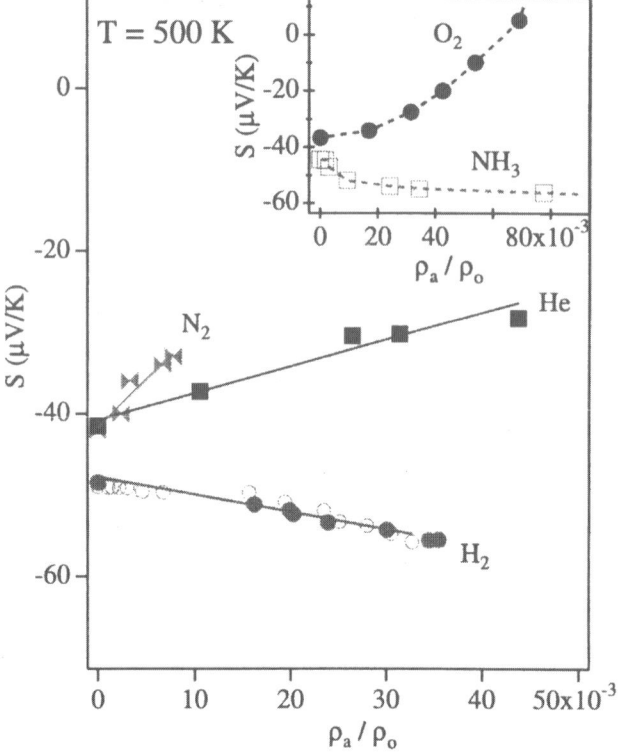

Figure 10.24. Nordheim-Gorter plots (*i.e.* S vs. ρ) showing the effect of gas adsorption on the electrical properties of the SWNT mat. The amount of the adsorbed gas is proportional to the value of ρ. A linear plot indicates physisorption while a curved plot is for chemisorption (inset). As S_0 varies slightly from sample to sample, different values for the intercepts on the S-axis are observed (After Adu *et al.*, 2001, copyright 2001 Elsevier. Reproduced with permission)

Figure 10.24 reproduces the Nordheim-Gorter plots for the isothermal adsorption of He, N_2 and H_2 on the SWNT mats at 500 K. Also included in the inset of Figure 10.24 are the corresponding plots for O_2 and NH_3. It becomes clear from Figure 10.24 that He, N_2 and H_2 were physisorbed on the SWNT surface whereas O_2 and NH_3 were chemisorbed. This is consistent with the charge transfer processes between the nanotube and O_2 and NH_3 discussed in proceeding subsections. Furthermore, the opposite signs for the slopes of the Nordheim-Gorter plots corresponding to H_2 and He or N_2 indicate a different energy dependence for the electron scattering rate for the H_2 and He or N_2 sites. Overall, it was clearly demonstrated that the Nordheim-Gorter plot underlies the basis for the utility of a SWNT thermoelectric "*nanonose*".

10.8.1.5 Carbon Nanotube Carbon Dioxide Sensors

Unlike all of the above carbon nanotube gas sensors based on SWNT(s), Ong and co-workers (Ong *et al.*, 2001; Varghese *et al.*, 2001) have devised a multi-wall carbon nanotube (MWNT) CO_2 sensor by monitoring changes in the MWNT permittivity with CO_2 exposure. Using a planar, inductor-capacitor resonant-circuit transducer (Ong and Grimes, 2000) with the configuration shown in Figure 10.25(a), these authors can remotely monitor changes in the MWNT complex permittivity (*i.e.* $\varepsilon_r' - j\varepsilon_r''$) with CO_2 exposure by measuring the corresponding changes in the resonant frequency of a loop antenna (Figure 10.25(b)).

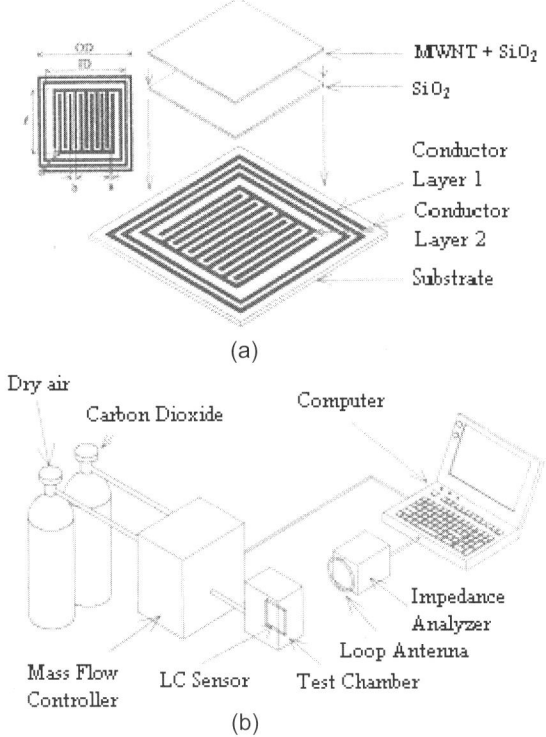

Figure 10.25. (a) Schematic drawing of the MWNT CO_2 sensor. A planar inductor-interdigital capacitor pair is defined on a printed circuit board. The capacitor is first coated with a protective SiO_2 layer followed by a layer of CO_2 responsive MWNT-SiO_2 composite; (b) the experimental set-up for the MWNT CO_2 sensor. The sensor is placed inside a sealed Plexiglas chamber and monitored via a loop antenna by measuring its impedance with an impedance analyzer (After Ong *et al.*, 2001; Varghese *et al.*, 2001, copyright 2001 Elsevier)

As CO_2 has a relatively low ε_r' (*ca.* 1) compared to the MWNT (*ca.* 15), the adsorption of CO_2 causes an overall decrease in ε_r' of the sensor. Meanwhile, the ε_r'', which is directly proportional to the conductivity of the MWNT ($\sigma = 2\pi f \varepsilon_0 \varepsilon_r''$, where f is the resonant frequency and $\varepsilon_0 = 8.854\times10^{-12}$ F/m is the free space permittivity), also decreases due to the reduced conductivity when the sensor is exposed to CO_2 (Marliere et al., 2000).

Figure 10.26(a) shows the typical responses in terms of ε_r' and ε_r'' of the MWNT CO_2 sensor as it is alternately cycled between dry air and pure CO_2. As can be seen in Figure 10.26, the change in the complex permittivity is highly reversible, with no observable hysteresis. The corresponding changes as a function of the volume fraction of CO_2 to dry air is given in Figure 10.26(b), which shows a symmetric step function, indicating that the sensor response is linear and highly reversible with either increasing or decreasing CO_2 concentrations. Besides, a response time of less than one minute may be obtained from Figure 10.26. These results make the MWNT CO_2 sensor very attractive for various long-term wireless monitoring applications, especially for sensing CO_2 in sealed food packages where the presence of a high level of CO_2 is often taken as an indication for spoilage.

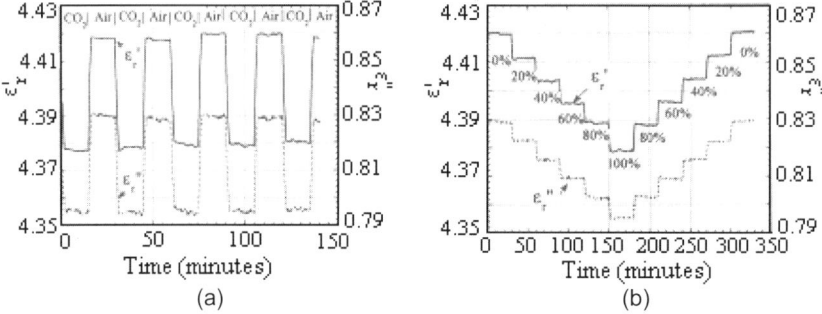

Figure 10.26. Measured ε_r' and ε_r'' values when the MWNT sensor is (a) cycled between pure CO_2 and dry air, and (b) exposed to CO_2 gas with concentrations varying from 0–100% (v/v) and then back to 0% (After Ong et al., 2001; Varghese et al., 2001, copyright 2001 Elsevier)

Most commercially available CO_2 sensors are based on the impedance measurements of a capacitor coated with a CO_2-responsive layer, including heteropolysiloxane (Endres et al., 1999), $BaTiO_3$ (Lee and Meyer, 2000), $CeO/BaCo_3/CuO$ (Matsubara et al., 2000), Ag_2SO_4 and Na_2CO_3 (Currie et al., 1999). Although these commercial products can offer a high degree of accuracy and reliability, they often require hard-wire connections between the sensor probe and detecting system. Unlike the nanotube CO_2 sensor, therefore, their commercial counterpart is not suitable for remote-sensing applications.

10.8.2 Carbon Nanotube Pressure and Temperature Sensors

Wood and Wagner (2000) have recently found that the disorder-induced Raman peak of SWNTs (D^* band, 2610 cm^{-1} in air) shifts dramatically upon immersion of the nanotubes in a liquid or measured in a diamond-anvil cell (Wood et al., 2000) with respect to the corresponding peak observed in air (Figure 10.27(a)). These authors, along with others (Lourie and Wagner, 1998), have further found that the Raman peak of the nanotubes embedded in polymer matrices also shifts significantly with temperature (Figure 10.27(b)). As shown in Figure 10.27(b), the wavenumber of the D^* band increases with decreasing temperature as the nanotubes experience compression. These changes are highly reversible, indicating the usefulness of carbon nanotubes as molecular sensors for pressure- and temperature-sensing.

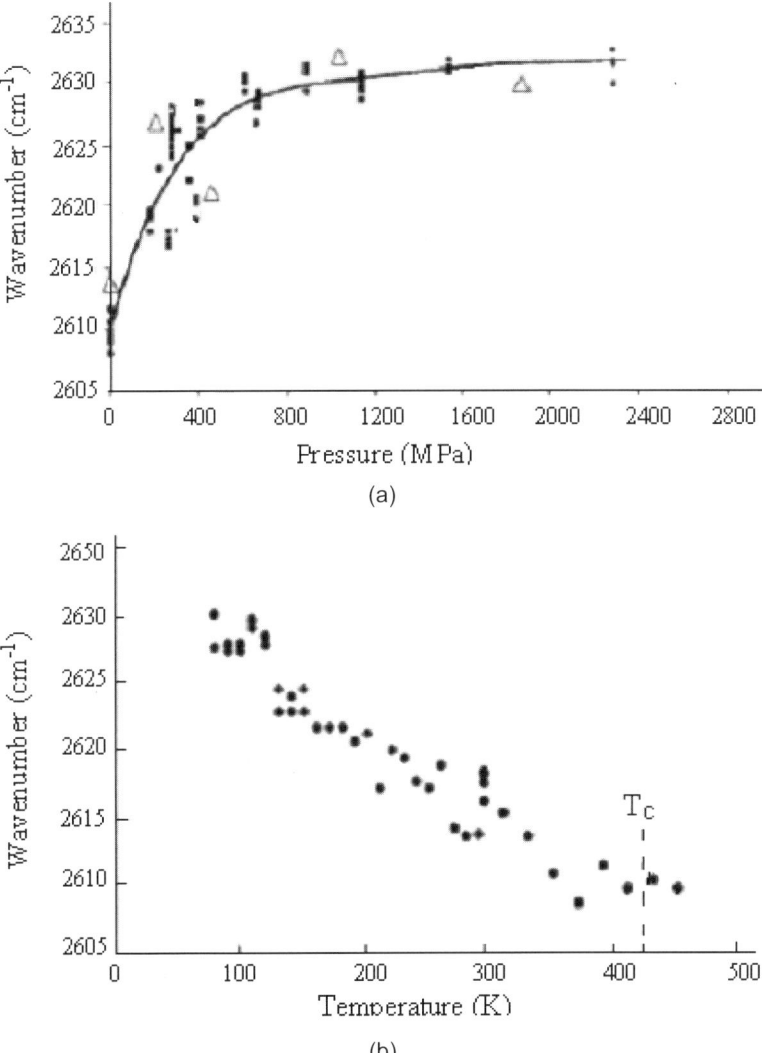

Figure 10.27. The peak position of the D^* Raman peak of (a) SWNTs as a function of the molecular (by immersion in liquids, black square) and macroscopic (using a diamond-anvil cell, open triangle) (After Wood et al., 2000); (b) SWNTs embedded in polycarbonate upon cooling (After Wood and Wagner, 2000, copyright 2000 American Institute of Physics)

10.8.3 Carbon Nanotube Chemical Force Sensors

Owing to their exceptional electrical and mechanical properties, carbon nanotubes (both SWNTs and MWNTs) have also been used as scanning probe microscope tips with a significantly better lateral resolution compared with commercial Si and SiN AFM tips (Dai *et al.,* 1996, Wong *et al.,* 1998a). Although these sharp nanotube tips offer advantages for high-resolution imaging, the unmodified nanotube tips, like most of the commercial tips, do not provide information about the chemical nature of the surface. Lieber and co-workers (Wong *et al.,* 1998b, 1998c) have recently addressed this point by covalently modifying the ends of both SWNT and MWNT tips with chemically and/or biologically active functionalities for chemically sensitive mapping at molecular resolution. For example, these authors have introduced –COOH groups at the opened end of an SWNT tip by acid oxidation (Figure 10.28) (Tsang *et al.,* 1994). They then demonstrated the possibility of chemically attaching an amine molecule, RNH_2 (*e.g.* benzylamine) onto the terminal –COOH group through the amide linkage.

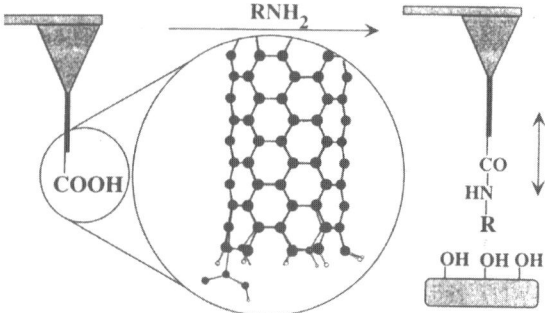

Figure 10.28. Schematic illustrating the SWNT tip modification via coupling an amine (RNH_2) onto a terminal –COOH functionality, and the use of the tip-modified nanotube as an AFM probe to sense specific interactions between the functional group (R) and surface –OH group (After *Wong et al.* 1998c, copyright 1998 The American Chemical Society. Reproduced with permission)

To demonstrate the chemically sensitive force mapping, Lieber and co-workers (Wong *et al.,* 1998b, 1998c) first used a –COOH-terminated SWNT AFM tip to image a micro-patterned surface consisting of 10 µm squares of a hexadecanethiol SAM (*i.e.* CH_3 rich surface) surrounded by a 16-mercaptohexadecanoic acid SAM (*i.e.* –COOH rich surface). As shown in Figure 10.29(a), the tapping mode image recorded in ethanol shows a greater phase lap for tip–COOH/sample–COOH *vs.* tip–COOH/sample–CH_3 regions due to the stronger adhesion force for the former. The corresponding image obtained with the benzylamine-modified SWNT AFM tip shows the patterned feature, but with the opposite phase contrast (Figure 10.29(b)) as this particular tip interacts more strongly with the –CH_3 rich surface in respect to the –COOH region. These results clearly demonstrate that chemically modified nanotube AFM tips can be used as sensing probes for chemically sensitive surface recognition.

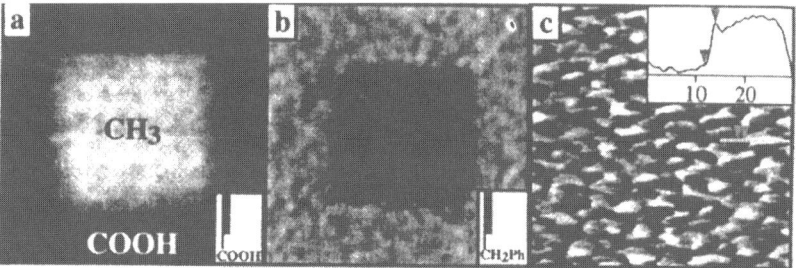

Figure 10.29. Tapping mode AFM images of a patterned sample with region-specific $-CH_3$ and $-COOH$ regions in ethanol recorded with (a) a $-COOH$-terminated SWNT tip; (b) a benzylamine-functionalized SWNT tip. Dark regions indicate greater phase lag and the images are 16μm × 16μm in size; (c) tapping mode phase image recorded on a partial bilayer structure, indicating a lateral chemical resolution of *ca.* 3 nm (After Wong *et al.* 1998c, copyright 1998 The American Chemical Society. Reproduced with permission)

10.8.4 Carbon Nanotube Resonator Mass Sensors

As is to be discussed in Chapter 11, de Heer and co-workers (Frank *et al.*, 1998) have measured the conductance of carbon nanotubes by replacing the tip of a scanning probe microscope with the nanotube, which forms an electrical circuit with a liquid metal. They revealed novel stepwise increases in current with increasing voltage, which was attributed to quantum effects.

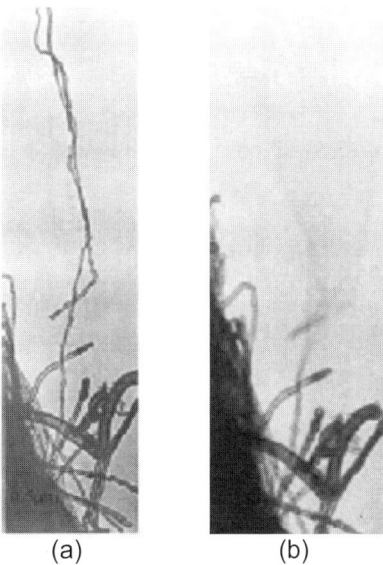

Figure 10.30. A carbon nanotube with another nanotube trapped on it (a) at the stationary and (b) the first resonance mode (After Gao R. *et al.*, 2000a, copyright 2000 The American Physical Society)

Using a similar approach, Gao et al. (2000a) have been able to measure mechanical properties of individual carbon nanotubes. In this case, a transmission electron microscopy (TEM) specimen holder was specially built for applying a voltage across the nanotube and its counter electrode so that the measurements can be done on a specific nanotube whose microstructure is determined by TEM. By applying an alternating voltage with a tunable frequency on the nanotube, resonance can be induced (Figure 10.30) from which certain mechanical properties of the nanotube can be determined. More interestingly, it was found that neither the point defects nor the volume defects caused any significant softening at the local region upon deformation, due to the collectively rippling deformation on the inner arc of a bent nanotube (Poncharal et al., 1999). Analogous to a pendulum, any object attached onto the nanotube could significantly affect the resonance frequency. The mass of the trapped object, down to femtogram (1 fg = 10^{-15} g), can thus be measured simply by monitoring changes in the resonance frequency.

10.8.5 Carbon Nanotube Glucose Sensors

As discussed in Chapter 5, Dai and co-workers (Gao et al., 2000b) have used the aligned carbon nanotubes produced from FePc to make novel conducting polymer-carbon nanotube (CP-NT) coaxial nanowires by electrochemically depositing a concentric layer of an appropriate conducting polymer uniformly onto each of the constituent aligned nanotubes (Figure 10.31).

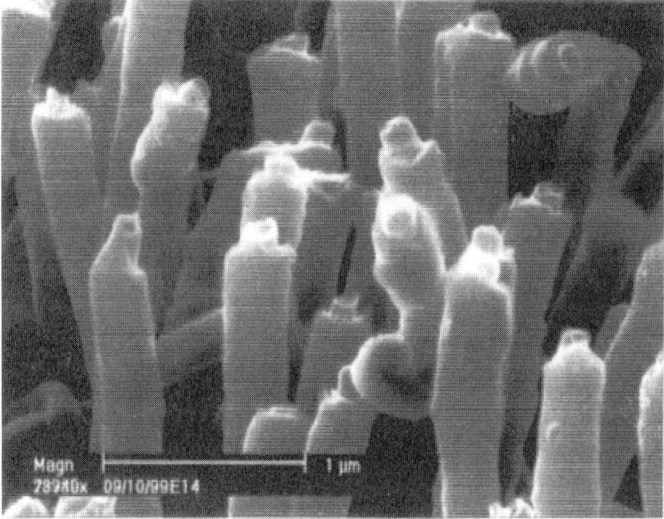

Figure 10.31. Typical SEM images of the CP-NT coaxial nanowires produced by cyclic voltammetry on the aligned carbon nanotube electrode, showing a thin layer of conducting polymer (polypyrrole) coating surrounding each of the constituent aligned carbon nanotubes (After Dai, 2001, copyright 2001 CSIRO Publishing)

The electrochemical performance of the aligned CP-NT coaxial nanowires was evaluated by carrying out cyclic voltammetry measurements. As for polyaniline films electrochemically deposited on conventional electrodes, the cyclic voltammetric response of the polyaniline-coated nanotube array in an aqueous solution of 1M H_2SO_4 (Figure 10.32(a)) shows oxidation peaks at 0.33 and 0.52 V (but with much higher current densities) (Sazou and Georgolios, 1997). This indicates that polyaniline films thus prepared are highly electroactive. As a control, the cyclic voltammetry measurement was also carried out on the bare aligned nanotubes under the same conditions (Figure 10.32(b)). In the control experiment, only capacitive current was observed with no peak attributable to the presence of any redox-active species.

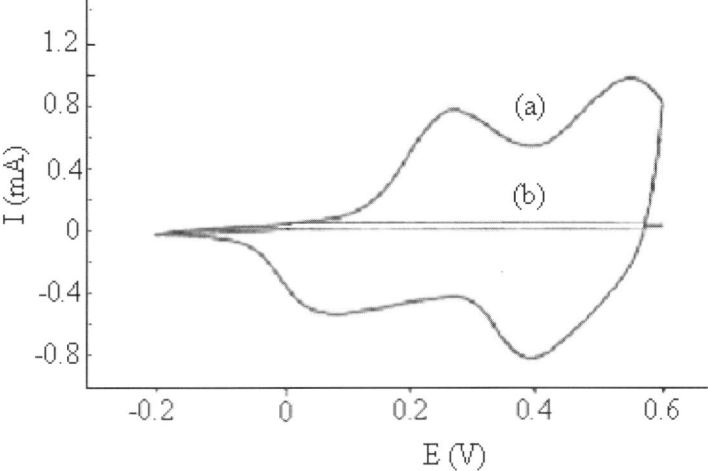

Figure 10.32. Cyclic voltammograms of (a) the polyaniline-coated CP-NT coaxial nanowires and (b) the bare aligned carbon nanotubes. Measured in an aqueous solution of 1M H_2SO_4 with a scan rate of 50 mV/s (After Gao et al., 2000b, copyright 2000 Wiley-VCH Verlag. Reproduced with permission)

The coaxial structure allows the nanotube framework to provide mechanical stability (Poncharal et al., 1999, Gao et al., 2000a) and efficient thermal/electrical conduction (Frank et al., 1998, Odom et al., 2000) to and from the conducting polymer layer. The large surface-interface area obtained for the nanotube-supported conducting polymer layer is an additional advantage for using them in sensing applications. In this context, Gao et al. (2003) have immobilized glucose oxidase onto the aligned carbon nanotube substrate by electrooxidation of pyrrole (0.1 M) in the presence of glucose oxidase (2 mg/ml) and $NaClO_4$ (0.1 M) in a pH 7.45 buffer solution. Then, the glucose oxidase-containing polypyrrole-carbon nanotube coaxial nanowires were used to monitor concentration change of hydrogen peroxide (H_2O_2) during the glucose oxidation reaction [Equation (10.1)] by measuring the increase in the electrooxidation current at the oxidative potential of H_2O_2 (i.e. the amperometric method).

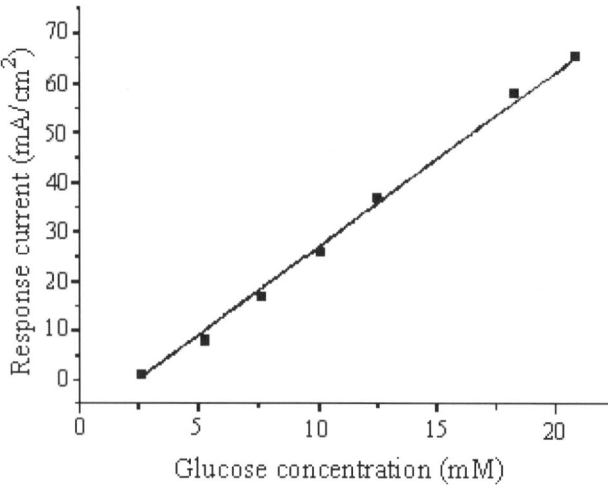

Figure 10.33. The dependence of electrooxidation current at the oxidative potential of H_2O_2 on the glucose concentration for the CP-NT coaxial nanowire sensor (After Gao *et al.* 2003, copyright 2003 Wiley-VCH Verlag)

As shown in Figure 10.33, a linear response of the electrooxidation current to the glucose concentration was obtained for the CP-NT nanowire sensor (Gao *et al.* 2003). The linear relationship extends to a glucose concentration as high as 20 mM, which is higher than the 15 mM typical limit used for the detection of blood glucose in practice (Yasuzawa and Kunugi, 1999). Furthermore, the amperiometric response was found to be about ten orders of magnitude higher than that of more conventional flat electrodes coated with glucose oxidase-containing polypyrrole films under the same conditions (Yasuzawa and Kunugi, 1999). The CP-NT nanowire sensors were also demonstrated to be highly selective, with their amperiometric responses being almost unchanged even in the presence of some interference species including ascorbic acid, urea and D-fructose. Therefore, the CP-NT nanowires could be used for making new glucose sensors with a high sensitivity, selectivity and reliability.

10.9 DNA Sensors

Watson and Crick's (1953) classical paper on the double helix structure of DNA spurred the age of DNA technology. A major feature of the Watson-Crick's model of DNA is that it provides a vision of how a base sequence of one strand of the double helix can precisely determine the base sequence of the partner strand for passing the genetic information in all living species (Chapter 1). The principle learned from this breakthrough has now been applied to the development of DNA sensors for DNA analysis and diagnosis, either *in vitro* or *in vivo*, through the very specific DNA pairing interaction.

10.9.1 DNA Sensors Based on Oligonucleotide-functionalized Polypyrroles

Based on their pioneering work on conducting polymers, Garnier et al. (1999) prepared oligonucleotide (OND)-substituted polypyrrole and measured its electrochemical response in various aqueous media containing either complementary or non-complementary OND targets. Their results showed that the cyclic voltammogram of the OND-functionalized polypyrrole did not change in the presence of a non-complementary OND in solution. In contrast, a significant change was observed upon the addition of a complementary OND target, indicating the occurrence of specific hybridization of the polypyrrole-grafted OND probe with its complementary OND analyte in solution. The observed biological recognition can be quantitatively determined by an amperometric analysis down to about 10^{-11} mol. The conjugated backbone thus acts as a macromolecular wire to transduce and amplify (Section 10.2.2.1) the biological information collected by the substituted OND groups into electrochemical signals. Therefore, this study provides a springboard to the development of highly sensitive biosensors from nucleotide modified conducting polymers.

10.9.2 DNA Diagnostic Biosensors

Without involving any conducting polymer, Berney et al. (2000) subsequently developed a sensor from a single-stranded oligonucleotide for specifying DNA sequences through the capacitance measurement. As an analyte DNA strand hybridized with its counterpart immobilized on a silicon surface, charge effects were induced, leading to changes in the dielectric properties of the bilayer and hence the capacitance. The response was fast, specific and required no addition of mediators to enhance or amplify the signal. The device can be optimized for the detection of complex sequences. Similarly, Hasebe et al. (1999) developed a mutagen sensor with the amperometric method by using a DNA-immobilized graphite felt as a sensing element.

10.9.3 DNA Sensor for Detection of Hepatitis B Virus

In a separate study, Hashimoto et al. (1998) used the photolithographic technique to construct a disposable-type DNA sensor for detecting hepatitis B virus (HBV) genome DNA. In a typical experiment, the DNA sensor was reacted with a HBV-DNA-inserted plasmid (pYRB259) immersed in a 100 µmol/l Hoechst 33258 solution and washed with a phosphate buffer. Electrochemical analyses revealed that the anodic peak current derived from the Hoechst 33258 bound to the formed hybrids on the electrode increased with increasing the concentration of pYRB259 in the range of 10^4 to 10^6 copy/ml. This finding prompted these authors to further evaluate the concentration of HBV-DNA extracted from patients' serums using the DNA sensor and competitive polymerase chain reaction (C-PCR). A correlation coefficient of 0.89 was obtained between the two assay methods. Repeating inter-assay measurements indicated an error range of 2–6% for the DNA sensor. Therefore, the microfabricated disposable DNA sensor was shown to be highly specific for quantitative detection of HBV-DNA in serum.

10.9.4 DNA Fluorescent Sensor for Lead Ions

Lead detection is of paramount importance to human health as even a low-level lead exposure (*e.g.* ≥ 480 nM in the blood) could lead to various adverse health effects (Needleman, 1992). In order to eliminate sophisticated equipment and/or sample treatment associated with most current methods for lead detection (anodic stripping, atomic absorption spectrometry, inductively coupled plasma mass spectrometry, *etc.*), Li J. and Lu (2000) developed a simple and inexpensive method for real-time detection of Pb^{2+} by choosing a deoxyribozyme for transducing signals through both molecular recognition of the metal ion and metal-induced hydrolytic cleavage. In particular, these authors chose an *in vitro*-selected deoxyribozyme (termed 17E) that is capable of cleaving a single RNA linkage within a DNA substrate (designated as 17DS) in the presence of Pb^{2+} ions (Figure 10.34(a)). They then labeled the 5'-end of the substrate with the fluorophore 6-carboxytetramethylrhodamine (TAMRA) and the 3'-end of the enzyme strand with a fluorescence quencher 4-(4'-dimethylaminophenylazo)benzoic acid (Dabcyl).

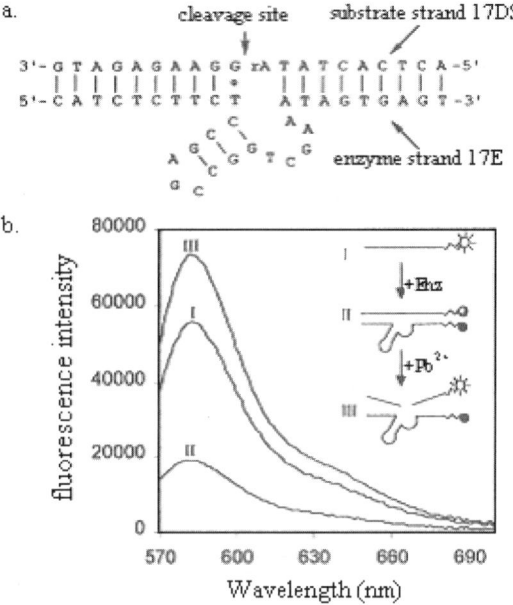

Figure 10.34. (a) Sequence and schematic representation of the deoxyribozyme/substrate complex. The cleavable substrate (Rh-17DS) is a DNA/RNA chimera in which rA represents a ribonucleotide adenosine; (b) Fluorescence spectra of the substrate (Rh-17DS) alone (I), after annealing by the deoxyribozyme (17E-Dy) (II), and after adding 500 nM $Pb(OAc)_2$ for 15 min (III) (After Li and Lu, 2000, copyright 2000 The American Chemical Society. Reproduced with permission)

As shown in Figure 10.34(b), the TAMRA-labeled substrate (Rh-17DS) displayed fluorescence broad emission band (curve I, Figure 10.34(b)) over 570–700 nm when excited at 560 nm. The fluorescence of Rh-17DS was quenched (curve II, Figure 10.34(b)) upon hybridization with a Dabcyl-attached enzyme strand (17E-Dy). After addition of Pb^{2+} ions, however, the fluorescence intensity increased significantly (curve III, Figure 10.34b) because the Dabcyl quencher was separated from the TAMRA fluorophore due to the cleavage of Rh-17DS strand. Due to the specificity towards metal ions associated with the enzymatic cleavage reaction, a selectivity of 80-fold for Pb^{2+} over other metal ions (Co^{2+}, Zn^{2+}, Mn^{2+}, Ni^{2+}, Cd^{2+}, Cu^{2+}, Mg^{2+}, Ca^{2+}, *etc.*), together with a quantifiable detection range from 10 nM to 4 µM for Pb^{2+}, was found. These findings clearly demonstrated that the Rh-17DS/17E-Dy is a promising sensing element for Pb^{2+} detection.

10.9.5 DNA Molecular Break Lights

A continuous assay for DNA cleavage by small molecules (*e.g.* the enediyne compound calicheamicin) has also recently been developed (Biggins *et al.*, 2000; Borman, 2000). This technique is based on hair-pin-shaped DNA (designated as *molecular break lights*), which can light up when a fragment breaks off in the presence of a DNA-cleaving agent (Figure 10.35).

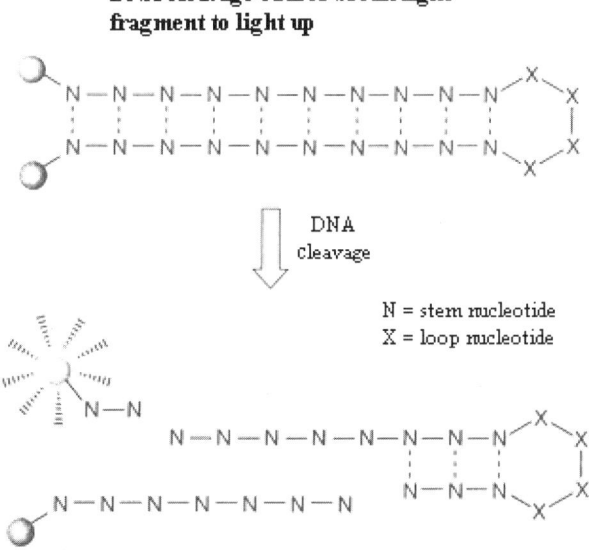

Figure 10.35. DNA cleavage causes break light fragment to light up (After Borman, 2000, copyright 2000 The American Chemical Society)

As can be seen in Figure 10.35, the molecular break lights are small hair-pin-shaped single-stranded oligonucleotides containing a fluorescent group at the 5′ end and a fluorescence quenching group at the 3′ end (for a detailed DNA structure, see Chapter 1). Also included into the molecular break light's stem is a DNA sequence susceptible to cleavage by a specific DNA-cleaving agent (Alcamo, 2001). The fluorescent group and its quencher are sufficiently close to each other in the as-prepared molecular break light to prevent the fluorescent group from lighting up. When a DNA-cleaving agent breaks the stem, however, the fluorophore-quencher pair becomes separated and the fluorophore illuminates. The intensity of the photoluminescence is directly proportional to the extent of DNA cleavage and hence serves as a measure for sensing the cleaving agent.

Like the molecular break light method, several other techniques, such as fluorescence correlation spectroscopy, fluorescence resonance energy transfer and radioactivity-based assays, have previously been used to detect DNA-cleaving agents (Alcamo, 2001). Comparing with these related techniques, however, the molecular break light method has many advantages, including lower interference from background fluorescence, greater simplicity, broader applicability and higher sensitivity. Furthermore, the molecular break light technique could be extended to an automated high-throughput drug discovery assay for screening novel protein-based or small-molecule-derived DNA-cleavage agents.

10.9.6 DNA Quartz Oscillators and Cantilevers

Apart from the aformentioned DNA sensors based on electrical and optical transducers, Nicolini *et al.* (1997) have deposited thin LB films containing single-stranded DNA onto quartz oscillators for sensing the presence of the complementary DNA sequences by monitoring mass increment arising from the specific hybridization with their immobilized counterparts. The validity of the nanogravimetric assay was confirmed by independent fluorescence measurements using 4′,6-diamidino-2-phenylindole (DAPI) and a CCD camera.

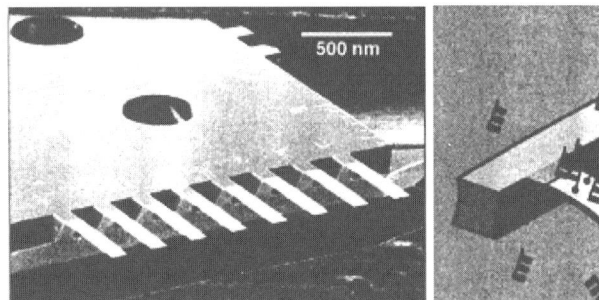

Figure 10.36. Cantilevers coated with single-stranded DNA, which bends upon binding to analytes. The deflection of the cantilever is detected with a laser beam (After Hellemans, 2001, copyright 2001 IBM Zurich)

In a separate but somewhat closely related study, biological samples were screened for the presence of particular genetic sequences using small cantilevers of the type used in atomic force microscopes (Wu G. *et al.*, 2001). The surface of each cantilever was coated with single-stranded DNA capable of binding to one particular target sequence. Binding with the analytes induces a surface stress, which bends the affected cantilever by nanometers (Figure 10.36) – not much, but enough to reveal that the bent cantilevers found their specific targets in a sample.

10.10 Sensor Arrays

Almost all of the chemical and biological sensors described in preceding sections are based on a "lock-and-key" principle, in which the selectivity of the analyte of interest is achieved through precise design of the receptor site. Such approaches are appropriate when a specific analyte is to be detected with controllable backgrounds and interferences. In these cases, an individual highly selective sensor is required for each analyte to be detected and this type of approach is not particularly useful for analyzing, classifying or assigning mixture samples (perfumes, drinks, mixtures of solvents, *etc.*).

The super performance of human olfactory systems (Figure 10.1) in odor detection and identification has stimulated considerable research effort to develop *sensor arrays* (also termed as artificial or electronic noses). Just as the human olfactory system contains many different olfactory receptor genes for simultaneously responding to various analytes, sensor arrays consist of many different sensors within each of them to respond to a number of different analytes (Figure 10.37).

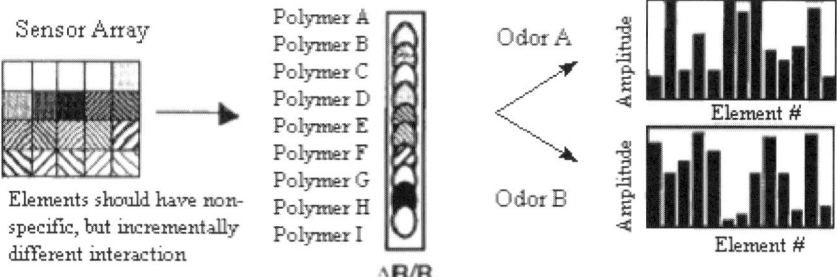

Figure 10.37. Schematic representation of a collection of incrementally different but non-specific sensors, used to generate a complex pattern (or fingerprint) characteristic of a given analyte (After Albert *et al.*, 2000, copyright 2000 The American Chemical Society. Reproduced with permission)

It is not necessary for the elements of such an array to be highly selective toward any given analyte. In order to cover the largest possible varieties of analytes in a mixture system, however, the collection of sensors in a sensor array should contain as much chemical/biological diversity as possible. Identification of a mixed sample by a sensor array is accomplished by a distinct pattern of responses produced over the collection of sensors in the array (Figure 10.37), from which a fingerprint characteristic of the given mixture system is obtained through signal-processing algorithms rather than from the response of a single sensor element. The advantage of this approach is that it can be used to sense complex mixture systems (*e.g.* cheeses, beers) without requiring separation of the mixture into individual constituent components prior to, or during, the analysis. However, this advantage could also be a drawback for the sensor array technology when the precise chemical composition of a complex mixture is required.

10.10.1 Conducting Polymer "Electronic Noses"

The development of conducting polymer sensor arrays, or *electronic noses*, based on substituted polypyrrole for gas and odor sensing was first reported by Persaud and Pelosi at the European Chemoreception Congress held in Lyon in 1984 and in a scientific paper in 1985 (Persaud and Pelosi, 1985). This pioneering work generated a great deal of interest among many researchers worldwide (see, for example, Amrani *et al.*, 1993, Hatfield *et al.*, 1994) and has led to the development of commercial "electronic nose" devices (Hatfield *et al.*, 1994).

The manufacture of polymer electronic noses is compatible with many advanced microfabrication technologies, including photolithography, ink-jet and micro-contact printing (Chapter 6). Figure 10.38 shows a schematic representation of a 32-sensor array formed on a ceramic substrate. As can be seen in Figure 10.38, the active sensing materials (*i.e.* chemoresistors) were formed between pairs of gold electrodes by electropolymerization of substituted heterocyclic monomers, such as pyrrole, aniline or thiophene. The resistance change of the conducting polymer "bridges" is measured as a change in voltage across the sensor at a constant current.

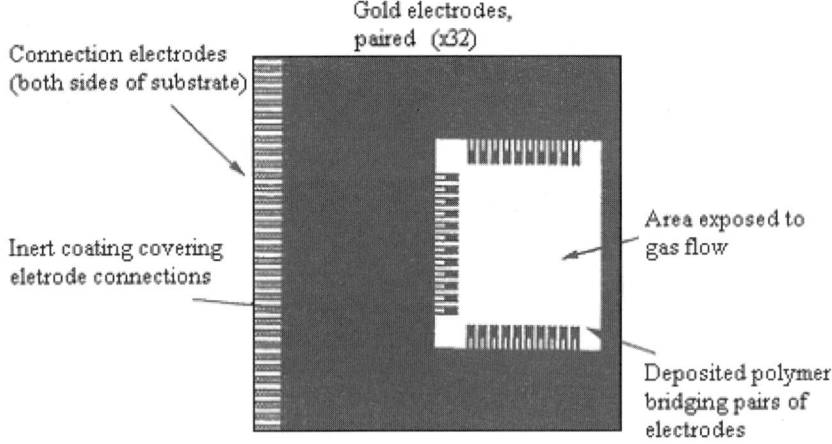

Figure 10.38. Schematic representation of a chip substrate showing the polymer deposited between pairs of electrodes (After Osada and De Rossi, 2000, copyright 2000 Springer-Verlag. Reproduced with permission)

Since the 1984 report on the first electronic nose based on polypyrrole, similar conjugated conducting polymer sensor arrays have been constructed for testing and sensing a wide range of mixture systems, including odor-sensing for gases (Schurmer *et al.*, 1991, Pearce *et al.*,1993), recognition of spoilage bacteria and yeasts in milk (Magan *et al.*, 2001), water quality control (Stuetz *et al.*, 1998, Di Francesco *et al.*, 2001), identification of fruit cultivar (Gelperin *et al.*, 1999), and characterization of wines (Baldacci *et al.*, 1998, Guadarrama *et al.*, 2000), and olive oil (Stella *et al.*, 2000, Aparicio *et al.*, 2000).

To circumvent the instability problem associated with most common conjugated conducting polymers (Chapter 2), conductively-filled conducting polymers that possess an electroactive component for signal transduction and insulating polymers to provide chemical diversity have been used in sensor arrays (Freund *et al.*, 1995, Lonergan *et al.*, 1996). In particular, Lonergan *et al.* (1996) developed an organic vapor sensor array, in which the individual sensor elements consist of carbon-black particles dispersed into insulating polymers. This approach offers advantages that the conductive element is very stable carbon-black, and that the chemical diversity required for the sensor array can be readily obtained by using different organic polymers as the insulating phase for the carbon-black-polymer composites. Sensor arrays were constructed by depositing thin films of such carbon-black-organic polymer composites across interdigitate electrodes with each sensing element containing the same carbon-black conducting phase but with a different organic polymer as the insulating phase. Upon exposure to a vapor mixture, the differing gas-solid partition coefficients for the various polymers of the sensor array produced a pattern of the swelling-induced resistance changes that can be used to classify vapors and vapor mixtures. It was found that this type of sensor array could be used to resolve common organic solvents, including molecules of different classes (*e.g.* aromatics against alcohols) as well as those within a particular class (*e.g.* benzene *vs.* toluene or methanol *vs.* ethanol).

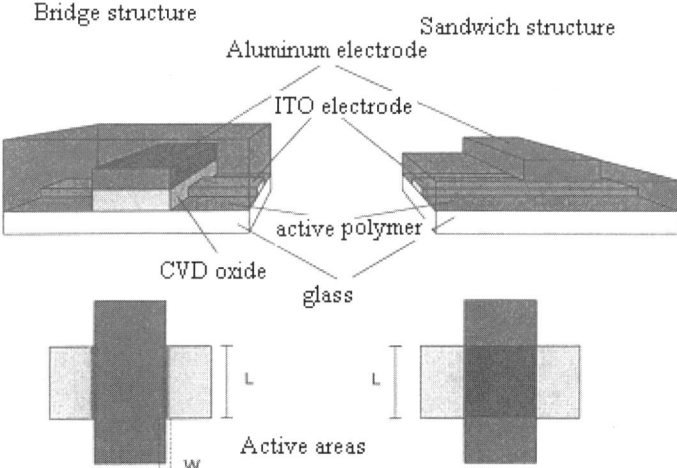

Figure 10.39. A schematic representation of bridge and sandwich structures of the conjugated polymer image sensor arrays (After Pede *et al.*, 1998, copyright 1998 Wiley-VCH Verlag. Reproduced with permission)

More interestingly, conjugated polymer sensor arrays, coupled with a photodiode, have also been used for optical imaging. In this regard, Pede *et al.* (1998) assembled a photodiode array by coating a bridge microstructure matrix with a conjugated polythiophene/C_{60} blend to form an array of pixels between perpendicular electrodes, in which both the upper and lower layers of electrodes were coated with a polymer layer and were separated by an insulating layer at the cross points (Figure 10.39). Then, the bridge devices were used as chemical sensors for detecting the distribution of iodine by using a polymer layer that can be oxidized by iodine. This bridge structure has also been used to drive an array of polymer electroluminescent diodes.

Using semiconducting polymers, Yu *et al.* (1998) have also developed large-area, full-color image sensors. These image sensors have a high photosensitivity, low dark current, large dynamic range, and can be fabricated by a simple coating process on flexible or curved substrates at a low temperature. In practice, full-color image sensing in the visible range has been demonstrated by combining a polymer detector array with a set of optical filters. Photosensors in other spectral ranges, UV or near-IR, can also be achieved through proper selection of the sensing material and/or optical filters.

10.10.2 DNA Arrays

The recent development of miniaturized *DNA arrays*, or *DNA chips*, has led to the revolution in the way in which many biomedical assays and tests are performed in both research and clinical diagnostic laboratories. For instance, DNA arrays used to study gene expression consist of a library of genes immobilized in a grid on a glass slide (Schena *et al.*, 1995). Each of the constituent sites of the grid contains DNA from a single gene that will bind to the messenger RNA (mRNA) produced by the gene concerned. Therefore, it is possible to obtain an instant visual read-out, revealing which genes are from a given tissue by tagging their mRNAs with fluorescent dyes prior to exposing the sample to the DNA chip. As a result, researchers who previously studied the activity of one gene at a time can now analyze the expression of thousands of genes simultaneously, for example, for tracking how patterns of gene expression change in diseases such as cancer (Marx, 2000).

Like the fabrication of electronic noses based on conducting polymers, DNA chips with large numbers of DNA test sites have been developed by photolithographic combinatorial synthesis (Chee *et al.*, 1996, Fodor *et al.*, 1991, 1993) and physical deposition methods (Eggers *et al.*, 1994, Lamture *et al.*, 1994). In the latter case, a robot may be used to dip a quill-like tip into a well containing a purified single-strand DNA chain and then spot down a tiny drop of the DNA onto a glass plate. After the robot washes and dries the tip, the same procedure is repeated with a different single-stranded DNA chain to build the array, with which many parallel DNA hybridization processes occur simultaneously on the immobilized oligonucleotide or other DNA-probing sequences pre-attached to the array surface.

So far, multicolor fluorescence has been used as the primary technology for the simultaneous analysis of multi-site hybridization processes on DNA chips. This detection technique is based on the different affinities of the hybridized fluorescent-labeled DNA strands from their unbound counterparts that can be washed away from the array surface. Fluorescence emission from the hybridized sample at each location on the chip is normally visualized either by confocal scanning microscopy or a CCD camera (Schena and Davis, 1998). Once the fluorescence emission from each of the microarray elements is converted into a digital output by the detection system, superimposing a grid over the microarray image and computing an average intensity value for each microarray element with automated software can obtain final quantitative results.

More recently, the so-called active microelectronic DNA arrays, in which electric fields are used to directly affect the hybridization reactions, have been developed. In a typical experimental set-up, a dc positive bias was applied to the individual microelectrodes beneath the selected test sites, causing rapid electrophoretic transport and concentration of negatively charged nucleic acid molecules at the selected microlocation test sites on the microelectronic array. This allows the immediate immobilization of nucleic acid probes (oligonucleotides, DNA, RNA, polynucleotides, *etc.*) by direct attachment (covalently or non-covalently) to the permeation layer overlaying the microelectrode. The target nucleic acids can then be transported, concentrated and hybridized to the pre-attached nucleic acid probes. Reversal of the electric field to negative bias then causes the rapid removal of the unhybridized DNA molecules. Furthermore, selective dehybridization of the DNA sequences from the attached complementary probe is possible by precise control of the electric field. Therefore, this technique provides a powerful and rapid approach to detect even a single base mismatch in target DNA sequences. These DNA chips consisting of electronically controlled arrays with independently addressable test sites represent one of the most advanced technologies in the array area. In view of the excellent electronic properties, large surface area and tiny probe size, the well-aligned carbon nanotube micro-patterns described in Chapter 6 could add additional advantages to this approach. For more detailed discussion on the fabrication and performance of active microelectronic arrays, interested readers are referred to the specialized book edited by Schena (1999).

10.10.3 Protein Arrays

Just as DNA arrays allow researchers to track the activity of thousands of genes at once, similar arrays of proteins, typically named as *protein arrays* or *proteome chips*, could be used to test the activity of potential drugs against thousands of protein targets simultaneously. While the fabrication of DNA chips has been an active research area for several years, the construction of protein chips is a more recent development. However, protein arrays might have more applications than DNA arrays. This is because it is proteins, not DNA or RNA, that carry out the vast majority of biochemical reactions in cells. Although proteome chips are more difficult to make than DNA chips, as proteins are harder to synthesize than DNA and it is easier for them to lose their activity upon being anchored down on solid surfaces (*i.e.* denaturation), certain functional protein arrays have now been made based on some clever designs (Arenkov *et al.,* 2000, MacBeath and Schreiber, 2000, Zhu *et al.,* 2000, 2001).

In order to stick proteins to the surface without denaturing, for instance, MacBeath and Schreiber (2000) firstly coated a layer of a protein called bovine serum albumin (BSA) on a glass slide to render it water-wettable. They then used a robot originally designed to synthesize DNA arrays to make the protein arrays containing billions of proteins become chemically linked to the BSA layer through chemical reactions between the lysine amino acids on the proteins and lysines in the BSA molecules. By so doing, these authors demonstrated without difficulty that their protein arrays could be used to test the actions of various proteins, as exemplified by the binding of a small, drug-like molecule bearing a fluorescent tag to particular proteins within such an array. These results could help reveal novel drug targets.

Along with MacBeath and Schreiber's work, a number of other versions of protein array technologies have appeared in recent years. To mention but a few examples, Fields and co-workers (Uetz *et al.*, 2000) have devised an array consisting of many tiny "test tubes" containing specially engineered yeast cells in a single plate, which allowed them to test the interactions of 6000 yeast proteins with many of the others all at once. In the meantime, Mirzabekov *et al.* (Arenkov *et al.*, 2000) reported a new technique for creating arrays of proteins immobilized inside tiny gel packets dotted across a surface. However, the study of protein function is only one of the many potential applications of protein array technology. It is also possible to array antibodies binding to specific proteins that could detect which proteins are actually being produced in various tissues, and hence provide clues to what causes various diseases. With so many microfabrication methods already reported and more to be developed, the DNA and protein arrays will surely be only the first few of many biochips to be developed for multitasking biomedical applications.

10.11 References

Adu, C.K.W., Sumanasekera, G.U., Pradhan, B.K., Romero, H.E., Eklund, P.C. (2001) *Chem. Phys. Lett.* **337**, 31.
Akagi, K, Piao, G., Kaneko, S., Sakamaki, K., Shirakawa, H., Kyotani, M. (1998) *Science* **282**, 1683
Albert, K.J., Lewis, N.S., Schauer, C.L., Sotzing, G.A., Stitzel, S.E., Vaid, T.P., Walt, D.R. (2000) *Chem. Rev.* **100**, 2595.
Albert, S.K., Remash, C., Murugesan, N., Gill, T.P.S., Periaswami, G., Kulkarni, S.D. in (1997) *Weld. Res. (Miami)*, **July**, American Welding Society.
Albrecht, M., Schlupp, M., Bargon, J., van Koten, G. (2001) *Chem. Commun.* 1874.
Albrecht, M., Gossage, R.A., van Koten, G., Spek, A.L. (1998) *Chem. Commun.* 1003.
Albrecht, M., Van, K.G. (1999) *Adv. Mater.* **11**, 171.
Alcamo, I.E. (2001) *DNA Technology*. 2nd Edition, Academic Press, California.
Amao, Y., Asai, K., Okura, I. (1999) *Bull. Chem. Soc. Jpn.* **72**, 2223, and references cited therein.
Amrani, M.E.H.; Ibrahim, M.S., Persaud, K.C. (1993) *Mat. Sci. Eng.* **C1**, 17.
Aparicio, R., Rocha, S.M., Delgadillo, I., Morales, M.T. (2000) *J. Agric. Food Chem.* **48**, 853.
Arenkov, P., Kukhtin, A., Gernmell, A., Voloshchuk, S., Chupeeva, V., Mirzabekov, A. (2000) *Anal. Biochem.* **278**, 123.
Baldacci, S., Matsuno, T., Toko, K., Stella, R., De Rossi, D. (1998) *Sens. Mater.* **10**, 185.

Barnard, R.D. (1972) *Thermoelectricity in Metal and Alloys*, John Wiley & Sons, Inc., New York.
Baughman, R.H., Shacklette, L.W. (1991) In *Science and Application of Conducting Polymers,* Salaneck, W.R., Clark, D.T., Samuelsen, E.J. (eds) Adam Hilger, New York.
Berney, H., West, J., Haefele, E., Alderman, J., Lane, W., Collins, J.K. (2000) *Sens. Actuators B* **68**, 100.
Biggins, J.B., Prudent, J.R., Marshall, D.J., Ruppen, M., Thorson, J.S. (2000) *Proc. Natl. Acad. Sci. USA* **97**, 13537.
Borman, S. (2000) *C&E News* **December 11**, 51.
Buttner, W., Maclay, G.J., Street, J.R. (1990) *Sens. Actuators B* **1**, 303.
Chee, M., Yang, R., Hubbell, E., Berno, A., Huang, X.C., Stern, D., Winkler, J., Lockhart, D.J., Morris, M.S., Fodor, S.P.A. (1996) *Science* **274**, 610.
Christensen, W.H., Sinha, D.N., Agnew, S.F. (1993) *Sens. Actuators B* **10**, 149.
Collins G.E., Buckley, L.J. (1996) *Synth. Met.* **78**, 93.
Collins, P.G., Bradley, K., Ishigami, M., Zettl, A. (2000) *Science* **286**, 1801.
Crawford, K.B., Goldfinger, M.B., Swager, T.M. (1998) *J. Am. Chem. Soc.* **120**, 5187.
Currie, J.F., Essalik, A., Marusic, J.C. (1999) *Sens. Actuators B* **59**, 235.
Czarnik, A.W. (1992) *Fluorescent Chemosensors for Ion and Molecule Recognition*, American Chemical Society, Washington, DC.
Dai, H., Hafner, J.H., Rinzler, A.G., Colbert, D.T., Smalley, R.E. (1996) *Nature* **384**, 147.
Dai, L. (1999) In *Semiconducting Polymers: Applications, Properties, and Synthesis*, Hsieh, B.R., Wei, Y. (eds) ACS Symp. Ser., 735, ACS, Washington DC.
Dai, L. (2001) *Aust. J. Chem.* **54**,11.
Dai, L., Mau, A.W.H. (2001) *Adv. Mater.* **13**, 899.
Dermody, D.L., Peez, R.F., Bergbreiter, D.E., Crooks, R.M. (1999) *Langmuir* **15**, 885.
Di Francesco, F., Lazzerini, B., Marcelloni, F., Pioggia, G. (2001) *Atmos. Eniron.* **35**, 1225.
Egashira, M., Shimizu, Y., Eshita, A. (1988) *J. Electrochem. Soc.* **135**, 2546.
Eggers, M., Hogan, M, Reich, R.K., Lamture, J., Ehrlich, D., Hollis, M., Kosicki, B., Powdrill, T., Beattie, K., Smith, S., Varma, R, Gangadharan, R., Mallik, A., Burke, B., Wallace, D. (1994) *Biotechniques* **17**, 516.
Eggins, B.R. (1996) *Biosensors: an Introduction*, John Wiley & Sons and B.G. Teubner Publishers, New York.
El-sherif, M.A., Yuan, J.M., MacDiarmid, A. (2000) *J. Intell. Mater. Sys. Struct.* **11**, 407.
Endres, H.-E., Hartinger, R., Schwaiger, M., Gmelch, G., Roth, M. (1999) *Sens. Actuators B* **57**, 83.
Fabre, B., Burlet, S., Cespuglio, R., Bidan, G. (1997) *J. Electroanal. Chem.* **426**, 75-83.
Favier, F., Walter, E.C., Zach, M.P., Benter, T., Penner, R.M. (2001) *Science* **293**, 2227.
Fodor, S.P.A., Read, J.L., Pirrung, M.C., Stryer, L., Lu, A.T., Solas, D. (1991) *Science* **251**, 767.
Fodor, S.P., Rava, P., Huang, X.C., Pease, A.C., Hotnies, C.P., Adams, C.L. (1993) *Nature* **364**, 555.
Foulkes, E. (1990) *Biological Effects of Heavy Metals*, CRC Press, Boca Raton, Florida.
Frank, S., Poncharal, P., Wang, Z.L., de Heer, W.A. (1998) *Science* **280**, 1744.
Freund, M.S., Lewis, N.S. (1995) *Proc. Natl. Acad. Sci. U.S.A.* **92**, 2652.
Gao, M., Dai, L., Wallace, G. (2003) *Electranaly.* **15**, 1089.
Gao, M., Huang, S., Dai, L., Wallace, G., Gao, R., Wang, Z. (2000b) *Angew. Chem. Int. Ed.* **39**, 3664.
Gao, R., Wang, Z.L., Bai, Z., de Heer, W.A., Dai, L. Gao, M. (2000a) *Phys. Rev. Lett.* **85**, 622.
Garnier, F., Korri-Youssoufi, H., Srivastava, P., Mandrand, B., Delair, T. (1999) *Synth. Met.* **100**, 89.

Gelperin, A., Dawson, J.L., Cazares, S.M., Seung, H.S. (1999) In *Electron. Noses Sens. Array Based Syst., Proc. Int. Symp. Olfaction Electron. Nose,* Hurst, W.J. (ed.), Technomic Publishing Co., Inc., Lancaster.

Gokel, G.W. (1991) *Crown Ethers and Cryptands*, The Royal Society of Chemistry, Cambridge.

Guadarrama, A., Fernandez, J.A., Iniguez, M., Souto, J., de Saja, J.A. (2000) *Anal. Chim Acta* **411**, 193.

Hasebe, Y., Arai, M., Yamauchi, T., Uchiyama, S. (1999) *Chem. Sens.* **15**, 163.

Hashimoto, K., Ito, K., Ishimori, Y. (1998) *Sens. Actuators B* **46**, 220.

Hatfield, J.V., Neaves, P.I., Hicks, P.J., Persaud, K.C., Travers, P. (1994) *Sens. Actuators B* **18**, 221.

Heil, C., Windscheif, G.R., Braschohs, S., Florke, J., Glaser, J., Lopez, M., Muller-Albrecht, J., Schramm, U., Bargon, J., Vogtle, F. (1999) *Sens. Actuators, B* **61**, 51.

Hellemans, A. (2000) *Science* 290, 1529.

Hoa, D.T., Kumar, T.N.S., Punekar, N.S., Srinivasa, R.S., Lal, R., Contractor, A.Q. (1992) *Anal. Chem.* **64**, 2645.

Huang, H., Dasgupta, P.K. (1991) *Anal. Chem.* **63**, 1570.

Hwang, B.J., Yang, J.Y., Lin, C.W. (2001) *Sens. Actuators B* **75**, 67.

Inoue, Y., Gokel, G.W. (1990) *Cation Binding by Macrocycles*, Marcel Dekker Inc., New York.

Jones, T.T., Chyan, O.M., Wrighton, M.S. (1987) *J. Am. Chem. Soc.* **109**, 5526, and references cited therein.

Kim, J., McQuade, D.T., McHugh, S.K., Swager, T.M. (2000) *Angew. Chem. Int. Ed.* **39**, 3868.

Kim, C., Park, E., Song, C.K., Koo, B.W. (2001) *Synth. Met.* 123, 493.

Koda, H., Nakajima, H., Sakurai, Y., Okamoto, A. (2000) *Chem. Sens. (Japanese)* **16** (Suppl. B), 106.

Kolla, P. (1997) *Angew. Chem. Int. Ed.* **36**, 800.

Kong, J., Chapline, M.G., Dai, H. (2001) *Adv. Mater.* **13**, 1384.

Kong, J., Franklin, N.R., Zhou, C., Chapline, M.G., Peng, S., Cho, K., Dai, H. (2000) *Science* **287**, 622.

Kovvali, A.S., Chen, H., Sirkar, K.K. (2000) *J. Am. Chem. Soc.* **122**, 7594.

Kranz, C., Wohlschlaeger, H., Schmidt, H.-L., Schuhmann, W. (1998) *Electroanalysis* **10**, 546.

Lam, M.H.W., Lee, D.Y.K., Man, K.W., Lau, C.S.W. (2000) *J. Mater. Chem.* **10**, 1825.

Lamture, J.B., Beattie K.L., Burke, B.E., Eggers, M.D., Ehrlich, D.J., Fowler, R., Hollis, M.A., Kosicki, B.B., Reich, R.K., Smith, S.R., Varma, R.S., Hogan, M.E. (1994) *Nucleic Acids Res.* **22**, 2121

Lee, M.S., Meyer, J.U. (2000) *Sens. Actuators B* **68**, 293.

Li, C.Z., He, H.X., Bogozi, A., Bunch, J.S., Tao, N.J. (2000) *Appl. Phys. Lett.* **76**, 1333.

Li, J., Lu, Y. (2000) *J. Am. Chem. Soc.* **122**, 10466. and references cited therein.

Lonergan, M.C., Severin, E.J., Doleman, B.J., Beaber, S.A., Grubbs, R.H., Lewis, N.S. (1996) *Chem. Mater.* **8**, 2298.

Lourie, O., Wagner, H.D. (1998) *J. Mater. Res.* **13**, 2418.

MacBeath, G., Schreiber, S.L. (2000) *Science* **289**, 1760.

MacDiarmid, A.G. (1997) *Synth. Met.* **84**, 27.

Magan, N., Pavlou, A., Chrysanthakis, I. (2001) *Sens. Actuators B* **72**, 28.

Mandelis, A., Christofides, C. (1993) *Physics, Chemistry and Technology of Solid State Gas Sensor Devices*, John Wiley & Sons, Inc., New York.

Marliere, C. Poncharal, P., Vaccarini, L., Zahab, A. (2000) *Mater. Res. Soc. Sym. Proc.* **539**, 173.

Marsella, M.J., Swager, T.M. (1993) *J. Am. Chem. Soc.* **115**, 12214.

Marx, J. (2000) *Science* **289**, 1670.

Matsubara, S., Kaneko, S., Morimoto, S., Shimizu, S., Ishihara, T., Takita, Y. (2000) *Sens. Actuators B* **65**, 128.
Matsue, T., Nishizawa, M., Sawaguchi, T., Uchida, I. (1991) *Chem. Commun.* 1029.
Matsue, T., Yamada, H., Chang, H.-C., Uchida, I., Nagata, K., Tomita, K. (1990) *Biochim. Biophys. Acta* **29**, 1038.
Maureen, R.A. (1997) *C&E News* **March 10**, 14.
McQuade, D.T., Hegedus, A.H., Swager, T.M. (2000) *J. Am. Chem. Soc.* **122**, 12389.
Miller, L.L., Kunugi, Y., Canavesi, A., Rigaut, S., Moorefield, C.N., Newkome, G.R. (1998) *Chem. Mater.* **10**, 1751.
Munstedt, H. (1986) *Polymer* **27**, 899.
Nakatsuji, Y., Kita, K., Inoue, H., Zhang, W., Kida, T., Ikeda, I. (2000) *J. Am. Chem.* **122**, 6307.
Needleman, H.L. (1992) *Human Lead Exposure*, CRC Press, Boca Raton, Florida.
Nicolini, C., Erokhin, V., Facci, P., Guerzoni, S., Ross, A., Paschkevitsch, P. (1997) *Biosens. Bioelectron.* **12**, 613.
Nishizawa, M., Matsue, T., Uchida, I. (1992) *Anal. Chem.* **64**, 2642.
Nosaka, T., Sakurai, Y., Natsukawa, K., Yotsuya, T., Nishida, K., Akae, N. (1999) *Adv. Sci. Technol.* (Faenza, Italy), **26** (Solid State Chemical and Biochemical Sensors), 255.
Odom, T.W., Huang, J.-L., Kim, P., Lieber, C.M. (2000) *J. Phys. Chem. B* **104**, 2794, and references cited therein.
Ogura, K., Kokura, M., Nakayama, M. (1995) *J. Electrochem. Soc.* **142**, L152.
Ogura, K., Shiigi, H., Nakayama, M. (1996) *J. Electrochem. Soc.* **143**, 2925.
Ong, K.G., Grimes, C.A. (2000) *Smart Mater. Struct.* **9**, 421.
Ong, K.G., Kichambare, P.D., Grimes, C.A. (2001) *Sensors and Actuators A* **93**, 33.
Osada, Y., De Rossi, D.E., (Eds) (2000) *Polymer Sensors and Actuators*, Springer, Berlin.
Otagawa, T., Madou, M., Wing, S., Rich-Alexander, J., Kusanagi, S., Fujioka, T., Yasuda, A. (1990) *Sens. Actuators B* **1**, 319.
Oubda, T., Zhao, P., Nauer, G.E. (1998) *Polym. News* **23**, 331.
Pearce, T.C., Gardner, J.W., Friel, S., Bartlett, P.N., Blair, N. (1993) *Analyst* **118**, 371.
Pede, D., Smela, E., Johansson, T., Johansson, M., Inganaes, O. (1998) *Adv. Mater.* **10**, 233.
Persaud, K.C., Pelosi, P. (1985) *Trans. Am. Soc. Artif. Organs* 31, 297.
Poncharal, P., Wang, Z.L., Ugarte, D., de Heer, W.A. (1999) *Science* **283**, 1513, and references cited therein.
Prodi, L., Bargossi, C., Montalti, M., Zaccheroni, N., Su, N., Bradshaw, J.S., Izatt, R.M., Savage, P.B. (2000) *J. Am. Chem. Soc.* **122**, 6769.
Qiu, H., Wan, M., Matthews, B., Dai, L. (2001) *Macromolecules* **34**, 675.
Radeva, E., Georgiev, V., Spassov, L., Koprinarov, N., Kanev, St. (1997) *Sens. Actuators B.* **42**, 11.
Rouhi, A.M. (2001) *C&E News* **June 11**, 29, and references cited therein.
Sadaoka, Y., Sakai, Y. (1986) *J. Mater. Sci.* **21**, 235.
Sakai, Y., Sadaoka, Y., Matsuguchi, M., Yokouchi, H., Tamai, K. (1995) *Mater. Chem. & Phys.* **42**, 73.
Sazou, D., Georgolios, C. (1997) *J. Electroanal. Chem.* **429**, 81.
Schena, M. (Ed.) (1999) *DNA Microarrays: A Practical Approach,* Oxford University Press, Oxford.
Schena, M., Davis, R.W. (1998) In *PCR Methods Manual,* Innis, M., Gelfand, D., Sninsky, J. (eds), Academic Press, San Diego.
Schena, M., Shalon, D., Davis, R.W., Brown, P.O. (1995) *Science* **270**, 467.
Schurmer, H.V., Corcoran, P., Gardner, J.W. (1991) *Sens. Actuators* **4**, 29.
Shimizu, Y., Komori, K., Egashira, M. (1989) *J. Electrochem. Soc.* **136**, 2256.
Sophie, D.-C., Pierre-Yves, S., Marc, D. (2000) *Polym. Mater. Sci. Eng.* **83**, 498.
Stella, R., Barisci, N., Serra, G., Wallace, G.G., De Rossi, D. (2000) *Sens. Actuators B* **63**, 1.
Swager, T.M. (1998) *Acc. Chem. Res.* **31**, 201.

Stuetz, R.M., Fenner, R.A., Engin, G. (1998) *Water Res.* **33**, 453.
Sumanasekera, G.U., Adu, C.A.K., Fang, S., Eklund, P.C. (2000) *Phys. Rev. Lett.* **85**, 1096.
Talaie, A. (1997) *Polymer* **38**, 1145.
Tanaka, F., Kawai, T., Kojima, S., Yoshino, K. (1999) *Synth. Met.* **102**, 1358.
Thess, A., Lee, R., Nikolaev, P., Dai, H., Petit, P., Robert, J., Xu, C., Lee, Y.H., Kim, S.G., Rinzler, A.G., Colbert, D.T., Scuseria, G.E., Tomanek, D., Fischer, J.E., Smalley, R.E. (1996) *Science* **273**, 483.
Torsi, L., Pezzuto, M., Siciliano, P., Rella, R., Sabbatini, L., Valli, L., Zambonin, P.G. (1998) *Sens. Actuators* B **48**, 362.
Tsang, S.C., Chen, Y.K., Harris, P.J.F., Green, M.L.H. (1994) *Nature* **372**, 159.
Uetz, P., Giot, L., Cagney, G., Mansfield, T.A., Judson, R.S., Knight, J.R., Lockshon, D., Narayan, V., Srinivasan, M., Pochart, P., Qureshi-Emili, A., Li, Y., Godwin, B., Conover, D., Kalbfleisch, T., Vijayadamodar, G., Yang, M., Johnston, M., Fields, S., Rothberg, J.M. (2000) *Nature* **403**, 623.
Varghese, O.K., Kichambre, P.D., Gong, D., Ong, K.G., Dickey, E.C., Grimes, C.A. (2001) *Sensors and Actuators B* **81**, 32.
Ward, M.D., Buttry, D.A. (1990) *Science* **249**, 4972.
Watson, J.D., Crick, F.H.C. (1953) *Nature* **171**, 737.
Williams, K.A., Eklund, P.C. (2000) *Chem. Phys. Lett.* **320**, 352.
Wong, S.S., Harper, J.D., Lansbury, P.T. Jr., Lieber, C.M. (1998a) *J. Am. Chem. Soc.* **120**, 603.
Wong, S.S., Joselevich, E., Woolley, A.T., Cheung, C., Lieber, C.M. (1998b) *Nature* **394**, 52.
Wong, S.S., Woolley, A.T., Joselevich, E., Cheung, C., Lieber, C.M. (1998c) *J. Am. Chem. Soc.* **120**, 8557
Wood, J.R., Wagner, H.D. (2000) *Appl. Phys. Lett.* **76**, 2883.
Wood, J., Zhao, Q., Frogley, M.D., Meurs, E.R., Prins, A.D., Peijs, T., Dunstan, D.J., Wagner, H.D. (2000) *Phys. Rev. B, Condens. Matter. Mater. Phys.* **62**, 7571.
Wu, G., Datar, R.H., Hansn, K.M., Thundat, T., Cote, R.J., Majumdar, A. (2001) *Nat. Biotechnol.* 19, 856.
Wu, X.Q., Shimizu, Y., Egashira, M. (1989) *J. Electrochem. Soc.* **136**, 2892.
Xin, Y., Hirata, M., Yosomiya, R. (1987) *Proc. Transducers* **87**, 669.
Yan, H., Lu, J.T. (1989) *Sens. Actuators* **19**, 33.
Yang, J.-S., Swager, T.M. (1998a) *J. Am. Chem. Soc.* **120**, 5321.
Yang, J.-S., Swager, T.M. (1998b) *J. Am. Chem. Soc.* **120**, 11864.
Yasuzawa, M., Kunugi, A. (1999) *Electrochem. Commun.* **1**, 459.
Yu, G., Wang, J., McElvain, J., Heeger, A.J. (1998) *Adv. Mater.* **10**, 1431.
Zhu, H., Klemic, J., Chang, S., Bertone, P., Casamayor, A., Klemic, K.G., Smith, D., Gerstein, M., Reed, M.A., Snyder, M. (2000) *Nature Genet.* **26**, 283.
Zhu, H., Bilgin, M., Bangham, R., Hall, D., Casamayor, A., Bertone, P., Lan, N., Jansen, R., Bidlingmaier, S., Houfek, T., Mitchell, T., Miller, P., Dean, R.A., Gerstein, M., Snyder, M. (2001) *Science* **293**, 2101.

Chapter 11

Actuators and Nanomechanical Devices

11.1 Introduction

The human muscle plays an important role in grasping and manipulating objects in our daily life. Grasping and manipulating objects of a small size down to the nanometer scale, however, has been a big challenge. Piezo-electric materials have long been known to be able to generate mechanical force when an electrical voltage is applied, or *vice versa*. As seen in Chapter 3, macromolecules can also respond to an external stimulus, whether it is an optical, electrical, mechanical or environmental stimulation. These intelligent macromolecules could provide useful actuations by cleverly designing and constructing them into appropriate working devices.

Based on the electrochemical doping and de-doping of the conducting polymers (Chapter 2), simple cantilever actuators for the direct conversion of electrical energy to mechanical energy have been developed (Baughman *et al.*, 1996). By analogy, smart actuation devices have also been made with some biomacromolecules, such as DNA and proteins. Carbon nanotubes are the most recent additions to the growing list of electroactive actuator materials. With the rapid development in nanoscience and nanotechnology, various actuation devices at the nanometer scale have been devised from carbon nanotubes since Baughman and co-workers reported the first carbon nanotube actuator in 1999 (Baughman *et al.*, 1999). Furthermore, various robotic systems based on fullerene C_{60} and carbon nanotubes have also been reported (Bar-Cohen, 2001). Owing to their unusual shape and size, these carbon nanoactuators and robotic systems are able to manipulate nano-size objects in a controlled, reproducible and reversible manner.

In this chapter, the experimental results from actuators and smart robotic devices, based on conducting polymers, biomacromolecules, fullerenes and carbon nanotubes, are summarized, along with some suggestions for the direction of future research.

11.2 Conducting Polymer Actuators

Based on the large dimensional changes associated with the electrochemical doping and de-doping processes, a variety of conjugated conducting polymers, including polypyrrole, polyaniline, polyalkylthiophene and polyarylvinylene, have been used to make the so-called electromechanical or electrochemomechanical actuators (also called artificial muscles). The word "electrochemomechanical" refers to an electrochemical reaction (*e.g.* redox or doping/de-doping of conducting polymers) that leads to a mechanical response (*e.g.* bending).

In conducting polymer actuators, the dimensional changes are normally caused by the volume changes arising from the inclusion or elimination of dopant counter ions (*i.e.* anions or cations), together with the associated solvating species. Reversible electromechanical actuators must consist of at least three elements, an anode, a cathod, and a separating electrolyte. Conducting polymers can be used either as the anode, cathode, or both.

In 1987, Baughman and co-workers (Baughman *et al.*, 1990) demonstrated the first so-called unimorph actuator based on conjugated conducting polymers by sputter-coating a thin layer of gold on one side of a film strip of a copolymer of 3-methylthiophene and 3-*n*-octylthiophene. Then, these authors immersed the polymer cathode and spatially separated a lithium counter electrode in an $LiClO_4$/propylene carbonate electrolyte for electrochemical doping of the cathode with ClO_4^-. This caused an expansion of the conducting polymer film relative to the gold coating, leading to the bending of the actuator strip so that the gold-coated surface became concave. Electrochemical de-doping was found to reverse the bending process.

Similarly, Otero *et al.* (1993) reported a conducting polymer bilayer actuator by laminating a freestanding polypyrrole (Ppy) film onto one side of a commercially available, double-sided adhesive tape or other flexible thin film with a Pt wire for electrical contact. The relative expansion or contraction of the polymer film, due to electrochemical or chemical redox reactions, with respect to the other layers leads to bending of the layered structure.

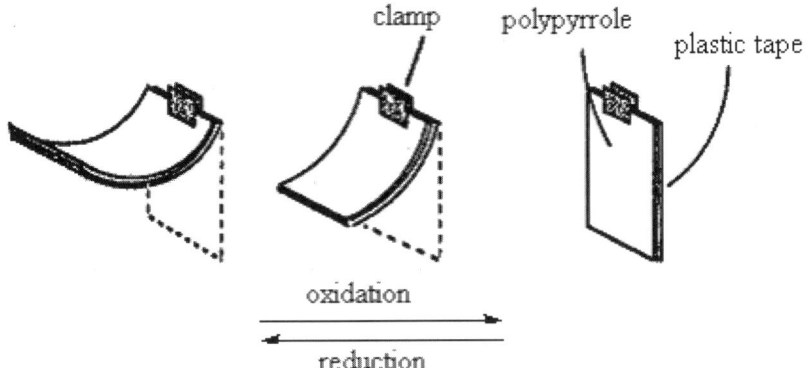

Figure 11.1. Schematic representation of artificial muscles based on conducting polymers (After Otero *et al.*, 1993, copyright 1993 Elsevier. Reproduced with permission)

The position of the free end can be reversibly bended up to 90° by applying various potentials on the bilayer actuator, which is used as the working electrode in a electrochemical cell with Pt counter and SCE reference electrodes and an $LiClO_4$ electrolyte. More interestingly, the actuator can be held at any specific physical position upon removal of the applied potential with a very predictable correlation between the value of applied potential and the angle of bending.

Similar unimorph conjugated conducting polymer actuators with special features have also been reported by others. For example, Pei and Inganäs *et al.* (1992, 1993) demonstrated that unimorph conducting polymer actuators made from either polypyrrole/Au/polyethylene, poly(3-octylthiophene)/polypyrrole or poly(3-octylthiophene)/polyethylene sandwiches can be used to convert chemical energy directly into mechanical energy. Using photoresist technology, Inganäs and co-workers (Smela *et al.*, 1993) have also down-sized their unimorph actuators to as small as 10 × 20 µm and have fabricated simultaneously thousands of such microactuators on a single silicon wafer by microlithographic techniques. Later, these authors (Smela *et al.*, 1995) constructed conducting polymer unimorph hinges from a polypyrrole layer bound to a layer of gold to rotate paddles in solution (Figure 11.2(a)). It was demonstrated that a conducting polymer unimorph hinge of dimension *ca.* 90 × 90 µm thus prepared can rotate paddles as large as 900 × 900µm, equivalent to move a 30 × 30m plate in water by a human arm. Furthermore, these microscopic hinged conducting polymer actuator structures can be reversibly assembled and disassembled from a flat array into a cube (Figure 11.2(b)).

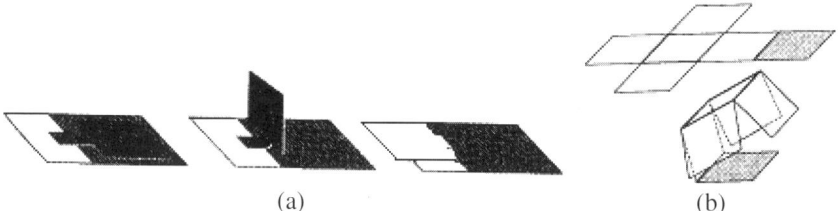

(a)   (b)

Figure 11.2. (a) The actuation paddles consist of rigid square plates with surface dimensions of 90 µm on bilayer hinges of 30 µm by 30 µm. From left to right: paddles lie flat at 0°; paddles are perpendicular to the surface around 90°; paddles lie flat at 180°. The position of the paddles is controlled by the applied voltage; (b) schematic representation of actuators with three consecutive rigid elements forming the closed cubic box (After Smela *et al.*, 1995, copyright 1995 American Association for the Advancement of Science)

It is worth pointing out that these unimorph or bilayer conducting polymer actuators can operate in aqueous media only, and not in organic solvents including acetonitrile (in which polypyrrole was synthesized) and propylene carbonate. This may be attributed to an increased ion solvation, greater affinity of the PPy for the ions, and more effective swelling of PPy in aqueous solutions, coupled with possible retention of organic solvents in the reduced or de-doped PPy films.

Nevertheless, suitable encapsulation (Baughman, 1996; Madden *et al.*, 2000) and the use of a solid polymeric electrolyte (Otero *et al.*, 1997) have made the bilayer actuators operatable in air. The working principle for these bimorph actuators is given in Figure 11.3 (Baughman, 1996).

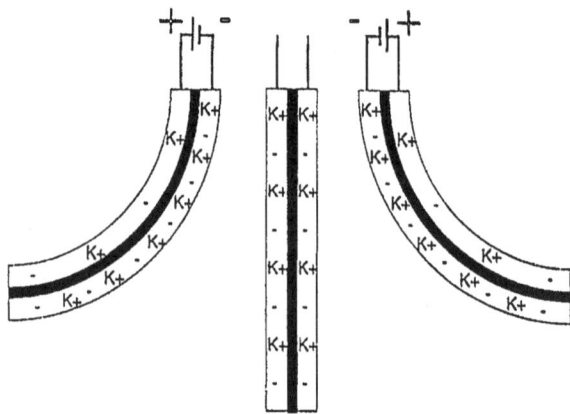

Figure 11.3. Schematic representation of three states during the electromechanical cycle of a bimorph actuator. Both electrodes have the same concentration of dopant (K) when the cantilever is undistorted, and electrochemical transfer of dopant between electrodes causes bending either to the right or to the left (After Baughman, 1996, copyright 1996 Elsevier. Reproduced with permission)

A working device is shown in Figure 11.4, in which a thin layer of gel electrolyte separates two gold-supported polypyrrole strips. Upon applying an external voltage on the two gold electrodes, one polymer layer expands while the other contracts due to the differential in diffusion of the counter ions into the two electrochemically doped polypyrrole layers.

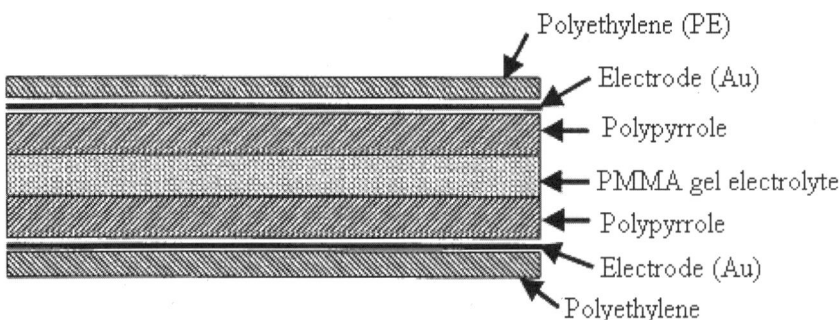

Figure 11.4. Cross-section view of an encapsulated bilayer actuator of 40 mm length and 10 mm width. The thickness of the PMMA gel, Au, polypyrrole and PE are 100, 0.1, 40 and 20 μm, respectively. The actuation performance of the encapsulated bilayer actuator is shown in Figure 11.5 (After Madden *et al.*, 2000, copyright 2000 Elsevier. Reproduced with permission)

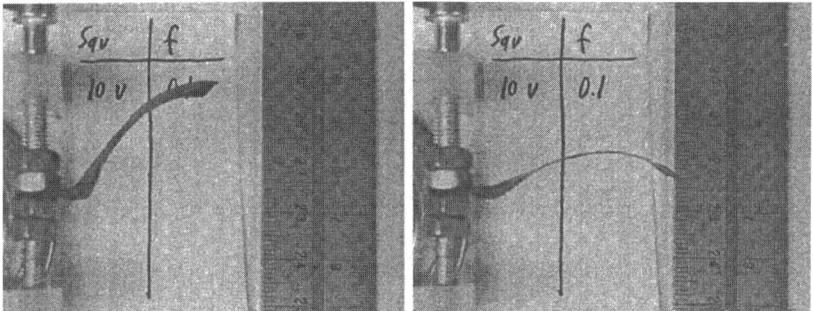

Figure 11.5. Photographs showing the maximum extent of deflection of the encapsulated bilayer actuator upon activation with a ±5 V square wave at 0.1 Hz in air (After Madden *et al.*, 2000, copyright 2000 Elsevier. Reproduced with permission)

Various actuators based on conjugated conducting polymers with a wide range of novel features have been reported (Baughman, 1996). In what follows, some cleverly designed conjugated conducting polymer actuators with interesting features will be spotlighted. As no attempt has been made to provide a comprehensive literature survey, the examples presented are not exhaustive.

11.2.1 Self-powered Actuators

In 1996, Baughman reported a conducting polymer actuator that can power itself through many mechanical work cycles (Baughman, 1996). As schematically shown in Figure 11.6, this self-powered actuator consists of three film-strip electrodes (they can be made from the same conjugated conducting polymer) that are separated by two solid-state electrolyte layers. The central electrode B can have a higher initial dopant concentration than either electrode A or electrode C.

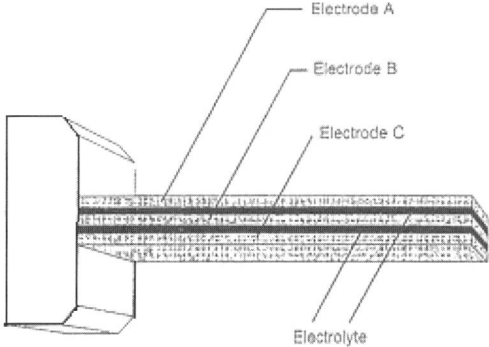

Figure 11.6. Schematic representation of a self-powered conducting polymer actuator that is capable of exhibiting reversible mechanical deformations (After Baughman, 1996, copyright 1996 Elsevier. Reproduced with permission)

To start with, the author connected one of the outer electrodes (*e.g.* A) and the central electrode to partially charge the electrode A. The electrochemical transfer of dopant species between the electrode B and A caused the bending of the cantilever towards the electrode C. Disconnecting the A and B electrodes and electrically connecting the B and C electrodes can reverse this deformation. The electrical connections could be achieved using a light beam if the two electrolyte layers contain a percolated photoconductor (Baughman, 1996).

11.2.2 Conducting Polymer Microtweezers

Another good example of bimorph actuators reported also by Baughman and co-workers (Baughman and Shacklette, 1991) is a conducting polymer microtweezer that could be reversibly opened and closed (Figure 11.7).

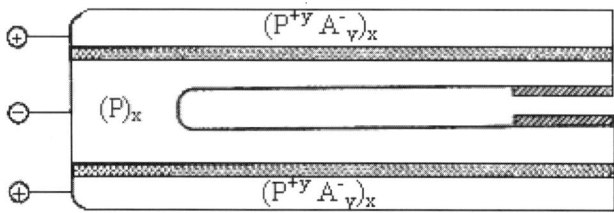

Figure 11.7. Schematic representation of paired bimorph conducting polymer actuators as a microtweezer (After Baughman and Shacklette, 1991, copyright 1991 Adam Hilger)

As shown in Figure 11.7, the microtweezer consists of paired bimorph conducting polymer (*e.g.* polypyrrole) actuators, in which each of the bimorph conducting polymer layers are separated by a layer of either liquid or solid electrolyte. Electrochemical transfer of dopant from the outer layer to the inner layer of each bimorph causes the opening of the tweezers. By reversing the biases of the applied voltage, the tweezers close. A few volts of the applied voltage are sufficient for operating the conducting polymer microtweezer whereas over 100 V is required to achieve the same opening/closing dimensions (*ca.* 200 µm) for a piezo-electric microtweezer.

Recently, Jager *et al.* (2000) have developed microactuators based on polypyrrole-gold bilayers, which can be used in an aqueous environment to pick up, lift, move and place micrometer-sized objects over a distance of 200 to 250 µm. These microactuators consist of three individually addressable actuation components made from polypyrrole/Au bilayers to act as an elbow, a wrist and a hand with fingers, respectively. To demonstrate the actuation performance, these authors first moved the arm with opened fingers on to the glass bead (Figure 11.8(A)). They then closed the fingers to grab the glass bead, lifted it from the surface, and moved the arm completely perpendicular to the surface (shown in gray in Figure 11.8(B)). After having positioned the hand at the base of the robot arm (Figure 11.8(C)), they finally opened the fingers to release the bead and retracted the arm completely (Figure 11.8(D)).

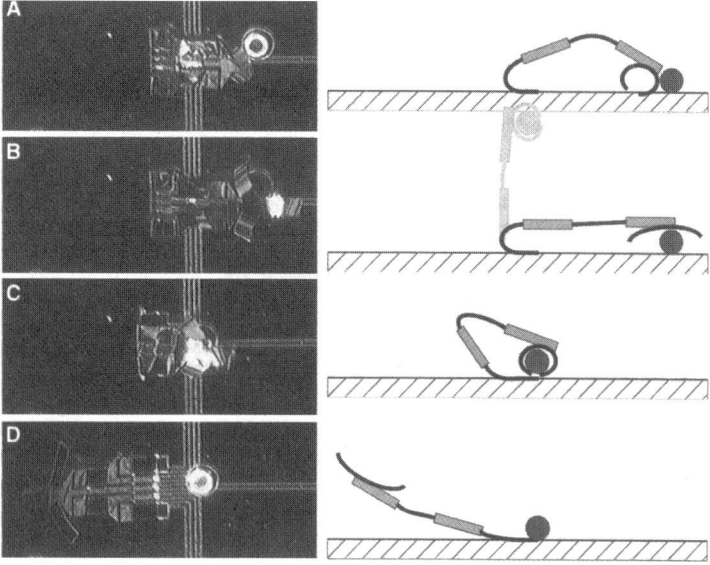

Figure 11.8. A sequence of pictures (*left*) showing the grabbing and lifting of a 100-μm glass bead and schematic drawings of the motion (*right*) (After Jager *et al.*, 2000, copyright 2000 American Association for the Advancement of Science)

The ability to work in an aqueous environment and their ability to be scaled down to micrometer size makes these microactuators based on the polypyrrole-gold bilayers ideal, for example, for single-cell manipulation and characterization.

Based on the same principle, these authors have more recently developed conducting polymer microactuators for three-dimensional movement control (Jager *et al.,* 2001). In particular, they used a sequence of individual actuators in a cascading manner so that the movement of the first actuator moves the base of the second actuator. As a result, three-dimensional movements can be readily achieved.

Actuators based on conjugated conducting polymers are capable of producing at least 10 times more force per unit cross-sectional area than skeletal muscle, with strains typically between 1% and 10%. The high force-generation capabilities, high work densities per cycle (*i.e.* the integral of stress with respect to strain over a cycle), and low operation voltages make conducting polymer actuators very attractive for use in robotics, prosthetics and microelectrochemomechanical systems (MEMS), and many others, in place of conventional electrostatic and piezo-electric actuators. Before large-scale commercial applications can be realized, however, some of the major problems associated with conducting polymer actuators must be solved, including the limitations of the cycle life, energy conversion efficiency, and their relatively slow speed.

11.3 Actuators Based on Composites of Ion-exchange Polymers and Metals

Ion-exchange polymer-metal composites (IPMC) are highly active actuation materials (Shahinpoor et al., 1999; Tadokoro et al., 2000; Oguro et al., 2000). They show a large deformation in response to a low applied voltage. IPMC actuators operate best in a humid environment, but can also be made as self-contained, encapsulated actuators to operate in dry environments. A typical IPMC is a perfluorinated polymeric ion-exchange membrane (IEM) chemically composited with a noble metal (e.g. Au or Pt), as shown in Equation (11.1).

$$[-(CF_2-CF_2)_n-(CF-CF_2)_m-] \\ \quad\quad\quad\quad\quad | \\ \quad\quad\quad\quad O-CF-CF_2-O-CF_2-O-CF_2-SO_3^-....M^+ \\ \quad\quad\quad\quad | \\ \quad\quad\quad\quad CF_3$$

(11.1)

Most ion-exchange polymeric membranes are hydrophilic and capable of adsorbing large amounts of polar solvents (e.g. H_2O). Pt ions dispersed throughout the hydrophilic regions of the polymer can be reduced into the corresponding metal atoms, leading to the formation of electrodes with a dendritic structure. The application of a voltage on two electrodes in close proximity to the membrane walls could cause the transport of ions within a solution through the membrane. Due to the ionic nature in the microscopic structure, IPMC also has the ability to shift its mobile ions of the same charge polarity within the membrane under an electric field. Thus bending can occur due to differential contraction and expansion of the outer most remote regions of a strip if an electric field is imposed across its thickness as shown in Figure 11.9.

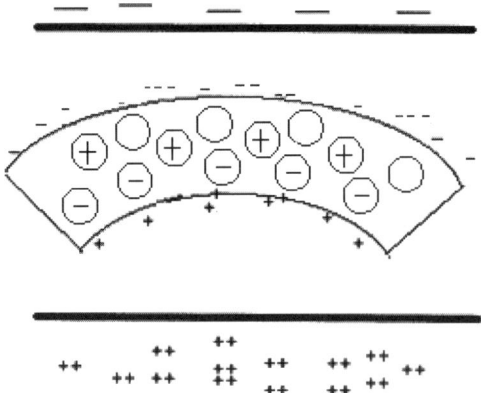

Figure 11.9. Schematic representation of the charge redistribution in an ionic polymer membrane under an electric field (After Shahinpoor et al., 1999, copyright 1999 The American Chemical Society. Reproduced with permission)

Based on the above principle, Shahinpoor and co-workers (Shahinpoor *et al.*, 1999; Tadokoro *et al.*, 2000; Oguro *et al.*, 2000) constructed an IPMC gripper that acts as a miniature low-mass robotic arm (Figure 11.10). The IPMC fingers are connected back-to-back via the electrical wiring, which allows the fingers to bend either inward or outward, and hence close or open the gripper, upon electrical activation.

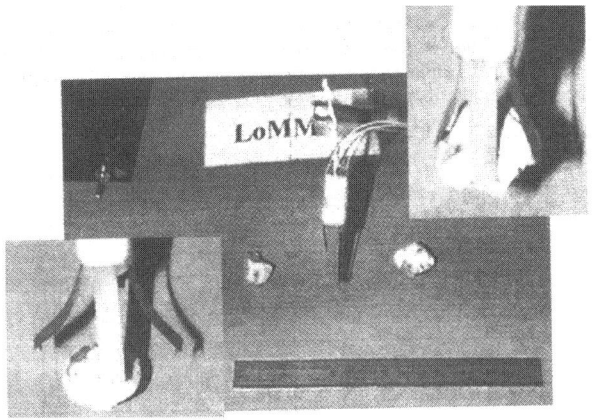

Figure 11.10. Photograph of an IPMC gripper lifting a 10.3 g rock under 5 V, 25 mW activation using four 0.1 g fingers made of perfluorinated IPMC (After Shahinpoor *et al.*, 1999, copyright 1999 The American Chemical Society. Reproduced with permission)

By constructing IPMC strips into the configuration shown in Figure 11.11, both linear- and platform-type actuators can be made for dynamic operation.

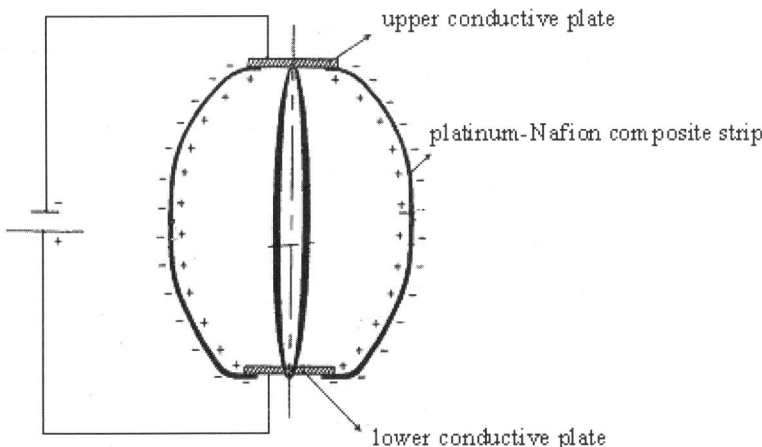

Figure 11.11. A typical linear-type robotic actuator made with IPMC legs (After Shahinpoor *et al.*, 1999, copyright 1999 The American Chemical Society. Reproduced with permission)

The dynamic deformation characteristics associated with the IPMC actuators allow the construction of various smart devices, such as noiseless swimming robotic structures (Figure 11.12(a)) and wing-flapping flying machines (Figure 11.12(b)).

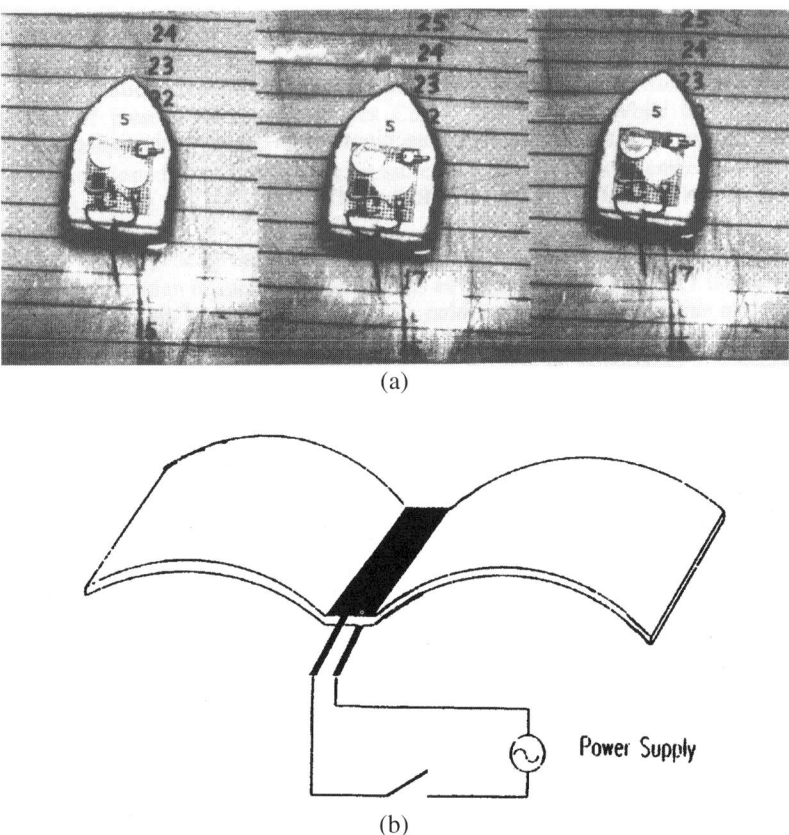

(a)

(b)

Figure 11.12. (a) Photograph of robotic swimmer with muscle undulation frequency of 3 Hz; (b) schematic representation of wing-flapping flying machines (After Shahinpoor *et al.*, 1999, copyright 1999 The American Chemical Society. Reproduced with permission)

Conversely, it is also possible to shift the mobile charges by bending the IPMC membrane. In this case, the mobile ions will shift toward the favored region where opposite charges are available. The deficit in one charge and excess in the other can be translated into a voltage gradient that can be easily sensed by a low-power amplifier and meter. This leads to a linear relationship between the voltage output and imposed quasi-static displacement of the tip of the IPMC strip, as shown in Figure 11.13. Therefore, IPMC could also be used for sensing active materials (Chapter 10).

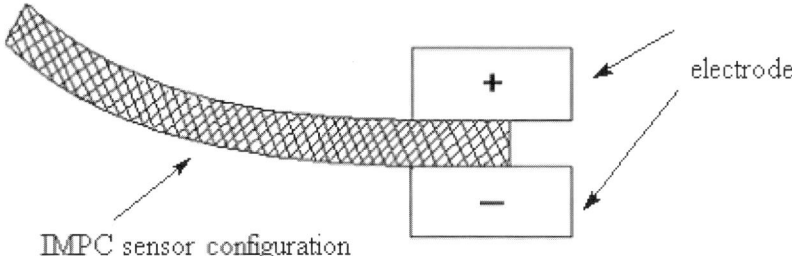

Figure 11.13. A simple IPMC sensor placed between two electrodes (After Shahinpoor *et al.*, 1999, copyright 1999 The American Chemical Society)

11.4 Responsive Polymer Actuators

For all of the actuators described above, the actuation effect results from a balanced expansion and contraction process. The expansion is normally caused by the incorporation of dopant counter ions via electrochemical doping or solvating species into the conducting polymer (IPMC) matrix under an electrical field. On the other hand, the expulsion of dopant counterions by de-doping or solvating species from IPMC by reversing the bias of the electrical field often leads to the contraction process. Consequently, the actuation processes described above are due to bulk phenomena, rather than being an intrinsic property of the individual polymer chains themselves (Marsella and Reid, 1999). The doping and de-doping of a single, isolated conjugated conducting polymer chain would not in itself be expected to elicit a significant muscle-like expansion or contraction. However, a large variety of intelligent polymers have been shown to demonstrate dramatic changes in their molecular conformations in response to environmental stimuli, including temperature, pH, light, ionic strength, as well as magnetic and electric fields (Chapter 3). Poly(*N*-isopropylacrylamide), NIPAAm, which effectively changes its molecular conformation in response to temperature changes, has been widely studied as a prototype polymer for other stimuli-responsive polymers (Rao *et al.*, 2000). Polymers of acrylate derivatives having an aliphatic tertiary amino group have been shown to be responsive to the change in pH. These findings allow the development of various actuators from responsive polymers.

Indeed, small particles of cross-linked poly(*N*-isopropylacrylamide) hydrogels and copolymers with basic or acidic groups have been simply filled into a cylinder to act as polymer actuators for flow control (Arndt *et al.*, 2000). Owing to the responsive nature of the filled hydrogel particles, the flow rate and the pressure drop for the flow through the cylinder depend on temperature, pH and the solvent properties. A photo-cross-linkable responsive polymer coating could offer many possibilities in the microfabrication of micro/nanodevices.

Using photopolymerization techniques for the preparation of asymmetric stimuli-responsive polymer bilayer structures, Li F.-M. *et al.* (1999) studied the thermo- and pH-responsive behavior of poly(*N,N*-dimethylaminoethyl methacrylate), poly(DMAEMA), and hydrogels based on the copolymers of DMAEMA and butyl methacrylate (BMA). As the degree of protonation of the tertiary amino group of DMAEMA changes with changing pH value, the highly protonated tertiary amino group in acidic medium will significantly increase the polymer hydrophilicity. While the pH-sensitive behavior is due to the existence of the tertiary amino group, the thermo-responsive behavior arises from the balance of hydrophilicity and hydrophobicity. Figure 11.14 shows a schematic representation and photographs of the thermo- and pH-responsive changes of an asymmetric bilayer sheet made from DMAEMA-BMA copolymer.

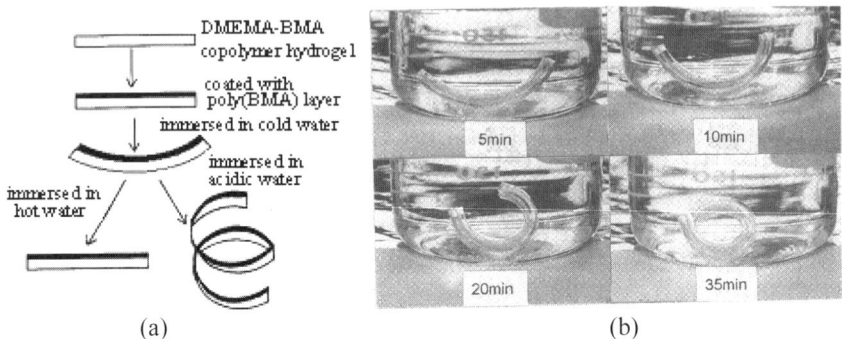

Figure 11.14. (a) Schematic illustration of thermo- and pH-responsive changes of an asymmetric bilayer sheet of DMAEMA-BMA copolymer (top layer: DMAEMA/BMA = ¼; bottom layer: DMAEMA/BMA = 1/1, by molar ratio); (b) photographs showing the time dependence of the shape change for the asymmetric DMAEMA-BMA bilayer sheet in deionized water of pH 5.0 (After Li F.-M. *et al.*, 1999, copyright 1999 The American Chemical Society. Reproduced with permission)

As can be seen in Figure 11.14(b), the asymmetric DMAEMA-BMA bilayer sheet becomes curved upon immersion into the deionized water. Since the outer layer is more hydrophilic than the inner layer, the sheet bends gradually until a circle is formed. The response dynamic depends on the compositions of the two layers and the pH value of the medium. Reversible thermo and pH responses have also been observed.

In a separate, but somewhat related, experiment, Hu *et al.* (1995) prepared bigel strips of the temperature-responsive (N-isopropylacrylamide), NIPA, gel and acetone concentration-responsive polyacrylamide, PAAM, gel. They then constructed three actuator structures based on the modulated gels: a bigel strip, a shape memory gel and a gel "hand". Like Figure 11.14, the PAAM/PAAM-NIPA interpenetrating network, IPN, bigel bends almost to a circle in response to a temperature increase or an increase in acetone concentration (Figure 11.15).

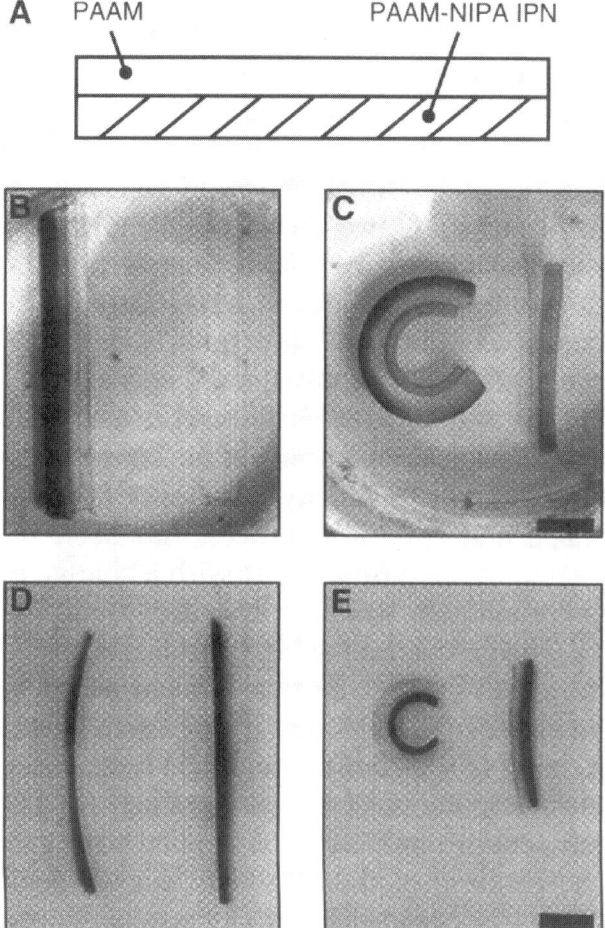

Figure 11.15. (A) Schematic representation of the bigel strip with one PAAM gel (empty area) modulated by the interpenetration of one side of the PAAM gel with the NIPA gel to form a PAAM-NIPA interpenetrating polymer network (shaded area); (B and C) photographs showing the bending of the bigel in water at a temperature of (B) 30.0 °C and (C) 37.8 °C; (D and E) photographs showing the bending of the bigel in a water-acetone mixture at an acetone concentration of (D) 20% and (E) 45% by weight. The gels at the right of pictures (B and C) and (D and E) are a pure NIPA and pure PAAM gel, respectively, as a reference. Scale bar is 5 mm (After Hu *et al.*, 1995, copyright 1995 American Association for the Advancement of Science)

The driving force for the above actuation processes lies in the fact that the volume of the ionic NIPA gel, containing 8 mM sodium acrylate in this particular case, shrinks drastically at temperatures higher than 37 °C, while the volume of the PAAM gel does not. However, the volume of the PAAM gel shrinks much more in acetone-water mixtures (> 34% by weight) than does the NIPA gel.

These authors went on further to demonstrate a shape memory gel by patternwise modulating the PAAM gel with the NIPA gel at four locations so that the bigel can bend only at these modulated sites as the temperature is increased (Figure 11.16).

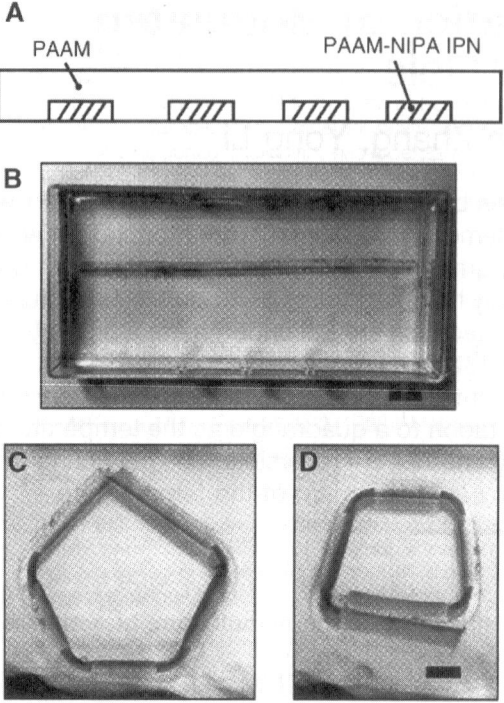

Figure 11.16. (A) Schematic representation of the shape memory gel produced by modulated interpenetration of one side of the PAAM gel with the NIPA gels at four locations (shaded areas). The separation distance between adjacent NIPA gel pieces is about 12 mm; (B through D) photographs showing the gel shape at various temperatures: (B) 22 °C, (C) 39 °C and (D) 41°C. Scale bar is 15 mm in (B) and 6 mm in (D); the length scales in (C) and (D) are the same (After Hu *et al.*, 1995, copyright 1995 American Association for the Advancement of Science)

As seen in Figure 11.16, the gel bends at the modulated sites to form a pentagon (39 °C, Figure 11.16(C)) and then a quadrangle (41 °C, Figure 11.16(D)) as the temperature increases. The change from the pentagon to the quadrangle is due to the higher shrinkage ratio of NIPA at 41 °C than at 39 °C. Consequently, the bigel bends more up to about 90° (*i.e.* quadrangle) at 41 °C than at 39 °C (about 70° bending to form a pentagon). It is interesting to find that the transitions between different shapes are reversible. Furthermore, a variety of other transitions, such as from a straight line to a sinusoidal form, from a long strip to a spiral, and from a flat sheet to a hollow tube, are readily achievable at various temperatures by proper design and control of the modulation patterns.

More interestingly, the same authors further demonstrated a gel "hand" consisting of two bigel strips having the same structure as that shown in Figure 11.15(A). By immersing the gel "hand" in water at room temperature (Figure 11.17(A)), the "fingers" of the gel "hand" are open and an object is released. By increasing the temperature to 35 °C (Figure 11.17(B)), the "fingers" are closed to grasp the object.

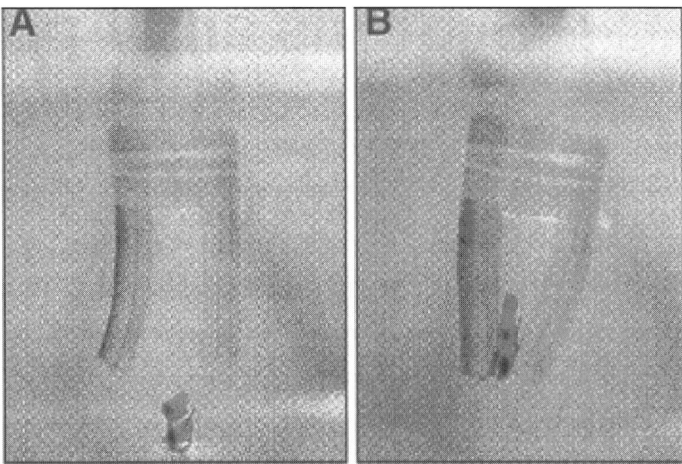

Figure 11.17. Photographs showing a gel "hand" made from two bigel strips that have the same structure as that shown in Figure 11.15(A) and are tied together at one end with a spacer. The NIPA-modulated sides are facing each other. (A) At 22 °C, the "fingers" are open and an object is released; (B) At 35 °C, the "fingers" are closed and the object is caught (After Hu *et al.*, 1995, copyright 1995 American Association for the Advancement of Science)

The availability of a large number of stimuli-responsive polymers (Chapter 3), together with various modification and modulation methods (Samra *et al.*, 2000), implies there is considerable room for making a wide variety of actuators based on functional polymers.

11.5 Carbon Nanotube Actuators

As early as in 1996, Baughman (1996) demonstrated the inherent limitations associated with doping-induced actuation in non-covalent directions for conducting polymers and predicated that actuation in covalently bonded directions would provide superior performance in terms of cycle life, work-density-per-cycle, and operating temperature range. Furthermore, Baughman proposed that non-Faradaic electrochemical charging of electrodes with large surface areas would produce usable dimensional changes for actuation in covalently bonded directions. The

476 Intelligent Macromolecules for Smart Devices

availability of materials that combined both high surface area and a covalently bonded network was realized with the ready availability of single-wall nanotubes.

Carbon nanotubes are the most recent additions to the growing list of actuator materials, since Baughman's group reported the first carbon nanotube actuator based on double-layer (non-Faradaic) charge injection in 1999 (Baughman *et al.* 1999). Recent reports have described several different actuation mechanisms for carbon nanotubes, including actuation by double-layer charge injection (Baughman *et al.*, 1999), electrostatic actuation (Kim and Lieber, 1999), and photo-thermal actuation (Zhang and Iijima, 1999). Double-layer charge injection is a particularly promising mechanism for nanotube actuation, wherein supercapacitor charging results in relatively large changes in covalent bond lengths. The achievable actuator strain, coupled with the spectacular mechanical properties of carbon nanotubes, means that these materials potentially offer a level of actuator performance exceeding that of all other actuator materials in terms of work-per-cycle and stress generation.

To construct the first carbon nanotube actuator, Baughman and co-workers (Baughman, *et al.*, 1999) used SWNT mats (buckypapers) prepared according to the process developed by Smalley and co-workers (Wildoer *et al.*, 1998). As is the case for constructing conducting polymer actuators, Baughman *et al.* stuck a strip of the buckypaper on each side of a piece of double-stick tape and immersed it in salt-water electrolytes. By applying a DC voltage to the SWNT electrodes, the bimorph structure was found to bend. Reversing the polarity caused the bimorph to bend in the opposite direction (Figure 11.18). The deflection is caused by an unequal expansion of the charge-carrying strips. Although both of the buckypapers elongate, one of the stretches expands more than the other and causes curling around its shorter partner. Tensile measurements on unsupported bucky paper provided axial strains of up to 0.2% while switching the electrochemical potential between +0.5 V and –1.0 V (*vs.* SCE).

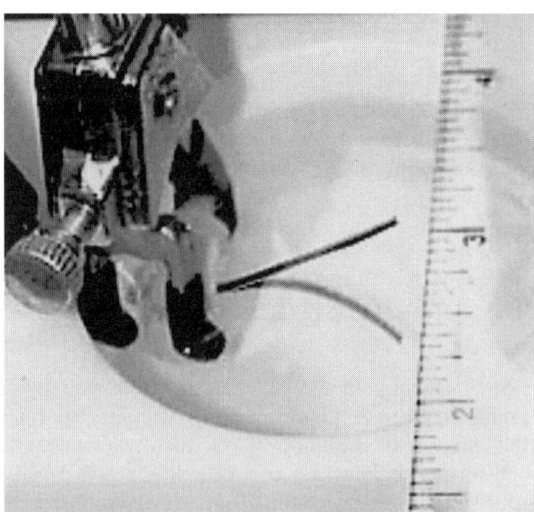

Figure 11.18. Time-lapse photograph showing bending of a SWNT bimorph structure upon the application of +/–1.5 V   (After Spinks *et al.*, 2001, copyright 2001 SPIE Press. Reproduced with permission)

Figure 11.19 shows that the actuator strain is approximately parabolic, with a minimum at around +0.5 V. Some hysteresis is observed, but this largely disappears when the actuator strain is plotted against the charge stored per carbon atom, rather than electrochemical potential applied. The maximum strain-voltage coefficient of 0.11%/V occurs between –0.2 V and –0.8 V (*vs.* SCE). This value is close to that predicted (0.096%/V) from the dimensional changes for the in-plane direction of graphite (Baughman *et al.*, 1996). The actuator strain sharply decreases with increasing scan rate, although actuator response could be observed at > 1 kHz (Baughman *et al.*, 1999).

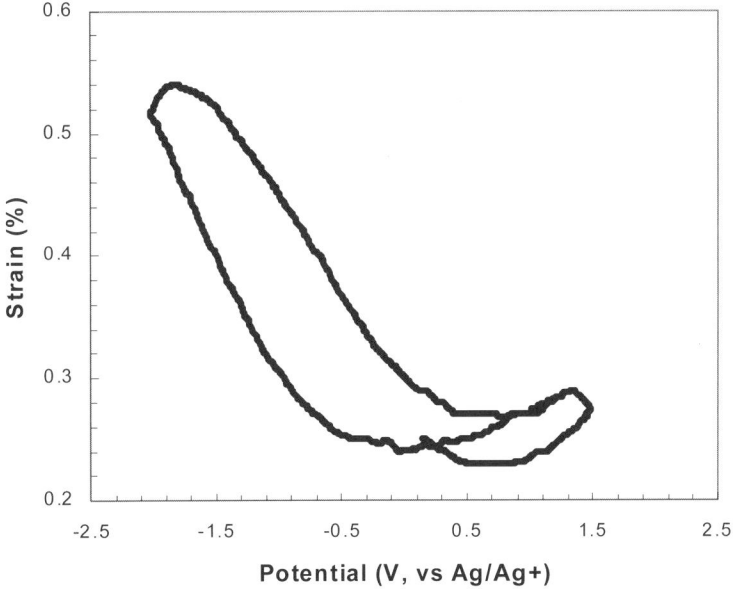

Figure 11.19. Actuator strain of carbon NT mat as a function of applied electrochemical potential (*vs.* SCE, scan rate = 50 mV/s) (After Spinks *et al.*, 2001, copyright 2001 SPIE Press. Reproduced with permission)

The available data predicted that nanotube actuators operating through non-Faradaic charging should eventually exceed all other actuator materials in terms of stress-generation and work-density-per-cycle capabilities. Very fast switching speeds and high cycle life are also expected from non-Faradaic charging.

To date it has been difficult to study the actuation in MWNTs due to the difficulty in producing macroscopic sheets with adequate mechanical properties. However, the synthesis method giving "forests" of MWNTs and the methods developed to transfer such arrays to other substrates (Huang *et al.*, 1999; Yang *et al.*, 1999; Dai and Mau, 2000; Huang *et al.*, 2000; Li D.-C. *et al.*, 2000; Chen and Dai, 2000) have enabled the demonstration of MWNT actuation. MWNT arrays were prepared on a flexible gold foil and immersed in 1M $NaNO_3$ electrolyte. The application of a potential of +1 V or –1 V versus SCE produced bending of the unimorph structure in the opposite direction. Higher voltages (+/–2 V) produced larger deformations.

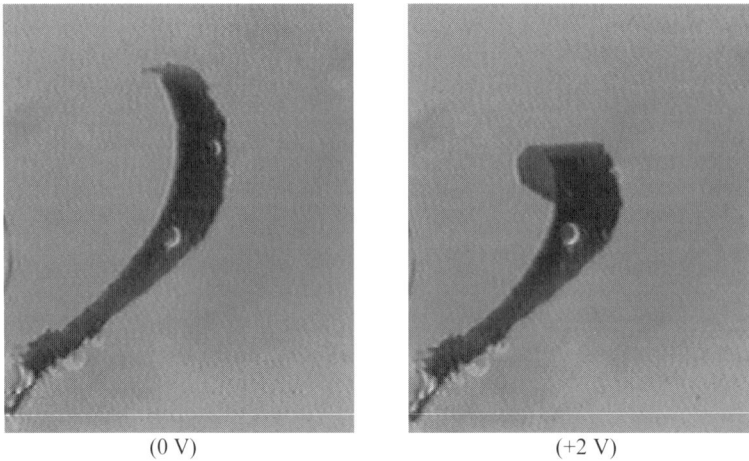

(0 V)   (+2 V)

Figure 11.20. Bending of aligned MWNT arrays on a gold foil by application of +2 or –2 V (*vs.* SCE) (After Spinks *et al.,* 2001, copyright 2001 SPIE Press. Reproduced with permission)

The gravimetric capacitance for the forests of MWNTs is about 30 F/g, which is similar to that for the SWNT bucky paper. However, the above cantilever-based actuator response for the forests of MWNTs is believed to be electrostatic. Electrostatic repulsion between different tubes in the nanotube forest would cause repulsion between these tubes. Since the nanotubes are firmly anchored to the substrate, the effect of this repulsion is a bending of the sheet, so that the separation between unanchored nanotube ends can be increased. While scientifically interesting, this electrostatic mechanism is much less promising for applications than the quantum chemical mechanism.

11.6 Smart Electromechanical Devices Based on Carbon Nanotubes

11.6.1 Carbon Nanotube Quantum Resistors and Nanoresonators

Frank *et al.* (1998) have measured the conductance of carbon nanotubes by replacing the tip of a scanning probe microscope with the nanotube, which forms an electrical circuit with a liquid metal (Figure 11.21). They revealed novel stepwise increases in current with increasing voltage, which was attributed to quantum effects.

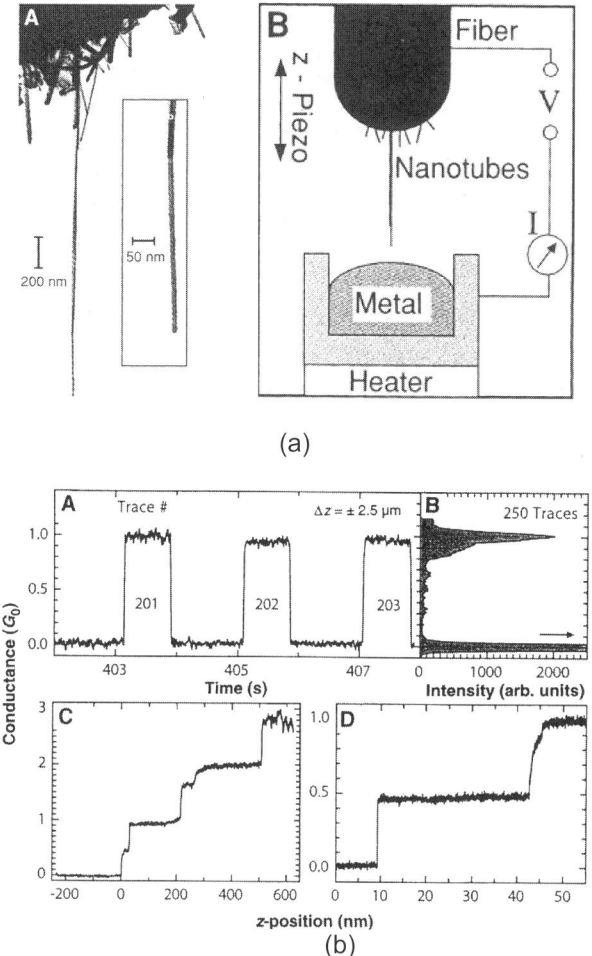

Figure 11.21. (a) (A) TEM image of the end of a nanotube fiber produced by the carbon-arc discharge method. The fiber shown is ~1 mm long and 0.05 mm at the tip, from which protrude several long and straight nanotubes. The long nanotube is 2.2 mm long and 14 nm wide. The inset shows the end of the longest tube under higher magnification; it is boundled together with another one that terminates 400 nm before the first one. (B) Schematic diagram of the measurement of the electrical properties of an individual carbon nanotube. The nanotube contact is lowered under scanning probe microscopic control to a liquid metal surface. After contact is established, the current is measured as the fiber is moved into the liquid metal, so that the conductance can be determined as a function of the position of the nanotube contact; (b) (A) conductance of a nanotube contact that is moved at constant speed into and out of the mercury contact as a function of time. The cycle is repeated to show its reproducibility; cycles 201 through 203 are displayed as an example. (B) Histogram of the conductance data of all 250 traces in the sequence. (C) A trace of a nanotube contact with two major plateaus, each with a minor pre-step. This trace is interpreted as resulting from a nanotube that is bundled with a second one [as in the inset of (a)(A)]. (D) The same as (C) under a higher magnification (After Frank *et al.*, 1998, copyright 1998 American Association for the Advancement of Science)

Using a similar approach, Wang and co-workers (Gao *et al.*, 2000) have been able to measure mechanical properties of individual carbon nanotubes. In this case, a transmission electron microscopy (TEM) specimen holder was specially built for applying a voltage across the nanotube and its counter electrode so that the measurements can be done on a specific nanotube whose microstructure is determined by TEM. By applying an alternating voltage with a tunable frequency on the nanotube, resonance can be induced (Figure 11.22(a)) from which certain mechanical properties of the nanotube can be determined. For example, the bending modulus of the MWNT produced by pyrolysis of FePc (Chapter 6) was determined to be about 30 GPa.

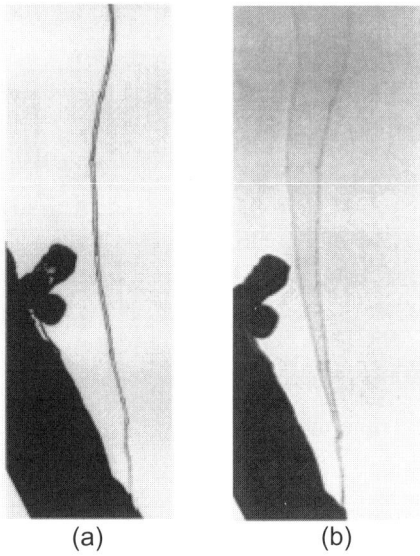

Figure 11.22. (a) A carbon nanotube at the stationary and (b) the first resonance mode; (c) a carbon nanotube with another nanotube trapped on it at the stationary and (c) the first resonance mode (After Gao *et al.*, 2000, copyright 2000 American Physical Society. Reproduced with permission)

More interestingly, it was found that neither the point defects nor the volume defects caused any significant softening at the local region upon deformation, due to the collectively rippling deformation on the inner arc of a bent nanotube (Poncharal *et al.*, 1999). The carbon nanotubes produced by pyrolysis usually have a high density of point defects. Figure 11.23(a) shows a TEM image of a nanotube that exhibits a neck structure at the middle of its body. From the shape of the vibration of the nanotube, there is no abrupt change at the defect point and the vibration curve is smooth. Another experiment shown in Figure 11.23(b) is on the electrostatic deflection of the nanotube when a constant voltage is applied across the electrodes. The nanotube showed a smooth deflection without visible change in its shape near the volume defect. Therefore, the volume defect seems to not introduce any significant softening at the local region, due, most probably, to the collectively rippling deformation on the inner arc of the bend nanotube (Poncharal *et al.*,1999); therefore, the vibration of the entire system could still be described by the elastic theory.

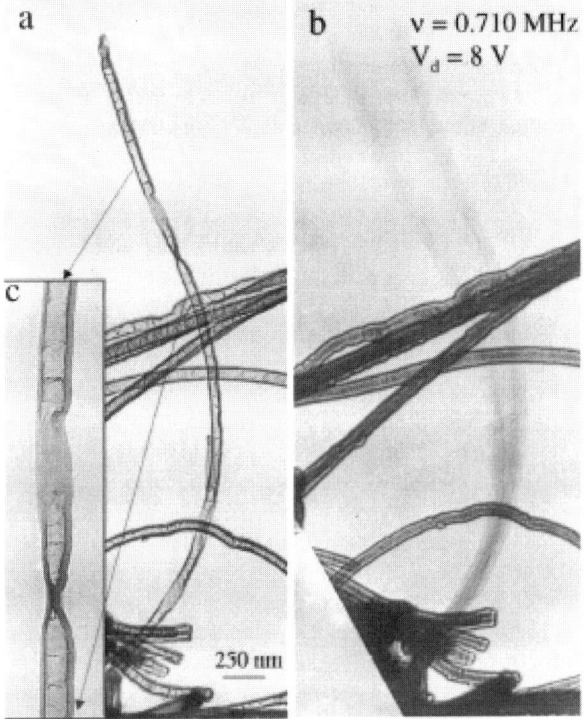

Figure 11.23. A carbon nanotube at (a) the stationary and (b) the first resonance mode; (c) shows the presence of a neck-like volume defect along the body of the nanotube (After Gao *et al.*, 2000, copyright 2000 American Physical Society. Reproduced with permission)

11.6.2 Carbon Nanotube Nanoprobes

In 1981, Gerd Binnig and Heinrich Rohrer of the IBM Zurich Research Laboratory invented the scanning tunneling microscope (STM). Their accomplishment was recognized in their receipt of the 1986 Nobel Prize in Physics (http://www.nobel.se/physics/laureates/1986). STM detects small currents that pass between the microscope's tip and the sample under examination, allowing researchers to see substances at the atomic level (Wiesendanger and Güntherodt, 1992). This success was followed by the development of other scanning probe devices, such as atomic force microscope (AFM) and scanning tunneling spectroscopy (STS) (Magonov and Whangbo, 1996). The working principle of the AFM is similar to that of an old-fashioned phonograph, in which a tiny probe setting on a cantilever is brought into direct contact (contact mode) with or close, but without contact (tipping mode), to the sample. As the tip moves across the sample's surface, the cantilever bends and its deflection is measured by reflecting a beam of laser light off the top of the cantilever, allowing variations in vertical surface topography to be detected.

It was the invention of STM and its variants that catalyzed the rapid advances in the field of nanoscience and nanotechnology over the past decade or so. Further advances in this field will continue to rely on the development of new analytical tools, such as nanotweezers, nanopipettes and nanopens, for making, manipulating and probing structures at the nanometer scale.

Commercial or custom-built scanning probe microscopes can now be found in every major research institution around the world for measuring electrical properties across nanoscopic objects, forces between molecules, and surface topographies, as well as manipulating matter at the scale of individual atoms. The more recently developed scanning instruments with multiple probe-tips, each tip of which can be run independently or in parallel, has further boosted the throughput of STM and AFM devices. The high-aspect ratio of carbon nanotubes, coupled with the intrinisically small diameter at the nanometer scale, as well as excellent electrical and mechanical characteristics, makes them ideal as scanning probe-tips. This opened up one of the most promising near-term applications of carbon nanotubes, which is being recognized by commercial AFM vendors.

Since the first report on the use of carbon nanotubes as AFM probes by Smalley and co-workers (Dai et al., 1996), it has been demonstrated that the carbon nanotube tips, as small as 0.4 nm in diameter, could offer a much higher image resolution than conventional silicon AFM probes. The latter typically has a tip size between two and 30 nanometers wide. Furthermore, the long length of carbon nanotubes allows the tracing of rough surfaces with steep and deep features, while their extraordinary mechanical strength and elastically buckling capability make carbon nanotube scanning probes very robust with a rapid recovery of the structural integrity after deformation.

Figure 11.24(a) shows AFM images of a 280 nm line/space array of photoresist polymer pattern (300 nm thick) on a silicon substrate acquired with a conventional pyramid-shaped silicon probe (left) and MWNT tip (right) (Nguyen et al., 2001). As can be seen in Figure 11.24, the AFM image obtained with a regular silicon probe shows artifacts of sloping sides for the resist lines due to the pyramidal shape of the probe. In contrast, the corresponding image for the same polymer pattern recorded with a micron-long MWNT probe shows no detectable artifact because of the tiny tip size and round shape. Consistent with the SEM image (Figure 11.24(b)), the AFM image obtained with the MWNT tip reveals photoresist lines with nearly vertical walls. The slight asymmetry in the profile at the top of the MWNT image has been attributed to the fact that the feedback mechanism and gain of the AFM have not been optimized for profilometry and thus the shape of the lines is dependent on scan direction. Compared with conventional silicon probes, therefore, the carbon nanotube probes have immense potential in probing 3D structures with nanoscopic dimensions. This, coupled with chemical modification, means that carbon nanotube probes can even provide chemically sensitive surface recognition, as described in Chapter 10.

Actuators and Nanomechanical Devices   483

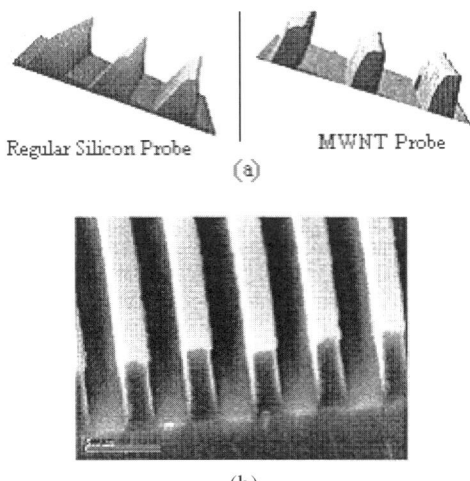

Figure 11.24. (a) AFM images of a 280 nm line/space array of photoresist (300 nm thick) on a silicon substrate acquired with a regular silicon probe (*left*) and a MWNT probe (*right*); (b) The corresponding SEM image taken for the same photoresist pattern (After Nguyen *et al.*, 2001, copyright 2001 Institute of Physics Publishing. Reproduced with permission)

11.6.3 Carbon Nanotube Nanotweezers

As can be concluded from the above discussion, advances in scanning probe techniques could revolutionize the field of nanotechnology in the same way, if not more, as the conventional microscopic techniques have changed our macroscopic world. For the full potential of nanotechnology to be realized, however, the development of new tools, such as tweezers, pipettes, and pens – analogs for macroscopic tools – is critical. In this regard, Kim and Lieber (1999) reported the first "nanotube nanotweezers" for manipulation and interrogation of substances at the nanoscopic scale.

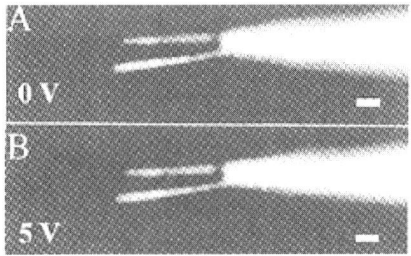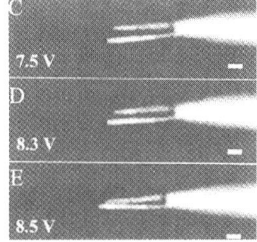

Figure 11.25. Dark-field optical micrographs of the nanotube nanotweezers (A through E) showing the electromechanical response to potentials of 0, 5, 7.5, 8.3, and 8.5 V, respectively. Although the nanotweezer arms are substantially smaller (50 nm in diameter) than optical wavelengths, they scatter sufficient light to be readily observed in the dark-field images. Scale bar is 1 μm (After Kim and Lieber, 1999, copyright 1999 American Association for the Advancement of Science)

To construct the nanotube nanotweezers, Kim and Lieber (1999) first patterned two electrodes on opposite sides of a glass pipette probe. They then attached a carbon nanotube to each of the electrodes (Figure 11.25). As seen in Figure 11.25, the electromechanical response of the two nanotube tweezer arms to an applied voltage with an appropriate bias can cause the free ends of the nanotubes to close. The application of an appropriate voltage with an opposite bias across the two tweezer arms will open them. Figure 11.26 shows the sequential processes of using the nanotube nanotweezer for manipulating polystyrene nanoclusters.

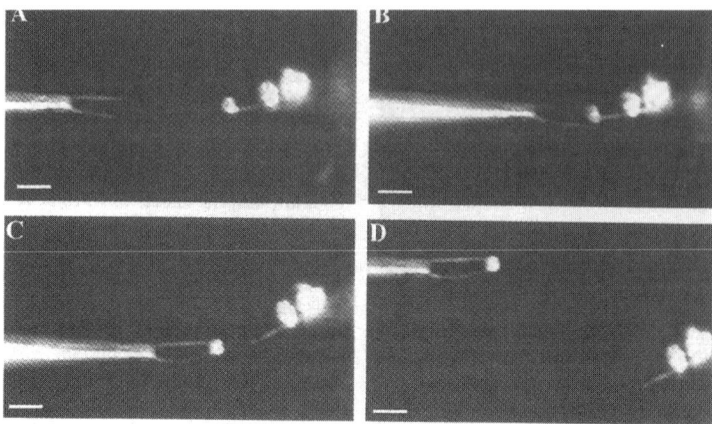

Figure 11.26. Dark-field optical micrographs showing the sequential process of nanotube nanotweezer manipulation of polystyrene nanoclusters containing fluorescent dye molecules. (A) Approach of the nanotweezers to nanoclusters. (B) Alignment of the tweezer arms on a small cluster. A voltage was applied to nanotweezer arms on the nanocluster, and then the nanotweezers and cluster were moved away from the sample support (C and D). Scale bars are 2 μm (After Kim and Lieber, 1999, copyright 1999 American Association for the Advancement of Science)

The above optical micrographs clearly show that the nanotube nanotweezers can be used to manipulate individual nanostructures. Since the tweezer arms are also individually addressable electrodes, it is possible to use them as a two-probe STM device for grasping small substances and for electrical measurements across them simultaneously. In addition, many other applications for the nanotube nanotweezers can be expected, including their use for manipulating biological structures on surfaces or even within cells.

11.6.4 Carbon Nanotube Bearings, Switches and Gears

Apart from the tools for nanoscale manipulation discussed above, another important step towards the advancement of nanotechnology is to develop building blocks for nanodevices in a controlled manner. Using a manipulator inside a high-resolution electron microscope, Cumings and Zettl (2000) at the University of California, Berkeley, peeled off a few outer layers of an MWNT end tip, exposing the inner core tube(s). They then attached the nanomanipulator to the core tubes to move it back and forth so as to obtain a telescoping structure as shown in Figure 11.27.

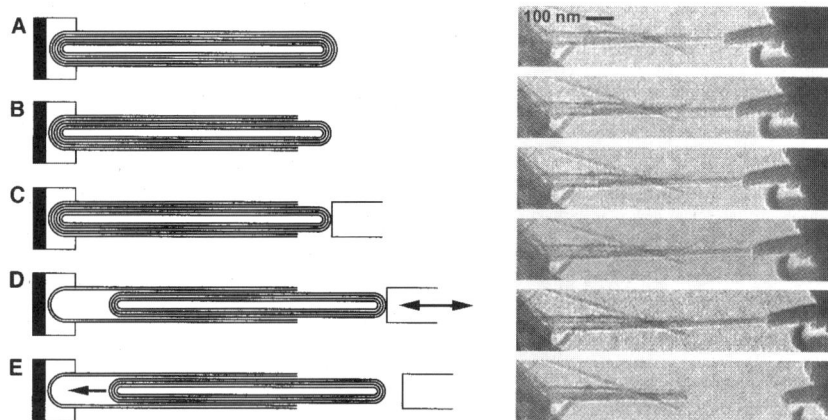

Figure 11.27. (a) Schematic representation of carbon nanotube bearing action. (A to C) the process of opening the end of an MWNT (A) to expose the core tubes (B), and attaching the nanomanipulator to the core tubes (C). (D and E) Two different classes of subsequent experiments performed. In (D), the nanotube is repeatedly telescoped while observations for wear are performed. In (E), the core is released and pulled into the outer-shell housing by the attractive van der Waals force; (b) selected frames of a video recording of the *in situ* telescoping of an MWNT. In the first five frames, the core nanotubes are slowly withdrawn to the right. In the sixth image, which occurred one video frame after the core was released, the core has fully retracted into the outer nanotube housing as a result of the attractive van der Waals force (After Cumings and Zettl, 2000, copyright 2000 American Association for the Advancement of Science)

In high-resolution transmission electron microscopy, the carbon nanotube under the nanomanipulation appears like a nanobearing with the outer layer(s) forming the sleeve and the inner ones the shaft. It was found that the inner core could spin around its long axis within the outer tubes with no wear or fatigue, as the surfaces are atomically perfect and the spacing between the shaft and the sleeve is just the van der Waals distance in graphite with no room for grit to form between them. The nanotube bearings, therefore, show great promise as low-friction, low-wear bearings in the coming world of nanomachines and/or microelectricalmechanical systems (MEMS) where friction is a big problem.

On the other hand, Cumings and Zettl (2000) demonstrated that MWNTs with their inner tube(s) pulled out can act as a nanoswitch. This is because lowering the van der Waals energy by increasing the tube-tube contact area could act as a constant force on the extruded tube(s). As the friction forces between the inner and outer tubes are very low, the extruded inner tube(s) can telescope back into the housing in the order of 10^{-9} seconds. The fast reaction switch, coupled with the electrically conducting nature of carbon nanotubes, provides the possibility of electronic control of back-and-forth movement. The reported nanotube nanobearings and nanoswitches opened up novel applications of carbon nanotubes in manufacturing future nanomachines.

Computer simulations have also demonstrated the possibility of formation of other carbon nanotube-based building blocks of molecular machines, including molecular gears, racks and pinions, though the syntheses of such structures still present problems.

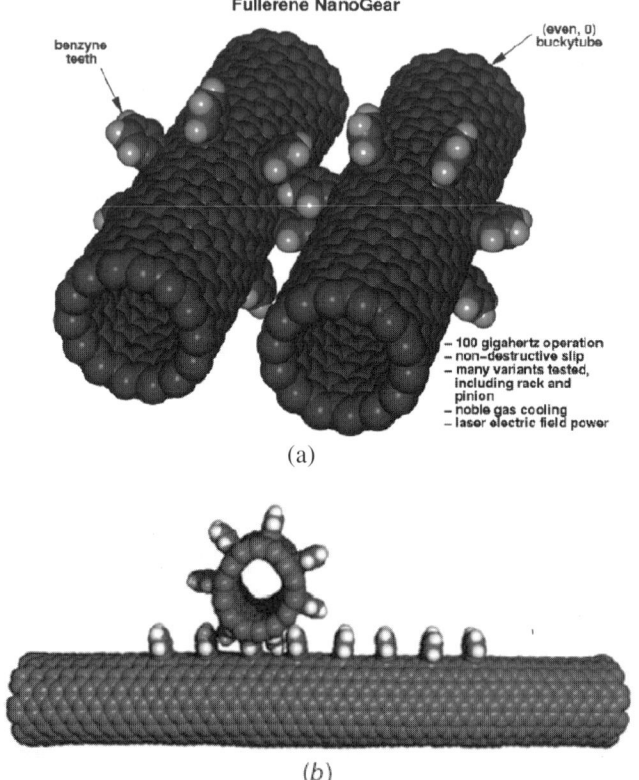

Figure 11.28. (a) Carbon nanotube gears made from carbon nanotubes and benzene teeth; (b) carbon nanotube rack and pinion, in which teeth are spaced two rings apart on the gear but three rings apart on the shaft (After Globus *et al.* 1998, copyright 1998 Institute of Physics Publishing. Reproduced with permission)

These researchers also investigated a rack-and-pinion system by carrying out computer simulation on a molecular gear consisting of a pair of (9,9) SWNTs with teeth spaced every three rings, rather than every other ring (Figure 11.28(b)). The large opening between the teeth allows such a rack-and-pinion system to convert rotation of the gear into linear motion of the (9,9) nanotube, or *vice versa*.

11.7 C_{60} Abacus and Fullerene Vehicles

Using a scanning tunneling microscopy (STM) tip, for example, Cuberes and co-workers (Cuberes *et al.*, 1996; Dagani, 1996) have successfully pushed C_{60} molecules back and forth along a one-atom-high step on a copper surface, demonstrating the possibility to construct a nanoscale C_{60} abacus (Figure 11.29).

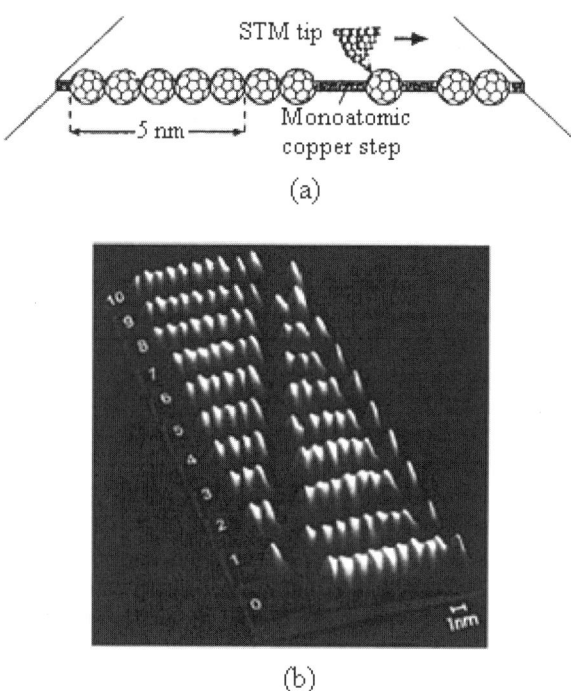

Figure 11.29. Sequence of STM images of the ten C_{60} molecules adsorbed on the Cu(111) step sites, plotted in a pseudo-three-dimensional illuminated presentation. Using the tip as a nanoactuator, the numbers 0 to 10 indicated in the figure have been sequentially computed and read using C_{60} molecules as counters. Top: the corresponding schematic representation of STM tip-aided repositioning of C adsorbed on the Cu(111) surface (After Cuberes *et al.*, 1996, copyright 1996 American Institute of Physics)

Other interesting, related developments include the use of fullerenes and carbon nanotubes for the generation and manipulation of nanoscale vehicles. Tour and co-workers have been trying to build a fleet of "nanotrucks" of 4 × 3 nm in size that run on buckyball "wheels" with a "chassis" consisting of a planar polycyclic molecule – the "loading bay" – situated between two axles supported by four C_{60} wheels (see Dagani, 1999). These authors expect to use a fleet of the C_{60} nanotrucks to deliver other molecular "cargos", to specific surface sites for assembling into nanoscale functional structures (*e.g.* nanoelectronic components).

11.8 Smart Devices Based on Biomolecules

11.8.1 Flagellar Motors

The cell is a self-replicating structure, in which a large number of important molecules of precise structures and functions, including DNA, RNA and proteins, are assembled through various chemical processes. They further spontaneously self-assemble or associate with other molecules into molecular machines. Among the many marvelous molecular machines employed by the cell, the flagellar motor is of particular interest as it is very similar to human-scale motors. The flagellar motor is a highly structured aggregate of proteins anchored in the membrane of many bacterial cells that provides rotary motion to turn the flagellar. The flagellar – a long whip-like structure that acts as the propeller can, in turn, propel the cells through water (Steinberg *et al.*, 1998; Goodsell *et al.*, 1996; Vall and Milligan, 2000).

11.8.2 DNA Switches

Recently, several interesting molecular machines based on DNA chains have also been reported. Mao *et al.* (1999) constructed DNA double-crossover (DX) molecules (Li X. *et al.*, 1996), in which two DNA double helices are joined to each other twice to form rigid molecules. These authors then attached two DX molecules to each of the ends of a longer DNA strand with a special mid-sequence that can switch conformation from the B conformation to the Z conformation (Rich *et al.*, 1984; Pohl and Jovin, 1972), or *vice versa*, in response to the solution conditions (Figure 11.30).

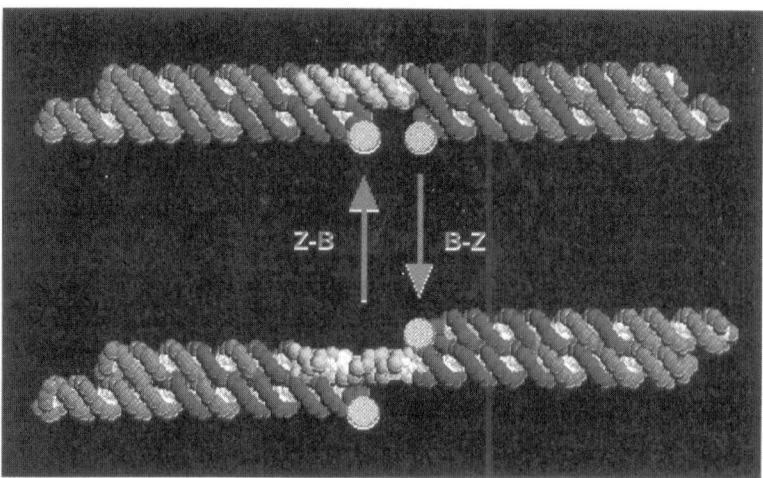

Figure 11.30. Schematic representation of a DNA-based nanoswitch. The two circles represent dyes whose separation distance changes with the conformational change of the middle DNA segment (After Mao *et al.*, 1999, copyright 1999 Macmillan Magazines Limited. Reproduced with permission)

In the B form, the two DX molecules lie on the same side of the longer DNA strand and on opposite sides for the Z form. The B–Z transition results in a rotary displacement of up to 6 nm, along with a 0.6 nm lengthening of the middle DNA segment. These changes can be monitored by measuring changes in the fluorescence of dyes attached to the free ends of the DX molecules. The above biomolecular smart systems, together with their counterparts based on fullerenes, carbon nanotubes and intelligent polymers discussed in this chapter, could provide various building blocks for the development of more complex smart devices of practical significance.

11.9 References

Amdt, K.-F., Kuckling, D., Richter, A. (2000) *Polym. Adv. Technol.* **11**, 496.
Bar-Cohen, Y. (ed.) (2001) *Electroactive Polymer Actuators as Artificial Muscles – Reality, Potential and Challenges*, SPIE Press, Bellingham.
Baughman, R.H. (1996) *Synth. Met.* **78**, 339.
Baughman, R.H., Cui, C., Zakhidov, A.A., Iqbal, Z., Barisci, J.N., Spinks, G.M., Wallace, G.G., Mazzoldi, A., De Rossi, D., Rinzler, A.G., Jaschinski, O., Roth, S., Kertesz, M. (1999) *Science* **284**, 1340.
Baughman, R.H., Shacklette, L.W., Elsenbaumer, R.L., Plichta, E.J., Becht, C. (1990) In *Conjugated Polymeric Materials: Opportunities in Electronics, Optoelectronics, and Molecular Electronics*, Brdas, J.L., Chance, R.R. (eds), Kluwer Scientific Publishers, Dordrecht.
Baughman, R.H., Shacklette, L.W. (1991) In *Science and Applications of Conducting Polymers*, Salaneck, W.R., Clark, D.T., Samuelsen, E.J. (eds), Adam Hilger, New York.
Chen, Q., Dai, L. (2000) *Appl. Phys. Lett.* **76**, 2719.
Cuberes, M.T., Schlittler, R.R., Gimzewski, J.K. (1996) *Appl. Phys. Lett.* **69**, 3016.
Cumings, J., Zettl, A. (2000) *Science* **289**, 602.
Dagani, R. (1996) *C&E News* **Dec. 2**, 20.
Dagani, R. (1999) *C&E News* **October**, 54.
Dai, H., Hafner, J.H., Rinzler, A.G., Colbert, D.T., Smalley, R.E. (1996) *Nature* **384**, 147.
Dai, L., Mau, A.W.H. (2000) *J. Phys. Chem.* **104**, 1891.
Frank, S., Poncharal, P., Wang, Z. L., de Heer, W.A. (1998) *Science* **280**, 1744.
Gao, R., Wang, Z.L., Bai, Z., de Heer, W.A., Dai, L., Gao, M. (2000) *Phys. Rev. Lett.* **85**, 622.
Globus, A., Bauschlicher, C.W., Jr., Han, J., Jaffe, R.L., Levit, C., Srivastava, D. (1998) *Nanotechnology* **9**, 192, and references cited therein.
Goodsell, D.S. (1996) *Our Molecular Nature: The Body's Motors, Machines, and Messages*, Copernicus, New York.
http://www.nobel.se/physics/laureates/1986
Hu, Z., Zhang, X., Li, Y. (1995) *Science* **269**, 525.
Huang, S., Dai, L., Mau, A.W.H., (1999) *J. Phys. Chem. B* **103**, 4223.
Huang, S., Mau, A.W.H., Turney, T.W., White, P.A., Dai, L. (2000) *J. Phys. Chem. B* **104**, 2193.
Jager, E.W.H., Inganäs, O., Lundström (2000) *Science* **288**, 2335.
Jager, E.W.H., Inganäs, O., Lundström, I. (2001) *Adv. Mater.* **13**, 76.
Kim, P., Lieber, C. M. (1999) *Science* **286**, 2148.

Kim, P., Odom, T.W., Huang, J., Lieber, C.M., (1999) In *AIP Conference Proceedings*, 486 (Electronic Properties of Novel Materials – Science and Technology of Molecular Nanostructures).
Li, F.-M., Chen, S.-J., Du, F.-S., Wu, Z.-Q., Li, Z.-C. (1999) In *Field Responsive Polymers*, ACS Symposium Series 726, Khan, I.M., Harrison, J.S. (eds), ACS, Washington DC.
Li, D.-C., Dai, L., Huang, S., Mau, A.W.H., Wang, Z.L. (2000) *Chem. Phys. Lett.* **316**, 349.
Li, X., Yang, X., Qi, J., Seeman, N.C. (1996) *J. Am. Chem. Soc.* **118**, 6131.
Madden, J.D., Cush, R.A., Kanigan, T.S., Hunter, I.W. (2000) *Synth. Met.* **113**, 185.
Magonov, S.N., Whangbo, M.-H. (1996) *Surface Analysis with STM and AFM: Experimental and Theoretical Aspects of Image Analysis*, VCH, Weinheim.
Mao, C., Sun, W., Shen, Z., Seeman, N.C. (1999) *Nature* **397**, 144.
Marsella, M.J., Reid, R.J. (1999) *Macromolecules* **32**, 5982.
Nguyen, C.V., Chao, K.-J., Stevens, R.M.D., Delzeit, L., Cassell, A., Han, J., Meyyappan, M. (2001) *Nanotechnology* **12**, 363.
Oguro, K., Asaka, K., Fujiwara, N., Onishi, K., Sewa, S. (2000) In *Electroactive Polymers*, Mater. Res. Soc. Symp. Proc., 600, MRS.
Otero, T.F., Rodriguez, J., Angulo, E., Santamaria, C. (1993) *Synth. Met.* **55–57**, 3713
Pei, Q., Inganäs, O. (1992) *Adv. Mater.* **4**, 277.
Pei, Q., Inganäs, O. (1993) *Synth. Met.* **55**, 3718.
Pohl, F.M., Jovin, T.M. (1972) *J. Mol. Biol.* **67**, 375.
Poncharal, P., Wang, Z.L., Ugarte, D., de Heer, W.A. (1999) *Science* **283**, 1513.
Rao, G.V.R., Lopez, G.P. (2000) *Adv. Mater.* **12**, 1692.
Rich, A., Nordheim, A., Wang, A.H.-J. (1984) *Annu. Rev. Biochem.* **53**, 791.
Samra, B.K., Galaev, Y.I., Mattiasson, B. (2000) *Angew. Chem. Int. Ed.* **39**, 2364.
Sansiñena, J.M., Olazábal, V., Otero, T.F., Polo da Fonseca, C.N., De Paoli, M.-A. (1997) *J. Chem. Soc., Chem. Commun.* 2217.
Shahinpoor, M., Bar-Cohen, Y., Xue, T., Harrison, J.S., Smith, J. (1999) In *Field Responsive Polymers*, ACS Symposium Series 726, Khan, I.M., Harrison, J.S. (eds), ACS, Washington DC.
Smela, E., Inganäs, O., Lundström, I. (1993) *J. Micrmech. Microeng.* **3**, 203.
Smela, E., Inganäs, O., Lundström, I. (1995) *Science* **268**, 1735
Spinks, G.M., Wallace, G.G., Baughman, R.H., L. Dai (2001) In *Electroactive Polymer Actuators as Artificial Muscles – Reality, Potential and Challenges*, Y. Bar-Cohen (ed.), SPIE Press: Bellingham, 2001.
Steinberg-Yfrach, G., Rigaud, J.-L., Durantini, E.N., Moore, A.L., Gust, D., Moore, A.L., Gust, D., Moore, T.A. (1998) *Nature* **392**, 479.
Tadokoro, S., Fukuhara, M., Bar-Cohen, Y., Oguro, K., Takamori, T. (2000) In *Electroactive Polymer Actuators and Devices*, SPIE-Procceding, 3987.
Vall, R.D., Milligan, R.A. (2000) *Science* **288**, 88.
Wiesendanger, R., Güntherodt, H.-J. (1992) *Scanning Tunneling Microscopy II: Further Applications and Related Scanning Techniques*, Springer-Verlag, Berlin.
Wildoer, J.W.G., Venema, L.C., Rinzier, A.G., Smalley, R.E., Dekker, C. (1998) *Nature* **391**, 59.
Yang, Y., Huang, S., He, H., Mau, A.W.H., Dai, L. (1999) *J. Am. Chem. Soc.* **121**, 10832.
Zhang, Y., Iijima, S. (1999) *Phys. Rev. Lett.* **82**, 3472.

Index

A

addition reaction, 134–148, 191, 217, 384

addition polymerization, 6

acid oxidation, 184, 185, 194, 250, 442

actuator, 3, 42, 81, 172, 203, 395, 432, 461–78, 487

aligned carbon nanotube, 184, 191–195, 229–251, 302–304, 338, 339, 378, 379, 413, 444, 445, 450

amperometric sensor, 412, 413

anode, 34, 62, 63, 166, 236, 279, 286, 300, 333–337, 359–368, 378, 362

atomic force microscopy (AFM), 2, 30, 16–163, 178, 186, 187, 190, 220, 250, 251, 309, 310, 341–343, 442, 443, 481–483

arc-discharge, 165

B

band gap, 44, 48, 82, 98, 187, 188, 224, 307, 308, 321, 360, 363, 369, 371–373

battery, 69, 333–336

B:C:N nanotube, 252

biochromism, 111

biofuel cell, 336

biosensor, 111, 191, 337, 405–407, 410, 413, 422, 447

bimorph actuator, 464, 466

bottom-up, 2, 30, 100

C

C_{60}, 1–6, 27–29, 67, 68, 117–119, 133–150, 157, 158, 181, 215, 253, 330–333, 359, 381–385, 406, 422, 425, 428–431, 454, 461, 487

carbon dioxide sensor, 439

carbon nanotube, 1–3, 6, 27–29, 75, 97, 112, 157–196, 203, 216, 228–251, 280, 302, 304, 305, 321, 322, 333, 337–345, 358, 359, 374, 375, 378–380, 385, 386, 406, 413, 425, 431–446, 461, 475–487, 489

carbon nanotube actuator, 461, 475–478

carbon nanotube sensor, 431–446

charge transfer polymer, 41, 42, 63, 67, 422

cathode, 34, 166, 229, 236, 249, 286, 302, 333, 334, 337, 358–368, 378, 380, 462

chemical doping, 46, 48

chemical force sensor, 442

chemical modification, 60, 100, 172, 186, 192, 194, 195, 210, 304, 307, 373, 378, 382
chemical sensor, 369, 372, 405, 406, 423, 431, 454
chemical vapor deposition (CVD), 236, 238, 247, 249
CO sensor, 427
coil–globule transition, 12
color tuning, 53, 307, 357, 369, 372, 373
conducting polymer, 3, 42-76, 99, 100, 103, 149, 195, 196, 204, 206, 214–218, 221, 222, 225, 226, 228, 253, 271, 275–288, 303, 304, 321–326, 332–336, 357, 369, 384, 394, 406–413, 422, 423, 425, 444, 445, 447, 452–454, 461–463, 465–467, 471, 475, 476
conductively filled polymer, 41, 42, 72, 75, 423–425
conjugated polymer, 27–29, 42–63, 65, 68, 76, 82–84, 86, 105, 109, 111, 126, 144, 149, 190, 203–219, 222, 224, 225, 279, 286–289, 306, 307, 321–323, 331, 357, 359, 364, 366, 368, 369, 372–374, 376–378, 382–385, 406–422, 425, 453, 454
convergent growth, 122, 125
copolymerization, 54–57, 89, 144
critical micelle concentration (CMC), 131
critical solution temperature, 16–18
cyclic voltammetry, 99, 444, 445
cycloaddition, 125, 136, 137, 139, 142–144, 148, 181

D

dendrimer, 3, 5, 6, 58, 59, 83, 95, 106, 112, 117–133, 148, 150, 176, 368, 387–392, 406, 426-428
dendrimer sensor, 426–428
dendritic block copolymer, 132

dimerization, 44, 139, 426
display, 3, 27, 29, 33, 42, 172, 203, 205, 214, 222, 228, 229, 247, 249, 301, 357–359, 372, 375, 376, 378, 379, 392, 393, 394–396, 431
divergent approach, 119, 121, 122, 132
deoxyribonucleic (DNA), 1, 3, 19–21, 31, 32, 81, 189, 321, 346, 353, 406, 446–451, 454–456, 461, 488, 489
DNA array, 454-456
DNA computing, 346, 351–353
DNA diagnostic biosensor, 447
DNA fluorescent sensor, 448
DNA molecular break light, 449
DNA sensor, 21, 446–451
DNA switch, 488
doping, 43–51, 54, 56, 58, 61, 62, 66, 67, 76, 82, 100, 149, 150, 207–209, 213, 215, 217–219, 225, 226, 285, 306, 324, 325, 332–334, 339, 368, 407, 437, 461, 462, 471, 475
Durham polyacetylene, 61

E

electrical transducer, 406,
electrically responsive polymer, 98
electrochemical doping, 47, 48, 62, 100, 368, 461, 462, 471
electrochromic display, 392, 393
electrochromic window, 392, 393
electroluminescence (EL), 28, 51, 58, 205, 215, 219, 224, 286, 289, 307, 357, 359, 369, 370, 372, 375, 381, 454
electromagnetic shielding, 322
electronic nose, 406, 451–454
electron emitter, 172, 229, 236, 358, 368
electronic paper, 392, 395, 396
electropolymerization, 60, 62, 76, 103, 221, 226, 275, 305, 422, 452
electrospinning, 282–286, 303

end-adsorption, 289

F
field-effect transistor, 48, 203, 326, 329, 339, 396
fluorescence, 126, 144, 182, 183, 187, 188, 212, 218, 219, 279, 347, 377, 381, 389, 391, 392, 413, 416, 417–422, 448–450, 454, 455, 489
fullerene, 1–3, 5, 6, 27–29, 41, 67, 68, 117–119, 133–150, 157, 158, 166, 181, 215, 228, 331, 333, 340, 358, 359, 381, 38-2385, 387, 422, 428–431, 461, 487, 489
fullerene vehicle, 487
fullerene charm bracelets, 141, 145
fullerene pearl necklaces, 144

G
gas sensor, 407, 423, 425–427, 430, 431, 435, 439
Gilch route, 52
glass transition temperature, 69
glucose sensor, 424, 444, 446

H
Hepatitis B, 447
highest occupied molecular orbital (HOMO), 101, 286, 333, 359, 360, 362, 363, 383
horizontally aligned carbon nanotube, 230, 232
humidity sensor, 423, 424, 428, 430
hydrocarbon vapor sensor, 407
hydrogen sensor, 433

I
I_2-induced conjugation, 55, 76, 144, 215
inorganic nanotube, 252
intelligent macromolecule, 1–4, 28, 29, 31
interpenetrating network, 6, 384, 472
iodine (vapor) sensor, 426
ion-exchange polymer, 468

ionically conducting polymer, 41, 42, 68–72
ionically responsive polymer, 97
ionochromism, 97, 98
isomerization, 57, 104, 106

L
Langmuir-Blodgett, 205, 206, 215
laser, 2, 109, 166, 168, 191, 221, 230, 247, 301, 322, 357, 359, 376–378, 431, 450, 482
light-harvesting, 387, 389
light-emitting diode (LED), 33, 66, 196, 205, 219, 286, 304, 307, 321, 322, 359–376
light-emitting electrochemical cell (LEC), 368–372
lower critical solution temperature (LCST), 17, 18, 89–92, 95
lowest unoccupied molecular orbital (LUMO), 67, 101, 134, 286, 333, 359, 360, 362, 363, 383

M
mask-induced self-assembling (MISA), 219
mechanochemical reaction, 185
microcontact printing (μCP), 221, 244, 245, 251, 395
micropatterned carbon nanotube, 203, 228–230, 233–236, 239–249
micropatterned conducting polymer, 215–228
micropatterning, 216, 230
molecular computing, 228, 229, 345, 347, 351
molecular wire, 215, 228, 229, 276, 345, 346, 347, 350, 351, 447
multilayer polymer, 60, 225, 287, 305, 306, 369, 373,
multi-wall carbon nanotube (MWNT), 157, 160, 161, 165, 166, 168, 172, 174, 175, 181, 193, 228, 229, 236, 238, 339, 380, 339, 340, 342, 377, 378, 477, 478, 480, 482, 483, 485

N

nanocircuit, 342
nanoelectronics, 321, 338, 349
nanofiber, 208, 265, 275, 279, 281, 282, 283, 284, 285, 300, 302, 303, 304, 309
nanofilm, 266, 286, 287, 288, 300, 305, 306, 309
nanoparticles 31, 32, 70, 90, 167, 170, 173, 236, 243, 247, 248, 265–275, 300, 301, 306–309
nanoprobe, 481
nanoporous, 184, 279, 300, 310
nanoresonator, 478
nanosphere, 90, 266, 268, 269, 272, 275
nanostructured polymer, 265, 300, 301, 309,
nanotubol, 231, 232
nanotweezer, 482–484
nanowire, 170, 195, 170, 195, 252, 265, 275–279, 282, 302, 304, 305
neutron reflectivity, 297–299, 367
nitrogen dioxide sensor, 432
non-carbon nanotube, 203, 230, 252
non-covalent, 29, 97, 98, 172, 188, 190, 249, 269, 270, 455, 475
non-linear optical, 42, 61, 205, 228, 359, 381, 382
non-redox doping, 48, 49
Fowler-Nordheim plot, 379

O

one-pot synthesis, 125, 126
optical limiter, 27, 327, 381, 382
oriented carbon nanotube, 238
oriented conjugated polymer, 206
oxidation, 43, 62, 98, 100, 101, 103, 167, 168, 173, 174, 184, 185, 192, 194, 195, 250, 252, 253, 275, 304, 305, 336, 366, 413, 426, 442, 445, 446
oxidized nanotube, 173–177, 234, 251, 337, 374
oxygen sensor, 434, 435

P

patterned emission, 219, 375
peptide nanotube, 253, 254
percolation threshold, 73, 74, 76, 423
perpendicularly aligned carbon nanotube, 236, 238, 239, 247, 249
phase separation, 16, 17, 143, 207, 223, 300, 306, 309, 332, 369
photobleaching, 108, 109
photochemical, 104, 136, 139, 140, 187, 217, 219, 221, 240, 242, 359, 383, 425
photo-doping, 48
photoelectrochromism, 103,
photolithographic patterning, 217, 242, 245
photoluminescence (PL), 67, 108, 214, 361, 450
photoresponsive polymer, 104, 105, 110, 384
photovoltaic cell, 3, 27, 144, 196, 286, 304, 322, 359, 381, 383–387, 390
pH-responsive polymer, 95, 289
plasma, 96, 140, 192, 193, 206, 216, 221, 225–228, 233, 238, 246, 247, 286, 287, 288, 304, 366, 448
plasma-enhanced chemical vapor deposition (PE-CVD), 238, 247
plasma patterning, 225, 247
plasma polymerization, 216, 225–228, 287, 288, 366
precursor polymer, 51, 61, 149, 184, 213, 214, 219, 226, 279, 288, 364, 365, 374, 375
polyacetylene, 27, 28, 42, 43, 45, 46, 47, 48, 51, 54, 55, 56, 57, 61, 76, 82, 83, 149, 206–208, 213, 215, 217, 224, 271, 279, 326–329, 334, 386, 407
poly(3-alkylthiophene) (P(3AT)), 53, 67, 332
polyaniline (PANI), 43, 45, 48, 49, 50, 51, 54, 76, 82, 83, 99–101, 103, 104, 138, 149, 150, 209, 213, 215, 217, 221, 222, 228,

252, 253, 275-279, 284-286, 287, 289, 306, 325, 333, 335, 368, 374, 407, 409, 410, 425, 445, 462
poly(ethylene oxide) (PEO), 69-71, 266, 270-272, 284-286, 292-294, 298, 299, 335, 368, 369, 393, 423, 425
polydimethylsiloxane (PDMS), 216, 244
polyisoprene, 55-58, 77, 84, 215, 217, 224
poly[2-methoxy-5-(2'-ethyl hexyloxy)-p-phenylene vinylene] (MEH-PPV), 52, 214, 286, 357, 368, 369, 375, 376, 382-385
poly(N-isopropylacrylamide) (PNIPAAm), 18, 88-90, 92, 95, 97
polymer brush, 32, 91, 97, 100, 215, 289, 290, 292, 294, 297, 298,
polymer electrolyte, 41, 68, 69, 70-72, 393, 394, 422
polymer nanotube, 253
polymerization, 6-9, 50, 51, 60, 61, 72, 76, 88, 89, 90, 97, 103, 123-125, 139, 140, 144-146, 150, 191, 206, 207, 210, 216, 217, 221, 223, 225, 226, 228, 233, 246, 247, 266-268, 273, 275, 279, 280, 281, 287, 288, 293, 297, 305, 366, 413, 422, 452, 471
poly(methyl methacrylate) (PMMA), 72, 146, 185, 219, 220, 266, 280, 281, 309, 310, 382, 383, 464
poly(p-phenylene vinylene) (PPV), 28, 51-53, 60, 126, 209, 211, 212, 213, 214, 219, 279, 286, 287, 289, 306, 307, 308, 357, 359-366, 368-376, 384
polypyrrole, 43, 45, 51, 54, 63, 76, 82, 99, 195, 213, 221, 226-228, 275, 279, 289, 303, 305, 306, 324, 325, 335, 336, 395, 407-411, 422, 425, 444-448, 452, 453, 462-464, 466, 467

polysaccharide, 25-27, 192
polystyrene (PS), 56, 58, 61, 62, 72, 76, 92, 93, 142-145, 224, 234, 266, 280, 283, 292, 293, 306, 309, 367, 376, 382, 383, 408, 430, 431, 484
polythiophene, 43, 45, 51, 53, 58, 86, 97, 98, 99, 112, 144, 215, 224, 253, 275, 279, 372, 375, 414, 454
poly[1-(trimethylsily)-1-propyne] (PTMSP), 83-85
potentiometric sensor, 412
protein, 1, 19, 21-25, 27, 29, 32, 33, 254, 406, 450, 455, 456
protein array, 455

Q

quantum efficiency, 287, 361-363, 368, 372, 384, 418
quantum resistor, 478
quantum (size) effect, 162, 443, 478
quantum yield, 126, 361, 377, 384, 388, 419

R

radius of gyration (R_g), 11, 12, 292
Raman spectroscopy, 251
random access memory, 229, 345
refractive index, 101, 294, 295, 377, 382
remote sensor, 411
resonator mass sensor, 443
responsive polymer, 3, 81-113, 289, 384, 394, 425, 471, 475
RNA, 1, 448, 454, 455

S

Schottky barrier diode, 326, 327, 328, 329
scanning tunneling microscope (STM), 2, 30, 31, 160, 161, 162, 163, 174, 215, 275, 276, 481, 482, 484, 487
second virial coefficient, 12, 16, 292
secondary doping, 50, 82, 215, 407

self-assemble, 21, 29, 48, 49, 101, 106, 188, 215, 216, 220, 222, 232, 234, 250, 251, 253, 270, 276, 280, 488
sensor, 3, 21, 29, 32, 81, 94, 98, 111, 112, 117, 172, 191, 196, 225, 228, 249, 268, 282, 294, 300, 304, 305, 337, 369, 372, 382, 389, 405–456, 471
sensor array, 405–407, 451, 452–454
shell-core, 270–273, 301
single-wall carbon nanotube (SWNT), 6, 158–161, 164, 166–168, 172–193, 228, 229, 232–236, 249–251, 339–345, 385, 431–443, 476, 478, 486
smart actuation device, 462
smart robotic device, 462
smart clothing, 33
smart drug, 33
smart electrochemical device, 476
smart surface, 32, 101, 102
smart textile, 425
smart window, 41, 101, 392, 393, 425
SO_2 gas sensor, 426
solvent-responsive polymer, 82–85
supercapacitor, 68, 333, 337, 446, 476
superconductivity, 41, 49, 67, 330, 331, 332, 333, 340
superconductor, 64, 68
supercritical fluid, 268, 269
surface force, 33, 290, 292, 294, 432
surface modification, 131, 172, 191 193, 216, 221, 225, 288
switch, 97, 102, 110, 190, 229, 344, 395, 411, 484–486, 488

T
temperature-responsive polymer, 86, 91, 289, 472
temperature sensor, 425, 440, 441
template synthesis, 166, 184, 236
thermoelectric power, 434–437
thermoelectric nanonose, 435, 438
TNT sensor, 418
top-down, 2, 30
total external reflection, 294

U
ultrahigh-molecular-weight polyethylene (UHMW-PE), 214
unimorph actuator, 462, 463
upper critical solution temperature (UCST), 17, 18

V
van der Waals, 25, 139, 188, 189, 269, 388, 435, 437, 485,
voltammetric sensor, 413

W
Wessling route, 51, 52
Wittig-type coupling, 54

X
X-ray crystal structure, 134
X-ray diffraction, 174, 211, 212, 307
X-ray photoelectron spectroscopy (XPS), 174, 181, 364, 365

Y
Young's modulus, 163

Z
Zwitterionic group, 293